Nonlinear Computational Solid Mechanics

Nonlinear Computational Solid Mechanics

Jamshid Ghaboussi
David A. Pecknold
Xiping Steven Wu

CRC Press
Taylor & Francis Group
Boca Raton London New York

CRC Press is an imprint of the
Taylor & Francis Group, an **informa** business

CRC Press
Taylor & Francis Group
6000 Broken Sound Parkway NW, Suite 300
Boca Raton, FL 33487-2742

First issued in paperback 2019

ISBN-13: 978-1-4987-4612-0 (hbk)
ISBN-13: 978-0-367-87524-4 (pbk)

Library of Congress Cataloging-in-Publication Data

Names: Ghaboussi, J., author. | Pecknold, D. A. W., author. | Wu, Xiping S., 1962- author.
Title: Nonlinear computational solid mechanics / by Jamshid Ghaboussi, David A. Pecknold and Xiping S. Wu.
Description: Boca Raton : CRC Press, [2017] | Includes bibliographical references and index.
Identifiers: LCCN 2016059346| ISBN 9781498746120 (hardback : alk. paper) |
ISBN 9781498746137 (ebook)
Subjects: LCSH: Mechanics, Applied--Mathematics. | Nonlinear mechanics--Matematics. |
Solids--Mathematical models. | Materials--Mathematical models.
Classification: LCC TA350 .G454 2017 | DDC 620.1001/511352--dc23
LC record available at https://lccn.loc.gov/2016059346

Visit the Taylor & Francis Web site at
http://www.taylorandfrancis.com

and the CRC Press Web site at
http://www.crcpress.com

To Our Families

Contents

Preface

Computational mechanics is well established and widely used in diverse fields of engineering and science. Nevertheless, it is a comparatively recent field that is still rapidly developing. Vast improvements in computational capabilities have made possible the routine computational simulation of mechanical and structural systems that would have been unimaginable only a few decades ago. Recent advances in information technology and bioscience have also led to the development of soft computing systems that provide new directions in computational mechanics.

There are now a number of excellent and widely used commercially available general-purpose software packages for simulating well into the nonlinear range the response of diverse types of structural and mechanical systems. Extensive material model and finite element libraries are a typical feature of these packages. In addition, in-house proprietary software systems for nonlinear analysis remain in use in many industries.

The justification and verification of analysis result from large-scale simulations that have become even more critical than the modeling and analysis itself. Engineering practitioners need a thorough understanding of the underlying fundamentals on which these computational tools are based in order to utilize them most effectively in their work. This book is intended to provide much of that necessary background in an accessible form.

The book can serve as a reference for engineers, analysts, and software developers in practice, as well as a graduate course text. Graduate students in various engineering and science disciplines can benefit from the relatively broad coverage of topics, including an introduction to information-based material modeling (in Chapter 12, “Soft Computing in Computational Mechanics”).

We assume that the reader is familiar with the basics of linear finite element analysis; that of course requires a working knowledge of the fundamentals of mechanics and of linear elasticity theory. We do not therefore cover in detail basic linear finite element theory; that would require an entire book in itself, and there are many excellent references available.

We start with an overview of linear and nonlinear mechanics and some simple examples of nonlinear behavior in Chapter 1. This is followed by three chapters (Chapters 2 through 4) discussing the fundamentals of nonlinear continuum mechanics, including the treatment of large displacements and strains (geometric nonlinearity), definitions of stresses and strains and their rates, and the principle of virtual work. Chapters 5 through 7 describe constitutive laws governing stress–strain relations, including material nonlinearity: Chapter 5 covers linear and nonlinear elastic material properties, Chapter 6 discusses stress invariants and material testing, and Chapter 7 covers elastoplastic material models. In Chapter 8, we discuss applications involving solid continuum mechanics, specifically the total Lagrangian formulation and the updated Lagrangian formulation.

The treatment of large rotations in three dimensions, required for nonlinear applications involving structural finite elements (i.e., beams, plates, and shells) with nodal rotational degrees of freedom, is presented in Chapter 9. Formulations for structural elements, including beams, plates, and shells, are covered in Chapter 10, before the discussion of incremental-iterative numerical solution methods for computational simulation in Chapter 11.

With the exception of this latter presentation of incremental-iterative Newton–Raphson methods, we do not attempt to delve deeply into the large inventory of numerical algorithms that are the cornerstones of numerical computational mechanics. That also requires an entire book in itself (see Ghaboussi and Wu, *Numerical Methods in Computational Mechanics*, CRC Press, Taylor & Francis Group 2016).

Throughout the book, we consistently present simple examples to illustrate and clarify the basic concepts and mechanics fundamentals that are being discussed.

Authors

Jamshid Ghaboussi is Emeritus Professor in the Department of Civil and Environmental Engineering at the University of Illinois at Urbana–Champaign. He received his doctoral degree from the University of California at Berkeley. He has more than 40 years of teaching and research experience in computational mechanics and soft computing with applications in structural engineering, geomechanics, and biomedical engineering. He has published extensively in these areas and is the inventor of five patents, mainly in the application of soft computing and computational mechanics. He is the coauthor of the book *Numerical Methods in Computational Mechanics* (CRC Press, Taylor & Francis Group) and author of the book *Soft Computing in Engineering* (CRC Press, Taylor & Francis Group). In recent years, he has been conducting research on complex systems and has coauthored the book *Understanding Systems: A Grand Challenge for 21st Century Engineering* (World Scientific Publishing).

David A. Pecknold is Emeritus Professor in the Department of Civil and Environmental Engineering at the University of Illinois at Urbana–Champaign. He received his PhD degree in structural engineering from the University of Illinois.

He has more than 40 years of teaching, consulting, and university and industrial research experience in structural mechanics and dynamics. He has conducted research and published technical papers on the strength of tubular joints in offshore structures, fatigue, and fracture in railway tank cars, modeling of damage and failure in high-performance structural composites, inelastic response of structures to earthquakes and other dynamic loadings, and nonlinear constitutive models for reinforced concrete and other engineering materials.

He has served as an industrial consultant on many projects, including studies of progressive collapse of offshore oil production platforms under wave and wind loads; investigations of structural vibrations in long-span floors, sports facilities, and other large structures; behavior of railway and subway tunnels; and failure in pressure vessels and containments, large-diameter flexible piping systems, and long-span roof systems.

Xiping Steven Wu received his PhD degree from the University of Illinois at Urbana–Champaign in 1991 in structural engineering. His professional career includes academic teaching and research, applied research for frontier development of oil and gas resources, and project delivery and management. He is a principal engineer in civil and marine engineering with Shell International Exploration & Production, Inc. He has more than 25 years of experience in onshore and offshore structural engineering, arctic engineering, deepwater floating systems, plant design, liquefied natural gas development, and capital project management. He has published more than 40 research papers in computational mechanics, numerical methods, arctic engineering, structural integrity and reliability, and applications of soft computing to civil engineering. He is the coauthor of the book *Numerical Methods in Computational Mechanics* (CRC Press, Taylor & Francis Group).

Chapter 1

Introduction

1.1 LINEAR COMPUTATIONAL MECHANICS

All solid structural systems are inherently nonlinear to some extent. In some cases, it may be useful to idealize such a system as linear, at least over a limited range of loading. Linear models of structural systems play an important role in the analysis and design of engineered structural systems; however, it is important to remember that they only approximate the actual behavior of the nonlinear system.

The rigorous definition of linearity involves the principle of superposition. Linear systems are defined as those that obey the principle of superposition: the response of the system to the sum of multiple loadings is the sum of the responses to each individual loading. The principle of superposition does not apply to nonlinear systems, which has some important consequences that will be discussed later.

Linear computational models are based on many simplifying assumptions, involving idealizations of material behavior, external loadings, and structural geometry. In linear computational models, we assume that displacements and strains are infinitesimal and stress–strain relations are linear and elastic. There is also an implied assumption that overall characteristics of the boundary conditions do not change during the analysis. This assumption has some important and fundamental consequences:

- The geometry of the structural system is assumed not to change because the displacements are infinitesimal. Equilibrium is expressed in the undeformed geometry.
- Consequently, there is no need to distinguish between a material particle and the point in space that it occupies.
- The definitions of strain and stress are unique.
- Strains are linearly related to displacements.
- Stresses are linearly related to forces.
- The response of the structural system to any applied force or displacement boundary condition is unique. This is of course a consequence of the assumption of linearity; in real structural systems, multiple solutions may be possible because of nonlinearity, as will be discussed in detail in later chapters.

Material behavior in linear systems is assumed to be linearly elastic; that is, stresses are linearly related to strains. This assumption is often reasonably accurate for small strains. Many materials behave almost linearly for small changes in stresses and strains. Linear elastic material behavior is also reversible; removing the applied stress or strain returns the system to its original condition. Stress–strain relations for real materials are often highly complex, nonlinear, and irreversible.

1.2 NONLINEAR COMPUTATIONAL MECHANICS

When we relax the assumption of infinitesimal displacements and strains, we have *geometric nonlinearity*. When we relax the assumption of linear stress–strain relations, we have *material nonlinearity*. These are the two primary sources of nonlinearity in structural systems. There are other sources of nonlinearity that can result from changes in the boundary conditions, contacts and separations between different parts of the structural system, and externally imposed changes in the geometry of the system, for example, resulting from construction or dismantling of a structural system.

Finite displacements and strains change the geometry of a structural system. Consequently, we must distinguish between material particles and the points in space that they occupy at a particular time. The motion of a material particle is described by the succession of points in space that it occupies. In addition, in geometrically nonlinear problems the definitions of stress and strain are not unique, and depend on the frame of reference that is employed.

Two reference systems for a solid continuum are discussed in this book:

- The *total Lagrangian (TL) formulation*, in which the initial undeformed configuration is the reference geometry to which strains and stresses are referred. In this formulation, the strains are *Green strains* and the corresponding work-conjugate stresses are *second Piola–Kirchhoff stresses*.
- The *updated Lagrangian (UDL) formulation*, in which the reference system is the current configuration to which the strains (*Almansi strains*) and the corresponding stresses (*Cauchy stresses*, often termed *true stresses*) are referred.

For both TL and UDL formulations, it is necessary to define *strain rates* and *stress rates* in order to eventually arrive at incremental equilibrium equations for a finite element system. Again, because of geometric nonlinearity, the definitions of strain rate and stress rate are not unique. These are presented and discussed in detail in Chapters 2 and 3.

A formulation in terms of (time) rates of stress and strain is necessitated by the rate form in which most nonlinear material models appear.

There are many types of nonlinear constitutive (stress–strain) models. Some are *nonlinear elastic*, which may be useful for modeling rubber-like materials and some biological materials (as discussed in Chapter 5). The stress–strain relations in such models are nonlinear but reversible; that is, when applied stresses are removed, the system returns to its original configuration, dissipating no energy in the process. Others are elastoplastic (Chapter 7), in which the relationship between stress rates and strain rates is nonlinear and often irreversible (i.e., dissipative).

For finite element systems that contain *structural elements* (e.g., beam, plate, and shell) that necessarily have *rotations* in addition to displacements as nodal degrees of freedom, we discuss (Chapter 9) a third type of formulation, that is, a *corotational (CR) formulation*, that resembles in some respects a UDL formulation. In this moving-coordinate type of formulation, the overall displacement and rotation of an individual element defines its updated coordinate system with respect to which local deformations are assumed to be small.

In this book, we have taken the general approach of discussing the fundamentals of continuum mechanics and constitutive laws used in nonlinear computational mechanics before we embark on finite element formulations (Chapters 8 through 10) and nonlinear equation solution techniques (Chapter 11) for various nonlinear problems. In the final chapter (Chapter 12), we present and discuss a new approach, *information-based modeling*, for characterizing material behavior based on information-rich tests.

1.3 NONLINEAR BEHAVIOR OF SIMPLE STRUCTURES

We now discuss the behavior of some simple structures from an intuitive perspective. Our aim is to illustrate some important aspects of nonlinear structural behavior and discuss the main underlying reasons for such behavior. We consider the effects of geometric nonlinearity in some simple structures that can be modeled with plane beam elements.

A two-dimensional beam element is shown in Figure 1.1. The six degrees of freedom of this element can be divided into two groups. Degrees of freedom 1 and 4 are associated with the axial mode of deformation, and degrees of freedom 2, 3, 5, and 6 are associated with the bending mode of deformation. In a linear analysis in which displacements and strains are infinitesimal and the geometry is assumed not to change, these two modes of deformation are uncoupled. We can observe this uncoupling in the element *linear elastic* stiffness matrix **K** that relates the displacement vector **u** to the load vector **p**:

$$\mathbf{K}\mathbf{u} = \mathbf{p}, \tag{1.1}$$

$$\mathbf{K} = \begin{bmatrix} \frac{AE}{l} & 0 & 0 & -\frac{AE}{l} & 0 & 0 \\ & \frac{12EI}{l^3} & \frac{6EI}{l^2} & 0 & -\frac{12EI}{l^3} & \frac{6EI}{l^2} \\ & & \frac{4EI}{l} & 0 & -\frac{6EI}{l^2} & \frac{2EI}{l} \\ & & & \frac{AE}{l} & 0 & 0 \\ & \text{Symm.} & & & \frac{12EI}{l^3} & -\frac{6EI}{l^2} \\ & & & & & \frac{4EI}{l} \end{bmatrix} \tag{1.2}$$

where l = length of the beam element, A = cross-sectional area, I = moment of inertia of the cross section, and E = Young's modulus.

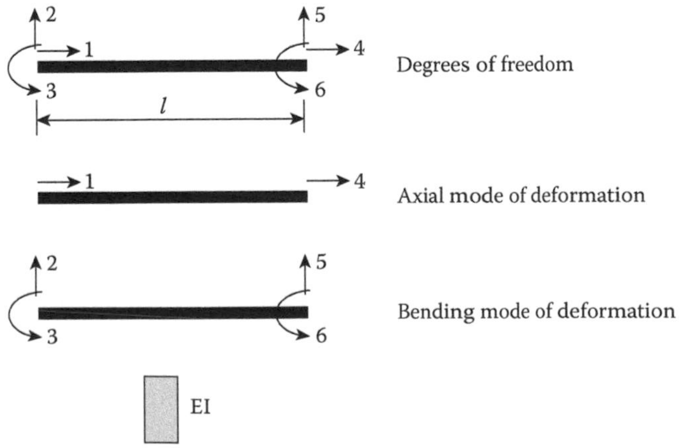

Figure 1.1 Two-dimensional beam element and modes of deformation.

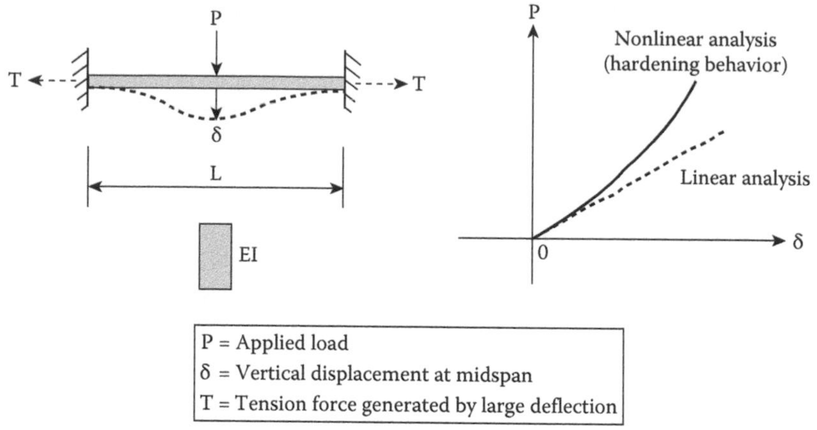

Figure 1.2 Linear and nonlinear behavior of a single-span fixed beam.

The coupling terms between the two modes of deformation in the element stiffness matrix in Equation 1.2 are zero. When displacements and strains are assumed to be finite, that is, we have geometric nonlinearity, the two modes of deformation are coupled, as we will see in Chapter 10. The interchange of strain energy between flexural and axial modes of deformation is responsible for some important aspects of nonlinear structural behavior that we will briefly describe in this section and discuss in detail in later chapters.

Consider the large displacement behavior of the single-span beam shown in Figure 1.2. According to linear analysis, its response is purely in bending. When displacements are small, this is a reasonable approximation. However, as the applied load is increased, the beam centerline begins to stretch and an increasing proportion of the strain energy created by the load is therefore stored in axial deformation (i.e., as membrane strain energy). We can intuitively recognize that this gradual transition from flexure to the stiffer axial extension mode of deformation is responsible for the nonlinear P versus δ response illustrated in Figure 1.2. (Try flexing a thin plastic ruler vs. stretching it.)

The axial extension generates a corresponding axial tension force. The calculated stiffening P–δ response shown in the figure results from taking account of the axial tension force acting through the vertical deflection.

Unlike the stiffening behavior of the fixed-fixed beam, the arch structure shown in Figure 1.3 exhibits softening load–deflection behavior. We recognize the underlying cause of this softening behavior as a gradual transition from a primarily axial to a primarily bending mode of deformation. That is, initially the arch responds essentially in axial compression (except near the concentrated load and near the fixed supports); as the load and displacements increase, an increasing proportion of the strain energy created by the load is stored as flexural strain energy. We see that the geometric nonlinearity is primarily responsible for this nonlinear (in this case, softening) behavior.

We mentioned earlier that the uniqueness of solution that is characteristic of linear structural systems does not apply to nonlinear systems. Nonuniqueness is an important aspect of the behavior of nonlinear structural systems. Here, we present a simple example in which the nonuniqueness is caused again by the interaction between axial and bending modes of deformation.

A sudden transition from one mode of deformation to another is usually associated with a *bifurcation* in the displacement response of the structure (Chapter 11). A simple example of this type of behavior is exhibited by the column shown in Figure 1.4. A "perfect" (i.e., initially straight) column under axial load undergoes only axial deformation; the lateral displacement δ

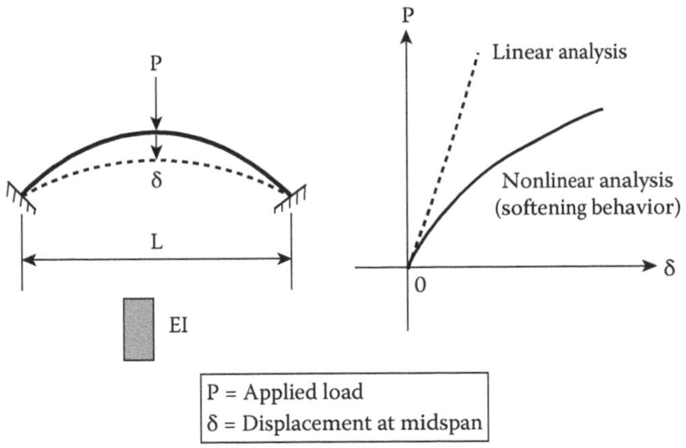

Figure 1.3 Large displacement behavior of an arch structure.

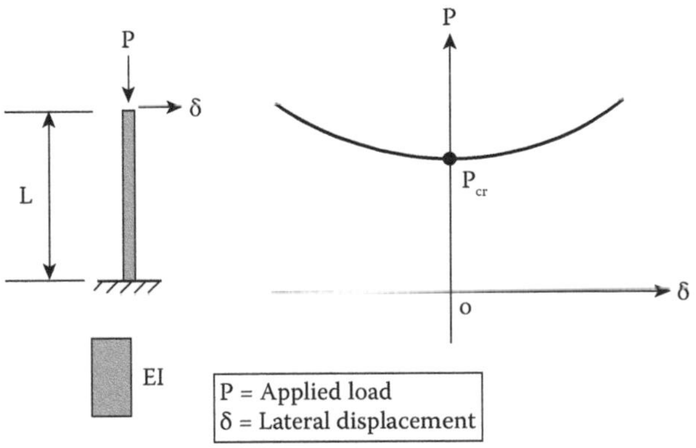

Figure 1.4 Bifurcation of a perfect column under axial load.

at the top of the column remains zero (i.e., there is no bending). When the axial load reaches the critical buckling load P_{cr}, suddenly three solution paths are available to the structure. In addition to the (unstable) equilibrium path involving only axial deformation ($\delta = 0$), two other solution paths (illustrated in Figure 1.4) are possible that involve significant bending deformations and lower strain energy.

Sudden transitions or bifurcations can occur in perfect structures (Chapter 11). For our simple column, it is sufficient to say that "perfect" means that it is straight, made of homogeneous material, and acted upon by a purely axial load. The absence of any of these three conditions will make the structure or its loading "imperfect" and may significantly affect its response.

The column shown in Figure 1.5 has a *geometric imperfection*. That is, it is not perfectly straight and vertical. The response of this column may be similar to the response of the perfect column (depending on the magnitude of the geometric imperfection), except that there is no longer a discrete bifurcation point. The transition to a bending mode of deformation is gradual, but is present from the beginning of the loading.

We conclude this introduction to the nonlinear behavior of simple structures with a brief discussion of *snap-through* buckling of an arch. (Bifurcation and snap-through are discussed in more detail in Chapter 11.)

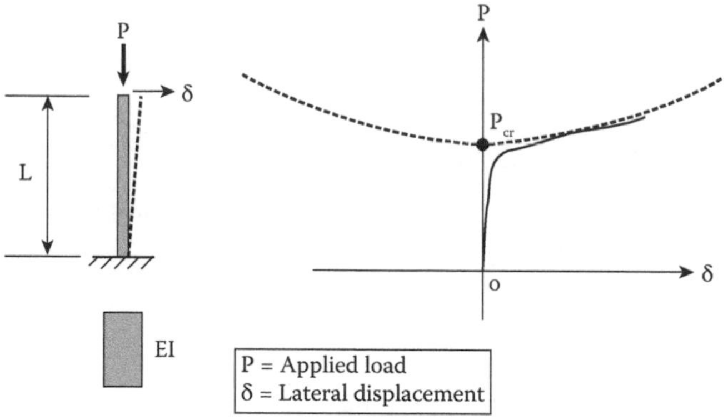

Figure 1.5 Buckling of an imperfect column under axial load.

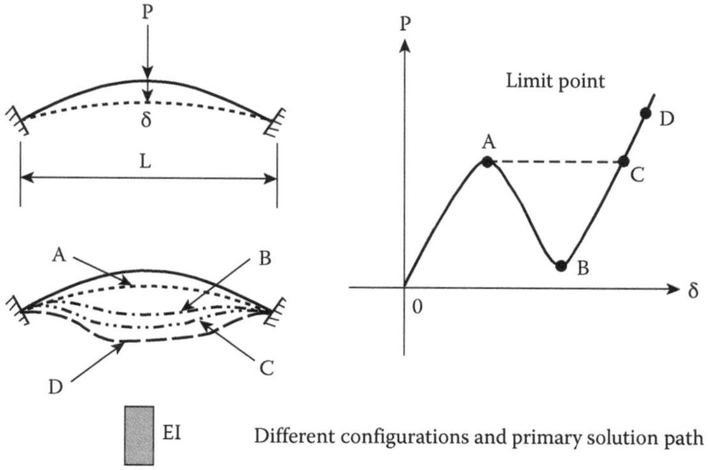

Figure 1.6 Large deformation behavior of an arch under concentrated load.

Consider the arch shown in Figure 1.6, under a concentrated load. We have already seen that under increasing load, the structure softens; that is, the slope of the load–displacement curve, which is a measure of the tangent stiffness, decreases. At some point (point A in Figure 1.6), the slope of this curve will become zero. This is a very special point called a *limit point.* The existence of a limit point on the load–displacement curve is closely associated with the *snap-through* buckling. If we try to increase the load beyond point A, the displacement will suddenly (and dynamically) increase to point C, and the structure will follow the CD portion of the load–displacement curve. We can say that the structure has snapped from configuration A to configuration C. The actual load–displacement curve beyond point A has a negative slope. In order to follow this portion of the load–displacement curve, we must start decreasing the load beyond point A, until another limit point, B, is reached.

Chapter 2

Kinematics of nonlinear deformation

In this chapter, a body is defined to be an assemblage of particles, occupying a region of space. The region of space occupied by a body is called a configuration. The motion is a one-parameter succession of configurations. The parameter is usually time, but in general can be a "time-like" parameter. This definition of configuration establishes a one-to-one correspondence $\overline{\mathbf{x}}$ between a particle and a point in space. If the particles of a body could be labeled individually, then the motion could be described by following the particles of a body. Since this is clearly not possible, the motion is instead described by specifying a particular configuration as the *reference configuration*. This reference configuration can be any configuration, including the original configuration. The term *original configuration*, which is used very often in practice, denotes the configuration at time zero and therefore is a special reference configuration. Any other configuration during a motion is called the current configuration or *deformed configuration*.

2.1 MOTION AND DEFORMATION OF LINE ELEMENTS

Following the general practice, however imprecise it may be, we will use the terms *reference configuration* and *deformed configuration*. However, this does not imply that the body is deformation-free at its original configuration. On the other hand, our deformed configuration may be the result of only a rigid body motion. As shown in Figure 2.1, we refer all configurations to a common Cartesian coordinate system. A *particle*, in the sense of a material particle, occupying a *point* in the reference configuration with the position vector $\mathbf{x}$, occupies another point in the deformed configuration with the position vector $\overline{\mathbf{x}}$. A point here refers to the instantaneous geometric location of a material particle; a body is the assemblage of infinitely many particles. The position vector $\mathbf{x}$ of a material particle in the reference configuration is represented by the coordinates (x_1, x_2, x_3) with respect to a Cartesian coordinate system, as shown in Figure 2.1. At a later time, $t = \tau$, the position of the particle is $\overline{\mathbf{x}}$, with coordinates $(\overline{x}_1, \overline{x}_2, \overline{x}_3)$ in the same Cartesian coordinate system. Since the components of these position vectors are referred to the same Cartesian coordinate system with Cartesian unit base vectors $\mathbf{e}_i$, then they can be expressed as

$$\mathbf{x} = x_i \mathbf{e}_i \tag{2.1}$$

$$\overline{\mathbf{x}} = \overline{x}_i \mathbf{e}_i \tag{2.2}$$

The motion is described by the following one-parameter set of continuous functions, relating the Cartesian components of the position vectors of points in the deformed configuration and the reference configuration.

$$\overline{x}_i = \overline{f}_i(x_j, \tau) \tag{2.3}$$

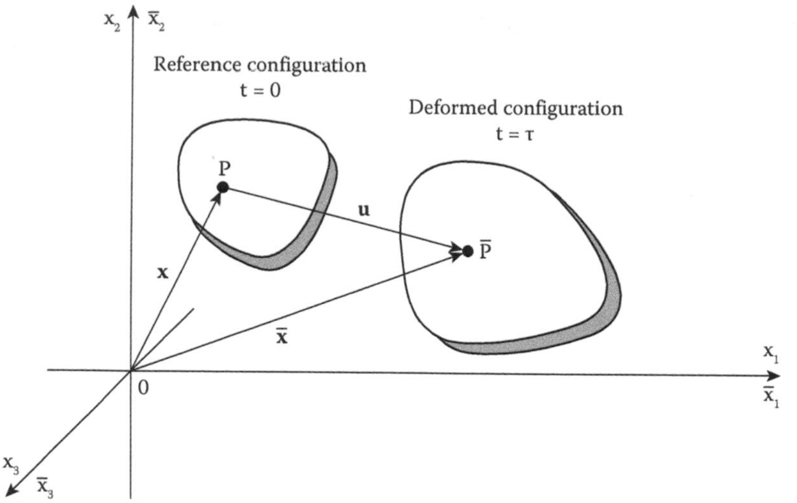

Figure 2.1 Configurations defined in a common Cartesian coordinate system.

This is called a *Lagrangian* formulation (the motion is described with respect to the reference configuration). This description of motion is invertible. Therefore, we also have

$$x_i = f_i(\bar{x}_j, \tau) \tag{2.4}$$

which is called a *Eulerian* formulation (the motion is described with respect to the deformed configuration).

Generally speaking, we call the Cartesian coordinate system defined by coordinate axes (x_1, x_2, x_3) the *material* coordinate system, and that defined by coordinate axes ($\bar{x}_1, \bar{x}_2, \bar{x}_3$) the *spatial* coordinate system. The spatial coordinate system identifies the position of the particle throughout the motion, whereas the material coordinate system "follows the particle."

As shown in Figure 2.1, a displacement vector **u** can be defined by the following equation:

$$\bar{\mathbf{x}} = \mathbf{x} + \mathbf{u} \tag{2.5}$$

In the reference configuration, the coordinate axes x_i form a Cartesian coordinate system. That is, if we were to trace the path formed by a series of material points by, for example, varying x_1 while holding x_2 and x_3 constant, that path would of course be a straight line. During the motion, that straight line along which only x_1 varies will, in general, become curved, as shown in Figure 2.2. Thus, the coordinate axes x_i will follow the motion, as though they were etched into the body at the reference configuration. This so-called *convected coordinate* system is a curvilinear coordinate system in the deformed configuration, as shown in Figure 2.2.

The base vectors of the convected coordinate system, which were the Cartesian base vectors in the reference configuration, are $\mathbf{G}_i$ in the deformed configuration. That is, we have $d\mathbf{x} = dx_i\, \mathbf{e}_i$ and

$$\begin{aligned} d\bar{\mathbf{x}} &= d\bar{x}_i \mathbf{e}_i = \left(\frac{\partial \bar{x}_i}{\partial x_j} dx_j\right) \mathbf{e}_i \\ &= \left(\frac{\partial \bar{x}_i}{\partial x_j} \mathbf{e}_i\right) dx_j = \mathbf{G}_j dx_j \end{aligned} \tag{2.6}$$

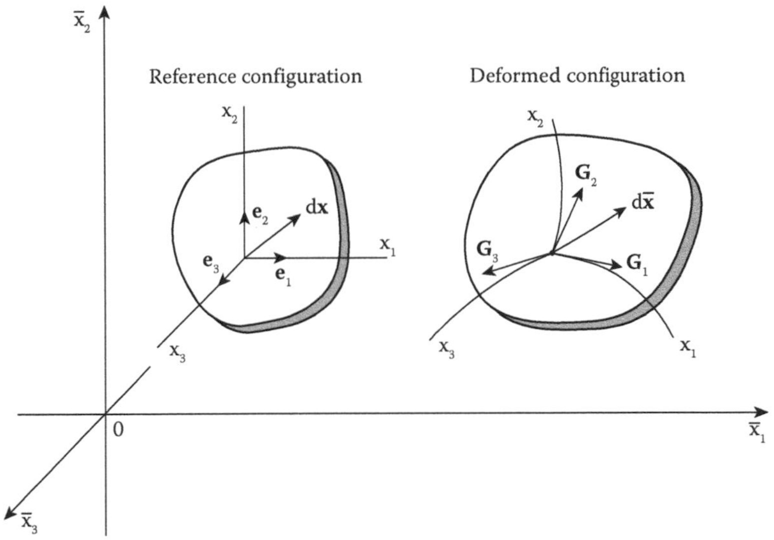

Figure 2.2 Transformation of Cartesian base vectors in reference configuration to convected base vectors in deformed configuration.

where

$$\mathbf{G}_i = \frac{\partial}{\partial x_i}(\bar{\mathbf{x}}) = \frac{\partial \bar{x}_j}{\partial x_i}\mathbf{e}_j \tag{2.7}$$

Note that $\partial \bar{x}_j / \partial x_i$ are the Cartesian components of the *convected base vectors* $\mathbf{G}_i$. Referring to Figure 2.2, we thus have the relations that

$$d\mathbf{x} = dx_i \; \mathbf{e}_i \tag{2.8}$$

and

$$d\bar{\mathbf{x}} = dx_i \; \mathbf{G}_i \tag{2.9}$$

Therefore, the convected base vectors $\mathbf{G}_i$ contain all the information about the deformation in the neighborhood of a material particle.

The *deformation gradient* tensor **F** transforms an arbitrary line element from the reference configuration to the deformed configuration

$$d\bar{\mathbf{x}} = \mathbf{F}d\mathbf{x} \tag{2.10}$$

where

$$\mathbf{F} = \begin{bmatrix} \frac{\partial \bar{x}_1}{\partial x_1} & \frac{\partial \bar{x}_1}{\partial x_2} & \frac{\partial \bar{x}_1}{\partial x_3} \\ \frac{\partial \bar{x}_2}{\partial x_1} & \frac{\partial \bar{x}_2}{\partial x_2} & \frac{\partial \bar{x}_2}{\partial x_3} \\ \frac{\partial \bar{x}_3}{\partial x_1} & \frac{\partial \bar{x}_3}{\partial x_2} & \frac{\partial \bar{x}_3}{\partial x_3} \end{bmatrix} \tag{2.11}$$

It may be seen from Equation 2.9 that **F** contains the Cartesian components of the three convected base vectors G_i as its columns, that is,

$$\mathbf{F} = [G_1, G_2, G_3] \tag{2.12}$$

For example, the first column of **F** is $\{\partial\bar{x}_1/\partial x_1, \partial\bar{x}_2/\partial x_1, \partial\bar{x}_3/\partial x_1\}^T$ (which are the Cartesian components of G_1).

Also, the transpose of the deformation gradient tensor $\mathbf{F}^T$ (**F** is generally not symmetric) maps the Cartesian base vectors to the convected base vectors.

$$G_i = F_{ji}\ e_j \tag{2.13}$$

By substituting Equation 2.5 into Equation 2.7, we can determine the components of the convected base vector in terms of the components of the displacement vector as

$$\begin{aligned} G_i &= \frac{\partial(x_j + u_j)}{\partial x_i} e_j = (\delta_{ij} + u_{j,i})e_j \\ &= e_i + u_{j,i}\ e_j \end{aligned} \tag{2.14}$$

and obviously

$$F_{ij} = \delta_{ij} + u_{i,j} \tag{2.15}$$

where $u_{i,j} \equiv \partial u_i/\partial x_j$ are the components of the Lagrangian or material *displacement gradient* **G**. In Equation 2.15, δ_{ij} is the Kronecker delta symbol, which is equal to 1 if i = j, and zero otherwise. Equation 2.15 may be written as

$$\mathbf{F} = \mathbf{I} + \mathbf{G} \tag{2.16}$$

where **I** is the unit matrix (tensor) and

$$\mathbf{G} = \begin{bmatrix} \dfrac{\partial u_1}{\partial x_1} & \dfrac{\partial u_1}{\partial x_2} & \dfrac{\partial u_1}{\partial x_3} \\ \dfrac{\partial u_2}{\partial x_1} & \dfrac{\partial u_2}{\partial x_2} & \dfrac{\partial u_2}{\partial x_3} \\ \dfrac{\partial u_3}{\partial x_1} & \dfrac{\partial u_3}{\partial x_2} & \dfrac{\partial u_3}{\partial x_3} \end{bmatrix} \tag{2.17}$$

Although the Lagrangian deformation gradient tensor **G** (like the convected base vectors G_i) contains complete information about the deformation in the neighborhood of a material point, it is not subsequently convenient to work with because it is not symmetric.

The symmetric (Lagrangian) *metric tensor* **C** is preferable for specifying material constitutive relations, which will be discussed in Chapter 7. **C** relates the (squared) length $d\bar{s}^2$of a deformed line element in the deformed configuration to its (squared) length in the reference configuration. It is defined by

$$d\bar{s}^2 = d\bar{\mathbf{x}}^T d\bar{\mathbf{x}} = d\mathbf{x}^T(\mathbf{F}^T\mathbf{F})d\mathbf{x} = d\mathbf{x}^T(\mathbf{C})d\mathbf{x} \tag{2.18}$$

Thus,

$$\mathbf{C} = \mathbf{F}^T\mathbf{F} \tag{2.19}$$

which is obviously symmetric.

C can further be expressed in terms of the displacement gradient tensor using Equation 2.16 as

$$\mathbf{C} = \mathbf{I} + \mathbf{G} + \mathbf{G}^T + \mathbf{G}^T\mathbf{G} \tag{2.20}$$

The components of **C** can be expressed as inner (dot) products of the convected base vectors as

$$C_{ij} = \mathbf{G}_i \cdot \mathbf{G}_j \tag{2.21}$$

or in terms of the components of the displacement vector as

$$C_{ij} = \delta_{ij} + u_{i,j} + u_{j,i} + u_{m,i}u_{m,j} \tag{2.22}$$

The foregoing is a Lagrangian description of motion (with respect to the reference configuration). An alternative (Eulerian) description of motion that characterizes the deformation process involves using a set of base vectors $\overline{\mathbf{G}}_i$ defined in the reference configuration. During the motion, these base vectors transform to the Cartesian unit base vectors in the deformed configuration, as shown in Figure 2.3.

In this dual formulation, the definition of the various deformation measures employs the inverse $\overline{\mathbf{F}} \equiv \mathbf{F}^{-1}$ of the deformation gradient tensor, which maps line elements in the deformed configuration to line elements in the reference configuration, that is, from Equation 2.10,

$$d\overline{\mathbf{x}} = \mathbf{F}d\mathbf{x} \Rightarrow d\mathbf{x} = \mathbf{F}^{-1}d\overline{\mathbf{x}} \tag{2.23}$$

$$\begin{aligned} d\mathbf{x} = dx_i\mathbf{e}_i &= \left(\frac{\partial x_i}{\partial \overline{x}_j}d\overline{x}_j\right)\mathbf{e}_i \\ &= \left(\frac{\partial x_i}{\partial \overline{x}_j}\mathbf{e}_i\right)d\overline{x}_j = \overline{\mathbf{G}}_j d\overline{x}_j \end{aligned} \tag{2.24}$$

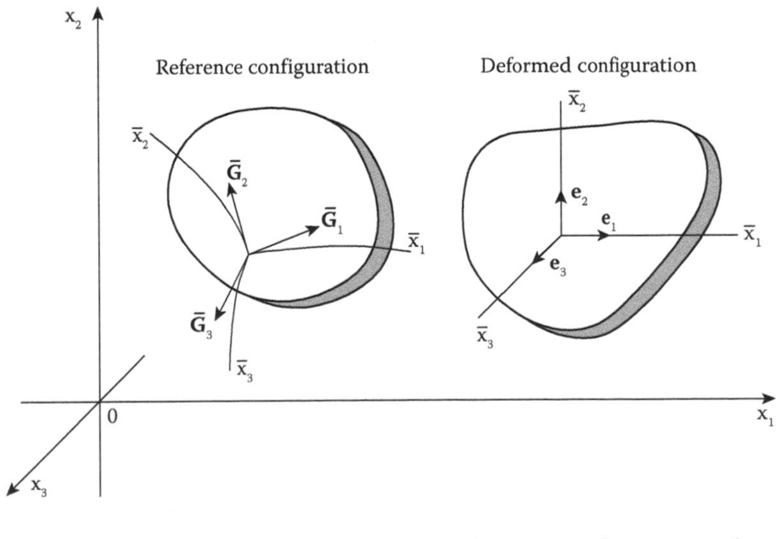

Figure 2.3 Transformation of base vectors in the reference configuration to Cartesian unit base vectors in the deformed configuration.

where

$$\overline{G}_i = \frac{\partial x_j}{\partial \overline{x}_i} e_j = \frac{\partial(\overline{x}_j - u_j)}{\partial \overline{x}_i} e_j = e_i - \overline{u}_{j,i}\ e_j \tag{2.25}$$

and $\overline{u}_{i,j}$ denotes $\partial u_i / \partial \overline{x}_j$.

The columns of $\overline{F}$ consist of the Cartesian components of the base vectors $\overline{G}_i$ (cf. Equation 2.12):

$$\overline{F} = \left[\overline{G}_1, \overline{G}_2, \overline{G}_3\right] \tag{2.26}$$

The Cartesian components of $\overline{F}$ are expressed in terms of displacement gradients by

$$\overline{F}_{ij} = \delta_{ij} - \frac{\partial u_i}{\partial \overline{x}_j} \tag{2.27}$$

or (cf. Equation 2.15)

$$\overline{F} = I - \overline{G} \tag{2.28}$$

where $\overline{G}$ is the Eulerian displacement gradient tensor (matrix) (cf. Equation 2.15):

$$\overline{G} = \begin{bmatrix} \frac{\partial u_1}{\partial \overline{x}_1} & \frac{\partial u_1}{\partial \overline{x}_2} & \frac{\partial u_1}{\partial \overline{x}_3} \\ \frac{\partial u_2}{\partial \overline{x}_1} & \frac{\partial u_2}{\partial \overline{x}_2} & \frac{\partial u_2}{\partial \overline{x}_3} \\ \frac{\partial u_3}{\partial \overline{x}_1} & \frac{\partial u_3}{\partial \overline{x}_2} & \frac{\partial u_3}{\partial \overline{x}_3} \end{bmatrix} \tag{2.29}$$

The symmetric Eulerian metric tensor $\overline{C}$ relates the (squared) length ds^2 of line elements in the reference configuration to their (squared) length in the deformed configuration (cf. Equation 2.16):

$$ds^2 = dx^T dx = d\overline{x}^T \overline{C} d\overline{x} \tag{2.30}$$

where (cf. Equation 2.17)

$$\overline{C} \equiv \overline{F}^T \overline{F} \tag{2.31}$$

With Equation 2.28, $\overline{C}$ can be expressed as

$$\overline{C} = I - \overline{G} - \overline{G}^T + \overline{G}^T \overline{G} \tag{2.32}$$

Although the Eulerian formulation appears to be not as practically useful as (total and updated) Lagrangian formulations in computational solid mechanics applications, in subsequent sections we will nevertheless briefly note the basic relationships of the Eulerian formulation.

2.2 DEFORMATION OF VOLUME AND AREA ELEMENTS

2.2.1 Volume elements

Consider a differential element of volume $dV = dx_1\ dx_2\ dx_3$, with sides lying along the Cartesian axes in the reference configuration, as shown in Figure 2.4. In the deformed configuration, the differential element, which was originally a cube, becomes a parallelepiped with volume $d\overline{V}$. The volume dV can also be written more generally as

$$\varepsilon_{ijk} dV = \mathbf{e}_i \cdot (\mathbf{e}_j \times \mathbf{e}_k) dx_1 dx_2 dx_3 \tag{2.33}$$

where ε_{ijk} (the permutation symbol) = 1 if i, j, k are cyclic and $\varepsilon_{ijk} = -1$ if i, j, k are anticyclic.

The parallelepiped in the deformed configuration has sides $\mathbf{G}_1 dx_1, \mathbf{G}_2 dx_2, \mathbf{G}_3 dx_3$, and its volume is given by

$$\varepsilon_{ijk}\ d\overline{V} = \mathbf{G}_i \cdot (\mathbf{G}_j \times \mathbf{G}_k) dx_1 dx_2 dx_3 \tag{2.34}$$

Referring to Equation 2.11, we recall that the Cartesian components of $\mathbf{G}_1$, $\mathbf{G}_2$, $\mathbf{G}_3$ form the columns of $\mathbf{F}$. From linear algebra, the scalar triple product of the three vectors $\mathbf{G}_1$, $\mathbf{G}_2$, $\mathbf{G}_3$ is therefore the determinant of $\mathbf{F}$ (or $\mathbf{F}^T$).

$$\mathbf{G}_i \cdot (\mathbf{G}_j \times \mathbf{G}_k) = \varepsilon_{ijk} |\mathbf{F}| \tag{2.35}$$

Therefore,

$$d\overline{V} = |\mathbf{F}| dV \tag{2.36}$$

As an aside, from the principle of the conservation of mass,

$$\overline{\rho}\ d\overline{V} = \rho\ dV \tag{2.37}$$

where ρ and $\overline{\rho}$ are, respectively, the mass densities in the reference and deformed configurations.

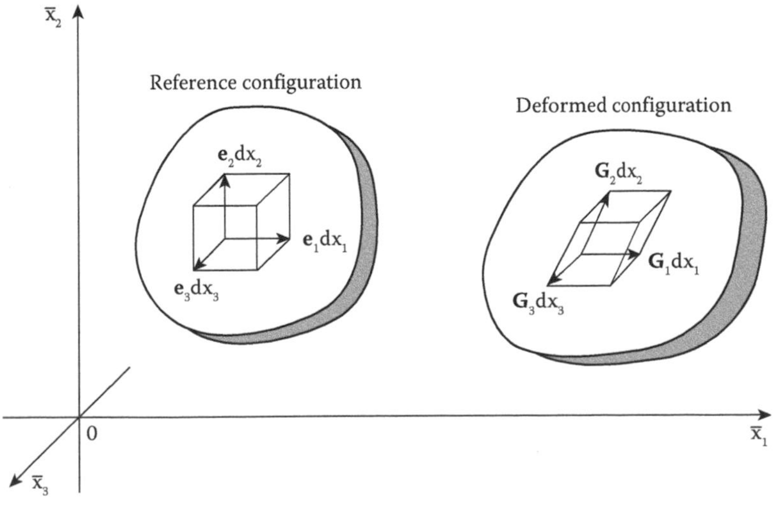

Figure 2.4 Deformation of a volume element.

2.2.2 Area elements

We now consider the deformation of an element of area during the motion that is important for definitions of stress, as well as for traction boundary conditions.

An element of area is a vector quantity, having both a scalar magnitude dA and an orientation, which can be specified by its unit normal vector **n**. As shown in Figure 2.5, the element of area $\mathbf{dA} = \mathbf{n}\,\mathrm{dA}$ in the reference configuration is deformed into $\mathbf{d\bar{A}} = \bar{\mathbf{n}}\,\mathrm{d\bar{A}}$ in the deformed configuration. We first express the element of area as the cross product of two vectors **da** and **db** such that $\mathbf{da} \times \mathbf{db} = \mathbf{dA}$.

These line elements are transformed into $\mathbf{d\bar{a}}$ and $\mathbf{d\bar{b}}$ in the deformed configuration such that

$$\mathbf{d\bar{A}} = \mathbf{d\bar{a}} \times \mathbf{d\bar{b}} \tag{2.38}$$

The vectors $\mathbf{d\bar{a}}$ and $\mathbf{d\bar{b}}$ can be represented by

$$\begin{cases} \mathbf{d\bar{a}} = \mathrm{d\bar{a}}_i \mathbf{e}_i = \left(\dfrac{\partial \bar{x}_i}{\partial x_j} \mathrm{da}_j \right) \mathbf{e}_i = \mathbf{G}_j \mathrm{da}_j \\ \mathbf{d\bar{b}} = \mathbf{G}_k \; \mathrm{db}_k \end{cases} \tag{2.39}$$

Use of Equations 2.39 in Equation 2.38 gives

$$\mathbf{d\bar{A}} = \mathbf{d\bar{a}} \times \mathbf{d\bar{b}} = (\mathbf{G}_j \times \mathbf{G}_k)\mathrm{da}_j \; \mathrm{db}_k \tag{2.40}$$

We now take the inner (dot) product of $\mathbf{G}_i$ with both sides of Equation 2.40 to give the relation

$$\mathbf{G}_i \cdot \mathbf{d\bar{A}} = \mathbf{G}_i \cdot (\mathbf{G}_j \times \mathbf{G}_k)\mathrm{da}_j \; \mathrm{db}_k \tag{2.41}$$

With Equation 2.35, Equation 2.41 becomes

$$\mathbf{G}_i \cdot \mathbf{d\bar{A}} = \varepsilon_{ijk} |\mathbf{F}| \mathrm{da}_j \; \mathrm{db}_k \tag{2.42}$$

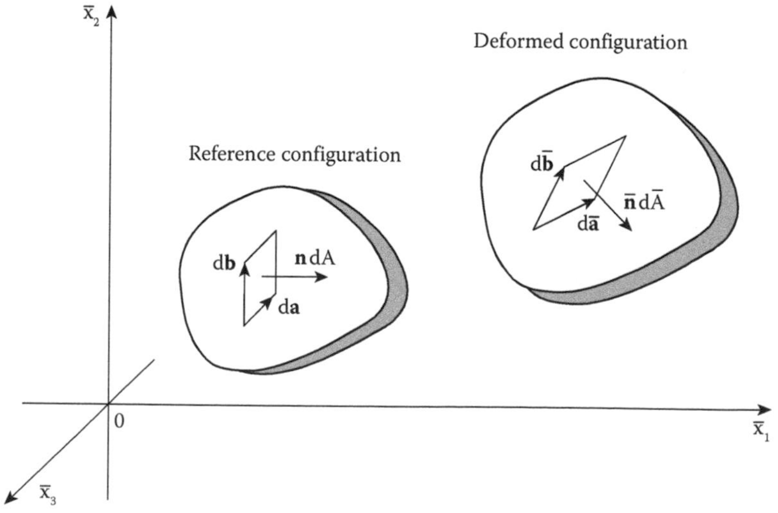

Figure 2.5 Deformation of an area element.

The left-hand side of Equation 2.42 can be written as

$$\mathbf{G}_i \cdot d\overline{\mathbf{A}} = F_{ki}\ \mathbf{e}_k \cdot (d\overline{A}_m \mathbf{e}_m) = F_{ki} d\overline{A}_k \quad (2.43)$$

now giving, with Equation 2.42,

$$F_{ki} d\overline{A}_k = |\mathbf{F}| dA_i \quad (2.44)$$

which, expressed in vector, rather than component, form gives finally

$$\mathbf{F}^T d\overline{\mathbf{A}} = |\mathbf{F}| d\mathbf{A} \quad (2.45)$$

which can also be written as

$$\mathbf{F}^T \overline{\mathbf{n}}\ d\overline{A} = |\mathbf{F}| \mathbf{n}\ dA \quad (2.46)$$

Equation 2.46 for the deformation of area elements is known as *Nanson's formula*.

Note that relations for the deformation of volume elements, Equation 2.34, and deformation of area elements, Equation 2.44, are easily adaptable to the Eulerian formulation, as well as the Lagrangian formulation.

2.3 STRAINS

In order to prescribe the constitutive relations of a material (i.e., to relate deformations to stresses), it is highly desirable to quantify local deformation in the neighborhood of a material point by means of a symmetric tensor, called the *strain* tensor. The most natural approach is to properly extend or generalize to the large deformation realm concepts of strain that are familiar and well established in conventional small deformation linear theory. There are several requirements that must be met in this process. These requirements are formalized in Section 2.4, but we will make use of them here in defining appropriate measures of strain (and later, stress).

One of the most fundamental and obvious requirements is that a valid measure of strain must be insensitive to rigid body motions. The conventional definition of strain in linear small displacement theory as the (normalized) change in the length of a line element does not satisfy this requirement when extended to large deformations and rotations.

2.3.1 Lagrangian (Green) strain

In the Lagrangian description of deformation, the metric tensor **C**, which measures the (squared) deformed length of an arbitrary line element, was defined in Equations 2.18 and 2.19, that is, $d\bar{s}^2 = d\mathbf{x}^T\mathbf{C}\ d\mathbf{x}$. Since $d\bar{s}$ and $d\bar{s}^2$ are scalar quantities, they are invariant under rigid body motions. However, **C** measures the deformed length of the line element, not the change in length due to the deformation. The Green strain does this as follows (Figure 2.6).

Consider the change in the (squared) length of the line element due to the deformation:

$$\begin{aligned} d\bar{s}^2 - ds^2 &= d\overline{\mathbf{x}}^T d\overline{\mathbf{x}} - d\mathbf{x}^T d\mathbf{x} \\ &= d\mathbf{x}^T (\mathbf{C} - \mathbf{I}) d\mathbf{x} \\ &= d\mathbf{x}^T (\mathbf{F}^T\mathbf{F} - \mathbf{I}) d\mathbf{x} \end{aligned} \quad (2.47)$$

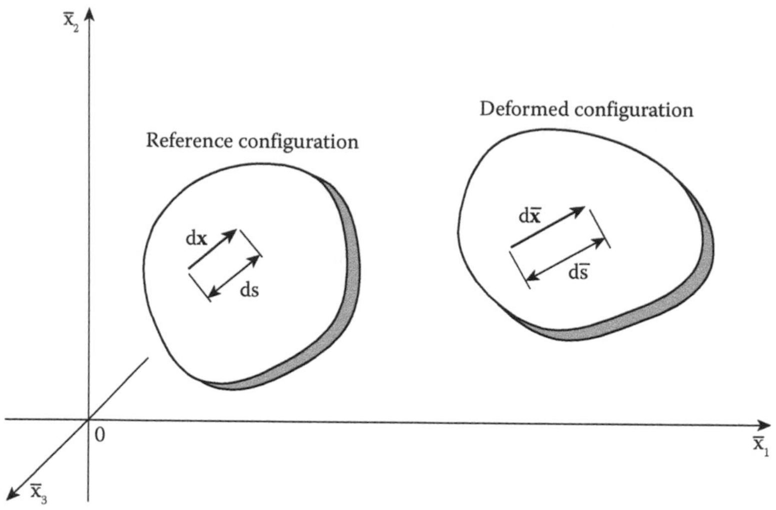

Figure 2.6 Definition of Green strain.

We now normalize by the reference length ds,

$$\begin{aligned}\frac{d\bar{s}^2 - ds^2}{ds^2} &= \frac{d\mathbf{x}^T}{ds}(\mathbf{F}^T\mathbf{F} - \mathbf{I})\frac{d\mathbf{x}}{ds} \\ &= \frac{d\mathbf{x}^T}{ds}(2\varepsilon)\frac{d\mathbf{x}}{ds}\end{aligned} \tag{2.48}$$

and multiply both sides of Equation 2.48 by $\frac{1}{2}$ to arrive at the Green (Lagrangian) definition of strain $\boldsymbol{\varepsilon}$ as

$$\frac{1}{2}\frac{(d\bar{s}^2 - ds^2)}{ds^2} \equiv \frac{d\mathbf{x}^T}{ds}\,\boldsymbol{\varepsilon}\,\frac{d\mathbf{x}}{ds} \tag{2.49}$$

where

$$\varepsilon = \frac{1}{2}(\mathbf{F}^T\mathbf{F} - \mathbf{I}) \tag{2.50}$$

The rationale for the incorporation of the $\frac{1}{2}$ factor in the definition will become clear later. Note that the unit vector $d\mathbf{x}/ds$ simply specifies the direction of an arbitrary line element. In terms of the displacement gradient $\mathbf{G}$, the Green strain tensor is

$$\varepsilon = \frac{1}{2}(\mathbf{C} - \mathbf{I}) = \frac{1}{2}(\mathbf{G} + \mathbf{G}^T + \mathbf{G}^T\mathbf{G}) \tag{2.51}$$

In terms of the Cartesian components of displacement u_i, the Green strain–displacement relations are

$$\varepsilon_{ij} = \frac{1}{2}(u_{i,j} + u_{j,i} + u_{m,i}u_{m,j}) \tag{2.52}$$

in which the linear and nonlinear components of the strains can be clearly identified. The linear portion of the Green strain ε_{ij}^{L} is

$$\varepsilon_{ij}^{L} = \frac{1}{2}(u_{i,j} + u_{j,i}) \tag{2.53}$$

which is the conventional definition of engineering strain for small deformations. This is the reason why the factor $\frac{1}{2}$ is introduced into the definition of Green strain.

In the two-dimensional case (plane strain), with indices (1, 2) replaced by indices (x, y), the Green strains are

$$\begin{aligned}
\varepsilon_{xx} &= \frac{\partial u_x}{\partial x} + \frac{1}{2}\left(\left(\frac{\partial u_x}{\partial x}\right)^2 + \left(\frac{\partial u_y}{\partial x}\right)^2\right) \\
\varepsilon_{yy} &= \frac{\partial u_y}{\partial y} + \frac{1}{2}\left(\left(\frac{\partial u_x}{\partial y}\right)^2 + \left(\frac{\partial u_y}{\partial y}\right)^2\right) \\
\varepsilon_{xy} &= \frac{1}{2}\left(\frac{\partial u_x}{\partial y} + \frac{\partial u_y}{\partial x} + \frac{\partial u_x}{\partial x}\frac{\partial u_x}{\partial y} + \frac{\partial u_y}{\partial x}\frac{\partial u_y}{\partial y}\right)
\end{aligned} \tag{2.54}$$

It is worth noting that while Green strain is given here in Cartesian tensor (or matrix) form, in finite element applications strain (and stress) tensors are for convenience converted to vector forms. For example, again in two dimensions (plane strain), the Green strain tensor is

$$\boldsymbol{\varepsilon} = \begin{bmatrix} \varepsilon_{xx} & \varepsilon_{xy} \\ \varepsilon_{yx} & \varepsilon_{yy} \end{bmatrix} \tag{2.55}$$

whereas the vector form of strain is

$$\mathbf{e} = \begin{Bmatrix} \varepsilon_{xx} \\ \varepsilon_{yy} \\ \gamma_{xy} \end{Bmatrix} \tag{2.56}$$

where $\gamma_{xy} \equiv \varepsilon_{xy} + \varepsilon_{yx} = 2\varepsilon_{xy}$.

2.3.1.1 Interpretation of Green strain components

We now consider some simple examples. Consider the rectangular plate shown in Figure 2.7.

The plate is stretched uniformly in the x_1 direction such that every material point with initial coordinate x_1 is displaced to $(1 + e)\,x_1$, that is,

$$\bar{x}_1 = (1 + e)x_1$$

$$\bar{x}_2 = x_2$$

$$\bar{x}_3 = x_3$$

The deformation gradient **F** is

$$\mathbf{F} = \begin{bmatrix} (1+e) & 0 & 0 \\ 0 & 1 & 0 \\ 0 & 0 & 1 \end{bmatrix} \tag{2.57}$$

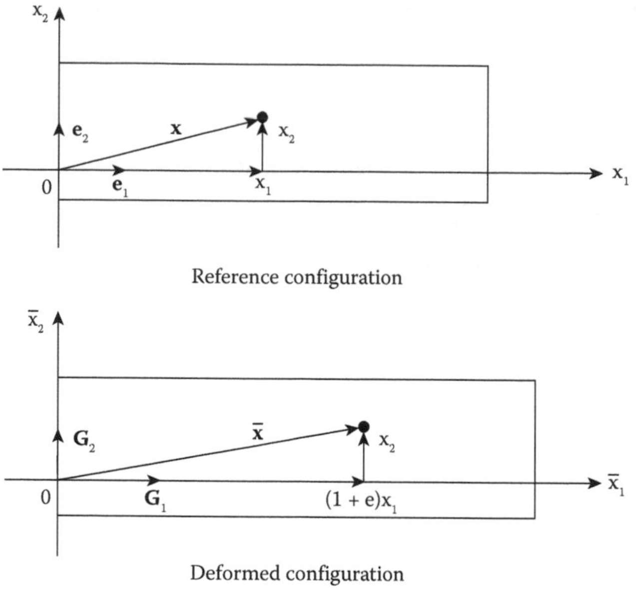

Figure 2.7 Uniaxially stretched plate.

The metric tensor $\mathbf{C}$ ($=\mathbf{F}^T\ \mathbf{F}$) is

$$C = \begin{bmatrix} (1+e)^2 & 0 & 0 \\ 0 & 1 & 0 \\ 0 & 0 & 1 \end{bmatrix} \tag{2.58}$$

and the (Lagrangian) Green strain is

$$\varepsilon = \frac{1}{2}(\mathbf{C}-\mathbf{I}) = \begin{bmatrix} e+\frac{1}{2}e^2 & 0 & 0 \\ 0 & 0 & 0 \\ 0 & 0 & 0 \end{bmatrix} \tag{2.59}$$

The Green strain $\varepsilon_{11} = e + \frac{1}{2}e^2$ measures the strain of a line element that lies along the x_1 axis in the reference configuration. All other strain components are zero, as expected. The linear portion of the Green strain is $\varepsilon^L_{11} = e$. For extensional deformations even as large as, say, 0.10, the Green measure of strain would exceed the linear approximation by only 5%.

Now consider the same rectangular plate stretched uniformly as before, but now also rotated about the x_3 axis through an angle θ. The deformed configuration is now shown in Figure 2.8.

The coordinates of a point in the deformed configuration are now

$$\bar{x}_1 = (1+e)\ \cos\ \theta\ x_1 - \sin\ \theta\ x_2$$

$$\bar{x}_2 = (1+e)\ \sin\ \theta\ x_1 + \cos\ \theta x_2$$

$$\bar{x}_3 = x_3$$

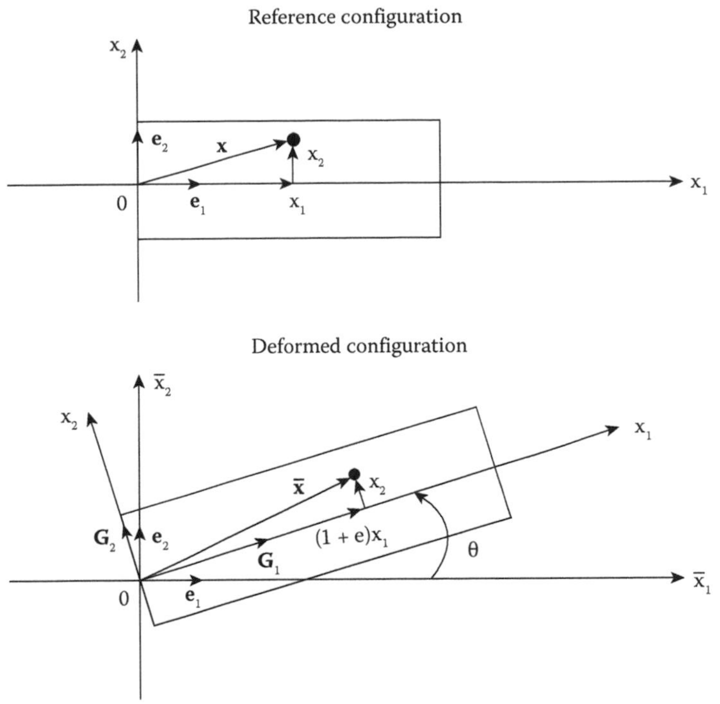

Figure 2.8 Plate stretched and rotated.

and the corresponding deformation gradient is

$$\mathbf{F} = \begin{bmatrix} (1+e)\cos\theta & -\sin\theta & 0 \\ (1+e)\sin\theta & \cos\theta & 0 \\ 0 & 0 & 1 \end{bmatrix} \tag{2.60}$$

The metric tensor **C** and the Green strain are as before:

$$\mathbf{C} = \begin{bmatrix} (1+e)^2 & 0 & 0 \\ 0 & 1 & 0 \\ 0 & 0 & 1 \end{bmatrix} \tag{2.61}$$

and

$$\varepsilon = \frac{1}{2}(\mathbf{C}-\mathbf{I}) = \begin{bmatrix} e+\frac{1}{2}e^2 & 0 & 0 \\ 0 & 0 & 0 \\ 0 & 0 & 0 \end{bmatrix} \tag{2.62}$$

Therefore, the metric tensor and the Green strain tensor are independent of finite rigid body rotations, as required for a valid deformation or strain measure.

2.3.1.2 *Diagonal components of Green strain*

The diagonal components of the Green strain tensor can each be interpreted in the same way as just shown for the uniaxially stretched plate. Under a general deformation, each of the material axes (x_1, x_2, x_3) in the neighborhood of a material point will be deformed and rotated.

A line element of length dx_1 lying along the x_1 axis in the reference configuration will be deformed into a line element (vector) $d\bar{x} = G_1 dx_1$. The normalized length of the stretched and rotated line element is

$$\frac{|d\bar{x}|}{dx_1} = \sqrt{G_1 \cdot G_1} = \sqrt{1 + 2\varepsilon_{11}} \tag{2.63}$$

The extension (change in length/original length) e_i of each of the axes x_i is then

$$e_i = \sqrt{1 + 2\varepsilon_{ii}} - 1 \quad i = 1, 2, 3 \text{ (no sum)} \tag{2.64}$$

2.3.1.3 *Off-diagonal components of Green strain*

In linear small deformation theory, (twice) the off-diagonal components of the strain tensor are physically interpreted as shear strains, the reduction in angle between pairs of lines (coordinate axes) originally perpendicular to each other. For example, considering the pair of axes x_1 and x_2, the shear strain $\gamma_{12} = 2\varepsilon_{12}$ is the reduction in the angle (from $\pi/2$) between the axes x_1 and x_2, under the assumption that the displacement gradients are negligible compared with unity.

In the Lagrangian nonlinear deformation theory, the off-diagonal Green strains can also be related to angle changes between pairs of (convected) material axes in the deformed configuration.

Consider the angle between the axes x_1 and x_2 in the deformed configuration, for example. The convected base vectors G_1 and G_2 contain all the relevant information.

The current angle between the convected axes in the deformed configuration is

$$\cos\theta = \frac{G_1 \cdot G_2}{|G_1||G_2|} \tag{2.65}$$

which can be expressed in terms of Green strains as

$$\sin\gamma = \frac{2\varepsilon_{12}}{(\sqrt{1 + 2\varepsilon_{11}})(\sqrt{1 + 2\varepsilon_{22}})} \tag{2.66}$$

in which γ ($= \pi/2 - \theta$) is the *change* in angle between axes x_1 and x_2 due to the deformation.

2.3.1.4 *Simple shear*

We now consider a plate undergoing a shearing deformation (shown in Figure 2.9) in which parallel planes normal to the x_2 axis simply slide a distance in the x_1 direction proportional to the x_2 coordinate.

The mapping from material to spatial coordinates is

$$\bar{x}_1 = x_1 + \kappa x_2$$
$$\bar{x}_2 = x_2$$
$$\bar{x}_3 = x_3$$

which gives the deformation gradient

$$\mathbf{F} = \begin{bmatrix} 1 & \kappa & 0 \\ 0 & 1 & 0 \\ 0 & 0 & 1 \end{bmatrix} \tag{2.67}$$

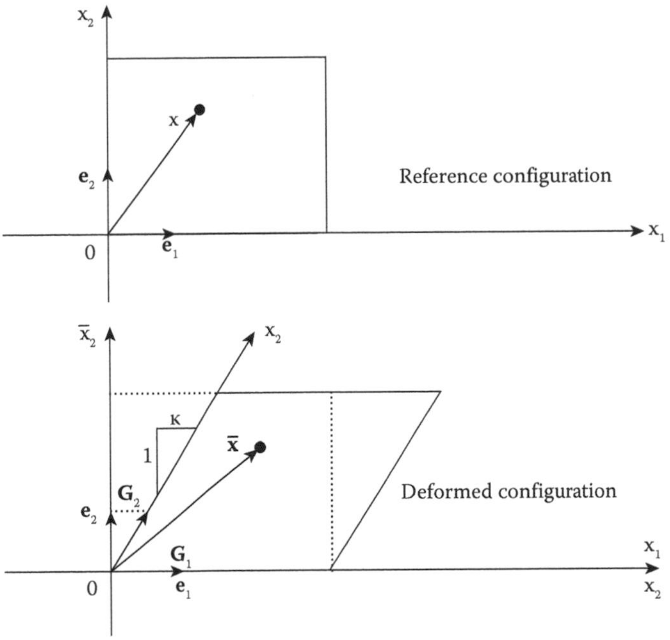

Figure 2.9 Simple shear.

The Green strain is

$$\varepsilon = \frac{1}{2}\begin{bmatrix} 0 & \kappa & 0 \\ \kappa & \kappa^2 & 0 \\ 0 & 0 & 0 \end{bmatrix} \tag{2.68}$$

We will return to this simple shear example later, but for now, we simply compare the geometry of the deformation with Equations 2.64 and 2.66.

Equation 2.64 gives for $e_2 (= |\bar{\mathbf{n}}| - |\mathbf{n}|$ in Figure 2.9), the (normalized) extension of a line originally in the x_2 direction, the value $\sqrt{1+\kappa^2} - 1$, which agrees with the geometry of the deformation shown. Equation 2.66 gives for the sine of the angle change between the x_1, x_2 axes the value $\sin\gamma = \kappa/\sqrt{1+\kappa^2}$, which again agrees with the geometry shown in Figure 2.10.

2.3.2 Eulerian (Almansi) strain

The Eulerian strain formulation starts from the same invariant definition of deformation $(d\bar{s}^2 - ds^2)$ as the Lagrangian formulation, but instead refers all quantities to the (current) deformed configuration. The base vectors $\overline{\mathbf{G}}_i$ defined in the reference configuration that are convected into the Cartesian unit base vectors in the deformed configuration (Figure 2.3) are defined in terms of the Eulerian deformation gradient $\overline{F}_{ij} (= \partial x_i/\partial \bar{x}_j)$ in Equations 2.24 and 2.25. The Eulerian (Almansi) strain tensor $\bar{\varepsilon}$ is defined by first expressing $(d\bar{s}^2 - ds^2)$ as

$$\begin{aligned} d\bar{s}^2 - ds^2 &= d\bar{\mathbf{x}}^T d\bar{\mathbf{x}} - d\mathbf{x}^T d\mathbf{x} \\ &= d\bar{\mathbf{x}}^T (\mathbf{I} - \overline{\mathbf{C}}) d\bar{\mathbf{x}} \end{aligned} \tag{2.69}$$

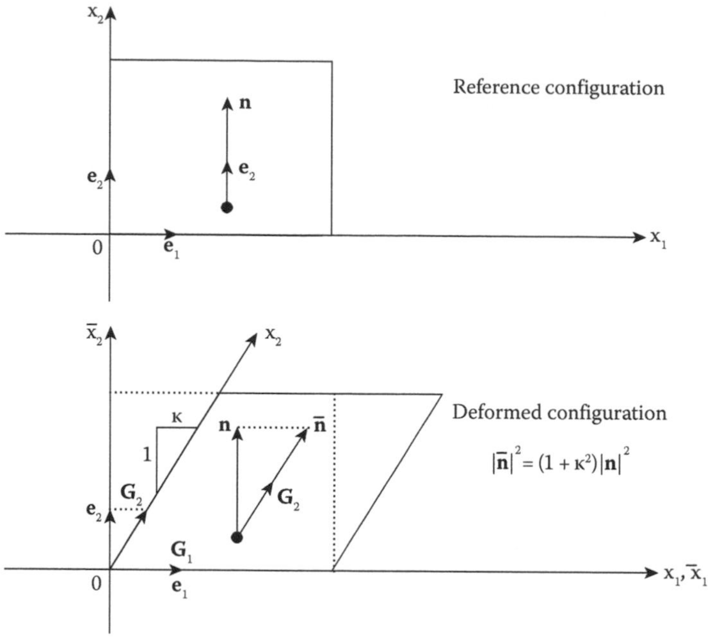

Figure 2.10 Geometry of simple shear deformation compared with Green strains.

where $\bar{\mathbf{C}} \equiv \bar{\mathbf{F}}^T\bar{\mathbf{F}}$, and then normalizing by the deformed length (cf. Equation 2.48):

$$\frac{1}{2}\left(\frac{d\bar{s}^2 - ds^2}{d\bar{s}^2}\right) = \frac{d\bar{\mathbf{x}}^T}{d\bar{s}}(\bar{\boldsymbol{\varepsilon}})\frac{d\bar{\mathbf{x}}}{d\bar{s}} \tag{2.70}$$

Note that the unit vector $\frac{d\bar{\mathbf{x}}}{d\bar{s}}$ in Equation 2.70 specifies a direction in the *deformed* configuration. Therefore, the diagonal components of the Almansi strain tensor measure extensions of line elements that are in the Cartesian $\bar{x}_1, \bar{x}_2, \bar{x}_3$ directions in the deformed configuration, and that were originally aligned along the base vectors $\bar{\mathbf{G}}_i$ in the reference configuration.

Using Equation 2.32, the Almansi strain tensor can be expressed in terms of the spatial displacemen gradient $\bar{\mathbf{G}}$ (cf. Equation 2.51) as

$$\bar{\boldsymbol{\varepsilon}} \equiv \frac{1}{2}(\mathbf{I} - \bar{\mathbf{C}}) = \frac{1}{2}(\bar{\mathbf{G}} + \bar{\mathbf{G}}^T - \bar{\mathbf{G}}^T\bar{\mathbf{G}}) \tag{2.71}$$

where

$$\bar{G}_{ij} = \frac{\partial u_i}{\partial \bar{x}_j} \equiv \bar{u}_{i,j} \tag{2.72}$$

In terms of the Cartesian components of displacement (cf. Equation 2.52), the components of the Almansi strain tensor are

$$\bar{\varepsilon}_{ij} = \frac{1}{2}(\bar{u}_{i,j} + \bar{u}_{j,i} - \bar{u}_{m,i}\bar{u}_{m,j}) \tag{2.73}$$

2.3.2.1 *Uniaxial stretching*

We again consider the plate shown in Figure 2.7. Recall that the mapping from reference to deformed configuration was

$$\bar{x}_1 = (1+e)x_1$$
$$\bar{x}_2 = x_2$$
$$\bar{x}_3 = x_3$$

which can be rewritten as

$$x_1 = \frac{1}{(1+e)}\bar{x}_1$$
$$x_2 = \bar{x}_2$$
$$x_3 = \bar{x}_3$$

corresponding to the Eulerian deformation gradient

$$\bar{\mathbf{F}} = \begin{bmatrix} \frac{1}{(1+e)} & 0 & 0 \\ 0 & 1 & 0 \\ 0 & 0 & 1 \end{bmatrix} \tag{2.74}$$

The Almansi strain can be expressed as

$$\bar{\varepsilon} = \frac{1}{2}(\mathbf{I} - \bar{\mathbf{F}}^T\bar{\mathbf{F}}) \tag{2.75}$$

The only nonzero component of $\bar{\varepsilon}$ is calculated as

$$\bar{\varepsilon}_{11} = \frac{1}{2}\left(1 - \frac{1}{(1+e)^2}\right) = \frac{e + \frac{1}{2}e^2}{(1+e)^2} \tag{2.76}$$

2.3.2.2 *Uniaxial stretching and rigid body rotation*

We now examine the plate shown in Figure 2.8, which is stretched and rotated in order to point out an important property of the Almansi strain tensor. The Lagrangian (material) deformation tensor **F** was given in Equation 2.58. The Eulerian (spatial) deformation gradient $\bar{\mathbf{F}}(\equiv \mathbf{F}^{-1})$ is calculated as

$$\bar{\mathbf{F}} = \begin{bmatrix} \frac{1}{(1+e)}\cos\theta & \frac{1}{(1+e)}\sin\theta & 0 \\ -\sin\theta & \cos\theta & 0 \\ 0 & 0 & 1 \end{bmatrix} \tag{2.77}$$

When we calculate the Almani strain tensor we find

$$\bar{\varepsilon} = \alpha \begin{bmatrix} c^2 & sc & 0 \\ sc & s^2 & 0 \\ 0 & 0 & 0 \end{bmatrix} \tag{2.78}$$

in which the abbreviations $c \equiv \cos\theta$ and $s \equiv \sin\theta$ are used and

$$\alpha = \frac{1}{(1+e)^2}\left(e + \frac{1}{2}e^2\right) \tag{2.79}$$

For the present, we note that the constant α appearing in Equation 2.79 is the Almansi strain component $\bar{\varepsilon}_{11}$ for the case when there is no rigid body rotation superimposed on the stretch. Conversely, for a pure rigid body motion (no stretch) $\bar{\mathbf{F}}^T\bar{\mathbf{F}} = \mathbf{I}$ and $\bar{\varepsilon} = \mathbf{0}$. We observe for future reference that if the spatial Cartesian coordinate system $\bar{x}_i$ were to be rotated through an angle θ about the $\bar{x}_3$ axis to create a new spatial Cartesian coordinate system, say $\bar{x}'_i$ (to which the Almansi strain was then referred), then the effect of the rotation on the Almansi strain tensor would be removed.

More generally, if the Cartesian reference frame is updated by rotating about the $\bar{x}_3$ axis by an arbitrary angle ϕ, the relation between the two Cartesian systems would be $\bar{\mathbf{x}}' = \mathbf{Q}\ \bar{\mathbf{x}}$, where

$$\mathbf{Q} = \begin{bmatrix} \cos\phi & \sin\phi & 0 \\ -\sin\phi & \cos\phi & 0 \\ 0 & 0 & 1 \end{bmatrix} \tag{2.80}$$

is orthogonal ($\mathbf{Q}^T\mathbf{Q} \equiv \mathbf{I}$). With respect to the updated Cartesian system, the Almansi strain is found from

$$\frac{1}{2}(d\bar{s}^2 - ds^2) = d\bar{\mathbf{x}}^T\bar{\varepsilon}\ d\bar{\mathbf{x}} = d\bar{\mathbf{x}}'^T(\mathbf{Q}\bar{\varepsilon}\ \mathbf{Q}^T)d\bar{\mathbf{x}}' \tag{2.81}$$

and therefore

$$\bar{\varepsilon}' = \mathbf{Q}\bar{\varepsilon}\ \mathbf{Q}^T \tag{2.82}$$

which is the general coordinate transformation for Cartesian tensors (matrices). We observe that if ϕ is selected to be equal to θ, the rigid body rotation of the plate in this example, then $\bar{\varepsilon}$ is, as expected, transformed to that calculated for the plate stretched without rotation.

Of course, the difficulty is that rotation in general would be calculated in response to loading, not a value specified *a priori* as in this simple example, and would therefore not be known in advance. However, nonlinear analyses have to be carried out numerically via incremental-iterative algorithms, so the (current) Cartesian reference system could be incrementally updated in the deformed configuration as the response is computed. This will be described subsequently when we discuss strain (and stress) rates.

2.3.2.3 Relation between Green strain and Almansi strain

The relation between the Lagrangian and Eulerian measures of strain will become important when strain (and stress) rates are discussed. To this end, consider the difference in (squared) length of a line element in the deformed and reference configurations $d\bar{s}^2 - ds^2$, which was fundamental in the definitions of Green and Almansi strains,

$$\frac{1}{2}(d\bar{s}^2 - ds^2) = d\mathbf{x}^T(\varepsilon)d\mathbf{x} = d\bar{\mathbf{x}}^T(\bar{\varepsilon})d\bar{\mathbf{x}}$$

Since this relation must be true for an arbitrary line element,

$$\varepsilon = \mathbf{F}^{\mathrm{T}}\bar{\varepsilon}\mathbf{F} \tag{2.83}$$

$$\bar{\varepsilon} = \bar{\mathbf{F}}^{\mathrm{T}}\varepsilon\bar{\mathbf{F}} \tag{2.84}$$

2.4 OBJECTIVITY AND FRAME INDIFFERENCE

We consider now, in general, the effect of a change of Cartesian reference system $(\bar{x}_1, \bar{x}_2, \bar{x}_3)$, specifically a rotation, on various vector and tensor measures, whether motion, deformation, strain, stress, or rates thereof. Suppose in the deformed configuration that we rotate the Cartesian spatial coordinate system to a new coordinate system, say $(\bar{x}'_1, \bar{x}'_2, \bar{x}'_3)$, as illustrated in Figure 2.11 and Equation 2.80 for a plane rotation about the $\bar{x}_3$ axis.

A general rotation in three dimensions is still expressed by means of an orthogonal rotation tensor (matrix) Q such that

$$\bar{\mathbf{x}}' = \mathbf{Q}\,\bar{\mathbf{x}} \tag{2.85}$$

where $\mathbf{Q}^{\mathrm{T}}\,\mathbf{Q} = \mathbf{I}$ and obviously, $\mathbf{Q}^{-1} = \mathbf{Q}^{\mathrm{T}}$. The rotation tensor Q may be constructed in various ways, one of which is via three successive planar rotations, say $\mathbf{Q} = \mathbf{Q}_1\,\mathbf{Q}_2\,\mathbf{Q}_3$ through Euler angles $(\boldsymbol{\varphi}_1, \boldsymbol{\varphi}_2, \boldsymbol{\varphi}_3)$, about each of the coordinate axes in turn.

Objectivity or frame indifference signifies the preservation of certain properties under an arbitrary coordinate rotation. For a vector quantity, objectivity means that the length of

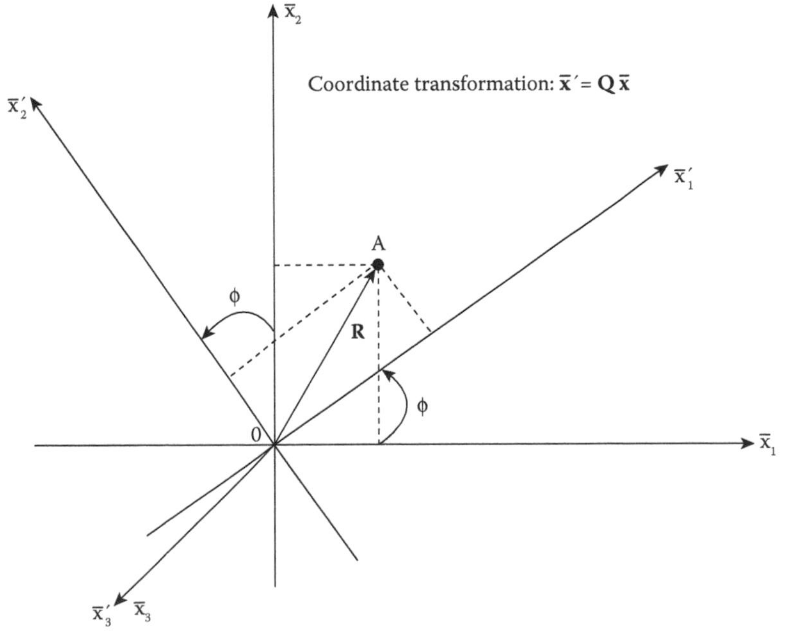

Figure 2.11 Rotation of Cartesian reference frame in two dimensions.

the vector is preserved under a coordinate rotation; that is, for a vector $\mathbf{a}$ that is rotated to $\mathbf{a}'$ via Equation 2.85,

$$\mathbf{a}'^T\mathbf{a}' = \mathbf{a}^T(\mathbf{Q}^T\mathbf{Q})\mathbf{a} = \mathbf{a}^T\mathbf{a} \tag{2.86}$$

which is satisfied for all orthogonal rotation tensors $\mathbf{Q}$.

For a second-rank tensor (matrix), the requirement for objectivity is that the coordinate transformation is a similarity transformation, as in Equation 2.82, that is,

$$\mathbf{A}' = \mathbf{Q}\ \mathbf{A}\ \mathbf{Q}^T \tag{2.87}$$

which preserves its eigenvalues, or principal values. Of course, the eigenvectors of $\mathbf{A}'$ are a rotated version of the eigenvectors of $\mathbf{A}$. To illustrate this, consider the spectral representation of the matrix $\mathbf{A}$ in terms of its eigenvalues and eigenvectors,

$$\mathbf{A} = \sum_i \lambda_i\ \boldsymbol{\varphi}_i\boldsymbol{\varphi}_i^T \tag{2.88}$$

in which $(\lambda_i, \boldsymbol{\varphi}_i)$ are the eigenpairs of $\mathbf{A}$, that is,

$$\mathbf{A}\boldsymbol{\varphi}_i = \lambda_i\ \boldsymbol{\varphi}_i \tag{2.89}$$

Applying the similarity transformation Equation 2.87, we find

$$\mathbf{A}' = \mathbf{Q}\ \mathbf{A}\ \mathbf{Q}^T = \sum_i \lambda_i\ \mathbf{Q}\ \boldsymbol{\varphi}_i\boldsymbol{\varphi}_i^T\mathbf{Q}^T \tag{2.90}$$

$$= \sum_i \lambda'_i\ \boldsymbol{\varphi}'_i\boldsymbol{\varphi}_i'^T \tag{2.91}$$

where $\lambda'_i = \lambda_i$ and $\boldsymbol{\varphi}'_i = \mathbf{Q}\ \boldsymbol{\varphi}_i$ (cf. Equation 2.85). Thus, the eigenvalues of $\mathbf{A}$ are preserved under the arbitrary similarity transformation. The eigenvectors of $\mathbf{A}$ are rotated, but their lengths are preserved.

2.4.1 Objectivity of some deformation measures

2.4.1.1 Deformation gradient

Consider first the effect of coordinate rotation on the deformation gradient $\mathbf{F}$. The Cartesian components of the transformed deformation gradient $\mathbf{F}'$ are given by

$$F'_{ij} = \frac{\partial \bar{x}'_i}{\partial x_j} = Q_{ik}\frac{\partial \bar{x}_k}{\partial x_j} \tag{2.92}$$

Therefore,

$$\mathbf{F}' = \mathbf{Q}\ \mathbf{F} \tag{2.93}$$

and

$$\bar{\mathbf{F}}' = \bar{\mathbf{F}}\ \mathbf{Q}^T \tag{2.94}$$

Therefore, neither the Lagrangian nor the Eulerian deformation gradient is objective.

2.4.1.2 *Metric tensor*

The Lagrangian and Eulerian metric tensors C and $\overline{C}$ are objective, and in fact are invariant under a coordinate rotation as shown by, for example,

$$C' = F'^T F' = F^T Q^T Q\, F = F^T F \tag{2.95}$$

as would be expected since line element lengths are preserved.

2.4.1.3 *Strain tensors*

The Lagrangian (Green) strain tensor $\varepsilon = \frac{1}{2}(C - I)$ is obviously objective, as is the Eulerian (Almansi) strain tensor $\overline{\varepsilon} = \frac{1}{2}(I - \overline{C})$.

2.5 RATES OF DEFORMATION

2.5.1 Velocity and velocity gradient

The *velocity* vector **v** is defined as the time rate of the position of a material particle **x** that is currently at position $\overline{x}$ in the deformed configuration, that is,

$$v \equiv \frac{\partial}{\partial t}\left[\overline{x}(x, t)\right] \tag{2.96}$$

with components

$$v_i = \frac{\partial}{\partial t}\left[\overline{x}_i(x, t)\right] \tag{2.97}$$

The *velocity gradient* tensor V is defined as the spatial gradient of the velocity vector in the deformed configuration. The components of the velocity gradient are

$$V_{ij} = \frac{\partial v_i}{\partial \overline{x}_j} \tag{2.98}$$

or

$$V = \begin{bmatrix} \frac{\partial v_1}{\partial \overline{x}_1} & \frac{\partial v_1}{\partial \overline{x}_2} & \frac{\partial v_1}{\partial \overline{x}_3} \\ \frac{\partial v_2}{\partial \overline{x}_1} & \frac{\partial v_2}{\partial \overline{x}_2} & \frac{\partial v_2}{\partial \overline{x}_3} \\ \frac{\partial v_3}{\partial \overline{x}_1} & \frac{\partial v_3}{\partial \overline{x}_2} & \frac{\partial v_3}{\partial \overline{x}_3} \end{bmatrix} \tag{2.99}$$

2.5.2 Deformation and spin tensors

The velocity gradient tensor is unsymmetric. It can be decomposed into the sum of the (symmetric) *deformation tensor* D and the (skew-symmetric) *spin* tensor W, defined as

$$D = \frac{1}{2}(V + V^T) \tag{2.100}$$

$$\mathbf{W} = \frac{1}{2}(\mathbf{V} - \mathbf{V}^T) \tag{2.101}$$

Thus, of course, $\mathbf{D}^T = \mathbf{D}$ and $\mathbf{W}^T = -\mathbf{W}$. It can be shown that the spin tensor $\mathbf{W}$ corresponds to a local rotation and that the deformation tensor $\mathbf{D}$ defines the local rate of deformation in the neighborhood of a material point. $\mathbf{D} = 0$ is necessary and sufficient to define a rigid body motion.

The components of the deformation tensor are

$$D_{ij} = \frac{1}{2}\left(\frac{\partial v_i}{\partial \overline{x}_j} + \frac{\partial v_j}{\partial \overline{x}_i}\right) \tag{2.102}$$

and the components of the spin tensor are

$$W_{ij} = \frac{1}{2}\left(\frac{\partial v_i}{\partial \overline{x}_j} - \frac{\partial v_j}{\partial \overline{x}_i}\right) \tag{2.103}$$

2.5.3 Deformation gradient rates

The (time) rate of the material deformation gradient $\mathbf{F}$ is determined as follows:

$$\begin{aligned} \frac{\partial}{\partial t} F_{ij} \equiv \dot{F}_{ij} &= \frac{\partial}{\partial t}\left(\frac{\partial \overline{x}_i}{\partial x_j}\right) = \frac{\partial}{\partial x_j}\left(\frac{\partial \overline{x}_i}{\partial t}\right) \\ &= \frac{\partial}{\partial \overline{x}_k}(\dot{\overline{x}}_i)\frac{\partial \overline{x}_k}{\partial x_j} \\ &= V_{ik}F_{kj} \end{aligned} \tag{2.104}$$

where

$$\dot{\overline{x}}_i \equiv \frac{\partial \overline{x}_i}{\partial t}$$

Thus,

$$\dot{\mathbf{F}} = \mathbf{V}\mathbf{F} \tag{2.105}$$

The time rate of the spatial gradient $\overline{\mathbf{F}}$ is determined as follows: Since $\mathbf{F}\overline{\mathbf{F}} \equiv \mathbf{I}$ and $\dot{\mathbf{I}} = 0$,

$$\begin{aligned} &\dot{\mathbf{F}}\overline{\mathbf{F}} + \mathbf{F}\dot{\overline{\mathbf{F}}} = 0 \\ &\dot{\overline{\mathbf{F}}} = -\overline{\mathbf{F}}\dot{\mathbf{F}}\overline{\mathbf{F}} = -\overline{\mathbf{F}}(\mathbf{V}\mathbf{F})\overline{\mathbf{F}} \end{aligned}$$

Therefore,

$$\dot{\overline{\mathbf{F}}} = -\overline{\mathbf{F}}\mathbf{V} \tag{2.106}$$

2.5.4 Rate of deformation of a line element

The rate of deformation of a line element $d\overline{\mathbf{x}}$ in the deformed configuration is found starting from Equation 2.10, that is, $d\overline{\mathbf{x}} = \mathbf{F}d\mathbf{x}$. Taking the time rate, we find

$$d\dot{\overline{\mathbf{x}}} = \dot{\mathbf{F}}d\mathbf{x} = \mathbf{V}\mathbf{F}d\mathbf{x} = \mathbf{V}d\overline{\mathbf{x}} \tag{2.107}$$

which can be expressed as

$$d\dot{\bar{x}} = (\mathbf{D} + \mathbf{W})d\bar{x}$$

2.5.5 Interpretation of deformation and spin tensors

The tensors **D** and **W** provide significant information about the rate of deformation in the neighborhood of a material point in the current deformed configuration.

Consider the three line elements (currently) lying along the Cartesian coordinate axes in the deformed configuration. For the line element lying along the $\bar{x}_1$ axis (i.e., $d\bar{x}_1 e_1$), we find from the general relation (Equation 2.100)

$$\begin{Bmatrix} d\dot{\bar{x}}_1 \\ d\dot{\bar{x}}_2 \\ d\dot{\bar{x}}_3 \end{Bmatrix} = \begin{bmatrix} \frac{\partial v_1}{\partial \bar{x}_1} & \frac{\partial v_1}{\partial \bar{x}_2} & \frac{\partial v_1}{\partial \bar{x}_3} \\ \frac{\partial v_2}{\partial \bar{x}_1} & \frac{\partial v_2}{\partial \bar{x}_2} & \frac{\partial v_2}{\partial \bar{x}_3} \\ \frac{\partial v_3}{\partial \bar{x}_1} & \frac{\partial v_3}{\partial \bar{x}_2} & \frac{\partial v_3}{\partial \bar{x}_3} \end{bmatrix} \begin{Bmatrix} d\bar{x}_1 \\ 0 \\ 0 \end{Bmatrix} = \begin{Bmatrix} \frac{\partial v_1}{\partial \bar{x}_1} \\ \frac{\partial v_2}{\partial \bar{x}_1} \\ \frac{\partial v_3}{\partial \bar{x}_1} \end{Bmatrix} d\bar{x}_1 \tag{2.108}$$

As shown in Figure 2.12, the line element is in general both stretched and rotated by the deformation. The extension of the line element in its original direction (i.e., $(\partial v_1/\partial \bar{x}_1)d\bar{x}_1 dt$) is given by $D_{11}d\bar{x}_1 dt$. Therefore, the diagonal terms of the deformation tensor **D** correspond to rates of elongation along the coordinate axes in the deformed configuration.

Note in Figure 2.12 that the line element $d\bar{x}_1 e_1$ also undergoes motion in the $\bar{x}_2$ and $\bar{x}_3$ directions. Line elements $d\bar{x}_2 e_2$ and $d\bar{x}_3 e_3$ undergo similar motions in all three coordinate directions, which leads to a combination of distortion (shear deformation) and rotation.

Figure 2.13 shows the deformations in the $\bar{x}_1$, $\bar{x}_2$ plane of the pair of line elements $d\bar{x}_1 e_1$ and $d\bar{x}_2 e_2$. The tensorial shear deformation is $1/2(\partial v_1/\partial \bar{x}_2 + \partial v_2/\partial \bar{x}_1)dt$, the average of the two angles indicated in Figure 2.13. These terms are given by D_{12} and D_{21} (Figure 2.14).

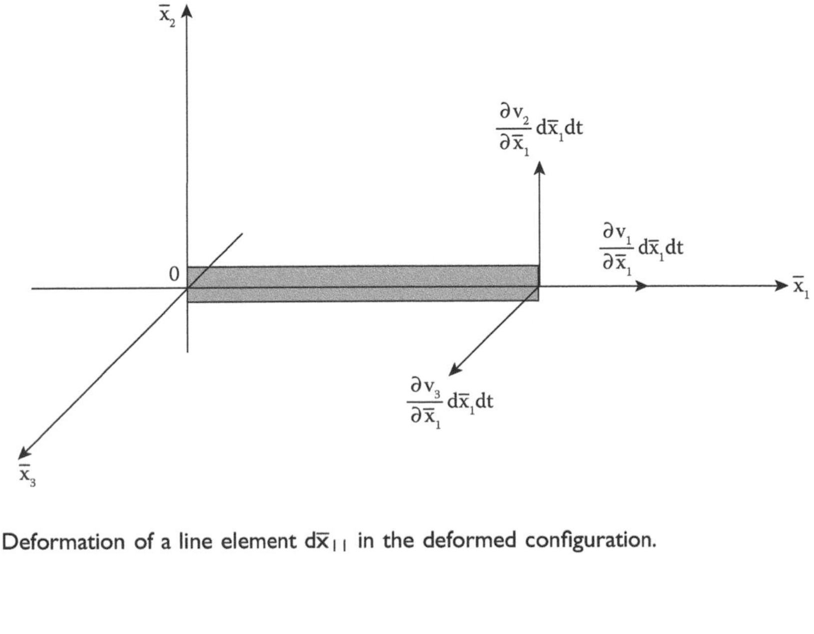

Figure 2.12 Deformation of a line element $d\bar{x}_{11}$ in the deformed configuration.

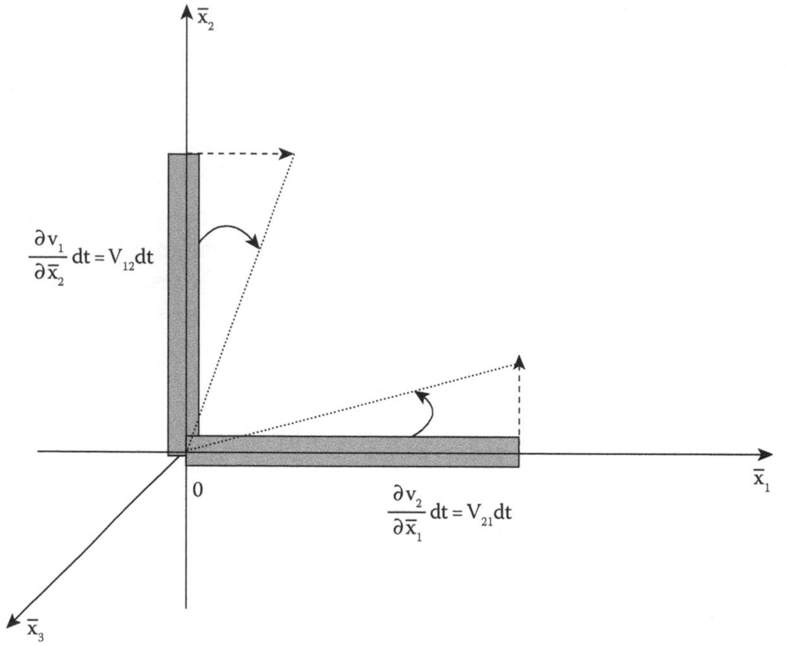

Figure 2.13 Deformation of a pair of orthogonal line elements in the deformed configuration.

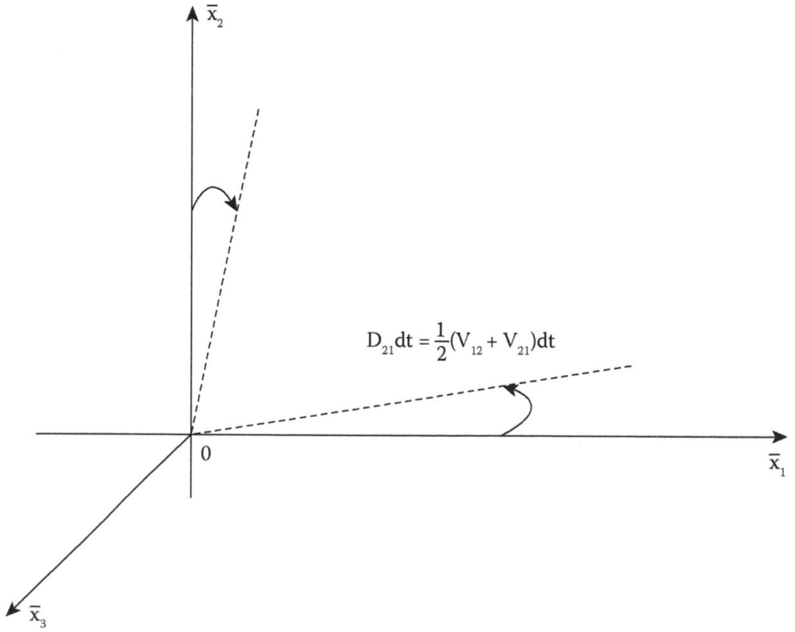

Figure 2.14 Shear deformation between a pair of orthogonal lines in the deformed configuration contributed by the off-diagonal terms of the deformation tensor D.

Note that small-angle approximations are used throughout since the deformations occur during an infinitesimal time interval dt. The elements of the deformation tensor **D** therefore correspond to familiar expressions for linear strain (rates).

Figure 2.15 illustrates the effect of the spin tensor **W** on the pair of orthogonal line elements, which is to rotate them without deformation. A skew-symmetric tensor (matrix) such as **W** can alternatively be represented as a vector, in this case the (rate of) rotation vector $\dot{w}$ shown in the figure.

The vector representation of the spin tensor **W** is defined in component form by

$$W_{ij} = -\varepsilon_{ijk}\dot{w}_k \tag{2.109}$$

The negative sign in Equation 2.109 is introduced to maintain the conventional "right-hand rule" sign convention for vector cross products. The resulting vector representation of the spin tensor is shown explicitly in Equation 2.110.

$$\mathbf{W} = \begin{bmatrix} 0 & -\dot{w}_3 & \dot{w}_2 \\ \dot{w}_3 & 0 & -\dot{w}_1 \\ -\dot{w}_2 & \dot{w}_1 & 0 \end{bmatrix} \tag{2.110}$$

The effect of the spin tensor on an arbitrary line element (i.e., $d\dot{\bar{x}} = \mathbf{W}d\bar{x}$) can be expressed in component form as

$$W_{ij}d\bar{x}_j = -\varepsilon_{ijk}\dot{w}_k d\bar{x}_j = \varepsilon_{ijk}\dot{w}_j d\bar{x}_k \tag{2.111}$$

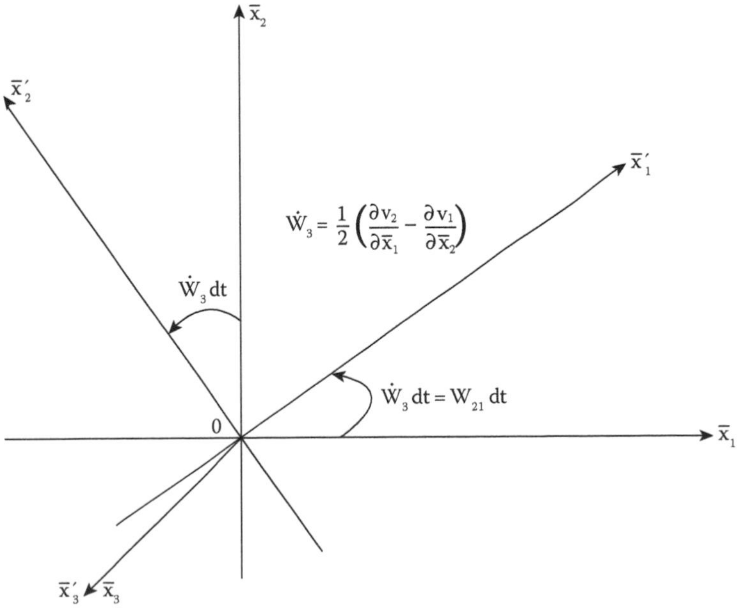

Figure 2.15 Local rotation in $\bar{x}_1, \bar{x}_2$ plane about $\bar{x}_3$ axis in the deformed configuration contributed by the spin tensor **W**.

Equation 2.111 represents a rigid body rotation (rate) of a line element about an axis defined by the vector $\dot{\mathbf{w}}$. The equation can also be put in the vector cross product form:

$$\mathbf{W}d\bar{\mathbf{x}} = \dot{\mathbf{w}} \times d\bar{\mathbf{x}} \tag{2.112}$$

2.5.6 Rate of change of a volume element

The rate of change of a volume element $d\bar{V}$ in the deformed configuration is easily found by noting that the deformation tensor $\mathbf{D}$ contains all the information about deformation rates in the deformed configuration. The spin tensor $\mathbf{W}$, as seen in the previous section, simply rotates line elements as a rigid body. The current volume element $d\bar{V} = d\bar{x}_1\ d\bar{x}_2\ d\bar{x}_3$, and we have seen that the diagonal elements of the deformation tensor give the rate of elongation along each of the coordinate axes. Therefore, in an infinitesimal time interval dt, the volume becomes

$$d\bar{V} + d\dot{\bar{V}}dt = (1 + D_{11}dt)(1 + D_{22}dt)(1 + D_{33}dt)\ d\bar{x}_1 d\bar{x}_2 d\bar{x}_3$$

Subtracting $d\bar{V}$ from both sides of the expression and then linearizing in dt, we find

$$d\dot{\bar{V}} = D_{kk}\ d\bar{V} \equiv \mathrm{tr}(\mathbf{D})d\bar{V} \tag{2.113}$$

where $\mathrm{tr}(\mathbf{D}) = D_{kk}$ is the "trace" (or first invariant) of $\mathbf{D}$.

We also have, from Equation 2.36,

$$d\dot{\bar{V}} = |\dot{\mathbf{F}}|dV \tag{2.114}$$

Therefore,

$$|\dot{\mathbf{F}}| = \mathrm{tr}(\mathbf{D})|\mathbf{F}| \tag{2.115}$$

2.5.7 Rate of change of an area element

We start from Nanson's equation (2.45) and take the (material) time rate

$$\dot{\mathbf{F}}^T d\bar{\mathbf{A}} + \mathbf{F}^T d\dot{\bar{\mathbf{A}}} = |\dot{\mathbf{F}}|d\mathbf{A} \tag{2.116}$$

and using Equation 2.115, we next obtain

$$\dot{\mathbf{F}}^T d\bar{\mathbf{A}} + \mathbf{F}^T d\dot{\bar{\mathbf{A}}} = \mathrm{tr}(\mathbf{D})|\mathbf{F}|d\mathbf{A} \tag{2.117}$$

Using Nanson's equation once again, in Equation 2.117,

$$\dot{\mathbf{F}}^T d\bar{\mathbf{A}} + \mathbf{F}^T d\dot{\bar{\mathbf{A}}} = \mathrm{tr}(\mathbf{D})\mathbf{F}^T d\bar{\mathbf{A}} \tag{2.118}$$

Substituting $\dot{\mathbf{F}} = \mathbf{V}\mathbf{F}$ in Equation 2.118 and solving for $d\dot{\bar{\mathbf{A}}}$, we find

$$d\dot{\bar{\mathbf{A}}} = \bar{\mathbf{F}}^T\mathbf{F}^T\left[\mathrm{tr}(\mathbf{D})\mathbf{I} - \mathbf{V}^T\right]d\bar{\mathbf{A}} = (\mathbf{F}\bar{\mathbf{F}})^T\left[\mathrm{tr}(\mathbf{D})\mathbf{I} - \mathbf{V}^T\right]d\bar{\mathbf{A}} \tag{2.119}$$

$$d\dot{\bar{A}} = \left[\mathrm{tr}(D)I - V^T\right] d\bar{A} \tag{2.120}$$

which may be written, using $V = D + W$, as

$$d\dot{\bar{A}} = \left[W + \mathrm{tr}(D)I - D\right] d\bar{A} \tag{2.121}$$

2.6 STRAIN RATES

In this section, we define valid strain rates that can be used in the specification of material constitutive relations. We first examine the material time rates of the Lagrangian (Green) and Eulerian (Almansi) strain tensors.

2.6.1 Lagrangian (Green) strain rate

Starting from the definition of the Green strain tensor, that is, $\varepsilon = \frac{1}{2}(F^T F - I)$, and taking the time rate, we find

$$\begin{aligned} \dot{\varepsilon} &= \frac{1}{2}(\dot{F}^T F + F^T \dot{F}) \\ &= \frac{1}{2}(F^T V^T F + F^T V\ F) \\ &= F^T D\ F \end{aligned} \tag{2.122}$$

We note that the Green strain rate vanishes in a rigid body motion, for which $D = 0$ is a necessary and sufficient condition. Therefore, $\dot{\varepsilon}$ is objective and is a valid measure of local deformation.

2.6.2 Eulerian (Almansi) strain rate

The material rate of Almansi strain is determined in an analogous manner, starting from the definition. That is, $\bar{\varepsilon} = \frac{1}{2}\left(I - \bar{F}^T \bar{F}\right)$. Taking the time rate, we find

$$\begin{aligned} \dot{\bar{\varepsilon}} &= \frac{1}{2}(-\dot{\bar{F}}^T \bar{F} - \bar{F}^T \dot{\bar{F}}) \\ &= \frac{1}{2}(V^T \bar{F}^T \bar{F} + \bar{F}^T \bar{F} V) \\ &= D - V^T \bar{\varepsilon} - \bar{\varepsilon} V \end{aligned} \tag{2.123}$$

which is not an objective strain rate since the velocity gradient V is not objective. Thus, the material rate of Almansi strain is not suitable for specifying rate-dependent constitutive relations.

However, if we use the relation $V = D + W$ in Equation 2.123 and rearrange, we obtain

$$\dot{\bar{\varepsilon}} - W\bar{\varepsilon} + \bar{\varepsilon} W = D - D\bar{\varepsilon} - \bar{\varepsilon} D \tag{2.124}$$

Since the deformation tensor $D = 0$ in a rigid body rotation, the quantity of the left side of Equation 2.124 is a valid strain rate.

It is called the *Almansi strain rate* $\bar{\varepsilon}^{\nabla}$ and is defined by

$$\bar{\varepsilon}^{\nabla} = \dot{\bar{\varepsilon}} - \mathbf{W}\bar{\varepsilon} + \bar{\varepsilon}\mathbf{W} \tag{2.125}$$

2.7 DECOMPOSITION OF MOTION

The specification of realistic material constitutive relations is a crucial component of nonlinear computational mechanics. To this end, it is important to be able to remove the rotation component from the local deformation in the neighborhood of a material point, since it should not influence material response.

In the context of incremental-iterative nonlinear computation, the material point under consideration would often be a Gaussian integration or sampling point within a finite element, itself contained in a mesh of many elements. It must also be kept in mind that the "rigid body" rotation contained within the various deformation and strain measures varies in general from point to point within a solid body. In order to decompose the motion into true deformation and rotation, we use *polar decomposition.*

2.7.1 Polar decomposition

Polar decomposition is one of several well-known matrix factorizations. Any matrix $\mathbf{A}$ can be expressed as the product of an orthogonal matrix $\mathbf{R}$ (i.e., $\mathbf{R}^T\mathbf{R} = \mathbf{I}$) and a positive semidefinite matrix $\mathbf{U}$ (i.e., eigenvalues ≥ 0) as

$$\mathbf{A} = \mathbf{R}\mathbf{U} \tag{2.126}$$

known as the *right polar decomposition.* The same matrix can be expressed in the *left polar decomposition* as

$$\mathbf{A} = \mathbf{V}\mathbf{R} \tag{2.127}$$

where the same orthogonal matrix appears in both Equations 2.126 and 2.127, and it may be observed that $\mathbf{V}$ is related to $\mathbf{U}$ by a similarity transformation,

$$\mathbf{V} = \mathbf{R}\mathbf{U}\mathbf{R}^T \tag{2.128}$$

which preserves the eigenvalues of $\mathbf{U}$. Therefore, $\mathbf{V}$ is obviously also positive semidefinite. We can think of $\mathbf{V}$ as $\mathbf{U}$ rotated into a new Cartesian coordinate system determined by the rotation matrix $\mathbf{R}$. We will see next that when applied to the various deformation measures, both forms of the polar decomposition have physical interpretations.

2.7.2 Polar decomposition of deformation gradient

The right polar decomposition of the deformation gradient is

$$\mathbf{F} = \mathbf{R}\mathbf{U} \tag{2.129}$$

where $\mathbf{U}$ is called the *right stretch tensor.*

The left polar decomposition of the deformation gradient is

$$\mathbf{F} = \bar{\mathbf{V}}\mathbf{R} \tag{2.130}$$

where $\bar{\mathbf{V}}$ is called the *left stretch tensor.*

The deformation of an arbitrary line element can now be decomposed in two different ways. The first is

$$d\bar{\mathbf{x}} = \mathbf{F}d\mathbf{x} = \mathbf{R}(\mathbf{U}d\mathbf{x}) = \mathbf{R}d\bar{\mathbf{x}}' \tag{2.131}$$

in which the line element $d\mathbf{x}$ is first stretched via the right stretch tensor

$$d\bar{\mathbf{x}}' = \mathbf{U}d\mathbf{x} \tag{2.132}$$

and then rotated without deformation to its final position,

$$d\bar{\mathbf{x}} = \mathbf{R}d\bar{\mathbf{x}}' \tag{2.133}$$

The second alternative is

$$d\bar{\mathbf{x}} = \mathbf{F}d\mathbf{x} = \bar{\mathbf{V}}(\mathbf{R}d\mathbf{x}) = \bar{\mathbf{V}}d\mathbf{x}' \tag{2.134}$$

in which the line element is first rotated without deformation,

$$d\mathbf{x}' = \mathbf{R}d\mathbf{x} \tag{2.135}$$

and then deformed to its final position,

$$d\bar{\mathbf{x}} = \bar{\mathbf{V}}d\mathbf{x}' \tag{2.136}$$

The two intermediate configurations, $d\bar{\mathbf{x}}'$ and $d\mathbf{x}'$, are, of course, not the same. They are related by $d\bar{\mathbf{x}}' = \mathbf{R}^T\bar{\mathbf{V}}d\mathbf{x}'$.

Figure 2.16 illustrates the polar decomposition of the deformation gradient.

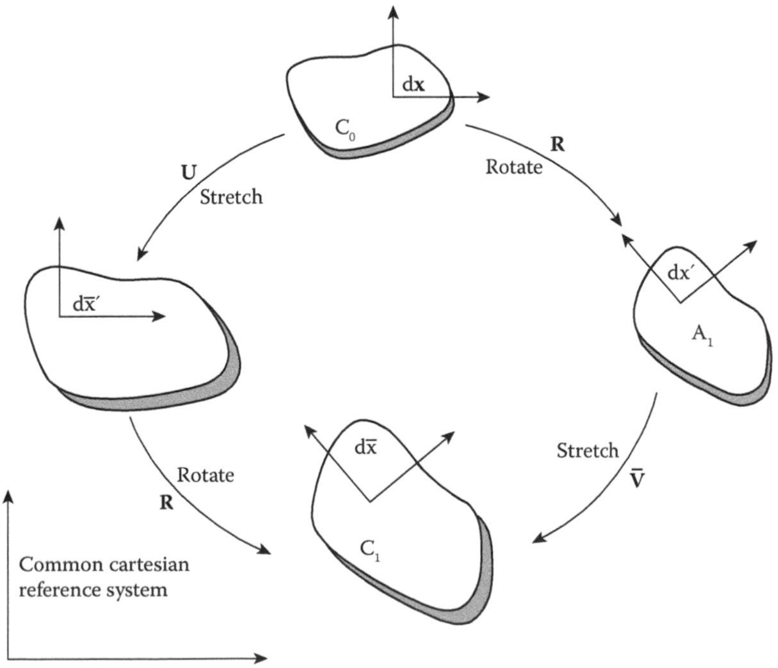

Figure 2.16 Polar decomposition of deformation gradient.

2.7.3 Computation of polar decomposition

2.7.3.1 *Right stretch tensor*

The right stretch tensor **U** can be extracted from the deformation gradient **F** as follows. We first observe that the Lagrangian metric tensor **C** (= $\mathbf{F}^T\mathbf{F}$) can be expressed in the form

$$\mathbf{F}^T\mathbf{F} = \mathbf{U}^T(\mathbf{R}^T\mathbf{R})\mathbf{U} = \mathbf{U}^T\mathbf{U} \tag{2.137}$$

Equation 2.137 may be written as

$$\mathbf{U}^T\mathbf{U} = \mathbf{C} \tag{2.138}$$

which can be solved for **U** as an eigenvalue problem.

If the eigenpairs of **C** are denoted by $(\lambda_i^C, \boldsymbol{\varphi}_i)$, then the eigenpairs of **U** are $(\lambda_i, \boldsymbol{\varphi}_i)$, where $\lambda_i = \sqrt{\lambda_i^C}$. **C** is positive definite, so all $\lambda_i^C > 0$.

The eigenvectors $\boldsymbol{\varphi}_i$ form an orthonormal set of vectors lying in the reference configuration (labeled C_0 in Figure 2.16).

Line elements in C_0 that lie along the special directions $\boldsymbol{\varphi}_i$ that are defined by the deformation remain unrotated during the motion $\mathbf{dx} \to \mathbf{d\bar{x}}'$, described by Equation 2.132. The unit vectors $\boldsymbol{\varphi}_i$ define the *material axes*. The eigenvalues of **U** (= $(\lambda_1, \lambda_2, \lambda_3)$) are termed the *principal stretches*.

Line elements lying in arbitrary directions in C_0 undergo both deformation and rotation during the motion $\mathbf{dx} \to \mathbf{d\bar{x}}'$.

During the next step, $\mathbf{d\bar{x}}' \to \mathbf{d\bar{x}}$, described by Equation 2.133, all line elements are rotated without length change.

2.7.3.2 *Left stretch tensor*

The left stretch tensor $\bar{\mathbf{V}}$ has the same eigenvalues as **U** since the two stretch tensors are related by a similarity transformation (cf. Equation 2.128). The eigenpairs of $\bar{\mathbf{V}}$ are therefore $(\lambda_i, \mathbf{R}\boldsymbol{\varphi}_i)$.

During the first step of the motion, $\mathbf{dx} \to \mathbf{dx}'$, described by Equation 2.135, all line elements in C_0 are rotated without deformation. During the next step of the motion, $\mathbf{dx}' \to \mathbf{d\bar{x}}$, described by Equation 2.136, line elements lying along the directions of the eigenvectors of $\bar{\mathbf{V}}$ (i.e., $\mathbf{R}\boldsymbol{\varphi}_i$, which are the rotated material axes in the intermediate configuration) are stretched, but undergo no further rotation.

Line elements $\mathbf{dx}'$ lying in arbitrary directions in the intermediate configuration undergo both stretching and additional rotation during the motion $\mathbf{dx}' \to \mathbf{d\bar{x}}$.

We see that the left and right polar decompositions are in a sense equivalent, and one or the other may be preferable from a computational standpoint, in different circumstances.

Both methods require the solution of a perhaps very large number of (3 × 3) eigenvalue problems (at each sampling point in a large mesh) during an incremental-iterative numerical simulation. Even though the numerical solution of such a small (3 × 3) eigenvalue problem can be carried out very efficiently, it may still be a significant contributor to overall computational effort.

2.7.4 Strains

The Green strain tensor $\boldsymbol{\varepsilon}$ can be expressed in terms of the right stretch tensor **U** using the polar decomposition of **F** as

$$\boldsymbol{\varepsilon} = \frac{1}{2}(\mathbf{F}^T\mathbf{F} - \mathbf{I}) = \frac{1}{2}(\mathbf{U}^T\mathbf{U} - \mathbf{I}) \tag{2.139}$$

We use the spectral representation of the right stretch tensor,

$$\mathbf{U} = \sum_i \lambda_i \boldsymbol{\varphi}_i \boldsymbol{\varphi}_i^T \tag{2.140}$$

to express Equation 2.139 as

$$\boldsymbol{\varepsilon} = \frac{1}{2}\sum_i (\lambda_i^2 - 1)\boldsymbol{\varphi}_i \boldsymbol{\varphi}_i^T \tag{2.141}$$

in which we use the identity for the unit tensor,

$$\mathbf{I} = \sum_i \boldsymbol{\varphi}_i \boldsymbol{\varphi}_i^T \tag{2.142}$$

Therefore, the principal Green strains ε_i are related to the principal stretches by

$$\varepsilon_i = \frac{1}{2}(\lambda_i^2 - 1) \tag{2.143}$$

The spectral representation of the left stretch tensor $\overline{\mathbf{V}}$ follows directly from that of $\mathbf{U}$ (Equation 2.140) and the tensor transformation $\overline{\mathbf{V}} = \mathbf{R}\ \mathbf{U}\ \mathbf{R}^T$ as

$$\overline{\mathbf{V}} = \sum_i \lambda_i \boldsymbol{\varphi}_i' \boldsymbol{\varphi}_i'^T \tag{2.144}$$

in terms of the rotated eigenvectors

$$\boldsymbol{\varphi}_i' = \mathbf{R}\boldsymbol{\varphi}_i \tag{2.145}$$

The Almansi strain tensor $\bar{\boldsymbol{\varepsilon}}$ can be similarly related to the right stretch tensor $\mathbf{U}$ and its eigenpairs as follows, in which we use Equation 2.140 and the symmetry of $\mathbf{U}$ (and $\mathbf{U}^{-1}$).

$$\bar{\boldsymbol{\varepsilon}} = \frac{1}{2}(\mathbf{I} - \overline{\mathbf{F}}^T\overline{\mathbf{F}}) = \frac{1}{2}(\mathbf{I} - \mathbf{R}\mathbf{U}^{-1}\mathbf{U}^{-1}\mathbf{R}^T) \tag{2.146}$$

$$\bar{\boldsymbol{\varepsilon}} = \frac{1}{2}\mathbf{R}(\mathbf{I} - \mathbf{U}^{-1}\mathbf{U}^{-1})\mathbf{R}^T \tag{2.147}$$

leading to

$$\bar{\boldsymbol{\varepsilon}}' = \frac{1}{2}(\mathbf{I} - \mathbf{U}^{-1}\mathbf{U}^{-1}) \tag{2.148}$$

where $\bar{\boldsymbol{\varepsilon}}$ and $\bar{\boldsymbol{\varepsilon}}'$ are related by the similarity transformation

$$\bar{\boldsymbol{\varepsilon}} = \mathbf{R}\ \bar{\boldsymbol{\varepsilon}}'\ \mathbf{R}^T \tag{2.149}$$

and they therefore have the same eigenvalues. The spectral representation of $\bar{\varepsilon}'$ is

$$\bar{\varepsilon}' = \frac{1}{2}\sum_i \left(1 - \frac{1}{\lambda_i^2}\right)\varphi_i\varphi_i^T \tag{2.150}$$

and therefore that of $\bar{\varepsilon}$ is

$$\bar{\varepsilon} = \frac{1}{2}\sum_i \left(1 - \frac{1}{\lambda_i^2}\right)\varphi_i'\varphi_i'^T \tag{2.151}$$

again in terms of the rotated eigenvectors φ_i' (Equation 2.145). Thus, the principal Almansi strains $\bar{\varepsilon}_i$ in terms of the principal stretches are

$$\bar{\varepsilon}_i = \frac{1}{2}\left(1 - \frac{1}{\lambda_i^2}\right) \tag{2.152}$$

2.7.5 Strain and deformation rates

2.7.5.1 *Green strain rate*

The material rate of Green strain is closely related to the deformation tensor **D**, as shown in Equation 2.122. If we now use the right polar decomposition of **F** in that relation, we obtain

$$\begin{aligned}\dot{\varepsilon} &= \mathbf{F}^T\mathbf{D}\ \mathbf{F} = \mathbf{U}^T(\mathbf{R}^T\mathbf{D}\ \mathbf{R})\mathbf{U} \\ &= \mathbf{U}^T\mathbf{d}\ \mathbf{U}\end{aligned} \tag{2.153}$$

in which the *unrotated deformation rate* **d** is defined by

$$\mathbf{d} = \mathbf{R}^T\mathbf{D}\ \mathbf{R}$$

2.7.5.2 *Material rotation rate*

We start from the deformation gradient rate $\dot{\mathbf{F}} = \mathbf{VF}$,

$$\mathbf{V} = \dot{\mathbf{F}}\mathbf{F}^{-1} \tag{2.154}$$

and introduce into it the left polar decomposition,

$$\mathbf{V} = (\dot{\bar{\mathbf{V}}}\mathbf{R} + \bar{\mathbf{V}}\dot{\mathbf{R}})\mathbf{F}^{-1} \tag{2.155}$$

Using $\mathbf{F}^{-1} = \mathbf{R}^T\bar{\mathbf{V}}^{-1}$ in Equation 2.155, we next obtain

$$\mathbf{V}\ \bar{\mathbf{V}} = \dot{\bar{\mathbf{V}}} + \bar{\mathbf{V}}(\dot{\mathbf{R}}\mathbf{R}^T) \tag{2.156}$$

Because $\mathbf{R}\mathbf{R}^T = \mathbf{I}$, therefore $\dot{\mathbf{R}}\mathbf{R}^T(= -\mathbf{R}\dot{\mathbf{R}}^T = -(\dot{\mathbf{R}}\mathbf{R}^T)^T)$ is skew symmetric. The skew-symmetric material rotation rate tensor $\mathbf{\Omega}$ is defined by $\dot{\mathbf{R}}\mathbf{R}^T = \mathbf{\Omega}$, yielding

$$\dot{\mathbf{R}} \equiv \Omega\mathbf{R} \tag{2.157}$$

$\mathbf{\Omega}$ represents the rate of rotation of the material axes, as distinguished from the spin tensor $\mathbf{W}$, which represents the rate of rotation of the Cartesian axes in the deformed configuration. Equation 2.156 now becomes

$$\mathbf{V}\ \overline{\mathbf{V}} = \dot{\overline{\mathbf{V}}} + \overline{\mathbf{V}}\mathbf{\Omega} \tag{2.158}$$

We wish to separate Equation 2.158 into symmetric and antisymmetric parts, and therefore we subtract from it its transpose, and then insert $\mathbf{V} = \mathbf{D} + \mathbf{W}$ to yield

$$\left[\mathbf{D}\overline{\mathbf{V}} - \overline{\mathbf{V}}\mathbf{D}\right] + \left[\mathbf{W}\overline{\mathbf{V}} + \overline{\mathbf{V}}\mathbf{W}\right] = \left[\overline{\mathbf{V}}\mathbf{\Omega} + \mathbf{\Omega}\overline{\mathbf{V}}\right] \tag{2.159}$$

Each of the three bracketed matrices in Equation 2.159 is antisymmetric, and therefore each can be represented by a vector, as shown previously in Equation 2.109, yielding a system of three linear equations for the vector representation of the material rotation rate $(\dot{\omega}_1, \dot{\omega}_2, \dot{\omega}_3)^T$ in terms of the easily calculated quantities $\mathbf{D}$, $\mathbf{W}$, and $\mathbf{V}$.

The vector representation of Equation 2.159 is

$$\left[\mathrm{tr}(\overline{\mathbf{V}})\mathbf{I} - \overline{\mathbf{V}}\right](\dot{\boldsymbol{\omega}} - \dot{\mathbf{W}}) = \dot{\mathbf{p}} \tag{2.160}$$

where $\mathbf{P} \equiv [\mathbf{D}\overline{\mathbf{V}} - \overline{\mathbf{V}}\mathbf{D}]$, and $P_{ij} = -\varepsilon_{ijk}\dot{p}_k$, yielding the following expression for the material rotation rate (Dienes 1979):

$$\dot{\boldsymbol{\omega}} = \dot{\mathbf{W}} + \left[\mathrm{tr}(\overline{\mathbf{V}})\mathbf{I} - \overline{\mathbf{V}}\right]^{-1}\dot{\mathbf{p}} \tag{2.161}$$

2.7.6 Simple examples

Now that the basic kinematic relations have been introduced, we return to two simple examples that were briefly discussed earlier, to examine them in more detail.

2.7.6.1 Plate stretched and rotated

The plate shown in Figure 2.8 is rotated and stretched. We now consider the uniaxial stretch (or elongation) e and rotation θ to be functions of the loading parameter t, with rates $\dot{e}$ and $\dot{\theta}$, respectively. The mapping from the reference configuration to the (current) deformed configuration is as before,

$$\begin{cases} \overline{x}_1 = (1+e)\cos\theta\ x_1 - \sin\theta\ x_2 \\ \overline{x}_2 = (1+e)\sin\theta\ x_1 + \cos\theta\ x_2 \\ \overline{x}_3 = x_3 \end{cases} \tag{2.162}$$

and the corresponding deformation gradient is

$$\mathbf{F} = \begin{bmatrix} (1+e)\cos\theta & -\sin\theta & 0 \\ (1+e)\sin\theta & \cos\theta & 0 \\ 0 & 0 & 1 \end{bmatrix} \tag{2.163}$$

The inverse of the deformation gradient is

$$\mathbf{F}^{-1} = \begin{bmatrix} \frac{1}{(1+e)}c & \frac{1}{(1+e)}s & 0 \\ -s & c & 0 \\ 0 & 0 & 1 \end{bmatrix} \tag{2.164}$$

in which c ≡ cosθ and s ≡ sinθ.

The deformation gradient rate is

$$\dot{\mathbf{F}} = \dot{e}\begin{bmatrix} c & 0 & 0 \\ s & 0 & 0 \\ 0 & 0 & 0 \end{bmatrix} + \dot{\theta}\begin{bmatrix} -(1+e)s & -c & 0 \\ (1+e)c & -s & 0 \\ 0 & 0 & 0 \end{bmatrix} \tag{2.165}$$

and the velocity gradient **V** is

$$\mathbf{V} = \dot{\mathbf{F}}\mathbf{F}^{-1} = \dot{\theta}\begin{bmatrix} 0 & -1 & 0 \\ 1 & 0 & 0 \\ 0 & 0 & 0 \end{bmatrix} + \frac{\dot{e}}{(1+e)}\begin{bmatrix} c^2 & sc & 0 \\ sc & s^2 & 0 \\ 0 & 0 & 0 \end{bmatrix} \tag{2.166}$$

which splits naturally into symmetric and antisymmetric parts because of the motion parameterization chosen.

The skew-symmetric portion is the spin tensor

$$\mathbf{W} = \dot{\theta}\begin{bmatrix} 0 & -1 & 0 \\ 1 & 0 & 0 \\ 0 & 0 & 0 \end{bmatrix} \tag{2.167}$$

and the symmetric portion is the deformation rate tensor

$$\mathbf{D} = \frac{\dot{e}}{(1+e)}\begin{bmatrix} c^2 & sc & 0 \\ sc & s^2 & 0 \\ 0 & 0 & 0 \end{bmatrix} \tag{2.168}$$

The (right) polar decomposition of **F** yields

$$\mathbf{R} = \begin{bmatrix} \cos\theta & -\sin\theta & 0 \\ \sin\theta & \cos\theta & 0 \\ 0 & 0 & 1 \end{bmatrix} \text{ and } \mathbf{U} = \begin{bmatrix} 1+e & 0 & 0 \\ 0 & 1 & 0 \\ 0 & 0 & 1 \end{bmatrix} \tag{2.169}$$

The left stretch tensor $\overline{\mathbf{V}} = \mathbf{R}\mathbf{U}\mathbf{R}^T$ is

$$\overline{\mathbf{V}} = \begin{bmatrix} 1+ec^2 & esc & 0 \\ esc & 1+es^2 & 0 \\ 0 & 0 & 1 \end{bmatrix} \tag{2.170}$$

and we note again that the left stretch $\overline{\mathbf{V}}$ and the deformation rate **D** are expressed in the deformed configuration.

The unrotated deformation rate is (cf. Equation 2.168)

$$\mathbf{d} = \frac{\dot{e}}{(1+e)}\begin{bmatrix} 1 & 0 & 0 \\ 0 & 0 & 0 \\ 0 & 0 & 0 \end{bmatrix} \tag{2.171}$$

We confirm from Equation 2.171 that d_{11}, for example, measures the logarithmic strain rate (i.e., change of length normalized by current length) of a line lying originally in the $\overline{x}_1$ direction in the reference configuration. At a "time" during the motion when $\theta = 90°$, $d_{11} = \dot{e}/(1+e)$, while $D_{22} = \dot{e}/(1+e)$ measures the logarithmic strain rate of the line in its current orientation, lying along the $\overline{x}_2$ axis.

We complete the example by noting the Green strain,

$$\varepsilon = \frac{1}{2}(U^T U - I) = \left(e + \frac{1}{2}e^2\right)\begin{bmatrix} 1 & 0 & 0 \\ 0 & 0 & 0 \\ 0 & 0 & 0 \end{bmatrix} \tag{2.172}$$

and the Green strain rate,

$$\dot{\varepsilon} = \dot{e}(1+e)\begin{bmatrix} 1 & 0 & 0 \\ 0 & 0 & 0 \\ 0 & 0 & 0 \end{bmatrix} \tag{2.173}$$

2.7.6.2 *Simple shear*

We now consider again the simple shear deformation shown in Figure 2.9. The mapping from material to spatial coordinates is

$$\begin{cases} \overline{x}_1 = x_1 + \kappa x_2 \\ \overline{x}_2 = x_2 \\ \overline{x}_3 = x_3 \end{cases} \tag{2.174}$$

which leads to the deformation gradient

$$F = \begin{bmatrix} 1 & \kappa & 0 \\ 0 & 1 & 0 \\ 0 & 0 & 1 \end{bmatrix} \tag{2.175}$$

The parameter κ is now considered to be a function of the loading parameter t, with rate $\dot{\kappa}$. The Green strain computed previously is

$$\varepsilon = \frac{1}{2}\begin{bmatrix} 0 & \kappa & 0 \\ \kappa & \kappa^2 & 0 \\ 0 & 0 & 1 \end{bmatrix} \tag{2.176}$$

The Green strain rate is found directly as

$$\dot{\varepsilon} = \frac{\dot{\kappa}}{2}\begin{bmatrix} 0 & 1 & 0 \\ 1 & 2\kappa & 0 \\ 0 & 0 & 0 \end{bmatrix} \tag{2.177}$$

The velocity gradient V is

$$V = \begin{bmatrix} 0 & \dot{\kappa} & 0 \\ 0 & 0 & 0 \\ 0 & 0 & 0 \end{bmatrix} \tag{2.178}$$

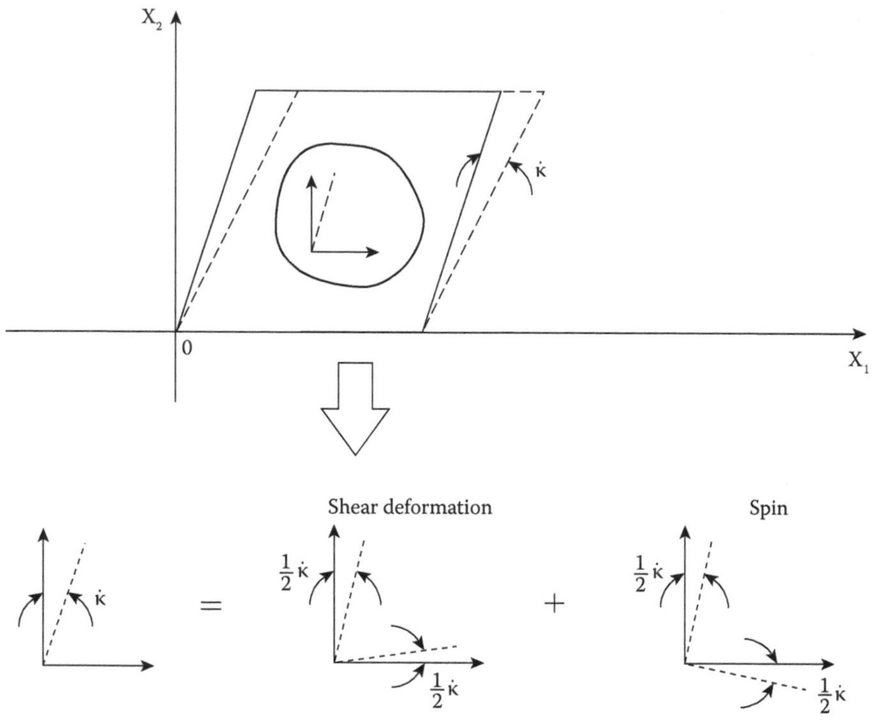

Figure 2.17 Deformation and spin rates in simple shear.

from which we find the deformation tensor **D** as

$$\mathbf{D} = \frac{\dot{\kappa}}{2}\begin{bmatrix} 0 & 1 & 0 \\ 1 & 0 & 0 \\ 0 & 0 & 0 \end{bmatrix} \tag{2.179}$$

and the spin tensor **W** as

$$\mathbf{W} = \frac{\dot{\kappa}}{2}\begin{bmatrix} 0 & 1 & 0 \\ -1 & 0 & 0 \\ 0 & 0 & 0 \end{bmatrix} \tag{2.180}$$

Figure 2.17 illustrates the total deformation rate as a combination of shear deformation rate **D** and spin **W**.

The right stretch tensor and principal stretches are found from an eigenvalue analysis of the Lagrangian metric tensor $\mathbf{C} = \mathbf{F}^T\mathbf{F} = \mathbf{U}^2$.

$$\mathbf{C} = \begin{bmatrix} 1 & \kappa & 0 \\ \kappa & 1+\kappa^2 & 0 \\ 0 & 0 & 1 \end{bmatrix} \tag{2.181}$$

The eigenvalues of **C** are easily calculated in this simple example. They are

$$\begin{cases} \lambda_1^C = 1 + \dfrac{\kappa^2}{2} + \kappa\sqrt{1 + \dfrac{\kappa^2}{4}} \\ \lambda_2^C = 1 \\ \lambda_3^C = 1 + \dfrac{\kappa^2}{2} - \kappa\sqrt{1 + \dfrac{\kappa^2}{4}} \end{cases} \tag{2.182}$$

The eigenvalues λ_i of the right stretch **U** are the square roots of the eigenvalues of **C**. That is,

$$\begin{cases} \lambda_1 = \sqrt{\lambda_1^C} \\ \lambda_2 = \sqrt{\lambda_2^C} \\ \lambda_3 = \sqrt{\lambda_3^C} \end{cases} \tag{2.183}$$

The major and minor principal stretches λ_1 and λ_3 associated with eigenvectors (i.e., directions) $\boldsymbol{\varphi}_1$ and $\boldsymbol{\varphi}_3$, respectively, lie in the $(\overline{x}_1, \overline{x}_2)$ plane. The intermediate principal stretch $\lambda_2 = 1$ throughout the motion since there is no deformation in the $\overline{x}_3$ direction. The principal stretches as functions of the deformation parameter κ are shown in Figure 2.18.

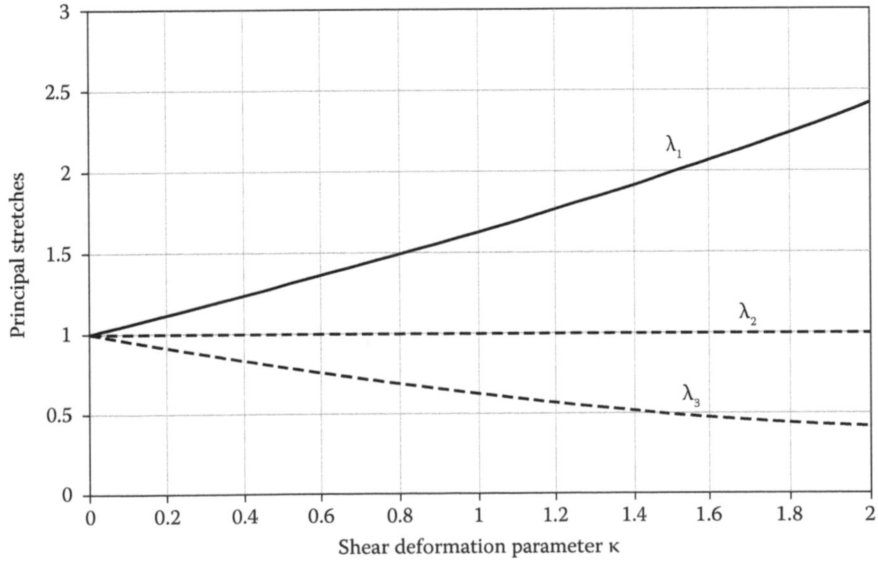

Figure 2.18 Principal stretches in simple shear.

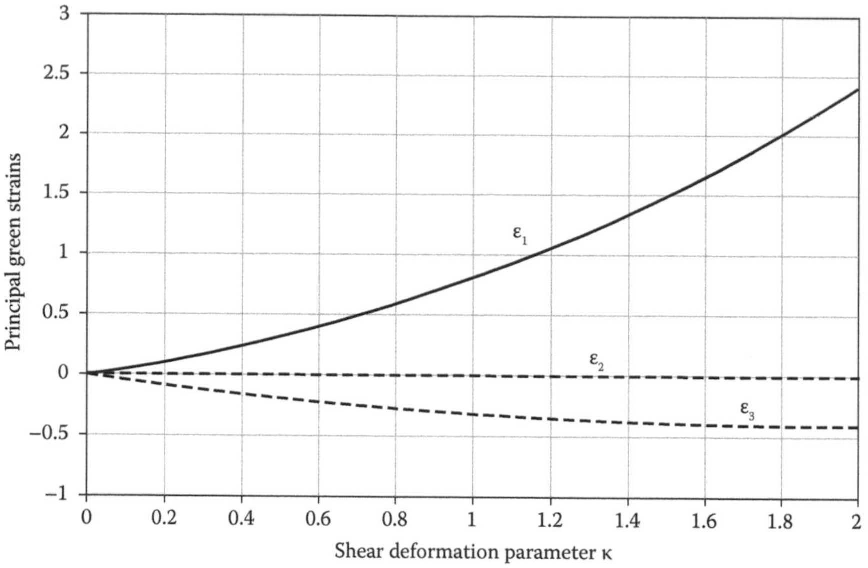

Figure 2.19 Principal Green strains in simple shear.

The principal Green strains are given by $\varepsilon_i = \frac{1}{2}(\lambda_i^2 - 1)$. They are shown in Figure 2.19. The rotation tensor **R** can be determined from the deformation gradient **F** in Equation 2.175 by

$$\mathbf{R} = \mathbf{F}\,\mathbf{U}^{-1} \tag{2.184}$$

where the inverse of the right stretch tensor can be constructed from its spectral representation as

$$\mathbf{U}^{-1} = \sum_i \frac{1}{\lambda_i} \varphi_i \varphi_i^T \tag{2.185}$$

These and subsequent calculations are performed numerically, and some further results are shown next.

Figure 2.20 shows the right polar decomposition of the motion at a deformation parameter $\kappa = 1$. The major and minor principal stretch directions (the eigenvectors of **U**) are also indicated. During the first step, as mentioned previously, line elements lying in the principal directions are stretched but not rotated; line elements lying in other directions are both stretched and rotated. In the second step, all line elements are rotated without deformation. The principal stretches shown in Figure 2.18 are (1.618, 1, 0.618) at $\kappa = 1$, corresponding to engineering principal strains (0.618, 0, −0.382). Of course, these are very large strains. However, very large plastic strain concentrations can, and do, develop in highly localized regions around sharp notches and cracks, for example.

In order to learn more about suitable stress and strain rate measures for computation, it is of interest to examine the behavior of the spin tensor in simple shear, as compared with the material rotation rate tensor. In this context, we recall that the spin tensor **W** was introduced into the Almansi strain rate in Equation 2.125.

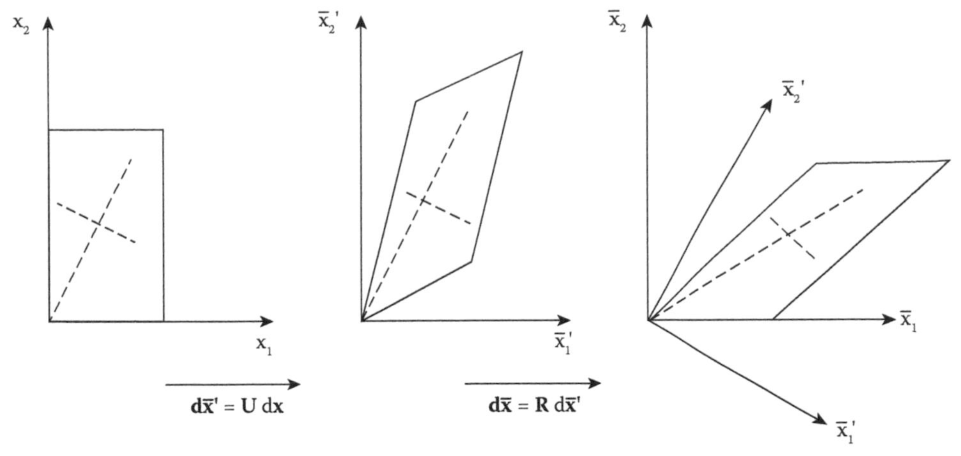

Figure 2.20 Polar decomposition of simple shear at deformation parameter κ = 1.

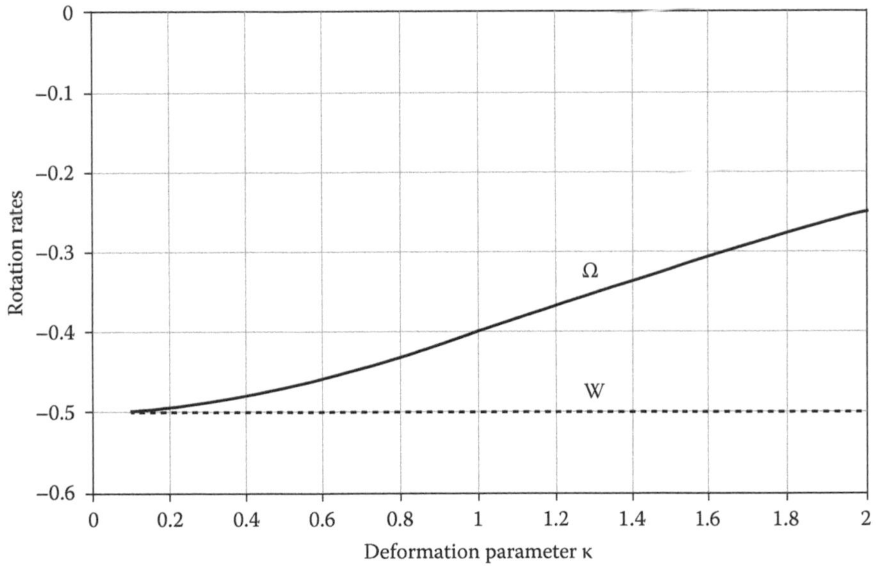

Figure 2.21 Material rotation rate and spin in simple shear.

We proceed numerically in order to exercise the algorithm summarized in Equation 2.161, although it is possible in this simple example to obtain a closed-form expression for the material rotation rate $\dot{\theta}$ in terms of the deformation rate parameter $\dot{\kappa}$ (Dienes 1979).

Figure 2.21 shows the material rotation (calculated from Equation 2.161) and spin, each as a coefficient of loading rate $\dot{\kappa}$. As we saw in Equation 2.180, in this example the spin tensor **W** has a constant value $\dot{w}_3 = -\dot{\kappa}/2$ (spin about the $\bar{x}_3$ axis), even as the deformation parameter continues to grow. The material rotation rate $\dot{\omega}_3$, in contrast, starts out initially equal to the

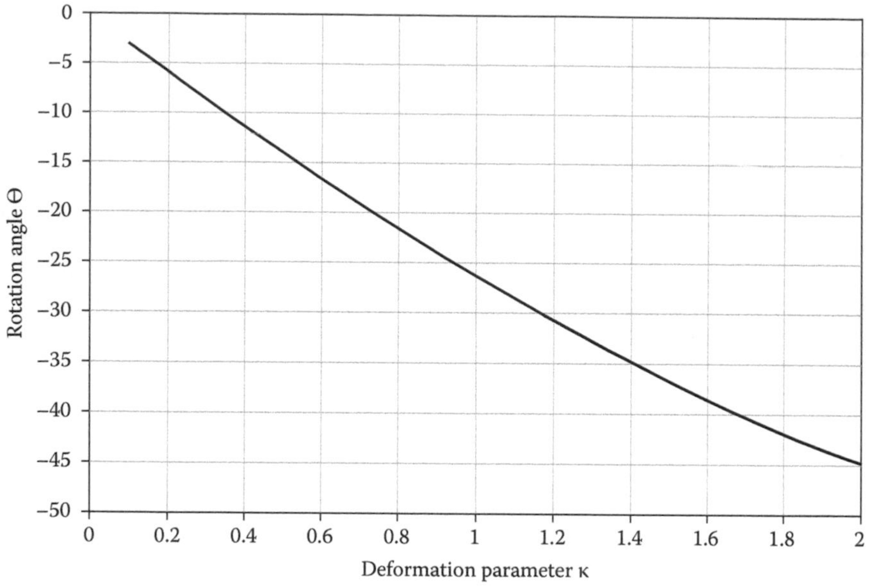

Figure 2.22 Rotation tensor (plane rotation angle θ) in simple shear.

spin $-\dot{\kappa}/2$, but decreases as the rate of rotation of the principal stretch directions slows down, reaching a value of $\dot{\omega}_3 = -0.400\dot{\kappa}$ at $\kappa=1$, and $\dot{\omega}_3 = -0.250\dot{\kappa}$ at $\kappa = 2$. The total rotation of the principal stretch axes reaches 26.6° at $\kappa = 1$ and 45° at $\kappa = 2$, as shown in Figure 2.22.

It seems reasonable that the material rotation rate tensor $\mathbf{\Omega}$ might be preferable to the spin tensor $\mathbf{W}$ in the development of suitable stress rates, strain rates, and constitutive relations since it should provide more relevant information about material response. The material rotation rate tensor $\mathbf{\Omega}$ will play a prominent role in all these areas.

Chapter 3

Stresses in deformable bodies

Stress is normally defined as the resultant force per unit area in the neighborhood of a particle. It is therefore a local measure of internal force around a particle that requires the assumption of local homogeneity in the material. Of course, real materials have microstructure, and they are far from homogeneous at the microscale. A familiar example is most metals, which have crystalline microstructures. Even though the assumption of local homogeneity is necessary in the definition of stress at the macroscale, it is useful to remember the limitations of this assumption as we approach the microscale.

3.1 TRACTION VECTOR ON A SURFACE

With the assumption of local homogeneity, we can define a resultant force vector $d\bar{f}$ in the deformed configuration acting on a plane with area $d\bar{A}$ and unit outward normal vector $\bar{n}$, as shown in Figure 3.1.

The *traction vector* (with dimensions of stress) is defined as the limit

$$\bar{t} = \lim_{d\bar{A} \to 0} \frac{d\bar{f}}{d\bar{A}} \tag{3.1}$$

and is, of course, dependent on the orientation of the surface; that is, it is a function of the normal vector $\bar{n}$.

3.2 CAUCHY STRESS PRINCIPLE

The Cauchy stress principle states, in essence, that the traction vector $\bar{t}$ is a *linear* function of the normal vector $\bar{n}$, or, in Cartesian component form,

$$\bar{t}_i = \bar{\sigma}_{ij}\bar{n}_j \tag{3.2}$$

where the transformation matrix (tensor) $\bar{\sigma}_{ij}$ consists of components that are independent of the normal vector $\bar{n}$.

This result may be stated in a more intuitive way as follows. In order to completely define the state of the internal forces in the neighborhood of a particle, we must obviously be able to determine the traction vector on any arbitrarily oriented surface at that location. Because any arbitrary normal vector (i.e., surface) can be expressed as a linear combination of three independent normal vectors, it is sufficient for this purpose to consider three non-coplanar surfaces.

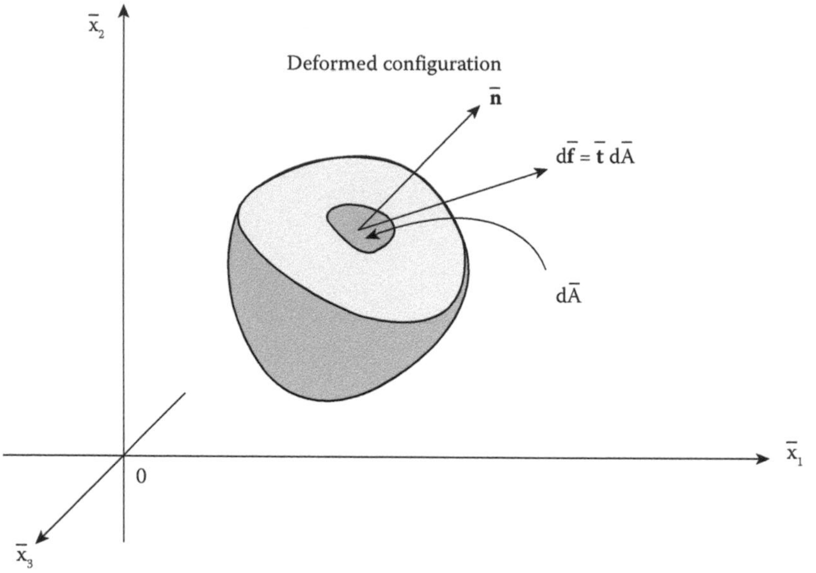

Figure 3.1 Traction vector in deformed configuration.

3.3 CAUCHY STRESS TENSOR

We normally consider the three surfaces that are orthogonal to the axes of the $\bar{x}_i$ Cartesian coordinate system, that is, which have unit normal vectors $(\bar{e}_1, \bar{e}_2, \bar{e}_3)$. The Cartesian components of these three traction vectors, that is, $\bar{\sigma}_{ij}$, are the nine components of the *Cauchy stress tensor* in the Cartesian coordinate system.

The Cartesian components of Cauchy stress in the deformed body are

$$\bar{\sigma} = \begin{bmatrix} \bar{\sigma}_{11} & \bar{\sigma}_{12} & \bar{\sigma}_{13} \\ \bar{\sigma}_{21} & \bar{\sigma}_{22} & \bar{\sigma}_{23} \\ \bar{\sigma}_{31} & \bar{\sigma}_{32} & \bar{\sigma}_{33} \end{bmatrix} \tag{3.3}$$

Therefore, the traction vector on the surface with unit normal $\bar{e}_1$ is

$$\bar{t}_1 = \begin{Bmatrix} \bar{\sigma}_{11} \\ \bar{\sigma}_{21} \\ \bar{\sigma}_{31} \end{Bmatrix} \tag{3.4}$$

The first subscript (row number) of the Cauchy stress indicates the direction in which the stress acts, and the second subscript indicates the coordinate plane on which it acts.

The force vector $d\bar{f}$ acting on the deformed surface is determined from the Cauchy stress by

$$d\bar{f} = \bar{\sigma}\ \bar{n}\ d\bar{A} \tag{3.5}$$

The Cauchy stress tensor is symmetric in most cases (unless, e.g., there are body force couples acting). In subsequent chapters, we will assume and make use of the symmetry of the Cauchy stress tensor, that is,

$$\bar{\sigma} = \begin{bmatrix} \bar{\sigma}_{11} & \bar{\sigma}_{12} & \bar{\sigma}_{13} \\ \bar{\sigma}_{12} & \bar{\sigma}_{22} & \bar{\sigma}_{23} \\ \bar{\sigma}_{13} & \bar{\sigma}_{23} & \bar{\sigma}_{33} \end{bmatrix} \tag{3.6}$$

but in this chapter, we continue to display stress tensors in their more general form, as in Equation 3.3.

In finite element applications, vector forms of stresses and strains are used, as we will see subsequently. The six independent stress components that constitute the "stress vector" are conventionally listed in the order

$$\{\bar{\sigma}\} = \begin{Bmatrix} \bar{\sigma}_{11} \\ \bar{\sigma}_{22} \\ \bar{\sigma}_{33} \\ \bar{\sigma}_{12} \\ \bar{\sigma}_{13} \\ \bar{\sigma}_{23} \end{Bmatrix} \tag{3.7}$$

which we will also use. We caution that some authors list the shear stresses in a different order from that shown in Equation 3.7.

It must be kept in mind that stresses (and strains) are truly tensor quantities, not vectors. As a result, their vector representations do not transform under coordinate rotations as tensors do.

In addition, as we will see, although stress tensors and strain tensors do transform the same way under coordinate rotations, their vector forms do not because of the different ways that, for very good reason, their vector representations are constructed from their tensor forms.

Cauchy stress defined in the deformed configuration and referring to the spatial coordinate system is often termed *true stress* or engineering stress. Other stress measures, to be defined subsequently, are not as easy to visualize or interpret as Cauchy stress.

The traction vector on any arbitrary plane within the material or on its surface can be determined in terms of the traction vectors $\bar{\mathbf{t}}_i$ on the three orthogonal Cartesian planes directly from the Cauchy stress principle, as shown in Equation 3.2. Equation 3.2 can also be expressed in the form

$$\bar{\mathbf{t}} = \bar{\mathbf{t}}_i \bar{n}_i \tag{3.8}$$

which is illustrated in Figure 3.2 for a two-dimensional case.

In Equation 3.8, $\bar{\mathbf{t}}$ is the traction vector on the arbitrary plane with normal vector $\bar{\mathbf{n}}$, which has Cartesian components $\bar{n}_i$. The traction vector $\bar{\mathbf{t}}_i$ contains the i-th column of the Cauchy stress tensor.

Equation 3.8 is displayed in expanded form below:

$$\bar{\mathbf{t}} = \begin{Bmatrix} \bar{\sigma}_{11} \\ \bar{\sigma}_{21} \\ \bar{\sigma}_{31} \end{Bmatrix} \bar{n}_1 + \begin{Bmatrix} \bar{\sigma}_{12} \\ \bar{\sigma}_{22} \\ \bar{\sigma}_{32} \end{Bmatrix} \bar{n}_2 + \begin{Bmatrix} \bar{\sigma}_{13} \\ \bar{\sigma}_{23} \\ \bar{\sigma}_{33} \end{Bmatrix} \bar{n}_3 \tag{3.9}$$

We emphasize here the distinction between $\bar{\mathbf{t}}_i$ and $\bar{t}_i$. $\bar{\mathbf{t}}_i$ is the traction *vector* on the coordinate plane with unit normal e_i, while $\bar{t}_i$ is the (scalar) Cartesian component of the traction vector $\bar{\mathbf{t}}$ acting on the plane with unit normal $\bar{\mathbf{n}}$ and area $d\bar{A}$.

Equation 3.8 is a vector equilibrium equation balancing the traction vectors acting on a four-sided body (in three dimensions). The orthogonal coordinate planes form three faces

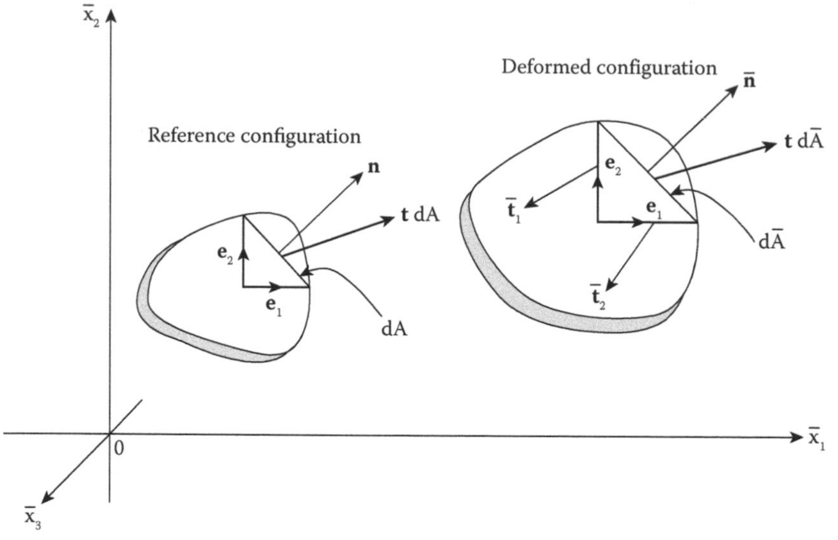

Figure 3.2 Cauchy stresses and traction vector for two-dimensional case.

of the body and are truncated by the fourth arbitrarily oriented face with orientation specified by the unit vector $\bar{\mathbf{n}}$.

In the two-dimensional case, we have the relation that

$$\bar{\mathbf{t}} = \begin{Bmatrix} \bar{\sigma}_{11} \\ \bar{\sigma}_{21} \end{Bmatrix} \bar{n}_1 + \begin{Bmatrix} \bar{\sigma}_{12} \\ \bar{\sigma}_{22} \end{Bmatrix} \bar{n}_2 \tag{3.10}$$

3.4 PIOLA–KIRCHHOFF STRESS TENSORS

Cauchy stress is a Eulerian description of stress, because it is defined in the deformed configuration and refers to the spatial coordinate system. A Lagrangian description, on the other hand, employs a stress tensor that refers to the material coordinate system in the reference configuration.

Consider the surface area $d\bar{A}$ with unit outward normal $\bar{\mathbf{n}}$ in the deformed configuration, as shown in Figure 3.3. The force vector on this surface is $d\bar{\mathbf{f}}$, and the traction vector is $\bar{\mathbf{t}}$. In the reference configuration, the surface area was dA with unit outward normal $\mathbf{n}$.

A Lagrangian stress tensor $\hat{\boldsymbol{\sigma}}$, called the *first Piola–Kirchhoff* stress tensor, is defined so that the force vector $d\bar{\mathbf{f}}$ in the deformed configuration is determined from stresses in the usual way, but now using areas and directions in the reference configuration, by

$$d\bar{\mathbf{f}} = \hat{\boldsymbol{\sigma}}\ \mathbf{n}\ dA \tag{3.11}$$

Comparing the expressions in Equations 3.5 and 3.11 for the force vector $d\bar{\mathbf{f}}$, we have

$$\bar{\boldsymbol{\sigma}}\ \bar{\mathbf{n}}\ d\bar{A} = \hat{\boldsymbol{\sigma}}\ \mathbf{n}\ dA \tag{3.12}$$

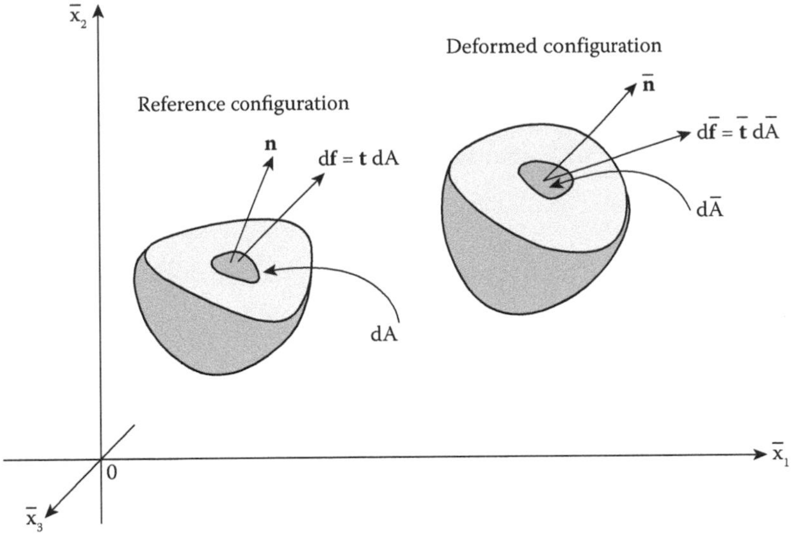

Figure 3.3 Force and traction vectors in reference and deformed configurations.

Using Nanson's equation in Equation 3.12, we obtain the relation between the first Piola–Kirchhoff stress and Cauchy stress as

$$\hat{\boldsymbol{\sigma}} = |\mathbf{F}|\bar{\boldsymbol{\sigma}}\ \bar{\mathbf{F}}^T \tag{3.13}$$

The force vector in the deformed configuration can also be related to a pseudo–traction vector $\hat{\mathbf{t}}$ defined with respect to the reference configuration by

$$d\bar{\mathbf{f}} \equiv \bar{\mathbf{t}}\ d\bar{A} = \hat{\mathbf{t}}\ dA \tag{3.14}$$

The pseudo–traction vector $\hat{\mathbf{t}}$ is related to pseudo–traction vectors $\hat{\mathbf{t}}_i$ acting on the Cartesian planes by relations completely analogous to Equations 3.8 and 3.2, that is,

$$\hat{\mathbf{t}} = \hat{\mathbf{t}}_i n_i = \hat{\sigma}_{ij} n_j \mathbf{e}_i \tag{3.15}$$

and we see that the pseudo–traction vector $\hat{\mathbf{t}}$ is related to the first Piola–Kirchhoff stress through a relation analogous to Equation 3.9 by

$$\hat{\mathbf{t}} = \begin{Bmatrix} \hat{\sigma}_{11} \\ \hat{\sigma}_{21} \\ \hat{\sigma}_{31} \end{Bmatrix} n_1 + \begin{Bmatrix} \hat{\sigma}_{12} \\ \hat{\sigma}_{22} \\ \hat{\sigma}_{32} \end{Bmatrix} n_2 + \begin{Bmatrix} \hat{\sigma}_{13} \\ \hat{\sigma}_{23} \\ \hat{\sigma}_{33} \end{Bmatrix} n_3 \tag{3.16}$$

Thus, the pseudo–traction vectors $\hat{\mathbf{t}}_i = \hat{\sigma}_{ki}\ \mathbf{e}_k$ in Equation 3.15 are the i-th columns of the first Piola–Kirchhoff stress tensor. The relation between the pseudo–traction vectors $\hat{\mathbf{t}}_i$ and the traction vectors $\bar{\mathbf{t}}_i$ acting on the Cartesian planes in the deformed configuration follows directly from Equation 3.13, since the vectors $\hat{\mathbf{t}}_i$ and $\bar{\mathbf{t}}_i$ form the columns of $\hat{\boldsymbol{\sigma}}$ and $\bar{\boldsymbol{\sigma}}$, respectively.

$$\hat{\mathbf{t}}_i = |\mathbf{F}|\bar{F}_{ij}\bar{\mathbf{t}}_j \tag{3.17}$$

It is clear from Equation 3.13 that the first Piola–Kirchhoff stress is not symmetric, and because of this, it is not suitable for general use in computational mechanics.

Therefore, we now consider a symmetric Lagrangian stress tensor, the *second Piola–Kirchhoff* (2PK) stress tensor $\boldsymbol{\sigma}$.

Consider now a pseudo–force vector $\mathbf{df}^0$ that is related to the 2PK stress tensor $\boldsymbol{\sigma}$ by a relation analogous to Equation 3.11, that is,

$$\mathbf{df}^0 = \boldsymbol{\sigma}\ \mathbf{n}\ \mathrm{dA} \tag{3.18}$$

A pseudo–traction vector $\mathbf{t}^0$ may now be defined in the usual way so that

$$\mathbf{df}^0 = \mathbf{t}^0 \mathrm{dA} \tag{3.19}$$

where

$$\mathbf{t}^0 = \boldsymbol{\sigma}\ \mathbf{n} = \sigma_{ij} n_j\ \mathbf{e}_i \tag{3.20}$$

This pseudo–traction vector $\mathbf{t}^0$ acting on the plane with original normal vector $\mathbf{n}$ and area dA can also be written as

$$\mathbf{t}^0 = \mathbf{t}_i^0 n_i \tag{3.21}$$

where

$$\mathbf{t}_i^0 = \sigma_{ki}\ \mathbf{e}_k \tag{3.22}$$

is the i-th column of the 2PK stress tensor. Equation 3.20 is written in expanded form as

$$\mathbf{t}^0 = \begin{Bmatrix} \sigma_{11} \\ \sigma_{21} \\ \sigma_{31} \end{Bmatrix} n_1 + \begin{Bmatrix} \sigma_{12} \\ \sigma_{22} \\ \sigma_{32} \end{Bmatrix} n_2 + \begin{Bmatrix} \sigma_{13} \\ \sigma_{23} \\ \sigma_{33} \end{Bmatrix} n_3 \tag{3.23}$$

The pseudo–force vector $\mathbf{df}^0$ is now transformed into the force vector $\mathbf{d\bar{f}}$ acting in the deformed configuration via the deformation gradient by

$$\mathbf{d\bar{f}} = \mathbf{F}\ \mathbf{df}^0 \tag{3.24}$$

Therefore, we have

$$\mathbf{d\bar{f}} = \mathbf{F}\ \boldsymbol{\sigma}\ \mathbf{n}\ \mathrm{dA} \tag{3.25}$$

Using Nanson's equation in Equation 3.25, we have

$$\mathbf{d\bar{f}} = \frac{1}{|\mathbf{F}|} \mathbf{F}\ \boldsymbol{\sigma}\ \mathbf{F}^{\mathrm{T}} \bar{\mathbf{n}}\ \mathrm{d\bar{A}} = \bar{\boldsymbol{\sigma}}\ \bar{\mathbf{n}}\ \mathrm{d\bar{A}} \tag{3.26}$$

from which follows the relation between the Cauchy stress tensor and the symmetric 2PK stress tensor:

$$\bar{\boldsymbol{\sigma}} = \frac{1}{|\mathbf{F}|} \mathbf{F}\ \boldsymbol{\sigma}\ \mathbf{F}^{\mathrm{T}} \tag{3.27}$$

Equation 3.27 can be inverted to give

$$\boldsymbol{\sigma} = |\mathbf{F}|\bar{\mathbf{F}}\ \bar{\boldsymbol{\sigma}}\ \bar{\mathbf{F}}^{T} \tag{3.28}$$

The same transformation shown in Equation 3.24 applies of course to the pseudo–traction vectors $\mathbf{t}^0$ and $\mathbf{t}^0_i$ producing traction vectors $\mathbf{t}$ and $\mathbf{t}_i$, respectively.

The traction vector $\mathbf{t}$ acts on an arbitrary plane in the deformed configuration, which had normal $\mathbf{n}$ and area dA in the reference configuration. The three traction vectors $\mathbf{t}_i$ act in the deformed configuration on planes that were Cartesian coordinate planes in the reference configuration.

In order to further interpret 2PK stresses, we return to the expression for the force vector (Equation 3.25) and write it in expanded form as

$$d\bar{f}_i\ \mathbf{e}_i = \left(\frac{\partial \bar{x}_k}{\partial x_j}\mathbf{e}_k\right)\sigma_{jm}n_m\ dA \tag{3.29}$$

We recognize the bracketed term in Equation 3.29 as the convected base vector $\mathbf{G}_j$, so that the force vector can be written as

$$d\bar{\mathbf{f}} = \sigma_{jk}n_k\mathbf{G}_j\ dA \tag{3.30}$$

Equation 3.30 determines the traction vector on an arbitrary plane. As a specific example, on a plane with *original* normal $\mathbf{n} = \{1, 0, 0\}^T$ and area dA, that is, the Cartesian coordinate plane perpendicular to the $\bar{x}_1$ axis, the force vector $d\bar{\mathbf{f}}$ acting on that plane (which now has normal $\bar{\mathbf{n}}$ and area $d\bar{A}$) in the deformed configuration is, in terms of 2PK stresses,

$$d\bar{\mathbf{f}} = \left[\sigma_{11}\mathbf{G}_1 + \sigma_{21}\mathbf{G}_2 + \sigma_{31}\mathbf{G}_3\right]dA \tag{3.31}$$

Therefore, the 2PK stresses are directed along the convected base vectors (which are not of unit length in general) in the deformed configuration as shown in Figure 3.4 for the $(\bar{x}_1, \bar{x}_2)$ plane.

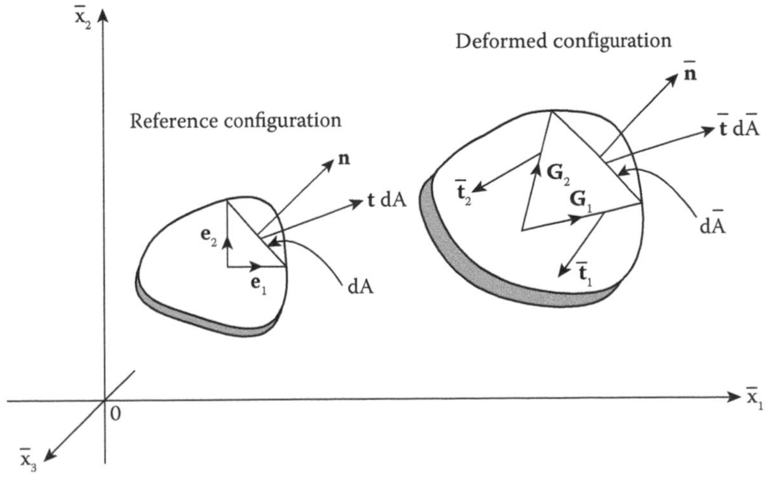

Figure 3.4 Traction vectors and 2PK stresses.

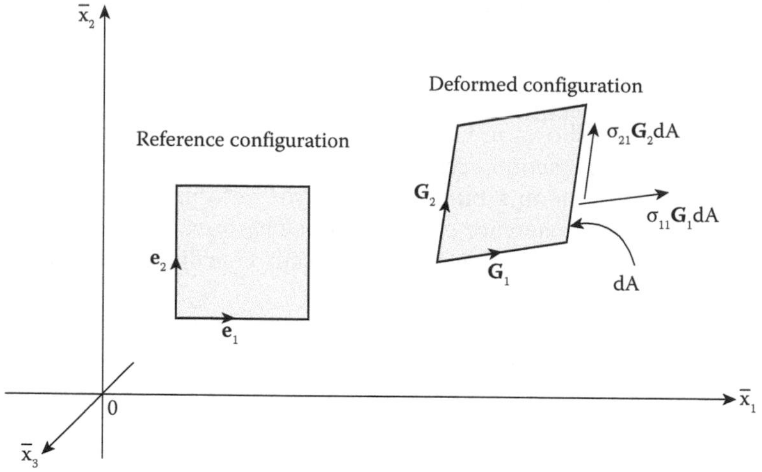

Figure 3.5 2PK stresses in deformed configuration.

Note again that the original element of area dA is used in these expressions for the force vector $d\bar{f}$. Explicit expressions for the traction vectors are

$$t = \sigma_{jk} n_k G_j \tag{3.32}$$

$$t_i = \sigma_{ji} G_j \tag{3.33}$$

Figures 3.4 and 3.5 illustrate these relations for the two-dimensional case.

3.5 STRESS RATES

Material constitutive relations in rate form are generally required for incremental-iterative nonlinear finite element analysis. That is, stress rates must be related to strain rates by material models. These models will typically be nonlinear; that is, the material properties may be stress, strain, history, and/or time dependent. As will be discussed subsequently, only *work-conjugate* stress rate–strain rate pairs can be coupled together for this purpose. We now consider several different rates of stress that have been developed.

3.5.1 Material rates of stress

The material rate of stress is simply defined as the (time) rate of change of stress at a given particle (i.e., for $\mathbf{x}$ = constant). The material rate of Cauchy stress is then

$$\dot{\bar{\sigma}} = \frac{\partial}{\partial t}\bar{\sigma} \tag{3.34}$$

and the material rate of symmetric Piola–Kirchhoff (2PK) stress is

$$\dot{\sigma} = \frac{\partial}{\partial t}\sigma \tag{3.35}$$

In order to determine whether these are valid stress rates, we must examine the behavior of these stress rates under an imposed rigid body rotation at the deformed configuration.

We can show as follows that under such a motion, $\dot{\boldsymbol{\sigma}}$ vanishes. Consider the force vector in the deformed configuration, that is,

$$d\bar{f} = \mathbf{F}\ \boldsymbol{\sigma}\ dA \tag{3.36}$$

We take the time (material) rate of both sides of Equation 3.36 to obtain

$$d\dot{\bar{f}} = \dot{\mathbf{F}}\ \boldsymbol{\sigma}\ dA + \mathbf{F}\ \dot{\boldsymbol{\sigma}}\ dA \tag{3.37}$$

In a rigid body motion from the deformed configuration, $\dot{\mathbf{F}} = \mathbf{W}\ \mathbf{F}$ and $d\dot{\bar{f}} = \mathbf{W}\ d\bar{f}$, the latter equality by virtue of the fact that lengths of vectors are preserved in rigid body (or coordinate) rotations. Thus, Equation 3.37 becomes

$$\mathbf{W}\ d\bar{f} = \mathbf{W}(\mathbf{F}\ \boldsymbol{\sigma}\ dA) + \mathbf{F}\ \dot{\boldsymbol{\sigma}}\ dA = \mathbf{W}\ d\bar{f} + \mathbf{F}\ \dot{\boldsymbol{\sigma}}\ dA \tag{3.38}$$

and consequently, $\dot{\boldsymbol{\sigma}} = 0$. We can therefore conclude that the material rate of 2PK stress is an objective stress rate.

Since Cauchy stress refers to the spatial coordinate system $\bar{\mathbf{x}}$, we can expect that its material rate is not objective, as we will confirm next.

3.5.2 Jaumann rate of Cauchy stress

In order to define valid stress rates involving Cauchy stress, we first establish the relation between the material rates of 2PK and Cauchy stress. We begin with Equation 3.28, which relates these two stress tensors, and take its material rate, to arrive at

$$\dot{\boldsymbol{\sigma}} = |\mathbf{F}|\bar{\mathbf{F}}\left[\dot{\bar{\boldsymbol{\sigma}}} - \mathbf{V}\bar{\boldsymbol{\sigma}} - \bar{\boldsymbol{\sigma}}\mathbf{V}^T + \mathrm{tr}(\mathbf{D})\,\bar{\boldsymbol{\sigma}}\right]\bar{\mathbf{F}}^T \tag{3.39}$$

in which we have used $|\dot{\mathbf{F}}| = \mathrm{tr}(\mathbf{D})|\mathbf{F}|$ and $\dot{\bar{\mathbf{F}}} = -\bar{\mathbf{F}}\ \mathbf{V}$.

We now evaluate Equation 3.39 for the special case of a rigid body rotation in which the velocity gradient tensor $\mathbf{V} = \mathbf{W}$ and $\mathbf{D} = 0$. Thus, we now have from Equation 3.39

$$0 = \dot{\bar{\boldsymbol{\sigma}}} - (\mathbf{W}\bar{\boldsymbol{\sigma}} + \bar{\boldsymbol{\sigma}}\mathbf{W}^T) \tag{3.40}$$

in a rigid body motion. Since $\dot{\bar{\boldsymbol{\sigma}}}$ is not zero in a rigid body motion, it is not an objective stress rate.

As the material rate of the Cauchy stress tensor is not objective, it cannot be used to formulate material constitutive laws that require the satisfaction of objectivity or frame indifference.

However, using Equation 3.40, we can define an objective rate of Cauchy stress, known as the *Jaumann rate*, as

$$\bar{\boldsymbol{\sigma}}^J = \dot{\bar{\boldsymbol{\sigma}}} - \mathbf{W}\bar{\boldsymbol{\sigma}} - \bar{\boldsymbol{\sigma}}\mathbf{W}^T \tag{3.41}$$

which does vanish under rigid body motion.

The Jaumann stress rate can be interpreted as referring to a set of orthogonal unit vectors (i.e., a Cartesian coordinate system) that coincides with the spatial Cartesian base vectors in the deformed configuration but rotates with the motion.

The Jaumann rate of Cauchy stress has in the past been widely used in nonlinear finite element computations. But, although it is objective, it has been found to give physically unrealistic results in the simple shear test problem at very large strain levels.

3.5.3 Truesdell rate of Cauchy stress

It is also possible to refer the stress rate to a set of convected base vectors that coincide with the Cartesian base vectors at the deformed configuration and follow not only the local rotation, but also the local deformation.

In order to do this, we return to Equation 3.39 and define the bracketed term therein as the *Truesdell stress rate*, that is,

$$\dot{\boldsymbol{\sigma}} = |\mathbf{F}|\ \overline{\mathbf{F}}\overline{\boldsymbol{\sigma}}^{Tr}\overline{\mathbf{F}}^{T} \tag{3.42}$$

where

$$\overline{\boldsymbol{\sigma}}^{Tr} = \dot{\overline{\boldsymbol{\sigma}}} - \mathbf{V}\overline{\boldsymbol{\sigma}} - \overline{\boldsymbol{\sigma}}\mathbf{V}^{T} + \mathrm{tr}(\mathbf{D})\,\overline{\boldsymbol{\sigma}} \tag{3.43}$$

By virtue of its definition, the Truesdell rate of Cauchy stress is obviously objective. We note also that the material rate of 2PK stress is related to the Truesdell rate of Cauchy stress in the same way as 2PK stress and Cauchy stress are related (viz. Equation 3.28).

Substituting $\mathbf{V} = \mathbf{D} + \mathbf{W}$ in Equation 3.43 and then using the definition of the Jaumann rate in Equation 3.41, we find the relation between the Jaumann and Truesdell rates as

$$\overline{\boldsymbol{\sigma}}^{Tr} = \overline{\boldsymbol{\sigma}}^{J} - \mathbf{D}\overline{\boldsymbol{\sigma}} - \overline{\boldsymbol{\sigma}}\mathbf{D} + \mathrm{tr}(\mathbf{D})\,\overline{\boldsymbol{\sigma}} \tag{3.44}$$

The expression for the Truesdell stress rate can also be determined in a direct manner, which may provide insight, as follows.

Consider configuration C_1 at time t and configuration C_2 at time $\tau = t + \Delta t$ in the deformed configuration. We denote the Cauchy stresses at C_1 and C_2 as $\overline{\boldsymbol{\sigma}}$ and $\overline{\boldsymbol{\sigma}}_\tau$, respectively, and

$$\overline{\boldsymbol{\sigma}}_\tau = \overline{\boldsymbol{\sigma}} + \Delta t\ \dot{\overline{\boldsymbol{\sigma}}} \tag{3.45}$$

Now we define a stress tensor $\boldsymbol{\sigma}_\tau$, which are the stresses at C_2 but referred to the Cartesian base vectors at C_1. The stress tensor $\boldsymbol{\sigma}_\tau$ is in fact the symmetric Piola–Kirchhoff stress at C_2, with C_1 considered the reference configuration. It can therefore be related to $\overline{\boldsymbol{\sigma}}_\tau$ through the following equation:

$$\boldsymbol{\sigma}_\tau = |\mathbf{F}_\tau|\overline{\mathbf{F}}_\tau\ \overline{\boldsymbol{\sigma}}_\tau\ \overline{\mathbf{F}}_\tau^{T} \tag{3.46}$$

The deformation gradient in Equation 3.46 is for the portion of the motion from C_1 to C_2. Since we have assumed these two configurations to be separated by an infinitesimal time increment Δt, the position vectors at C_1 and C_2 can be related by

$$\overline{\mathbf{x}}_\tau = \overline{\mathbf{x}} + \Delta t\ \mathbf{v} \tag{3.47}$$

in which $\mathbf{v}$ is the velocity vector at C_1.

Recall that the velocity vector is the material rate of the position vector of a particle in the deformed configuration. Therefore, $\Delta\overline{\mathbf{x}} \approx \Delta t\ \mathbf{v}$, and the deformation gradients can then be written as follows:

$$\mathbf{F}_\tau = \frac{\partial \overline{\mathbf{x}}_\tau}{\partial \overline{\mathbf{x}}} = \mathbf{I} + \Delta t\ \mathbf{V} \tag{3.48}$$

The determinant of the deformation gradient tensor is then

$$|\mathbf{F}_\tau| = 1 + \Delta t \ \mathrm{tr}(\mathbf{D}) \tag{3.49}$$

and the Eulerian deformation gradient tensor (to the first order in Δt) is

$$\overline{\mathbf{F}}_\tau = \mathbf{I} - \Delta t \ \mathbf{V} \tag{3.50}$$

Substituting these expressions in Equations 3.45 and 3.46, we obtain

$$\begin{aligned}\sigma_\tau &= \left[1 + \Delta t \ \mathrm{tr}(\mathbf{D})\right]\left[\mathbf{I} - \Delta t \ \mathbf{V}\right](\overline{\sigma} + \Delta t \ \dot{\overline{\sigma}})\left[\mathbf{I} - \Delta t \ \mathbf{V}^T\right] \\ &= \overline{\sigma} + \Delta t \ \left[\dot{\overline{\sigma}} - \mathbf{V}\overline{\sigma} - \overline{\sigma}\mathbf{V}^T + \mathrm{tr}(\mathbf{D})\,\overline{\sigma}\right] + O(\Delta t^2)\end{aligned} \tag{3.51}$$

The Truesdell stress rate at C_1 (at time t) is defined as

$$\overline{\sigma}^{Tr} = \lim_{\Delta t \to 0}\left[\frac{1}{\Delta t}(\sigma_\tau - \overline{\sigma})\right] \tag{3.52}$$

which yields the expression derived earlier in Equation 3.43.

3.5.4 Unrotated Cauchy stress and the Green–Naghdi rate

The Jaumann and Truesdell rates of Cauchy stress are both objective stress rates and therefore can be used in material models expressed in rate form, as is necessary in general for incremental-iterative nonlinear finite element analysis.

These stress rates both successfully address the nonobjectivity of the material rate of Cauchy stress by including terms involving the spin tensor **W** that compensate for the rigid body rotation component of the motion.

The Green–Naghdi rate of Cauchy stress achieves the same objective using the more physically meaningful material rotation rate Ω instead of the spin tensor. In order to provide the rationale for the Green–Naghdi rate, we first introduce the unrotated Cauchy stress.

3.5.4.1 Unrotated Cauchy stress

Using the right polar decomposition of the motion, Nanson's equation can be decomposed as

$$d\overline{\mathbf{A}} = |\mathbf{F}|\overline{\mathbf{F}}^T d\mathbf{A} = \mathbf{R}\ \left(|\mathbf{U}|\mathbf{U}^{-1}d\mathbf{A}\right) = \mathbf{R}\ d\overline{\mathbf{A}}' \tag{3.53}$$

where $d\overline{\mathbf{A}}'$ is the area element in the unrotated configuration $d\overline{\mathbf{x}}' = \mathbf{U}\ dx$. Thus, in the second stage of the motion ($d\overline{\mathbf{x}} = \mathbf{R}\ d\overline{\mathbf{x}}'$),

$$d\overline{\mathbf{A}} = \mathbf{R}\ d\overline{\mathbf{A}}' \tag{3.54}$$

The force vector in the deformed configuration is

$$d\overline{\mathbf{f}} = \overline{\sigma}\ d\overline{\mathbf{A}} = \overline{\sigma}\ \mathbf{R}\ d\overline{\mathbf{A}}' \tag{3.55}$$

and the force vector in the unrotated configuration is $d\overline{\mathbf{f}}' = \mathbf{R}^T d\overline{\mathbf{f}}$. Thus, Equation 3.55 becomes

$$d\overline{\mathbf{f}}' = \mathbf{R}^T d\overline{\mathbf{f}} = (\mathbf{R}^T \sigma\ \mathbf{R})\ d\overline{\mathbf{A}}' = \overline{\sigma}' d\overline{\mathbf{A}}' \tag{3.56}$$

where the *unrotated Cauchy stress* $\bar{\boldsymbol{\sigma}}'$ is

$$\bar{\boldsymbol{\sigma}}' = \mathbf{R}^{\mathrm{T}} \bar{\boldsymbol{\sigma}}\ \mathbf{R} \tag{3.57}$$

3.5.4.2 *Green–Naghdi rate*

The deformation rate **D** is related to the unrotated deformation rate **d** by

$$\mathbf{d} = \mathbf{R}^{\mathrm{T}} \mathbf{D}\ \mathbf{R} \tag{3.58}$$

and we recall that the unrotated deformation rate appears in the right polar decomposition of the deformation gradient (**F** = **RU**) prior to the rigid body rotation of all line elements to their final position in the deformed configuration. The principal material axes at this first stage in the motion have not yet been rotated—thus the terminology "unrotated" deformation rate.

Therefore, the pair $(\bar{\boldsymbol{\sigma}}', \mathbf{d})$ is defined in the unrotated configuration in the same coordinate system, as is the pair $(\bar{\boldsymbol{\sigma}}, \mathbf{D})$ in the deformed configuration. We now take the rate of both sides of Equation 3.57 to obtain

$$\dot{\bar{\boldsymbol{\sigma}}}' = \dot{\mathbf{R}}^{\mathrm{T}} \bar{\boldsymbol{\sigma}}\ \mathbf{R} + \mathbf{R}^{\mathrm{T}} \dot{\bar{\boldsymbol{\sigma}}}\ \mathbf{R} + \mathbf{R}^{\mathrm{T}} \bar{\boldsymbol{\sigma}}\ \dot{\mathbf{R}} \tag{3.59}$$

Then by inserting $\dot{\mathbf{R}} = \boldsymbol{\Omega}\ \mathbf{R}$ and noting that $\boldsymbol{\Omega}$ is skew symmetric ($\boldsymbol{\Omega}^{\mathrm{T}} = -\boldsymbol{\Omega}$), we arrive at

$$\dot{\bar{\boldsymbol{\sigma}}}' = \mathbf{R}^{\mathrm{T}} \left[\dot{\bar{\boldsymbol{\sigma}}} - \boldsymbol{\Omega} \bar{\boldsymbol{\sigma}} - \bar{\boldsymbol{\sigma}} \boldsymbol{\Omega}^{\mathrm{T}} \right] \mathbf{R} \tag{3.60}$$

and define the bracketed term in Equation 3.60 as the Green–Naghdi rate of Cauchy stress $\bar{\boldsymbol{\sigma}}^{\mathrm{GN}}$, where

$$\bar{\boldsymbol{\sigma}}^{\mathrm{GN}} = \dot{\bar{\boldsymbol{\sigma}}} - \boldsymbol{\Omega} \bar{\boldsymbol{\sigma}} - \bar{\boldsymbol{\sigma}} \boldsymbol{\Omega}^{\mathrm{T}} \tag{3.61}$$

Note that the material rate of unrotated Cauchy stress is related to the Green–Naghdi rate in the same way as the unrotated Cauchy stress is related to Cauchy stress (viz. Equation 3.57).

$$\dot{\bar{\boldsymbol{\sigma}}}' = \mathbf{R}^{\mathrm{T}} \bar{\boldsymbol{\sigma}}^{\mathrm{GN}} \mathbf{R} \tag{3.62}$$

We also note that the Green–Naghdi rate can be obtained from the Jaumann rate by simply substituting the material rotation rate $\boldsymbol{\Omega}$ for the spin tensor **W**.

3.6 EXAMPLES OF STRESS RATES FOR SIMPLE STRESS CONDITIONS

We now present several simple examples to provide insight into the physical meaning of the stress rates that have been discussed.

3.6.1 Uniaxial extension of an initially stressed body

Consider a plate undergoing pure uniaxial extension under a constant state of initial stress (or prestress), as shown in Figure 3.6.

Note that if the stress remains constant during a succession of deformed configurations, then that stress must necessarily be Cauchy stress. The constant Cauchy stress is assumed as

$$\bar{\boldsymbol{\sigma}} = \begin{bmatrix} a & b & 0 \\ b & c & 0 \\ 0 & 0 & 0 \end{bmatrix} \tag{3.63}$$

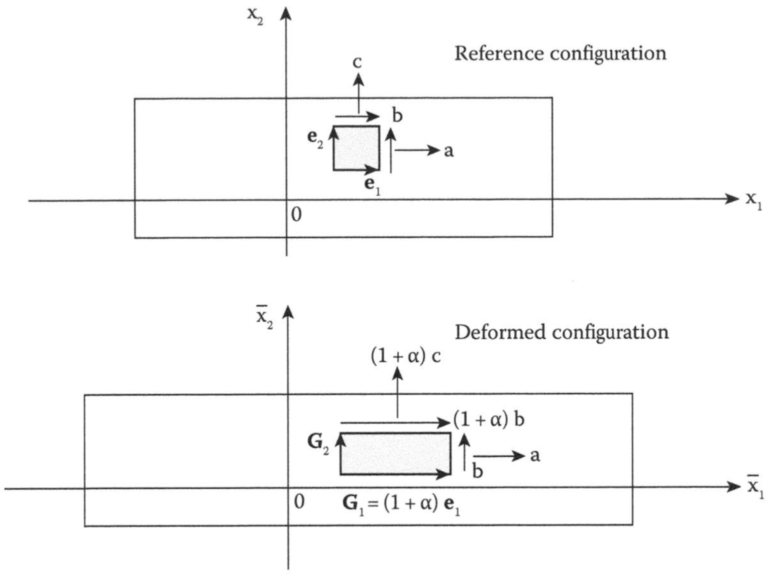

Figure 3.6 Tractions in a plate in pure extension under constant state of stress.

Therefore, the material rate of Cauchy stress in this example is zero.

$$\dot{\boldsymbol{\sigma}} = 0 \tag{3.64}$$

We may think of this example as a viscoelastic body undergoing creep deformation at a constant state of stress.

We recall that in pure uniaxial extension, the deformation gradient is of the form

$$\mathbf{F} = \begin{bmatrix} 1+\alpha & 0 & 0 \\ 0 & 1 & 0 \\ 0 & 0 & 1 \end{bmatrix} \tag{3.65}$$

The stretch $\alpha(t)$ is a time (load)-dependent parameter with rate $\dot{\alpha}$. The velocity vector is

$$\mathbf{v} = \begin{Bmatrix} v_1 \\ v_2 \\ v_3 \end{Bmatrix} = \begin{Bmatrix} \dot{\bar{x}}_1 \\ \dot{\bar{x}}_2 \\ \dot{\bar{x}}_3 \end{Bmatrix} = \begin{Bmatrix} \dot{\alpha} x_1 \\ 0 \\ 0 \end{Bmatrix} \tag{3.66}$$

The velocity gradient tensor is

$$\mathbf{V} = \dot{\alpha} \begin{bmatrix} \dfrac{1}{(1+\alpha)} & 0 & 0 \\ 0 & 0 & 0 \\ 0 & 0 & 0 \end{bmatrix} \tag{3.67}$$

The Eulerian deformation gradient tensor $\bar{\mathbf{F}}$ is obtained by inverting the Lagrangian deformation gradient tensor, giving

$$\bar{\mathbf{F}} = \begin{bmatrix} \dfrac{1}{1+\alpha} & 0 & 0 \\ 0 & 1 & 0 \\ 0 & 0 & 1 \end{bmatrix} \tag{3.68}$$

Using the relationship between the 2PK stresses and the Cauchy stresses, we obtain the 2PK stresses as

$$\boldsymbol{\sigma} = \begin{bmatrix} \dfrac{a}{1+\alpha} & b & 0 \\ b & c(1+\alpha) & 0 \\ 0 & 0 & 0 \end{bmatrix} \tag{3.69}$$

To examine the physical meaning of the symmetric Piola–Kirchhoff stresses, we first recall that these stresses refer to the convected base vectors $\mathbf{G}_1$ and $\mathbf{G}_2$, as shown in Figure 3.6. In this example, $\mathbf{G}_1$ is directed along the Cartesian base vector $\mathbf{e}_1$, but its length has increased to $(1 + \alpha)$, while $\mathbf{G}_2$ has remained equal to $\mathbf{e}_2$. The force vector on the face with normal $\mathbf{e}_1$ in terms of Cauchy stresses is

$$\bar{t}_1\, d\bar{A}_1 = \begin{Bmatrix} \bar{\sigma}_{11} \\ \bar{\sigma}_{21} \\ 0 \end{Bmatrix} d\bar{A}_1 = \begin{Bmatrix} a \\ b \\ 0 \end{Bmatrix}(1) \tag{3.70}$$

and the force vector on the face with normal $\mathbf{e}_2$ is

$$\bar{t}_2\, d\bar{A}_2 = \begin{Bmatrix} \bar{\sigma}_{12} \\ \bar{\sigma}_{22} \\ 0 \end{Bmatrix} d\bar{A}_2 = \begin{Bmatrix} b \\ c \\ 0 \end{Bmatrix}(1+\alpha) \tag{3.71}$$

as indicated in Figure 3.6. The same force vectors in terms of 2PK stresses are

$$t_1 dA_1 = \left(\sigma_{11} \begin{Bmatrix} (1+\alpha) \\ 0 \\ 0 \end{Bmatrix} + \sigma_{21} \begin{Bmatrix} 0 \\ 1 \\ 0 \end{Bmatrix} + \sigma_{31} \begin{Bmatrix} 0 \\ 0 \\ 1 \end{Bmatrix} \right) dA_1 = \begin{Bmatrix} \sigma_{11}(1+\alpha) \\ \sigma_{21} \\ 0 \end{Bmatrix}(1) \tag{3.72}$$

and

$$t_2 dA_2 = \left(\sigma_{12} \begin{Bmatrix} (1+\alpha) \\ 0 \\ 0 \end{Bmatrix} + \sigma_{22} \begin{Bmatrix} 0 \\ 1 \\ 0 \end{Bmatrix} + \sigma_{32} \begin{Bmatrix} 0 \\ 0 \\ 1 \end{Bmatrix} \right) dA_2 = \begin{Bmatrix} \sigma_{12}(1+\alpha) \\ \sigma_{22} \\ 0 \end{Bmatrix}(1) \tag{3.73}$$

Thus, comparing Equation 3.70 with Equation 3.72,

$$\sigma_{11} = \frac{a}{(1+\alpha)} \qquad \sigma_{21} = b \tag{3.74}$$

and Equation 3.71 with Equation 3.73,

$$\sigma_{12} = b \qquad \sigma_{22} = c(1+\alpha) \tag{3.75}$$

In the deformed configuration, the elements of area are $d\bar{A} = \{1+\alpha, 1, 1\}^T$, while in the reference configuration, the areas are $dA = \{1, 1, 1\}^T$. The convected base vectors have Cartesian components $G_1 = \{1+\alpha, 0, 0\}^T$, $G_2 = \{0, 1, 0\}^T$, and $G_3 = \{0, 0, 1\}^T$.

The Cauchy stresses act on the deformed area and are referenced to the Cartesian unit vectors, so, for example, on the face normal to e_2, although the Cauchy stresses are $\{a, b, 0\}^T$ *per unit area*, the force components are $\{a(1+\alpha), b(1+\alpha), 0\}^T$ acting on the area $d\bar{A}_2 = (1+\alpha)$ in the deformed configuration.

Conversely, the 2PK stresses are referred to the areas in the reference configuration (which in this example have been taken as unit values) and to the convected base vectors, so, for example, on the face normal to e_1, the force components are $\{\sigma_{11}(1+\alpha), \sigma_{21}, 0\}^T$. Thus, $\sigma_{11} = a/(1+\alpha)$ and $\sigma_{21} = b$ because G_1 (to which σ_{11} is referenced) has stretched by a factor $(1+\alpha)$ and G_2 has remained unit length.

Although the state of (Cauchy) stress is assumed to remain constant, since the convected base vectors change during the motion, the material rate of the symmetric Piola–Kirchhoff stress will not vanish. It is

$$\dot{\boldsymbol{\sigma}} = \dot{\alpha}\begin{bmatrix} -\dfrac{a}{(1+\alpha)^2} & 0 & 0 \\ 0 & c & 0 \\ 0 & 0 & 0 \end{bmatrix} \tag{3.76}$$

In pure extension, the spin tensor $\mathbf{W} = 0$; that is, there is no local rotation and the velocity gradient consists of local rate of deformation only. Consequently, the Jaumann stress rate in this example is equal to the material rate of Cauchy stress, which is zero for a constant state of (Cauchy) stress.

$$\bar{\boldsymbol{\sigma}}^J = \dot{\bar{\boldsymbol{\sigma}}} - \mathbf{W}\bar{\boldsymbol{\sigma}} - \bar{\boldsymbol{\sigma}}\mathbf{W}^T = 0 \tag{3.77}$$

The material rotation is also zero (the principal stretch axes do not rotate), so the Green–Naghdi stress rate is also zero.

$$\bar{\boldsymbol{\sigma}}^{GN} = \dot{\bar{\boldsymbol{\sigma}}} - \boldsymbol{\Omega}\bar{\boldsymbol{\sigma}} - \bar{\boldsymbol{\sigma}}\boldsymbol{\Omega}^T = 0 \tag{3.78}$$

However, because the deformation tensor is nonzero, the Truesdell stress rate is nonzero. It can be determined as

$$\bar{\boldsymbol{\sigma}}^{Tr} = \dot{\bar{\boldsymbol{\sigma}}} - \mathbf{V}\bar{\boldsymbol{\sigma}} - \bar{\boldsymbol{\sigma}}\mathbf{V}^T + \mathrm{tr}(\mathbf{D})\,\bar{\boldsymbol{\sigma}} \tag{3.79}$$

After carrying out the computation, we have

$$\bar{\boldsymbol{\sigma}}^{Tr} = \dot{\alpha}\begin{bmatrix} -\dfrac{a}{(1+\alpha)} & 0 & 0 \\ 0 & \dfrac{c}{(1+\alpha)} & 0 \\ 0 & 0 & 0 \end{bmatrix} \tag{3.80}$$

For a physical interpretation of the Truesdell stress rate, referring to Figure 3.7, we consider the deformed configuration at time t and the configuration at time t + Δt. The base vectors that were the Cartesian base vectors (e_1, e_2, e_3) at time t, following the motion, change into

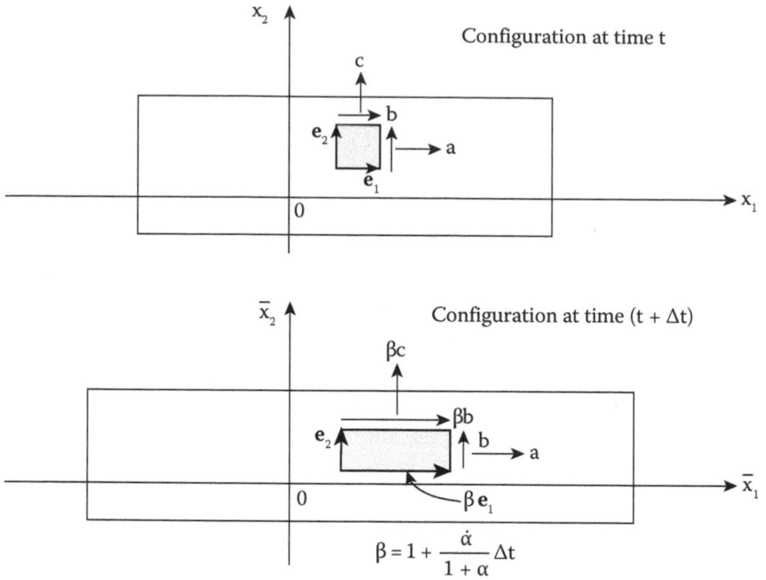

Figure 3.7 Physical interpretation of Truesdell stress rate in simple extension.

$(\beta e_1, e_2, e_3)$ at time $t + \Delta t$. This change is governed by the velocity gradient tensor (Equation 3.67). We therefore have

$$\beta = 1 + \frac{\partial v_1}{\partial \bar{x}_1}\Delta t = 1 + \frac{\dot{\alpha}}{1+\alpha}\Delta t \tag{3.81}$$

The Truesdell stress rate refers to the convected base vectors, which coincide with the Cartesian base vectors at the *deformed* configuration. They can be determined directly from

$$\bar{\boldsymbol{\sigma}}^{Tr} = \lim_{\Delta t \to 0}\left[\frac{1}{\Delta t}(\bar{\boldsymbol{\sigma}}_{t+\Delta t} - \bar{\boldsymbol{\sigma}}_t)\right] \tag{3.82}$$

Specifically, we obtain for the components in the (1, 2) plane

$$\begin{cases} \bar{\sigma}_{11}^{Tr} = \lim\limits_{\Delta t \to 0}\left[\dfrac{1}{\Delta t}\left(\dfrac{a}{\beta} - a\right)\right] = -\dfrac{a}{1+\alpha}\dot{\alpha} \\ \bar{\sigma}_{22}^{Tr} = \lim\limits_{\Delta t \to 0}\left[\dfrac{1}{\Delta t}(c\beta - c)\right] = \dfrac{c}{1+\alpha}\dot{\alpha} \\ \bar{\sigma}_{12}^{Tr} = \lim\limits_{\Delta t \to 0}\left[\dfrac{1}{\Delta t}\left(b\dfrac{\beta}{\beta} - b\right)\right] = 0 \end{cases} \tag{3.83}$$

3.6.2 Rigid body rotation of an initially stressed body

We now consider the rigid body rotation of a plate that carries an initial uniaxial stress that remains constant as the plate rotates, as shown in Figure 3.8.

The total motion could be thought of as a first stage (not shown) involving uniaxial stretching without rotation, resulting in the uniaxial stress shown, followed by a rigid body rotation

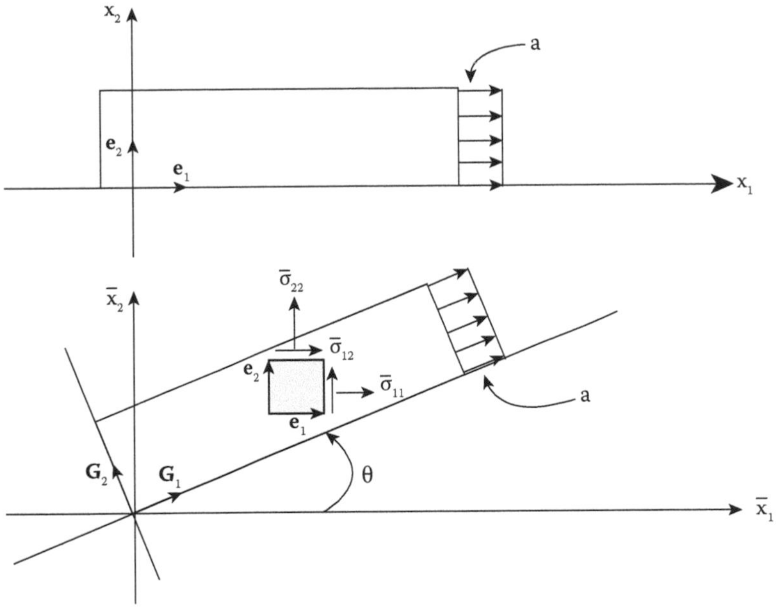

Figure 3.8 Rigid body rotation of an initially stressed body.

without further deformation, much like a right polar decomposition of the total motion. In this viewpoint, the reference configuration shown would not be an undeformed, unstressed configuration.

Therefore, the initial stress shown in the reference configuration could be considered either the Cauchy stress $\bar{\boldsymbol{\sigma}}$ or the unrotated Cauchy stress $\bar{\boldsymbol{\sigma}}'$. Which of the two viewpoints one adopted would depend on whether the subsequent motion involved deformation, as well as rotation. We prefer to think of the Cauchy stress in the reference configuration as the unrotated Cauchy stress, to distinguish it from the Cauchy stress in the rotating configuration, Therefore,

$$\bar{\boldsymbol{\sigma}}' = \begin{bmatrix} a & 0 & 0 \\ 0 & 0 & 0 \\ 0 & 0 & 0 \end{bmatrix} \tag{3.84}$$

During the rigid body rotation, there is no deformation; thus, the convected base vectors simply rotate and remain of unit length. However, the Cauchy stresses do change as the body rotates, as illustrated in Figure 3.8.

The deformation gradient for the rigid rotation is simply $\mathbf{F} = \mathbf{R}$, where the orthogonal matrix $\mathbf{R}$ is, in this case, a plane rotation about the $\bar{x}_3$ axis, that is,

$$\mathbf{R} = \begin{bmatrix} \cos\theta & -\sin\theta & 0 \\ \sin\theta & \cos\theta & 0 \\ 0 & 0 & 1 \end{bmatrix} \tag{3.85}$$

During the rigid body rotation, $|\mathbf{F}| = |\mathbf{R}| = 1$; that is, there is no volume change. The Cauchy stress tensor in the deformed (rotated) configuration is

$$\bar{\boldsymbol{\sigma}} = \mathbf{R}\,\bar{\boldsymbol{\sigma}}'\,\mathbf{R}^T \tag{3.86}$$

which gives

$$\bar{\sigma} = a \begin{bmatrix} c^2 & sc & 0 \\ sc & s^2 & 0 \\ 0 & 0 & 0 \end{bmatrix} \tag{3.87}$$

where $c \equiv \cos\theta$ and $s \equiv \sin\theta$. We note that when $\theta = 90°$, $\bar{\sigma}_{11} = 0$ and $\bar{\sigma}_{22} = a$, whereas initially (i.e., at $\theta = 0$) $\bar{\sigma}_{11} = a$ and $\bar{\sigma}_{22} = 0$, although the true axial stress acting on the plate has not changed with the rotation.

If we calculate the 2PK stress from the Cauchy stress in the rotated configuration,

$$\boldsymbol{\sigma} = |\mathbf{F}|\bar{\mathbf{F}}\ \bar{\boldsymbol{\sigma}}\ \bar{\mathbf{F}}^{\mathrm{T}} = |\mathbf{R}|\mathbf{R}^{\mathrm{T}}\bar{\boldsymbol{\sigma}}\ \mathbf{R} = \mathbf{R}^{\mathrm{T}}\left[\mathbf{R}\bar{\boldsymbol{\sigma}}'\mathbf{R}^{\mathrm{T}}\right]\ \mathbf{R} = \bar{\boldsymbol{\sigma}}' \tag{3.88}$$

thereby confirming that the 2PK stress remains constant during the rigid body rotation, as expected.

The (nonzero) material rate of Cauchy stress during the rigid body rotation is

$$\dot{\bar{\sigma}} = a\ \dot{\theta} \begin{bmatrix} -2sc & c^2 - s^2 & 0 \\ c^2 - s^2 & 2sc & 0 \\ 0 & 0 & 0 \end{bmatrix} \tag{3.89}$$

In this example, the velocity gradient $\mathbf{V} = \dot{\mathbf{F}}\mathbf{F}^{-1} = \dot{\mathbf{R}}\mathbf{R}^{\mathrm{T}} = \boldsymbol{\Omega}$ and the deformation rate

$$\mathbf{D} = \frac{1}{2}(\mathbf{V} + \mathbf{V}^{\mathrm{T}}) = \frac{1}{2}(\boldsymbol{\Omega} + \boldsymbol{\Omega}^{\mathrm{T}}) = 0$$

so the spin tensor

$$\mathbf{W} = \frac{1}{2}(\boldsymbol{\Omega} - \boldsymbol{\Omega}^{\mathrm{T}}) = \boldsymbol{\Omega}$$

the material rotation rate, which can be calculated explicitly as

$$\mathbf{W} = \dot{\mathbf{R}}\mathbf{R}^{\mathrm{T}} = \dot{\theta} \begin{bmatrix} -s & -c & 0 \\ c & -s & 0 \\ 0 & 0 & 0 \end{bmatrix} \begin{bmatrix} c & s & 0 \\ -s & c & 0 \\ 0 & 0 & 1 \end{bmatrix} = \dot{\theta} \begin{bmatrix} 0 & -1 & 0 \\ 1 & 0 & 0 \\ 0 & 0 & 0 \end{bmatrix} \tag{3.90}$$

As a consequence, the Jaumann, Green–Naghdi, and Truesdell stress rates are all equal. From Equations 3.89 and 3.90,

$$\begin{aligned} \mathbf{W}\bar{\boldsymbol{\sigma}} + \bar{\boldsymbol{\sigma}}\mathbf{W}^{\mathrm{T}} &= a\dot{\theta} \begin{bmatrix} 0 & -1 & 0 \\ 1 & 0 & 0 \\ 0 & 0 & 0 \end{bmatrix} \begin{bmatrix} c^2 & sc & 0 \\ sc & s^2 & 0 \\ 0 & 0 & 0 \end{bmatrix} + a\dot{\theta} \begin{bmatrix} c^2 & sc & 0 \\ sc & s^2 & 0 \\ 0 & 0 & 0 \end{bmatrix} \begin{bmatrix} 0 & 1 & 0 \\ -1 & 0 & 0 \\ 0 & 0 & 0 \end{bmatrix} \\ &= a\dot{\theta} \begin{bmatrix} -sc & -s^2 & 0 \\ c^2 & sc & 0 \\ 0 & 0 & 0 \end{bmatrix} + a\dot{\theta} \begin{bmatrix} -sc & c^2 & 0 \\ -s^2 & sc & 0 \\ 0 & 0 & 0 \end{bmatrix} = a\dot{\theta} \begin{bmatrix} -2sc & c^2 - s^2 & 0 \\ c^2 - s^2 & 2sc & 0 \\ 0 & 0 & 0 \end{bmatrix} \end{aligned} \tag{3.91}$$

Subtracting Equation 3.91 from Equation 3.89 to arrive at any or all three objective Cauchy stress rates gives, for this example,

$$\overset{\circ}{\bar{\sigma}}{}^{J} = \overset{\circ}{\bar{\sigma}}{}^{Tr} = \overset{\circ}{\bar{\sigma}}{}^{GN} = 0 \tag{3.92}$$

3.6.3 Simple shear of an initially stressed body

We now consider a case in which a plate carries a general two-dimensional state of (Cauchy) stress in the reference configuration and is subsequently subjected to a simple shear deformation that involves, as we have seen, both pure shear deformation and rotation of principal stretch axes. We assume in this example that the Cauchy stress remains constant during the subsequent shear deformation.

The Cauchy stress in the reference configuration is

$$\bar{\sigma} = \begin{bmatrix} a & b & 0 \\ b & c & 0 \\ 0 & 0 & 0 \end{bmatrix} \tag{3.93}$$

and its material (time) rate is consequently zero, that is,

$$\dot{\bar{\sigma}} = 0 \tag{3.94}$$

The kinematics of the subsequent simple shear motion has been presented in Section 2.7.6. We recall some of those details here. The parameter $\kappa(t)$ determines the motion by

$$\begin{cases} \bar{x}_1 = x_1 + \kappa\ x_2 \\ \bar{x}_2 = x_2 \\ \bar{x}_3 = x_3 \end{cases} \tag{3.95}$$

The Lagrangian convected base vectors are

$$\begin{cases} G_1 = e_1 \\ G_2 = \kappa e_1 + e_2 \\ G_3 = e_3 \end{cases} \tag{3.96}$$

and the deformation gradient and velocity gradient tensors are

$$F = \begin{bmatrix} 1 & \kappa & 0 \\ 0 & 1 & 0 \\ 0 & 0 & 1 \end{bmatrix} \tag{3.97}$$

$$V = \begin{bmatrix} 0 & \dot{\kappa} & 0 \\ 0 & 0 & 0 \\ 0 & 0 & 0 \end{bmatrix} \tag{3.98}$$

The inverse of the deformation gradient is

$$\bar{F} = \begin{bmatrix} 1 & -\kappa & 0 \\ 0 & 1 & 0 \\ 0 & 0 & 1 \end{bmatrix} \tag{3.99}$$

There is no volume change in simple shear ($|F| = 1$), and the 2PK stress tensor can be determined from the Cauchy stress tensor as follows:

$$\sigma = |F|\bar{F}\bar{\sigma}\bar{F}^T \;=\; \begin{bmatrix} 1 & -\kappa & 0 \\ 0 & 1 & 0 \\ 0 & 0 & 1 \end{bmatrix} \begin{bmatrix} a & b & 0 \\ b & c & 0 \\ 0 & 0 & 0 \end{bmatrix} \begin{bmatrix} 1 & 0 & 0 \\ -\kappa & 1 & 0 \\ 0 & 0 & 1 \end{bmatrix} \tag{3.100}$$

$$\sigma = \begin{bmatrix} a - 2b\kappa + c\kappa^2 & b - c\kappa & 0 \\ b - c\kappa & c & 0 \\ 0 & 0 & 0 \end{bmatrix} \tag{3.101}$$

In order to understand this result, it is instructive to determine the components of the 2PK stress tensor directly by balancing tractions on differential elements, as shown in Figure 3.9.

In Figure 3.9a, the Cauchy stresses are shown (Equation 3.93). In Figure 3.9b, the 2PK stresses are shown acting on the plane that had unit normal $(1, 0, 0)^T$, that is, $n_1 = 1$, in the reference configuration. Recall that the force vector in the deformed configuration in terms of 2PK stress is $d\bar{\mathbf{f}} = \mathbf{F}\ \boldsymbol{\sigma}\ dA$, which can be expressed as (viz. Equation 3.29)

$$d\bar{\mathbf{f}} = \bar{\mathbf{t}} d\bar{A} = \sigma_{jk} n_k \mathbf{G}_j dA \tag{3.102}$$

In Figure 3.9c, we consider a three-sided element. The force vector on the positive face is

$$d\bar{\mathbf{f}}_1 = \left[\sigma_{11}\ \mathbf{e}_1 + \sigma_{21}(\kappa\ \mathbf{e}_1 + \mathbf{e}_2)\right](1) \tag{3.103}$$

Note that even though the deformed area of the positive face is $d\bar{A}_1 = \sqrt{1+\kappa^2}$, the force in terms of 2PK stress is referenced to the original area of the face $dA_1 = 1$. The forces on the remaining two sides of the three-sided element are written in terms of Cauchy stresses. A force balance in the $\bar{x}_1$ direction yields

$$\sigma_{11} + \kappa\ \sigma_{21} = a - \kappa\ b \tag{3.104}$$

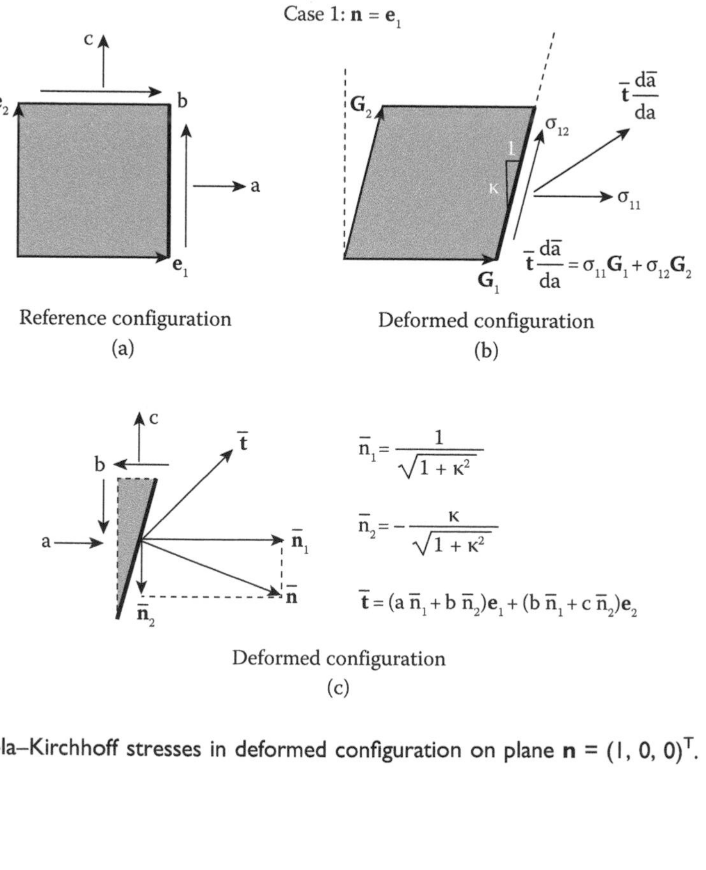

Figure 3.9 (a–c) Piola–Kirchhoff stresses in deformed configuration on plane $\mathbf{n} = (1, 0, 0)^T$.

$$\sigma_{21} = b - \kappa\ c \tag{3.105}$$

yielding

$$\begin{aligned} \sigma_{11} &= a - 2\kappa\ b + \kappa^2\ c \\ \sigma_{21} &= b - \kappa\ c \end{aligned} \tag{3.106}$$

Next, consider the plane that had unit normal $(0,1,0)^T$, that is, $n_2 = 1$, in the reference configuration, as shown in Figure 3.10. We again have three figures, showing Cauchy stresses in Figure 3.10a, 2PK stresses in Figure 3.10b, and both in Figure 3.10c. The force vector on the positive face is

$$d\bar{f}_2 = \left[\sigma_{12}\ e_1 + \sigma_{22}(\kappa\ e_1 + e_2)\right](1) \tag{3.107}$$

In Figure 3.10c, force balances in the $\bar{x}_1$ and $\bar{x}_2$ directions yield

$$\begin{cases} \sigma_{12} + \kappa\ \sigma_{22} = b \\ \sigma_{22} = c \end{cases} \tag{3.108}$$

from which we find $\sigma_{12} = b - \kappa\ c$, $\sigma_{22} = c$, confirming Equation 3.101 obtained by transforming the Cauchy stress tensor.

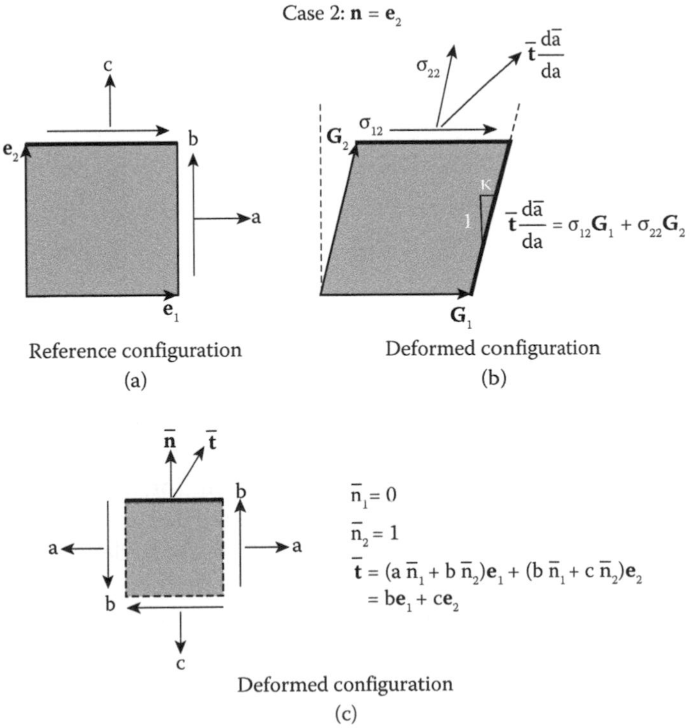

Figure 3.10 (a–c) Piola–Kirchhoff stresses in deformed configuration on plane $n = (0, 1, 0)^T$.

The material rate of 2PK stress is, from Equation 3.101,

$$\dot{\sigma} = \dot{\kappa}\begin{bmatrix} -2b + 2\kappa c & -c & 0 \\ -c & 0 & 0 \\ 0 & 0 & 0 \end{bmatrix} \tag{3.109}$$

which does not vanish even though the initial (Cauchy) stress remains constant during the deformation.

3.6.3.1 *Jaumann stress rate*

The spin tensor is

$$\mathbf{W} = \frac{\dot{\kappa}}{2}\begin{bmatrix} 0 & 1 & 0 \\ -1 & 0 & 0 \\ 0 & 0 & 0 \end{bmatrix} \tag{3.110}$$

and

$$\mathbf{W}\bar{\sigma} + \bar{\sigma}\mathbf{W}^T = \frac{\dot{\kappa}}{2}\begin{bmatrix} 0 & 1 & 0 \\ -1 & 0 & 0 \\ 0 & 0 & 0 \end{bmatrix}\begin{bmatrix} a & b & 0 \\ b & c & 0 \\ 0 & 0 & 0 \end{bmatrix} + \dot{\kappa}\begin{bmatrix} a & b & 0 \\ b & c & 0 \\ 0 & 0 & 0 \end{bmatrix}\begin{bmatrix} 0 & -1 & 0 \\ 1 & 0 & 0 \\ 0 & 0 & 0 \end{bmatrix} \tag{3.111}$$

Combining Equation 3.111 with Equation 3.93, we obtain for the Jaumann rate of Cauchy stress:

$$\bar{\sigma}^J = \frac{\dot{\kappa}}{2}\begin{bmatrix} -2b & a-c & 0 \\ a-c & 2b & 0 \\ 0 & 0 & 0 \end{bmatrix} \tag{3.112}$$

If, for example, the initial Cauchy stress was a pure shear, $a = c = 0$ and $b = \tau$, then the Jaumann rate would be simply

$$\bar{\sigma}^J = \dot{\kappa}\begin{bmatrix} -\tau & 0 & 0 \\ 0 & \tau & 0 \\ 0 & 0 & 0 \end{bmatrix} \tag{3.113}$$

For a physical interpretation of the Jaumann stress rate, we consider the deformed configuration at time t and a configuration at time $t + \Delta t$ that includes local rotation (via **W**) but not local deformation (via **D**) from the deformed configuration. In other words, we consider the effect on Cauchy stress of a local rigid body *rate* of rotation from the deformed configuration.

Figure 3.11a shows an element of (current) volume in the deformed configuration to which Cauchy stress is referred. Figure 3.11b shows the simple shear deformation. Figures 3.11c and 3.11d show the rate of deformation from the deformed configuration separated into its truly deformational (**D**) and rotational (**W**) components, as we have seen previously. It is Figure 3.11d that we now examine in more detail.

The Cartesian unit vectors e_1^*, e_2^* (and e_3^*) coincide with the Cartesian unit vectors at time t in the deformed configuration, but are convected (rotated) via the spin tensor through an infinitesimal angle $(-1/2\ \dot{\kappa}\Delta t)$ at time $t + \Delta t$. We temporarily set $\beta = \dot{\kappa}\Delta t/2$ for convenience and obtain

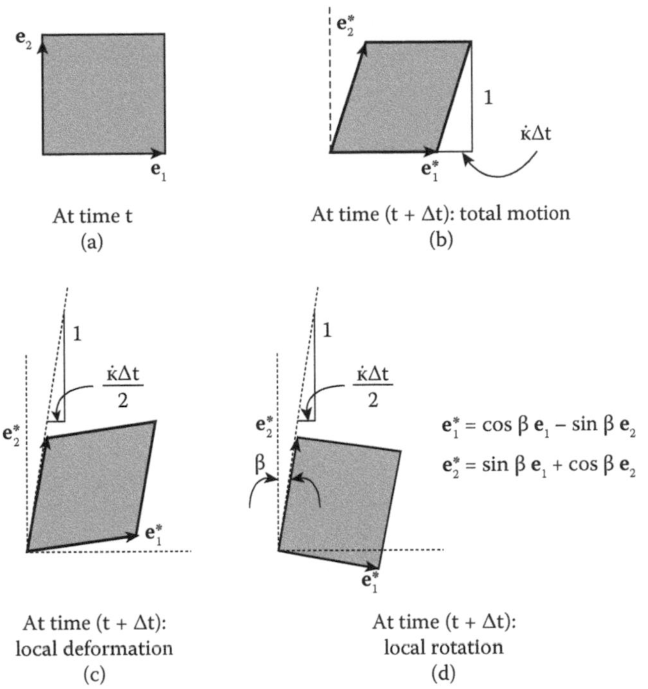

Figure 3.11 (a–d) Convected base vectors after local rotation from deformed configuration.

$$\begin{cases} e_1^* = \cos\beta\ e_1 - \sin\beta\ e_2 \\ e_2^* = \sin\beta\ e_1 + \cos\beta\ e_2 \\ e_3^* = e_3 \end{cases} \tag{3.114}$$

The force vector on the plane whose normal was e_1 at time t is

$$d\bar{f}_1 = \left[\sigma_{11}^* e_1^* + \sigma_{21}^* e_2^*\right]\ dA_1 \tag{3.115}$$

that is, the σ_{ij}^* are really 2PK stresses relative to the deformed configuration, and the e_i^* are "convected" base vectors from that same configuration as reference. If we use Equation 3.114 in Equation 3.115, we find

$$\begin{aligned} d\bar{f}_1 &= \sigma_{11}^*(\cos\beta\, e_1 - \sin\beta\ e_2) + \sigma_{21}^*(\sin\beta\, e_1 + \cos\beta\ e_2) \\ &= (\sigma_{11}^*\cos\beta + \sigma_{21}^*\sin\beta)e_1 + (-\sigma_{11}^*\sin\beta + \sigma_{21}^*\cos\beta)e_2 \end{aligned} \tag{3.116}$$

where dA_1 is taken as 1. On the other hand, the same force vector can be expressed in terms of Cauchy stresses as

$$d\bar{f}_1 = \bar{\sigma}\ d\bar{A} = \bar{\sigma}\ \bar{n}\ d\bar{A} \tag{3.117}$$

where $\bar{n} \equiv n^*$ and

$$n^* = \begin{Bmatrix} \cos\beta \\ -\sin\beta \\ 0 \end{Bmatrix} \tag{3.118}$$

We note again that the scalar area dA_1 does not change since only the rotational component of the motion is under consideration. We then find from Equation 3.117 the force vector in terms of Cauchy stress:

$$\mathbf{d}\bar{\mathbf{f}}_1 = \begin{bmatrix} a & b & 0 \\ b & c & 0 \\ 0 & 0 & 0 \end{bmatrix} \begin{Bmatrix} \cos\beta \\ \sin\beta \\ 0 \end{Bmatrix} = \begin{Bmatrix} a\cos\beta - b\sin\beta \\ b\cos\beta - c\sin\beta \\ 0 \end{Bmatrix} \tag{3.119}$$

From the two expressions for $\mathbf{d}\bar{\mathbf{f}}_1$ in Equations 3.116 and 3.119, we find

$$\begin{cases} \sigma_{11}^* + \dfrac{1}{2}\dot{\kappa}\Delta t\ \sigma_{21}^* = a - \dfrac{1}{2}\dot{\kappa}\Delta t\ b \\ \sigma_{21}^* - \dfrac{1}{2}\dot{\kappa}\Delta t\ \sigma_{11}^* = b - \dfrac{1}{2}\dot{\kappa}\Delta t\ c \end{cases} \tag{3.120}$$

Solving Equations 3.120 for σ_{11}^* and σ_{21}^* dropping terms in $(\dot{\kappa}\Delta t)^2$ compared with 1, we find

$$\begin{cases} \sigma_{11}^* = a - \dot{\kappa}\Delta t\ b \\ \sigma_{21}^* = b + \dfrac{\dot{\kappa}\Delta t}{2}(a - c) \end{cases} \tag{3.121}$$

If we apply the same process to determining the force vector $\mathbf{d}\bar{\mathbf{f}}_2$ acting on the plane whose normal was $\mathbf{e}_2$ at time t, we eventually find

$$\begin{cases} \sigma_{12}^* = b + \dfrac{\dot{\kappa}\Delta t}{2}(a - c) \\ \sigma_{22}^* = c + \dot{\kappa}\Delta t\ b \end{cases} \tag{3.122}$$

The components of the Jaumann stress rate tensor in the (1, 2) plane can now be determined from

$$\bar{\sigma}_{ij}^J = \lim_{\Delta t \to 0} [\sigma_{ij}^* - \bar{\sigma}_{ij}] \tag{3.123}$$

leading to

$$\begin{cases} \sigma_{11}^J = -\dot{\kappa} b \\ \sigma_{12}^J = \dfrac{\dot{\kappa}}{2}(a - c) \\ \sigma_{22}^J = \dot{\kappa} b \end{cases} \tag{3.124}$$

in agreement with Equation 3.112. The same results can also be obtained by balancing forces on free bodies in a manner similar to that used for verifying the 2PK stresses (Figures 3.9 and 3.10).

3.6.3.2 Truesdell stress rate

In the simple shear deformation, as we have seen, the velocity gradient tensor is

$$\mathbf{V} = \begin{bmatrix} 0 & \dot{\kappa} & 0 \\ 0 & 0 & 0 \\ 0 & 0 & 0 \end{bmatrix} \tag{3.125}$$

and the deformation rate tensor is

$$\mathbf{D} = \frac{\dot{\kappa}}{2}\begin{bmatrix} 0 & 1 & 0 \\ 1 & 0 & 0 \\ 0 & 0 & 0 \end{bmatrix} \tag{3.126}$$

Therefore, tr(**D**) = 0. The Truesdell rate can then be calculated as

$$\bar{\boldsymbol{\sigma}}^{Tr} = \dot{\bar{\boldsymbol{\sigma}}} - \mathbf{V}\bar{\boldsymbol{\sigma}} - \mathbf{V}^T\bar{\boldsymbol{\sigma}} = \dot{\kappa}\begin{bmatrix} -2b & -c & 0 \\ -c & 0 & 0 \\ 0 & 0 & 0 \end{bmatrix} \tag{3.127}$$

Truesdell stress can also be computed from the rate of 2PK stress by

$$\bar{\boldsymbol{\sigma}}^{Tr} = \frac{1}{|\mathbf{F}|}\mathbf{F}\,\dot{\boldsymbol{\sigma}}\,\mathbf{F}^T \tag{3.128}$$

For a physical interpretation of the Truesdell stress rate, we again consider the deformed configuration at times t and t + Δt, as shown in Figure 3.12. The base vectors at time t are ($\mathbf{e}_1$, $\mathbf{e}_2$, $\mathbf{e}_3$).

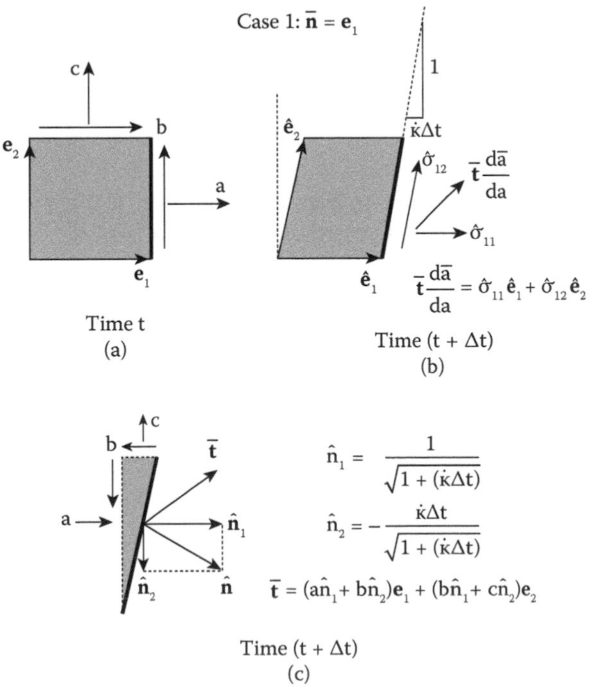

Figure 3.12 (a–c) Truesdell stress rate. Tractions on plane with normal $\mathbf{e}_1$ in deformed configuration.

At time $t + \Delta t$, the base vectors have been rotated and deformed by the simple shear motion to $(\hat{e}_1, \hat{e}_2, \hat{e}_3)$, and they no longer form an orthonormal basis. We first recall the basic relation for the convected Lagrangian base vectors in a motion $d\bar{x} = F\ dx$ as

$$G_i = F_{ki}\ e_k \tag{3.129}$$

and calculate their time rate as

$$\dot{G}_i = \dot{F}_{ki}\ e_k = (F_{mi} V_{km})\ e_k \tag{3.130}$$

Adapting Equation 3.130 to this example, we see that the time rate of the convected vectors $(\hat{e}_1, \hat{e}_2, \hat{e}_3)$ is

$$\dot{\hat{e}}_i = (F_{mi} V_{km})\ e_k \tag{3.131}$$

and that

$$\hat{e} = \left[I + (VF)^T \Delta t\right] e \tag{3.132}$$

Consequently, we find

$$\begin{cases} \hat{e}_1 = e_1 \\ \hat{e}_2 = \dot{\kappa}\Delta t\ e_1 + e_2 \\ \hat{e}_3 = e_3 \end{cases} \tag{3.133}$$

The Truesdell stress rate refers to these convected base vectors, and we see that it is essentially a 2PK material rate with the deformed configuration as its reference. The components of the Truesdell rate can be computed from

$$\bar{\sigma}_{ij}^{Tr} = \lim_{\Delta t \to 0} \frac{1}{\Delta t}\left[\hat{\sigma}_{ij} - \bar{\sigma}_{ij}\right] \tag{3.134}$$

where $\hat{\sigma}_{ij}$ are the components of stress at time $t + \Delta t$ along the convected base vectors $\hat{e}_i$ referenced to the areas at time t in the deformed configuration, and $\bar{\sigma}_{ij}$ are the Cauchy stresses at time t.

We first consider the plane whose normal vector was e_1 in the deformed configuration. At time $t + \Delta t$, the force vector on this plane (Figure 3.12b) is

$$d\bar{f}_1 = (\hat{\sigma}_{11}\hat{e}_1 + \hat{\sigma}_{21}\hat{e}_2) d\bar{A}_1 \tag{3.135}$$

where $d\bar{A}_1$ is taken as 1. Equation 3.135 can be expressed in terms of the base vectors at time t, using Equation 3.133, as

$$\begin{aligned} d\bar{f}_1 &= \hat{\sigma}_{11} e_1 + \hat{\sigma}_{21}(\dot{\kappa}\Delta t\ e_1 + e_2) \\ &= (\hat{\sigma}_{11} + \dot{\kappa}\Delta t\ \hat{\sigma}_{21})\ e_1 + \hat{\sigma}_{21} e_2 \end{aligned} \tag{3.136}$$

We now express the same force vector in terms of Cauchy stresses. Applying Nanson's formula,

$$d\hat{A} = |F| \bar{F}^T d\bar{A} \tag{3.137}$$

to this example, we find

$$d\hat{A} = \begin{bmatrix} 1 & 0 & 0 \\ -\dot{\kappa}\Delta t & 1 & 0 \\ 0 & 0 & 1 \end{bmatrix} \begin{Bmatrix} d\bar{A}_1 \\ 0 \\ 0 \end{Bmatrix} \tag{3.138}$$

and the force vector in terms of Cauchy stress is

$$d\bar{f}_1 = \sigma\ d\hat{A} = \begin{bmatrix} a & b & 0 \\ b & c & 0 \\ 0 & 0 & 0 \end{bmatrix} \begin{Bmatrix} 1 \\ -\dot{\kappa}\Delta t \\ 0 \end{Bmatrix} d\bar{A}_1 \tag{3.139}$$

Equating the two expressions for the force vector and solving for $\hat{\sigma}_{11}$ and $\hat{\sigma}_{21}$, we find

$$\begin{cases} \hat{\sigma}_{11} = a - 2b\ \dot{\kappa}\Delta t\ +\ c(\dot{\kappa}\Delta t)^2 \\ \hat{\sigma}_{21} = b - c\ \dot{\kappa}\Delta t \end{cases} \tag{3.140}$$

and applying Equation 3.134,

$$\begin{cases} \sigma_{11}^{Tr} = -2b\dot{\kappa} \\ \sigma_{21}^{Tr} = -c\dot{\kappa} \end{cases} \tag{3.141}$$

If we now apply the same process to the plane whose normal vector was e_2 in the deformed configuration at time t (Figure 3.13), we find

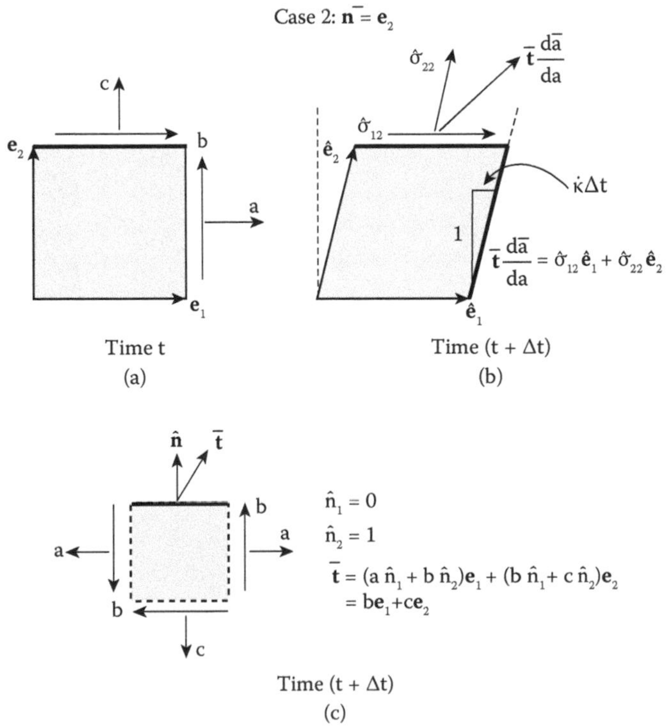

Figure 3.13 (a–c) Truesdell stress rate. Tractions on plane with normal e_2 in deformed configuration.

$$\begin{cases} \hat{\sigma}_{22} = c \\ \hat{\sigma}_{12} = b - c\ \dot{\kappa}\Delta t \end{cases} \tag{3.142}$$

From Equation 3.134 for $\Delta t \to 0$, we arrive at

$$\overline{\boldsymbol{\sigma}}^{Tr} = \dot{\kappa}\begin{bmatrix} -2b & -c & 0 \\ -c & 0 & 0 \\ 0 & 0 & 0 \end{bmatrix} \tag{3.143}$$

which agrees with Equation 3.127.

It is interesting to note the difference between the Jaumann and Truesdell stress rates for a pure shear deformation with initial Cauchy stress $\overline{\sigma}_{12} = \overline{\sigma}_{21} = \tau$.

$$\overline{\boldsymbol{\sigma}}^{Tr} = \dot{\kappa}\begin{bmatrix} -2\tau & 0 & 0 \\ 0 & 0 & 0 \\ 0 & 0 & 0 \end{bmatrix} \tag{3.144}$$

$$\overline{\boldsymbol{\sigma}}^{J} = \dot{\kappa}\begin{bmatrix} -\tau & 0 & 0 \\ 0 & \tau & 0 \\ 0 & 0 & 0 \end{bmatrix} \tag{3.145}$$

Chapter 4

Work and virtual work

In this chapter, we explore work and energy concepts that form the basis of the nonlinear finite element analysis methods that will be described in later chapters.

In Chapters 2 and 3, we presented the kinematics of nonlinear deformations and the several ways of describing internal forces in a deformed body, without yet connecting those forces (stresses) to the deformations that they produce.

The relations between stress and deformation are expressed through material (constitutive) models, which will most often be in rate form, that is, relating stress rates to strain rates. The basic requirements of objectivity and frame indifference for valid stress and strain rates have been discussed. However, it must still be established that a specific pair of stress and strain (rate) measures are compatible for use in a constitutive model; that is, they must be "work conjugate." We first present a fundamental tool, the *divergence theorem*, that will find much use throughout this book.

4.1 DIVERGENCE THEOREM

The divergence theorem (due to Gauss) is one of several well-known integral theorems. We state it briefly as follows.

For a vector **A** defined in a volume V and over its (complete and sufficiently smooth) enveloping surface S, the divergence theorem provides the following identity:

$$\int_S n_i A_i \, dS = \int_V \frac{\partial A_i}{\partial x_i} \, dV \tag{4.1}$$

where n_i is the unit outward normal to the surface and $\partial A_i / \partial x_i$ is the "divergence of **A**," often written as $\nabla \cdot \mathbf{A}$ or "div **A**."

4.2 STRESS POWER

Consider a small, but finite subvolume of the body (Figure 4.1), which we term a free body. This free body is of no particular specific shape. We shall denote its volume by $\overline{V}$ and its surface by $\partial\overline{V}$.

Surface forces $d\bar{\mathbf{f}}(\equiv \bar{\mathbf{t}}\ d\overline{A})$ act on each element of surface area $d\overline{A}$ that together make up the entire surface $\partial\overline{V}$ of the free-body volume $\overline{V}$. Body forces $\bar{\mathbf{b}}$ per unit volume of $\overline{V}$ also act in the interior of the free body. The motion and deformation of the material in the free body are specified in the usual way by the deformation gradient $d\bar{\mathbf{x}} = \mathbf{F}d\mathbf{x}$.

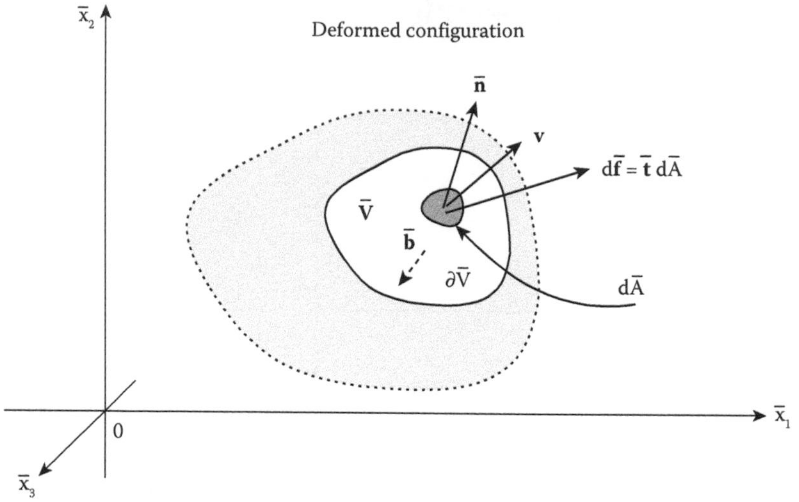

Figure 4.1 Tractions on the surface of a free body in the deformed configuration.

The rate of work (i.e., power) created by the surface force acting on the element of surface is $\mathbf{v}^T d\bar{\mathbf{f}}$, where $\mathbf{v}$ is the velocity vector at that point on the surface (i.e., on $d\bar{A}$). The rate of work done by the body force acting on a differential element of volume $d\bar{V}$ is $\mathbf{v}^T\bar{\mathbf{b}}\, d\bar{V}$. The power created by the forces on the entire free body during the motion (again, using the Cartesian component form for convenience) is then

$$\dot{W} = \int_{\partial\bar{V}} v_i\ \bar{t}_i\ d\bar{A} + \int_{\bar{V}} v_i\ \bar{b}_i\ d\bar{V} \tag{4.2}$$

The traction vector on the deformed surface can be represented in terms of Cauchy stress by $\bar{t}_i\ d\bar{A} = \bar{\sigma}_{ij}\ \bar{n}_j\ d\bar{A}$, giving

$$\dot{W} = \int_{\partial\bar{V}} v_i\ \bar{\sigma}_{ij}\ \bar{n}_j\ d\bar{A} + \int_{\bar{V}} v_i\ \bar{b}_i\ d\bar{V} \tag{4.3}$$

We transform the surface integral in Equation 4.3, using the divergence theorem, to get

$$\int_{\partial\bar{V}} \bar{n}_j(v_i\ \bar{\sigma}_{ij})\ d\bar{A} = \int_{\bar{V}} \frac{\partial}{\partial\bar{x}_j}(v_i\ \bar{\sigma}_{ij})d\bar{V} \tag{4.4}$$

and expand the volume integrand in Equation 4.4:

$$\int_{\bar{V}} \frac{\partial}{\partial\bar{x}_j}(v_i\ \bar{\sigma}_{ij})d\bar{V} = \int_{\bar{V}} \left[\frac{\partial v_i}{\partial\bar{x}_j}\ \bar{\sigma}_{ij} + v_i\frac{\partial\bar{\sigma}_{ij}}{\partial\bar{x}_j}\right] d\bar{V} \tag{4.5}$$

We now use Equation 4.5 in Equation 4.3 to arrive at

$$\dot{W} = \int_{\bar{V}} \left[\frac{\partial v_i}{\partial \bar{x}_j} \bar{\sigma}_{ij} + v_i \left(\frac{\partial \bar{\sigma}_{ij}}{\partial \bar{x}_j} + \bar{b}_i \right) \right] d\bar{V} \tag{4.6}$$

in which the first volume integral is

$$\dot{W}_I = \int_{\bar{V}} \frac{\partial v_i}{\partial \bar{x}_j} \bar{\sigma}_{ij} \; d\bar{V} = \int_{\bar{V}} \bar{\sigma}_{ij} V_{ij} \; d\bar{V} \tag{4.7}$$

and the second is

$$\dot{W}_E = \int_{\bar{V}} v_i \left(\frac{\partial \bar{\sigma}_{ij}}{\partial \bar{x}_j} + \bar{b}_i \right) d\bar{V} \tag{4.8}$$

The bracketed term in the integrand of Equation 4.8 is a local equilibrium equation,

$$\frac{\partial \bar{\sigma}_{ij}}{\partial \bar{x}_j} + \bar{b}_i = 0 \tag{4.9}$$

but we will generally not be concerned with enforcing pointwise equilibrium conditions, instead satisfying overall equilibrium in an integral or "weak" form over small regions (i.e., finite elements).

Equation 4.7 provides the result we seek,

$$\dot{W}_I = \int_{\bar{V}} \bar{\sigma}_{ij} V_{ij} \; d\bar{V} \tag{4.10}$$

so that the local (pointwise) rate of work of the internal stresses ("stress power") per unit deformed volume is

$$\dot{W}_I^o = \bar{\sigma}_{ij} V_{ij} = \bar{\sigma}_{ij} (D_{ij} + W_{ij}) \tag{4.11}$$

Since the Cauchy stress tensor is symmetric and the spin tensor is antisymmetric, the second term $\bar{\sigma}_{ij} W_{ij}$ vanishes identically and the rate of work of the internal stresses reduces to

$$\dot{W}_I^o = \bar{\sigma}_{ij} D_{ij} \tag{4.12}$$

Thus, we say that the Cauchy stress and the deformation rate tensor are *work conjugate* because their tensor product (often represented as $\bar{\boldsymbol{\sigma}} : \mathbf{D}$) produces a valid rate of work expression.

We can now find other work-conjugate stress and strain rate pairs by simply transforming Equation 4.12 appropriately.

4.2.1 2PK stress power

The relation between Cauchy stress and second Piola–Kirchhoff (2PK) stress that was developed in Chapter 3 is repeated below:

$$\bar{\boldsymbol{\sigma}} = \frac{1}{|\mathbf{F}|} \mathbf{F} \boldsymbol{\sigma} \mathbf{F}^T \tag{4.13}$$

We insert this relation in Cartesian component form into Equation 4.12 to get

$$\bar{\sigma}_{ij}D_{ij} = \frac{1}{|F|}F_{ik}\sigma_{km}F_{jm}D_{ij} = \sigma_{km}\left(\frac{1}{|F|}F_{ik}D_{ij}F_{jm}\right) \tag{4.14}$$

The bracketed term in Equation 4.14 can be written as

$$\frac{1}{|F|}\mathbf{F}^{T}\mathbf{D}\mathbf{F} = \frac{1}{|F|}\dot{\boldsymbol{\varepsilon}} \tag{4.15}$$

Therefore,

$$\int_{\bar{V}} \bar{\sigma}_{ij}D_{ij}\ d\bar{V} = \int_{\bar{V}} \sigma_{ij}\dot{\varepsilon}_{ij}\left(\frac{1}{|F|}d\bar{V}\right) = \int_{V} \sigma_{ij}\dot{\varepsilon}_{ij}\ dV \tag{4.16}$$

Thus, to evaluate the stress power for a volume of material, the integration is performed over the volume in the deformed configuration using the Eulerian work-conjugate pair $(\bar{\sigma}_{ij}, D_{ij})$ or, equivalently, over the volume in the reference configuration using the Lagrangian work-conjugate pair $(\sigma_{ij},\ \dot{\varepsilon}_{ij})$. That is,

$$\int_{\bar{V}} \bar{\sigma}_{ij}D_{ij}\ d\bar{V} = \int_{V} \sigma_{ij}\dot{\varepsilon}_{ij}\ dV \tag{4.17}$$

The latter formulation has some advantages for numerical finite element analysis.

4.2.2 Unrotated Cauchy stress power

We expect that the unrotated Cauchy stress $\bar{\boldsymbol{\sigma}}'$ and the unrotated deformation rate $\mathbf{d}$ also form a work-conjugate pair defined in the deformed configuration, since

$$\begin{aligned} \bar{\boldsymbol{\sigma}} &= \mathbf{R}\ \bar{\boldsymbol{\sigma}}'\ \mathbf{R}^{T} \\ \mathbf{D} &= \mathbf{R}\ \mathbf{d}\ \mathbf{R}^{T} \end{aligned} \tag{4.18}$$

To confirm this, we return to the tensor inner product $\bar{\boldsymbol{\sigma}} : \mathbf{D}$ and evaluate it using Equation 4.18 as follows. First, a tensor inner product, say $\mathbf{A}:\mathbf{B}$, can be expressed as

$$\mathbf{A} : \mathbf{B} \equiv \mathrm{tr}(\mathbf{A}\mathbf{B}^{T}) \tag{4.19}$$

We express Equations 4.18 in component form as

$$\begin{aligned} \bar{\sigma}_{ik} &= R_{il}\bar{\sigma}'_{lm}R_{km} \\ D_{jk} &= R_{jr}d_{rs}R_{ks} \end{aligned} \tag{4.20}$$

and apply the identity in Equation 4.19 to obtain

$$\delta_{ij}\bar{\sigma}_{ik}D_{jk} = \delta_{ij}(R_{il}\bar{\sigma}'_{lm}R_{km})(R_{jr}d_{rs}R_{ks}) = (\delta_{lr}\delta_{ms}\bar{\sigma}'_{lm}d_{rs}) = \bar{\sigma}'_{rm}d_{rm} \tag{4.21}$$

thereby confirming that

$$\bar{\boldsymbol{\sigma}} : \mathbf{D} = \bar{\boldsymbol{\sigma}}' : \mathbf{d} \tag{4.22}$$

4.3 VIRTUAL WORK

In the previous section, we utilized the concept of the rate of work done by a force acting through the velocity of its point of application (= $\mathbf{v}^T\mathbf{f}$). The velocity vector $\mathbf{v} = \dot{\bar{\mathbf{x}}} = \dot{\mathbf{u}}$, where $\mathbf{u} = \bar{\mathbf{x}} - \mathbf{x}$ is the displacement (vector) of the point of action of the force. Therefore, the rate of work done by the force can also be expressed as

$$\dot{W} = \dot{\mathbf{u}}^T\mathbf{f} \tag{4.23}$$

and in an infinitesimal time interval dt, the work done is

$$dW = \dot{W}dt = d\mathbf{u}^T\mathbf{f} \tag{4.24}$$

where the increment of displacement $d\mathbf{u} = \dot{\mathbf{u}}\ dt$.

We use the notation $\delta\mathbf{u}$ to denote a *virtual displacement* to distinguish it from the real displacement increment $d\mathbf{u}$ in Equation 4.24.

The term *virtual* as used here can be described as existing in essence or effect but not in actual fact.

The virtual work δW done by the force is then

$$\delta W = \delta\mathbf{u}^T\mathbf{f} \tag{4.25}$$

The virtual displacement $\delta\mathbf{u}$ can be thought of as a small perturbation from an equilibrium configuration in which the *force remains constant.*

4.4 PRINCIPLE OF VIRTUAL WORK

The *principle of virtual work* (PVW) is the fundamental basis for powerful and versatile approximate numerical analysis (e.g., finite element) procedures. It also provides a unified approach for formulating finite element models for all types of solid and structural components, and is equally applicable to nonlinear and linear analysis.

It is basically an alternative statement of equilibrium requirements (or perhaps, in a sense, a test of equilibrium) and can be viewed as a member of a large family of weighted-residual methods.

The PVW may be stated as follows: the virtual work done by a system of (internal and external) forces in equilibrium is zero for all possible kinematically admissible virtual displacements; that is, *if the system is in equilibrium, then* $\delta W = 0$ *for all possible kinematically admissible virtual displacements.*

It is actually the *converse of the principle of virtual work* (CPVW) that provides the important alternative to enforcement of local equilibrium in approximate analysis methods. The CPVW may be stated as follows: if the virtual work done by a system of (internal and external) forces is zero for all possible kinematically admissible virtual displacements, then the system of forces is in equilibrium; that is, *if* $\delta W = 0$ *for all possible kinematically admissible virtual displacements, then the system is in equilibrium.*

The logic of these two statements, without all the important qualifying conditions, may be summarized as

$$\boxed{VW = 0} \underset{PVW}{\overset{CPVW}{\rightleftarrows}} \boxed{EQM} \tag{4.26}$$

where VW = virtual work and EQM = equilibrium.

It is again emphasized that the forces, both internal (i.e., stresses) and external, are considered "frozen" in the equilibrium configuration that is being tested. The term *kinematically admissible* displacements means that the displacements that are used in the virtual work calculation must satisfy all the kinematic (i.e., motion) boundary conditions. The qualifier "all possible" displacements cannot be met, and in practice is taken to mean all possible displacements among the selected class of approximate displacements (e.g., the class of displacements defined by the chosen finite element shape functions). Note that there are no qualifying conditions related to satisfaction of traction-type boundary conditions on the surface of the body.

From this point on, we simply refer to PVW on the understanding that it is actually CPVW that is being applied.

4.4.1 Internal virtual work

For computational convenience, the virtual work of the system of internal and external forces is separated into two portions: internal virtual work δW_{int} (done by internal stresses) and external virtual work δW_{ext} (done by specified tractions) (i.e., loads) on the surface of the body or by body forces (e.g., gravity loads).

Therefore, the total virtual work is defined as

$$\delta W = \delta W_{int} + \delta W_{ext} \tag{4.27}$$

Thus, at equilibrium (when $\delta W = 0$), $\delta W_{int} = -\delta W_{ext}$. We caution that many authors use a different sign convention that would lead instead to internal and external virtual work being equated at equilibrium, rather than summed to zero, as in Equation 4.27.

Consider again the free body of a small volume of material in the deformed configuration shown in Figure 4.1. From Equation 4.2, we find the real work done during an infinitesimal time increment dt,

$$dW = \dot{W} dt = \int_{\partial\bar{V}} du_i\ \bar{t}_i\ d\bar{A} + \int_{\bar{V}} du_i\ \bar{b}_i\ d\bar{V} \tag{4.28}$$

where $v_i dt = \dot{u}_i dt = du_i$. The virtual work done on the free body during a virtual displacement δu_i from equilibrium in the deformed configuration is then

$$\delta W_{ext} = \int_{\partial\bar{V}} \delta u_i\ \bar{t}_i\ d\bar{A} + \int_{\bar{V}} \delta u_i\ \bar{b}_i\ d\bar{V} \tag{4.29}$$

As far as the free body is concerned, the traction on its surface is an external effect, so we choose to identify the surface integral in Equation 4.29 as external virtual work. We include the virtual work of the body force $\bar{b}_i$ in the external virtual work as a matter of convenience. When we express the traction vector in terms of Cauchy stress and apply the divergence theorem, we find

$$\delta W_{ext} = \int_{\bar{V}} \bar{\sigma}_{ij}\ \delta\left(\frac{\partial u_i}{\partial \bar{x}_j}\right) d\bar{V} + \int_{\bar{V}} \delta u_i \left(\frac{\partial \bar{\sigma}_{ij}}{\partial \bar{x}_j} + \bar{b}_i\right) d\bar{V} \tag{4.30}$$

The integrand in the second volume integral is zero (local equilibrium), and Equation 4.30 reduces to

$$\delta W_{ext} = \int_{\bar{V}} \bar{\sigma}_{ij}\ \delta\left(\frac{\partial u_i}{\partial \bar{x}_j}\right) d\bar{V} \tag{4.31}$$

Applying the PVW to the free body and using Equation 4.31, we find

$$\delta W_{int} + \int_{\overline{V}} \overline{\sigma}_{ij}\, \delta\left(\frac{\partial u_i}{\partial \overline{x}_j}\right) d\overline{V} = 0 \tag{4.32}$$

and therefore,

$$\delta W_{int} = -\int_{\overline{V}} \overline{\sigma}_{ij}\, \delta\left(\frac{\partial u_i}{\partial \overline{x}_j}\right) d\overline{V} \tag{4.33}$$

which can be expressed in a Lagrangian form as well, which we shall do in the next section.

We concede that the negative sign in the definition of the internal virtual work is merely a sign convention, but one that we believe is consistent. The following simple example may help to further explain this choice.

Example 4.1 Linear MDOF structural system

Consider a linear multi-degrees-of-freedom (MDOF) structural system with equilibrium equations

$$\mathbf{P} = \mathbf{KU} \tag{4.34}$$

where **P** is the vector of external and element equivalent nodal loads, **K** is the structure stiffness matrix, and **U** is the nodal displacement vector. The nodal equilibrium equations (4.34) can obviously be expressed as

$$\mathbf{P} - \mathbf{KU} = 0 \tag{4.35}$$

For a set of arbitrary nodal virtual displacements $\delta\mathbf{U}$, the virtual work done is

$$\delta\mathbf{U}^T[\mathbf{P} - \mathbf{KU}] = 0 \tag{4.36}$$

or

$$\delta\mathbf{U}^T\mathbf{P} - \delta\mathbf{U}^T\mathbf{KU} = 0 \tag{4.37}$$

Therefore,

$$\begin{aligned} &\delta W_{int} + \delta W_{ext} = 0 \\ &\delta W_{int} = -\,\delta\mathbf{U}^T\mathbf{KU} \\ &\delta W_{ext} = \delta\mathbf{U}^T\mathbf{P} \end{aligned} \tag{4.38}$$

The *internal resisting force vector* **I** is defined as

$$\mathbf{I} = \mathbf{KU} \tag{4.39}$$

and the nodal equilibrium equations can alternatively be expressed as

$$\mathbf{P} - \mathbf{I} = 0 \tag{4.40}$$

I_i, the i-th component of the internal resisting force vector **I**, is the force acting *on the elements* meeting at DOF U_i. The force from the elements that acts *on the node* at DOF i is therefore $-I_i$, and the equilibrium equation is $P_i - I_i = 0$.

For finite element models of continua, the internal resisting force vector is assembled from element internal resisting force vectors calculated from element stresses via Equation 4.33 or equivalent expressions.

For linear systems, the internal virtual work can be related to first-order increments of strain energy due to real nodal displacement increments.

4.4.2 Lagrangian form of internal virtual work

For incremental-iterative numerical solution methods based on either a total Lagrangian or updated Lagrangian formulation, internal virtual work is calculated in the reference configuration.

We start from the Lagrangian expression for stress power in terms of 2PK stress and Green strain rate (viz. Equation 4.17),

$$\dot{W}_I = \int_V \sigma_{ij}\dot{\varepsilon}_{ij}\, dV \tag{4.41}$$

and express the internal virtual work as

$$\delta W_{int} = -\int_V \sigma_{ij}\delta\varepsilon_{ij}\, dV \tag{4.42}$$

where the virtual (Green) strain is

$$\delta\varepsilon_{ij} = \frac{1}{2}(\delta u_{i,j} + \delta u_{j,i} + \delta u_{k,i}u_{k,j} + u_{k,i}\delta u_{k,j}) \tag{4.43}$$

and $u_{i,j} = G_{ij}$, the Lagrangian displacement gradient introduced in Chapter 2. The integrand in Equation 4.42 is then

$$\sigma_{ij}\delta\varepsilon_{ij} = \frac{1}{2}\sigma_{ij}(\delta u_{i,j} + \delta u_{j,i} + \delta u_{k,i}u_{k,j} + u_{k,i}\delta u_{k,j}) \tag{4.44}$$

Making use of the symmetry of the 2PK stress in Equation 4.44, we find that the Lagrangian expression for internal virtual work simplifies to

$$\delta W_{int} = -\int_V \sigma_{ij}(\delta_{ik} + u_{k,i})\delta u_{k,j}\, dV \tag{4.45}$$

4.5 VECTOR FORMS OF STRESS AND STRAIN

As mentioned previously, stress and strain tensors are often represented as vectors for convenience in numerical computation. The representation of the stress tensor as

$$\{\bar{\sigma}\} = \begin{Bmatrix} \bar{\sigma}_{11} \\ \bar{\sigma}_{22} \\ \bar{\sigma}_{33} \\ \bar{\sigma}_{12} \\ \bar{\sigma}_{13} \\ \bar{\sigma}_{23} \end{Bmatrix} \tag{4.46}$$

mandates that the strain tensor be represented as

$$\{\bar{\varepsilon}\} = \begin{Bmatrix} \bar{\varepsilon}_{11} \\ \bar{\varepsilon}_{22} \\ \bar{\varepsilon}_{33} \\ 2\bar{\varepsilon}_{12} \\ 2\bar{\varepsilon}_{13} \\ 2\bar{\varepsilon}_{23} \end{Bmatrix} \tag{4.47}$$

in order that the vector dot product of stress and strain give the same work consistent expression as the tensor inner product of stress and strain, that is,

$$\{\bar{\varepsilon}\}^{T}\{\bar{\sigma}\} \equiv \bar{\varepsilon} : \boldsymbol{\sigma} \tag{4.48}$$

Here, we temporarily depart from our standard bold-faced notation for both vectors and tensors, denoting the vector forms by $\{\cdot\}$ for clarity.

These vector representations of tensor quantities have consequences relating to coordinate system rotations. First, a coordinate rotation matrix, say $[T_\sigma]$, can be constructed to rotate the stress vector to a new Cartesian coordinate system by

$$\{\bar{\sigma}'\} = [T_\sigma]\{\bar{\sigma}\} \tag{4.49}$$

from the rotation tensor Q, to yield the same result as the equivalent tensor transformation,

$$\bar{\sigma}' = Q\sigma Q^{T} \tag{4.50}$$

but the matrix $[T_\sigma]$ is not orthogonal.

The second consequence is that because the vector representations have to be made differently for stress and strain in order to preserve work consistency (Equation 4.48), the strain vector requires a different coordinate transformation matrix. The invariance of the scalar product under a coordinate rotation requires

$$\begin{aligned}\{\bar{\varepsilon}'\}^{T}\{\bar{\sigma}'\} &= \{\bar{\varepsilon}\}^{T}\{\bar{\sigma}\} \\ &= \{\bar{\varepsilon}'\}^{T}[T_\sigma]\{\bar{\sigma}\}\end{aligned} \tag{4.51}$$

from which

$$\{\bar{\varepsilon}\} = [T_\sigma]^{T}\{\bar{\varepsilon}'\} \tag{4.52}$$

which may also be written as

$$\{\bar{\varepsilon}'\} = [T_\varepsilon]\{\bar{\varepsilon}\} \tag{4.53}$$

where $[T_\varepsilon] \equiv [T_\sigma]^{-T}$.

Chapter 5

Elastic material properties

5.1 INTRODUCTION

Constitutive laws are idealized mathematical models that are based primarily on observations of material behavior in specially designed experiments and/or under general loading conditions. These constitutive laws have to satisfy some fundamental laws of mechanics, known as *conservation laws*, in addition to approximating observed material response. We will discuss conservation laws of mechanics in more detail in Chapter 6 in the context of plasticity; at this point, we simply state the general requirement for all constitutive laws that the energy supplied to a material over complete loading and unloading cycles must be nonpositive. In other words, we cannot generate energy just by cyclically loading and unloading a material.

In the specific case of elastic material models, work done by stresses during loading is stored as elastic strain energy, which is fully recovered upon unloading, so that the work done over a complete loading and unloading cycle is zero.

Material tests are normally designed so that the state of stress and strain within a test sample is essentially uniform. That is, so that an isolated "material point" is represented in the test, and so that the results of the test can be readily interpreted in terms of the parameters of a specific mathematical material model.

However, material constitutive laws can also be determined from the behavior of systems (i.e., essentially more complex test specimens) subjected to general loading in which the stresses and strains within the system (specimen) are no longer spatially uniform. Naturally, developing a material model from such a test is a more complex task; we will address that possibility in Chapter 12.

Our first inclination is to think of elastic materials as behaving linearly. Although some elastic materials do behave almost linearly over a certain range of deformations, most truly elastic materials are inherently nonlinear. A material is said to exhibit elastic (not necessarily linear) behavior if there are no *permanent deformations* or residual stresses in any loading and unloading cycle; that is, the stress–strain relations follow the same paths under loading and unloading. A consequence of such a lack of permanent deformation is that there is no history dependency (i.e., memory) in elastic materials; that is, the current state of stress depends only on the current state of deformation.

In this chapter and Chapter 6, we discuss the "mathematical" constitutive laws that are commonly used in computational mechanics. It is important to mention that in recent years, information-based *soft computing methods* have also been used to capture material behavior. In these approaches, "neural networks" replace mathematical constitutive laws. This topic is discussed in Chapter 12 and in a book on soft computing (Ghaboussi 2017).

5.2 LINEAR ELASTIC MATERIAL MODELS

Linear elasticity is based on the assumption of infinitesimal displacements and strains, which leads to unique definitions of stresses and strains, also recognized as engineering stresses and strains. The material model is expressed through a generalized Hooke's law given in the following equation in its most general form.

$$\sigma_{ij} = D_{ijkl}\ \varepsilon_{kl} \tag{5.1}$$

The tensor of elastic constants D_{ijkl} is a fourth-order tensor that has certain symmetries because of the symmetries of σ_{ij} and ε_{ij}. Thus, follow the "minor" symmetries $D_{ijkl} = D_{jikl}$ and $D_{ijkl} = D_{ijlk}$. The elastic strain energy $2U = \sigma_{ij}\varepsilon_{ij} = D_{ijkl}\varepsilon_{kl}\varepsilon_{ij}$, which results in the "major" symmetries $D_{ijkl} = D_{klij}$. If Equation 5.1 is written in matrix–vector format, the resulting (6×6) symmetric elastic constitutive matrix has at most 21 independent material constants, which can describe different types of material anisotropy.

Isotropic elastic materials have no preferred material directions. That is, under an arbitrary rotation of reference axes, D_{ijkl} is invariant. This leads to the conclusion that for isotropic materials, there are only two independent material constants contained in D_{ijkl}. The fourth-order constant isotropic material elasticity tensor is

$$D_{ijkl} = \lambda\ \delta_{ij}\ \delta_{kl} + \mu(\delta_{ik}\ \delta_{jl} + \delta_{il}\ \delta_{jk}) \tag{5.2}$$

The material parameters λ and μ are known as Lamé constants. The stress–strain relations in terms of Lamé constants are given in the following equation:

$$\begin{aligned} \sigma_{ij} &= \left[\lambda\ \delta_{ij}\ \delta_{kl} + \mu(\delta_{ik}\ \delta_{jl} + \delta_{il}\ \delta_{jk})\right]\varepsilon_{kl} \\ &= \lambda\ \delta_{ij}\ \varepsilon_{kk} + \mu(\varepsilon_{ij} + \varepsilon_{ji}) \\ &= \lambda\ \delta_{ij}\ \varepsilon_{kk} + 2\mu\ \varepsilon_{ij} \end{aligned} \tag{5.3}$$

The two elastic constants can be expressed in several different forms. In engineering applications, we often use Young's modulus E and Poisson's ratio ν instead of the Lamé constants. The following equations relate these two sets of constants for isotropic elastic materials, where G ($\equiv\mu$) is the shear modulus.

$$\begin{cases} \lambda = \dfrac{\nu\ E}{(1+\nu)(1-2\nu)} \\ \mu = \dfrac{E}{2(1+\nu)} = G \end{cases} \tag{5.4}$$

Some rubberlike materials can undergo quite large elastic deformations. Metals, on the other hand, generally can undergo only very small elastic deformations before yielding plastically.

Once the assumption of infinitesimal deformations is removed, the definitions of stress and strain are no longer unique. In that case, the constitutive law must relate work-conjugate stress and strain pairs, as we have seen previously. In a Lagrangian formulation, this involves the second Piola–Kirchhoff (2PK) stress and Green strain, as shown in the following equation:

$$\boldsymbol{\sigma} = \mathbf{D}^e : \boldsymbol{\varepsilon} \tag{5.5}$$

We emphasize that the symmetric 2PK stress refers to the convected coordinates, and as such, it is not the "true" measure of stress that is familiar to engineers. True stress, or engineering stress, that is measured in laboratory material tests is in fact the Cauchy stress.

This is one aspect of the complications that arise from the nonuniqueness in the definition of stresses and strains.

There are three different general classes of elastic material models that exhibit, in varying degrees, properties that one might consider to characterize "elastic" behavior. These are (1) hyperelastic (Green elastic), (2) elastic (Cauchy elastic), and (3) hypoelastic material models.

In the following sections, we discuss these three types of elastic material models. We begin with Cauchy elastic material models.

5.3 CAUCHY ELASTIC MATERIAL MODELS

We start with a general nonlinear elastic stress–strain relation independent of stress and strain history.

$$\sigma_{ij} = f_{ij}(\varepsilon_{kl}) \tag{5.6}$$

We assume here that there exists a unique relationship between the current stresses and current strains. The above function can be expanded in polynomial form as

$$\sigma_{ij} = C_{ijkl}\ \varepsilon_{kl} + C_{ijklmn}\ \varepsilon_{kl}\ \varepsilon_{mn} + \cdots \tag{5.7}$$

This relation, although it appears very complicated, can be greatly simplified by introducing restrictions that are imposed by particular material symmetries.

For instance, for an isotropic material we have the following forms of tensor and matrix equations:

$$\sigma_{ij} = a_1\ \varepsilon_{ij} + a_2\ \varepsilon_{ik}\ \varepsilon_{kj} + a_3\ \varepsilon_{ik}\ \varepsilon_{kl}\ \varepsilon_{lj} + \cdots \tag{5.8}$$

$$\boldsymbol{\sigma} = a_1\ \boldsymbol{\varepsilon} + a_2\ \boldsymbol{\varepsilon}^2 + a_3\ \boldsymbol{\varepsilon}^3 + \cdots \tag{5.9}$$

where a_α, $\alpha = 1, 2, 3, \ldots$ are material constants, and $\boldsymbol{\sigma}$ and $\boldsymbol{\varepsilon}$ are 3 × 3 matrices. We can use the *Cayley–Hamilton* theorem to express this polynomial stress–strain relation in terms of the three strain invariants.

5.3.1 Characteristic polynomial of a matrix

The eigenvalues of a symmetric (3 × 3) matrix **A** are the roots of

$$\det(\mathbf{A} - \lambda\ \mathbf{I}) = 0 \tag{5.10}$$

Straightforward expansion of the determinant leads to the characteristic polynomial

$$\lambda^3 - I_1\lambda^2 + I_2\lambda - I_3 = 0 \tag{5.11}$$

The three roots of the cubic polynomial are the eigenvalues (λ_1, λ_2, λ_3). The invariants of the matrix **A** can be arranged in the form

$$\begin{cases} I_1 = \mathrm{tr}(\mathbf{A}) \equiv A_{kk} \\ I_2 = \dfrac{1}{2}(A_{kk}^2 - A_{ij}A_{ij}) \\ I_3 = \det(\mathbf{A}) \end{cases} \tag{5.12}$$

5.3.2 Cayley–Hamilton theorem

The Cayley–Hamilton theorem states that any symmetric second-order matrix satisfies its own characteristic equation. That is, in this case,

$$A^3 - I_1 A^2 + I_2 A - I_3\ \mathbf{I} = 0 \tag{5.13}$$

where **I** is the (3 × 3) identity matrix. The theorem is easily proved by using the spectral representation of the matrix, that is,

$$A = \sum_i \lambda_i\ \varphi_i \varphi_i^T \tag{5.14}$$

Therefore, according to the Cayley–Hamilton theorem, the strain tensor satisfies the following characteristic equation:

$$\boldsymbol{\varepsilon}^3 - I_1\ \boldsymbol{\varepsilon}^2 + I_2\ \boldsymbol{\varepsilon} - I_3\ \mathbf{I} = 0 \tag{5.15}$$

where the three strain invariants are

$$\begin{cases} I_1 = \mathrm{tr}(\boldsymbol{\varepsilon}) = \delta_{ij}\ \varepsilon_{ij} = \varepsilon_{kk} \\ I_2 = \dfrac{1}{2}\left(\varepsilon_{kk}^2 - \varepsilon_{ij}\ \varepsilon_{ij}\right) \\ I_3 = |\boldsymbol{\varepsilon}| = \dfrac{1}{6}\left(e_{ijk}\ e_{lmn}\ \varepsilon_{il}\ \varepsilon_{jm}\ \varepsilon_{kn}\right) \end{cases} \tag{5.16}$$

and e_{ijk} is the permutation symbol that was defined earlier.

5.3.3 General polynomial form for isotropic Cauchy elastic materials

By using the Cayley–Hamilton theorem recursively, we can express $\boldsymbol{\varepsilon}^3$ and all higher integral powers of $\boldsymbol{\varepsilon}$ in terms of $\boldsymbol{\varepsilon}$ and $\boldsymbol{\varepsilon}^2$. Therefore, the general polynomial stress–strain relation can be expressed in the following matrix or tensor form:

$$\boldsymbol{\sigma} = \alpha_0\ \mathbf{I} + \alpha_1\ \boldsymbol{\varepsilon} + \alpha_2\ \boldsymbol{\varepsilon}^2 \tag{5.17}$$

$$\sigma_{ij} = \alpha_0\ \delta_{ij} + \alpha_1\ \varepsilon_{ij} + \alpha_2\ \varepsilon_{ik}\ \varepsilon_{kj} \tag{5.18}$$

where the scalars α_0, α_1, and α_2 are polynomial functions of the strain invariants I_1, I_2, and I_3.

As an example, we consider an isotropic linear (i.e., $\alpha_2 = 0$) elastic material. Let $\alpha_1 = \beta_2$ and $\alpha_0 = \beta_0 + \beta_1 I_1$, where β_0, β_1, and β_2 are material constants. With these material constants, the above stress–strain relation can be rewritten in the following form:

$$\sigma_{ij} = (\beta_0 + \beta_1\ I_1)\delta_{ij} + \beta_2\ \varepsilon_{ij} \tag{5.19}$$

If the initial state is stress-free, then $\beta_0 = 0$, which leads to the following stress–strain relations:

$$\begin{aligned} \sigma_{ij} &= \beta_1\ I_1\ \delta_{ij} + \beta_2\ \varepsilon_{ij} \\ &= \beta_1\ \varepsilon_{kk}\ \delta_{ij} + \beta_2\ \varepsilon_{ij} \end{aligned} \tag{5.20}$$

We recognize that this relation (with $\beta_1 = \lambda$ and $\beta_2 = 2\mu$) is simply the generalized Hooke's law for linear isotropic elastic materials.

It is often convenient to express Equation 5.20 in an alternate form by setting

$$\begin{cases} \beta_2 = 2\mu = 2G \\ \beta_1 = \lambda = K - \dfrac{2G}{3} \end{cases} \tag{5.21}$$

where the bulk modulus K is the elastic constant relating the hydrostatic stress $p \equiv \sigma_{kk}/3$ to the volumetric strain $I_1 = \varepsilon_{kk}$. This substitution yields

$$\begin{aligned} \sigma_{ij} &= K\ I_1\ \delta_{ij} + 2G\left(\varepsilon_{ij} - \frac{1}{3}\ I_1\ \delta_{ij}\right) \\ &= K\ \varepsilon_{kk}\ \delta_{ij} + 2G\ e_{ij} \end{aligned} \tag{5.22}$$

where $e_{ij} \equiv \varepsilon_{ij} - \frac{1}{3}\ \delta_{ij}\ \varepsilon_{kk}$ is the deviatoric strain tensor.

The isotropic stress–strain relations can then be conveniently uncoupled as the sum of deviatoric (shear) and volumetric responses, that is,

$$\begin{cases} s_{ij} = 2G\ e_{ij} \\ p = K\ \varepsilon_{kk} \end{cases} \tag{5.23}$$

where $s_{ij} \equiv \sigma_{ij} - p\ \delta_{ij}$ is the deviatoric stress tensor.

In this example, we used the simplifying assumptions that the material is isotropic, linearly elastic, and initially stress-free. In general, Cauchy elastic materials can be anisotropic and nonlinear and may carry initial stress.

It is known that under some multiaxial loading and unloading stress cycles, Cauchy elastic material models may generate spurious energy, resulting in violation of energy conservation laws (Simo and Pister 1984). We will address this aspect subsequently.

5.4 HYPERELASTIC MATERIAL MODELS

The work done by the stresses in deforming a material volume element over a complete stress–strain cycle can be expressed in either a Lagrangian form or an equivalent Eulerian form, as we have seen in Chapter 4.

5.4.1 Strain energy density potential

For an elementary reference volume dV, the Lagrangian form of the work done (in terms of 2PK stress and Green strain) by the stresses over an arbitrary strain path from ε_a to ε_b is

$$\Delta W = \int_{\varepsilon_a}^{\varepsilon_b} \sigma_{ij} d\varepsilon_{ij} \tag{5.24}$$

In order that the integral be path independent, the stress must be derivable from a potential function, that is,

$$\sigma_{ij} = \frac{\partial W}{\partial \varepsilon_{ij}} \tag{5.25}$$

Equation 5.25 is the definition of a *Green elastic* material, which has come to be known as a *hyperelastic* material.

As a result,

$$\Delta W = \int_{\varepsilon_a}^{\varepsilon_b} \frac{\partial W}{\partial \varepsilon_{ij}} d\varepsilon_{ij} = W(\boldsymbol{\varepsilon}_b) - W(\boldsymbol{\varepsilon}_a) \tag{5.26}$$

Therefore, the work done over a complete strain cycle is zero. Consequently, a hyperelastic material model is conservative (neither creating nor dissipating energy) even for very large (essentially unlimited) elastic deformations.

We note that hyperelastic material models provide a unique one-to-one relationship between current total stress and strain, so that all hyperelastic materials are also (conservative) Cauchy elastic models.

5.4.2 Deformation invariants

For an isotropic hyperelastic material, the strain energy density function $W(\varepsilon_{ij})$ must be a function of the three invariants of the Green strain tensor.

$$W = W(I_1, I_2, I_3) \tag{5.27}$$

Therefore, the hyperelastic stress–strain relation for isotropic materials can be expressed in the form

$$\sigma_{ij} = \frac{\partial W}{\partial I_1}\frac{\partial I_1}{\partial \varepsilon_{ij}} + \frac{\partial W}{\partial I_2}\frac{\partial I_2}{\partial \varepsilon_{ij}} + \frac{\partial W}{\partial I_3}\frac{\partial I_3}{\partial \varepsilon_{ij}} \tag{5.28}$$

The derivatives of the invariants with respect to the Green strain tensor are

$$\begin{cases} \dfrac{\partial I_1}{\partial \varepsilon_{ij}} = \delta_{ij} \\[2ex] \dfrac{\partial I_2}{\partial \varepsilon_{ij}} = I_1\delta_{ij} - \varepsilon_{ij} \\[2ex] \dfrac{\partial I_3}{\partial \varepsilon_{ij}} = \varepsilon_{ik}\varepsilon_{kj} - I_1\varepsilon_{ij} + I_2\delta_{ij} \end{cases} \tag{5.29}$$

The last equality in Equation 5.29 is obtained using the identities

$$I_3 \equiv \frac{1}{6}\left[e_{ijk}e_{lmn}\varepsilon_{il}\varepsilon_{jm}\varepsilon_{kn}\right] \tag{5.30}$$

$$e_{ijk}e_{lmn} \equiv \delta_{il}\delta_{jm}\delta_{kn} + \delta_{im}\delta_{jn}\delta_{kl} + \delta_{in}\delta_{jl}\delta_{km} - \delta_{im}\delta_{jl}\delta_{kn} - \delta_{il}\delta_{jn}\delta_{km} - \delta_{in}\delta_{jm}\delta_{kl} \tag{5.31}$$

With these results, Equation 5.28 becomes

$$\sigma_{ij} = \left[\frac{\partial W}{\partial I_1} + I_1\frac{\partial W}{\partial I_2} + I_2\frac{\partial W}{\partial I_3}\right]\delta_{ij} + \left[-\frac{\partial W}{\partial I_2} - I_1\frac{\partial W}{\partial I_3}\right]\varepsilon_{ij} + \frac{\partial W}{\partial I_3}\varepsilon_{ik}\varepsilon_{kj} \tag{5.32}$$

The equation above is the most general form of hyperelastic stress–strain relation for an isotropic material. Even if $\partial W/\partial I_3 = 0$ for a particular model, the stress–strain relation may still be highly nonlinear because of the dependence of the strain energy on the strain invariants.

As a simple, but important, example, we consider a strain energy density function,

$$W = \alpha \ I_1^2 + \beta \ I_2 \tag{5.33}$$

where α and β are (constant) material parameters.

The resulting stress–strain relation is

$$\sigma_{ij} = (2\alpha + \beta)I_1 \ \delta_{ij} - \beta \ \varepsilon_{ij} \tag{5.34}$$

If we set $2\alpha + \beta = \lambda$ and $\beta = -2\mu$, where (λ, μ) are the Lamé elastic constants, we obtain

$$\sigma_{ij} = \lambda \ I_1 \ \delta_{ij} + 2\mu \ \varepsilon_{ij} \tag{5.35}$$

which are the familiar classical linear elastic stress–strain relations.

Since we have assumed material isotropy, the corresponding elastic constitutive tensor that produces the stress–strain relations of Equation 5.35 must be

$$D^e_{ijkl} = \lambda \ \delta_{ij}\delta_{kl} + \mu(\delta_{ik}\delta_{jl} + \delta_{il}\delta_{jk}) \tag{5.36}$$

Therefore, a constant fourth-order isotropic constitutive tensor that relates 2PK stress to Green strain is a consistent (i.e., conservative) hyperelastic material model. Since D^e_{ijkl} is constant, we note here that it can also be used to relate the material *rates* of 2PK stress and Green strain, that is,

$$\dot{\sigma}_{ij} = D^e_{ijkl}\dot{\varepsilon}_{kl} \tag{5.37}$$

Of course, a linear hyperelastic material model will not be appropriate for materials that can undergo large elastic deformations. We briefly mention here some well-known hyperelastic material models for such materials.

5.4.3 Rubber and rubberlike materials

The primary characteristic of natural rubbers and many rubberlike materials is that they are *incompressible* or nearly incompressible, as well as being capable of sustaining large elastic deformations.

Strain energy density functions for such materials are often conveniently expressed as functions of the Lagrangian metric tensor $\mathbf{C}$ ($\equiv \mathbf{F}^T\mathbf{F} = \mathbf{I} + 2\varepsilon$) rather than the Green strain tensor, because the incompressibility constraint is more compactly expressed in that form. For instance, the stress–strain relations for neo-Hookean material and Mooney (or Mooney–Rivlin) material are derived from the following strain energy density functions:

$$W = c(I_{1c} - 3) \tag{5.38}$$

$$W = c_1(I_{1c} - 3) + c_2(I_{2c} - 3) \tag{5.39}$$

These material models are often used to describe very large elastic deformations in materials, such as elastomers (e.g., rubbers), which are nearly incompressible.

The internal (incompressibility) constraint $I_{3c} = 1$ is treated by augmenting the strain energy density function by a Lagrange multiplier term as follows:

$$W^* = W + \frac{p}{2}(I_{3c} - 1) \tag{5.40}$$

The Lagrangian multiplier p is the hydrostatic stress that enforces the constraint. We note as an aside that the volumetric distortion contribution to the strain energy for a linear elastic material is (1/2)pθ, where θ is the dilatation (i.e., linearized volume change). There is no kinematic information from which to determine the hydrostatic stress. It must be determined from boundary or other conditions. For example, if the model is being applied to a thin elastic sheet for which plane stress is a reasonable approximation, p can be determined by setting the normal stress, say σ_{33}, equal to zero.

The invariants of the Green strain tensor can be expressed in terms of the invariants of the Lagrangean metric tensor as follows:

$$\begin{cases} I_1 = \delta_{ij}\ \varepsilon_{ij} = \dfrac{1}{2}(I_{1c} - 3) \\ I_2 = \dfrac{1}{2}\left(\varepsilon_{kk}^2 - \varepsilon_{ij}\ \varepsilon_{ij}\right) = \dfrac{1}{4}(I_{2c} - 2\ I_{1c} + 3) \\ I_3 = \left|\varepsilon\right| = \dfrac{1}{6}\left(e_{ijk}\ e_{lmn}\ \varepsilon_{il}\ \varepsilon_{jm}\ \varepsilon_{kn}\right) = \dfrac{1}{8}(I_{3c} -\ I_{2c} + I_{1c} - 1) \end{cases} \tag{5.41}$$

These relationships can be inverted to give expressions for the invariants of the Lagrangian metric tensor in terms of those of the Green strain tensor as follows:

$$\begin{cases} I_{1c} = 3 + 2I_1 \\ I_{2c} = 3 + 4I_1 + 4I_2 \\ I_{3c} = 1 + 2I_1 + 4I_2 + 8I_3 \end{cases} \tag{5.42}$$

It is evident from the third equation above that the incompressibility constraint is much more conveniently expressed in terms of I_{3c}.

5.4.3.1 Ogden hyperelastic model

Ogden (1972) proposed a three-term hyperelastic strain energy density function expressed in terms of principal stretches (λ_1, λ_2, λ_3) of the form

$$W = \sum_{r=1}^{3} \frac{\mu_r}{\alpha_r}\left[\lambda_1^{\alpha_r} + \lambda_2^{\alpha_r} + \lambda_3^{\alpha_r}\right] \tag{5.43}$$

and calibrated it using experimental results obtained by Treloar (1944) for thin vulcanized rubber sheets (for principal stretches exceeding 7). The incompressibility constraint is written as $\lambda_1\ \lambda_2\ \lambda_3 = 1$, so that

$$W^* = W + p(\lambda_1\lambda_2\lambda_3 - 1) \tag{5.44}$$

where we have dispensed with the ½ factor on the Lagrange multiplier term for strain energies formulated in terms of stretches, rather than Green strain.

For this isotropic model, the 2PK principal stresses (σ_1, σ_2, σ_3) are coaxial with the principal stretch axes, and are given by

$$\sigma_i = \frac{\partial W^*}{\partial \varepsilon_i} = \left[\frac{\partial W}{\partial \lambda_i} + p\frac{\partial(\lambda_1\lambda_2\lambda_3)}{\partial \lambda_i}\right]\frac{d\lambda_i}{d\varepsilon_i}\text{(no sum)} \tag{5.45}$$

Since $\varepsilon_i = (\lambda_i^2 - 1)/2$, $d\lambda_i/d\varepsilon_i = 1/\lambda_i$. Therefore,

$$\sigma_i = \frac{1}{\lambda_i}\left[\frac{\partial W}{\partial \lambda_i} + p\frac{\partial(\lambda_1\lambda_2\lambda_3)}{\partial \lambda_i}\right] \text{(no sum)} \tag{5.46}$$

The Cauchy stress is in general given by

$$\bar{\boldsymbol{\sigma}} = \frac{1}{|\mathbf{F}|}\mathbf{F}\ \boldsymbol{\sigma}\ \mathbf{F}^T = \mathbf{R}\left(\frac{1}{|\mathbf{U}|}\mathbf{U}\ \boldsymbol{\sigma}\ \mathbf{U}^T\right)\mathbf{R}^T = \mathbf{R}\ \bar{\boldsymbol{\sigma}}'\ \mathbf{R}^T \tag{5.47}$$

We note therefore that in this case, the principal axes of the *unrotated* Cauchy stress coincide with the principal stretch axes. On the other hand, the principal axes of Cauchy stress coincide with the principal axes of the left stretch tensor $\overline{\mathbf{V}}$ and the Almansi strain tensor (in the notation of Chapter 2, $\boldsymbol{\varphi}_i' = \mathbf{R}\ \boldsymbol{\varphi}_i$).

We now decompose Equation 5.47 using spectral representations to find the principal values of unrotated Cauchy stress as

$$\bar{\sigma}_i' = \lambda_i \frac{\partial W}{\partial \lambda_i} + p \text{ (no sum)} \tag{5.48}$$

in which we have utilized the incompressibility constraint $|\mathbf{U}| \equiv \lambda_1\lambda_2\lambda_3 = 1$. The principal values of Cauchy stress are of course the same as those of the unrotated Cauchy stress, so that we have finally

$$\bar{\sigma}_i = \lambda_i \frac{\partial W}{\partial \lambda_i} + p \text{ (no sum)} \tag{5.49}$$

The principal Cauchy stresses for the Ogden model are

$$\begin{cases} \bar{\sigma}_1 = \sum_{r=1}^{3} \mu_r \lambda_1^{\alpha_r} + p \\ \bar{\sigma}_2 = \sum_{r=1}^{3} \mu_r \lambda_2^{\alpha_r} + p \\ \bar{\sigma}_3 = \sum_{r=1}^{3} \mu_r \lambda_3^{\alpha_r} + p \end{cases} \tag{5.50}$$

Ogden found that the following parameter values

$$\begin{cases} \alpha_1 = 1.3 \\ \alpha_2 = 5.0 \\ \alpha_3 = -2.0 \end{cases} \tag{5.51}$$

$$\begin{cases} \mu_1 = 6.3 \\ \mu_2 = 0.012\ (\text{kg/cm}^2) \\ \mu_3 = -0.1 \end{cases} \tag{5.52}$$

gave excellent agreement with Treloar's experimental data. We now apply this to equibiaxial plane stress, as did Ogden in his 1972 paper.

For equibiaxial plane stress, $\overline{\sigma}_1 = \overline{\sigma}_2 = \overline{\sigma}$ and $\overline{\sigma}_3 = 0$, $\lambda_1 = \lambda_2 = \lambda$, and due to the incompressibility constraint $\lambda_3 = 1/(\lambda_1\lambda_2) = 1/\lambda^2$. From Equation 5.50,

$$\overline{\sigma}_3 = 0 = \sum_{r=1}^{3} \mu_r \lambda_3^{\alpha_r} + p \tag{5.53}$$

and with this value of the hydrostatic stress, we then find the relation between Cauchy stress and stretch for equibiaxial tension as

$$\overline{\sigma} = \sum_{r=1}^{3} \mu_r \left[\lambda^{\alpha_r} - \lambda^{-2\alpha_r}\right] \tag{5.54}$$

In order to calibrate his model from the test results, Ogden expressed this relationship in the same form in which the test results were presented, as a nominal tension stress f (tension force/original cross-sectional area) versus stretch λ. The nominal tension stress was defined as $f = \overline{\sigma}\lambda h/h_0 = \overline{\sigma}/\lambda$, where h = (sheet) thickness and = initial thickness. Therefore,

$$f = \frac{\overline{\sigma}}{\lambda} = \sum_{r=1}^{3} \mu_r \left[\lambda^{\alpha_r - 1} - \lambda^{-2\alpha_r - 1}\right] \tag{5.55}$$

Figure 5.1 shows the relationship between the nominal tension stress f and the stretch λ. We also show in Figure 5.1 a similar relation for uniaxial tension, again in terms of nominal rather than true (Cauchy) tension stress. For uniaxial tension, $\overline{\sigma}_2 = \overline{\sigma}_3 = 0$ and $\lambda_2 = \lambda_3 = 1/\sqrt{\lambda}$. The nominal stress versus stretch relation for unixial tension is

$$f = \frac{\overline{\sigma}}{\lambda} = \sum_{r=1}^{3} \mu_r \left[\lambda^{\alpha_r - 1} - \lambda^{-(1+\alpha_r/2)}\right] \tag{5.56}$$

The relations shown in Figure 5.1 are of course analytical, but they very accurately represent Treloar's test data. In Figure 5.2, we show the underlying relationship between

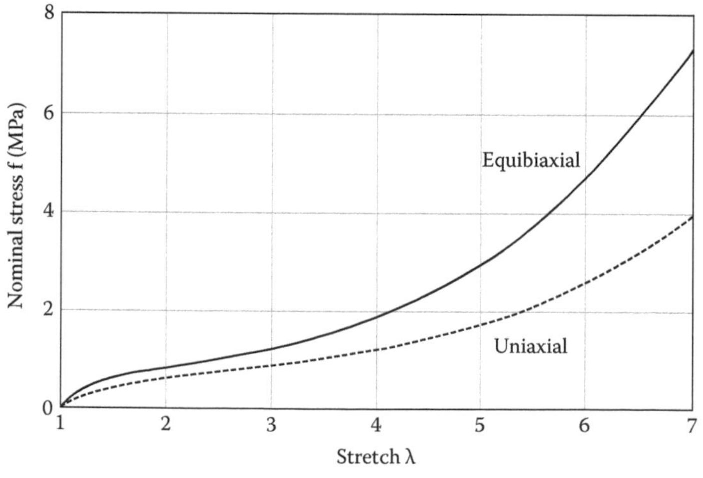

Figure 5.1 Nominal stress (MPa) vs. stretch.

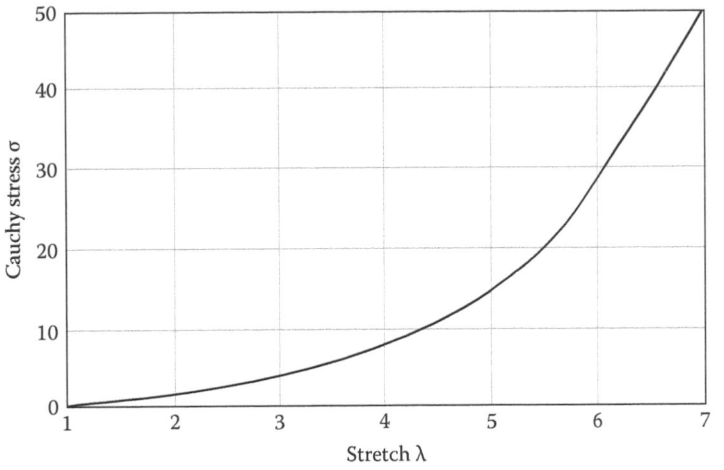

Figure 5.2 True (Cauchy) stress (MPa) vs. stretch—equibiaxial tension.

true (Cauchy) stress and the stretch, that is, Equation 5.54, from which the "test" data in Figure 5.1 is obtained.

It is evident that the "structural response" of the sheet shown in Figure 5.1 is more complex than the underlying material behavior that generates it, shown in Figure 5.2. This is a very simple example of utilizing a test of a (slightly in this case) more complex structural response to determine underlying material behavior, which we alluded to earlier in this chapter.

Treloar also reported data for a third type of test for the same material in which the in-plane lateral stretch was restrained, that is, $\lambda_2 = 0$, effectively producing a state of unequal biaxial tension in which shear stress is superimposed on equibiaxial tension. Ogden also employed these latter tests in his calibration.

We remark that this early hyperelastic model was calibrated for a specific rubber material that was tested in plane stress (i.e., thin membranes), and for biaxial *tension* combinations.

There are, of course, many applications in which compressive stresses are important. Treloar observed that the equibiaxial response in tension (e.g., in the [1, 2] plane) is equivalent to compression response in the three-direction, as demonstrated in Equation 5.57.

$$\begin{bmatrix} \bar{\sigma} & 0 & 0 \\ 0 & \bar{\sigma} & 0 \\ 0 & 0 & 0 \end{bmatrix} + \begin{bmatrix} -\bar{\sigma} & 0 & 0 \\ 0 & -\bar{\sigma} & 0 \\ 0 & 0 & -\bar{\sigma} \end{bmatrix} = \begin{bmatrix} 0 & 0 & 0 \\ 0 & 0 & 0 \\ 0 & 0 & -\bar{\sigma} \end{bmatrix} \tag{5.57}$$

$$\text{Equibiaxial tension} + \text{hydrostatic compression} = \text{uniaxial compression}$$

Since the material is incompressible, the hydrostatic compression produces no deformation. Therefore, the uniaxial response in compression ($0 < \lambda \leq 1$) is identical to the equibiaxial tension response ($1 < \lambda$). In this case, we set $\lambda_3 = \lambda$, $\lambda_1 = \lambda_2 = 1/\sqrt{\lambda}$ and determine p from $\bar{\sigma}_1 = 0$. This leads to the uniaxial true stress versus stretch relation shown in Figure 5.3, which is valid for $0 < \lambda$.

$$\bar{\sigma} = \sum_{r=1}^{3} \mu_r \left[\lambda^{\alpha_r} - \lambda^{-\alpha_r/2}\right] \tag{5.58}$$

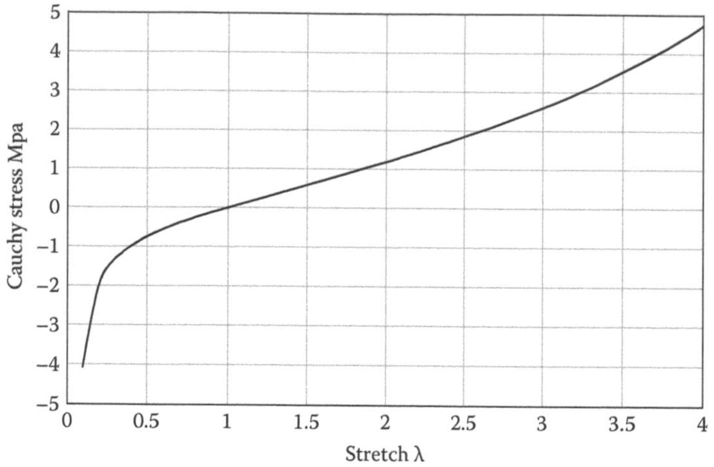

Figure 5.3 True uniaxial stress vs. stretch for $0 < \lambda$.

5.4.3.2 Balloon problem

We now consider an interesting and instructive problem that has attracted the attention of many researchers. We call it simply the "balloon problem." In its simplest form, which we consider here, it involves the inflation of a spherical balloon made of a thin rubber membrane. Analysis of this problem is relevant to the performance of high-altitude weather balloons (Alexander 1971), for example. The most interesting aspect of the balloon response as it is inflated is the appearance of a tension instability that is reminiscent of the snap-through of shallow arches. As is the case with arches, sometimes bifurcations into unsymmetric (in this case nonspherical) shapes can occur. This phenomenon has been observed experimentally and reported by Alexander.

We shall not, however, pursue that aspect of the problem here and shall assume that the balloon maintains its perfectly spherical shape during the different stages of inflation out to quite large strains (i.e., stretches). We also assume that an unlimited air supply is available so that pressure can be assumed to be independent of balloon volume.

A closely allied problem is that of a thin, initially flat membrane inflated to a spherical "cap" shape. This has in fact been the most common experimental method for measuring the equibiaxial tension response of rubber membranes (e.g., Treloar 1944; Rivlin and Saunders 1951; Oden 1972).

We follow Ogden and apply his proposed material model to this idealized problem in order to illustrate the phenomenon. A perfectly spherical balloon under internal (gauge) pressure is shown in Figure 5.4.

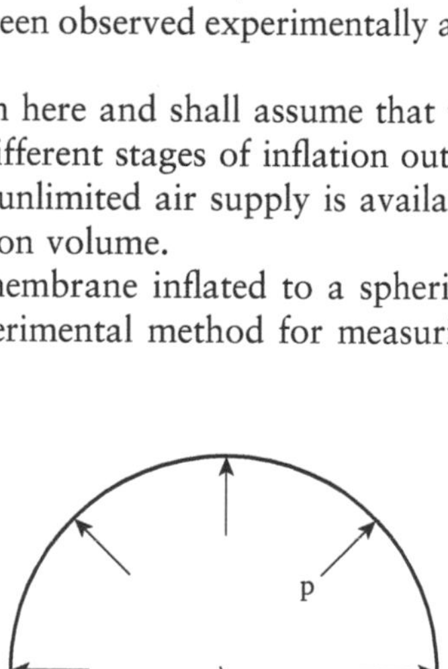

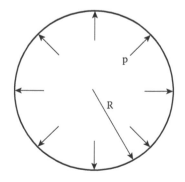

Figure 5.4 Inflation of a spherical balloon.

The balloon has radius R_0 and thickness h_0 in the unstretched configuration. As the balloon is inflated, the membrane surface stretches isotropically so that all surface dimensions increase by the stretch ratio λ Therefore, an element of surface area dA_0 in the unstretched configuration expands to an area $\lambda^2 dA_0$. Because the rubber material is incompressible, the deformed thickness reduces to $h \approx h_0 / \lambda^2$. Likewise, the radius in the deformed configuration expands to $R = \lambda R_0$.

To equilibrate the internal pressure p (above atmospheric) in the deformed configuration requires an isotropic membrane tension T per unit length of deformed midsurface:

$$T = pR/2 = \lambda p R_0/2 \tag{5.59}$$

In terms of the isotropic in-plane Cauchy membrane stress $\bar{\sigma}$, the membrane tension is

$$T = \bar{\sigma} h = \bar{\sigma} h_0/\lambda^2 \tag{5.60}$$

Therefore,

$$\frac{pR}{h_0} = \frac{2\bar{\sigma}}{\lambda^2} = \frac{2}{\lambda^2}\sum_{r=1}^{3} \mu_r \left[\lambda^{\alpha_r} - \lambda^{-2\alpha_r}\right] \tag{5.61}$$

The response can be nondimensionalized by introducing a relation between the shear modulus G and the Ogden parameters. From Equation 5.54, the initial slope of the biaxial tension stress versus stretch curve is $3\sum_{r=1}^{3} \alpha_r \mu_r$. For a linear elastic incompressible material, the corresponding slope is 6G. Therefore, we may set

$$2G = \sum_{r=1}^{3} \alpha_r \ \mu_r \tag{5.62}$$

from which we find for Ogden's material parameters the value G = 4.225 kg/cm^2 = 0.4143 MPa. The same relation results from equating initial slopes for uniaxial tension.

Figure 5.5 shows normalized inflation pressure pR_0/Gh_0 versus stretch for the balloon. The figure depicts a phenomenon familiar to anyone who has "blown up" a balloon. At first, more effort (i.e., more pressure) is required to inflate the balloon, but then at a certain point the balloon continues to expand much more easily, with reduced effort.

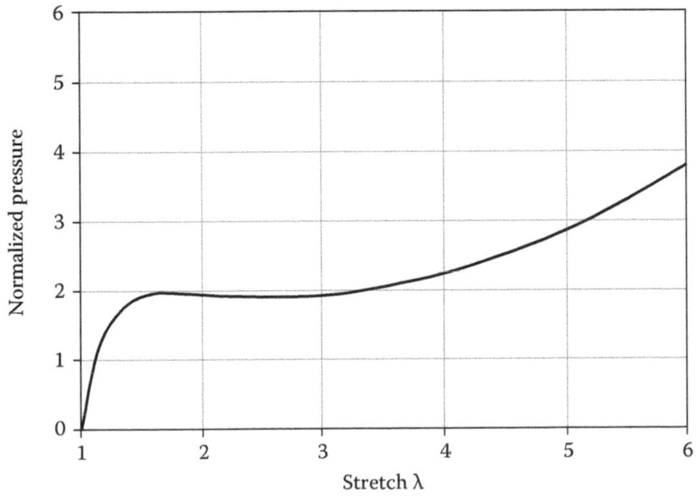

Figure 5.5 Inflation of a spherical balloon (Ogden material model).

The limit point, at which $dp/d\lambda = 0$ occurs, is at $\lambda \approx 1.8$ for this particular constitutive model, and does not vary much for many other well-known hyperelastic models that have been applied to this simple problem.

We remark that the falling pressure versus stretch relation beyond the limit point is unstable until the subsequent stiffening begins (at $\lambda \approx 2.6$ in this case). The pressure would have to be carefully reduced to follow the unstable portion of the response curve. This again reminds one of the snap-through of a shallow arch. The shapes of the resulting pressure versus stretch response curves according to different models differ somewhat, but virtually all exhibit the limit point behavior shown in Figure 5.5.

A notable exception is a simple model proposed by Gent (1996, 1999), which is essentially a modification of the neo-Hookean model (Equation 5.38) to include a factor that asymptotically limits the maximum stretch (which apparently may be on the order of $\lambda_m \approx 10$). The physical basis for this idealization is that the network of long-chain rubber molecules (whose extension via straightening and subsequent recovery on unloading is responsible for the material's elasticity) eventually reaches a maximum extension and essentially then "locks." For reasonable values of limiting stretch [expressed in terms of a limiting value of $(I_{1c} - 3)$], Gent's model also predicts the limit-point behavior displayed in Figure 5.5. Beyond the limit point, his material model has a much more rapidly stiffening response to increasing stretch and can predict a "snap-through" transition to a second stable configuration, which is consistent with observation.

To explore how (or if) the limit point phenomenon depends on the particular characteristics of the rubber material from which the balloon is made, we select a very simple (hypothetical) linear relation between Cauchy stress and natural strain for equibiaxial tension, that is,

$$\bar{\sigma} = 6G \ln \lambda \tag{5.63}$$

The initial slope $d\bar{\sigma}/d\lambda = 6G$ (when $\lambda = 1$) is chosen to match that of a linear elastic incompressible isotropic material. The shape of this material response curve is shown in Figure 5.6. It does not resemble the response of a rubberlike material, which typically exhibits stiffening with increasing stretch in equibiaxial tension.

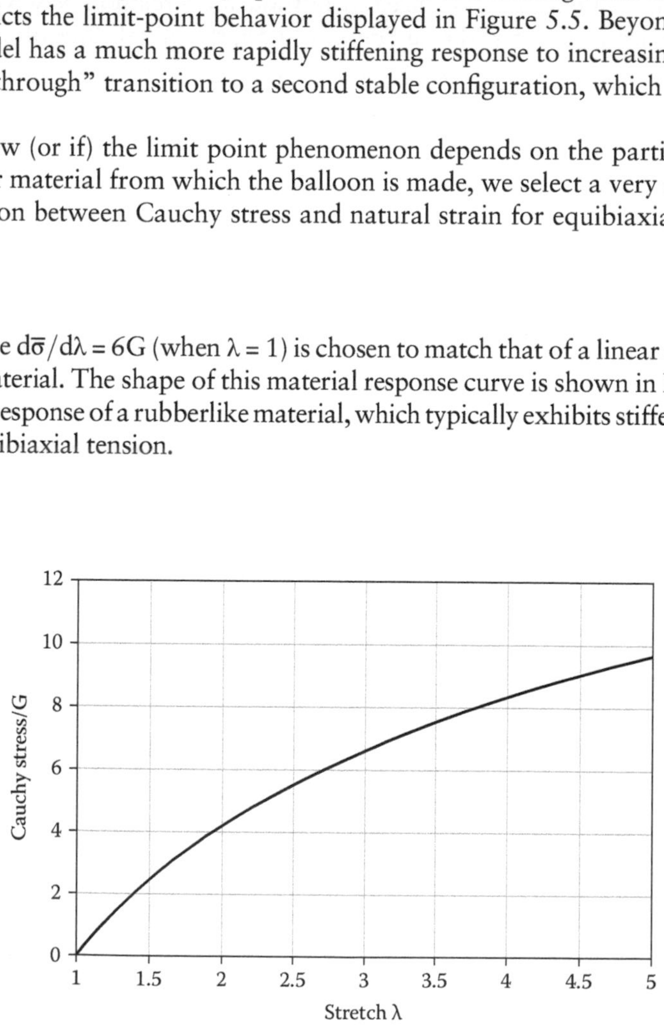

Figure 5.6 Hypothetical material response for balloon material.

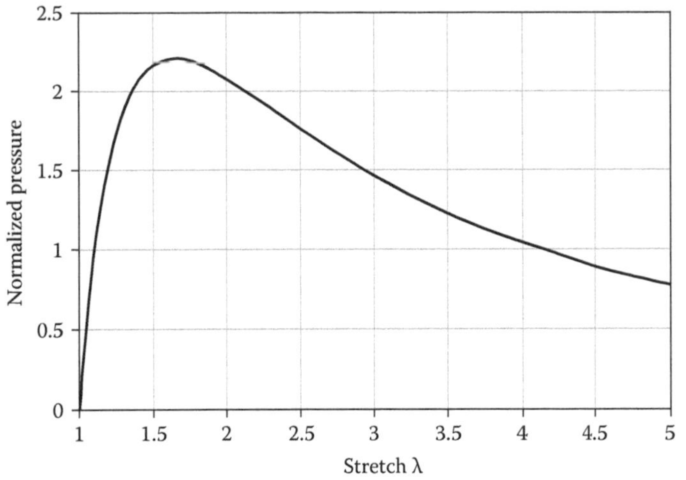

Figure 5.7 Inflation of spherical balloon—hypothetical material model.

The pressure versus stretch response relation for the balloon made of this hypothetical material with the constitutive relation Equation 5.63 is

$$\frac{pR_0}{Gh_0} = 12\frac{\ln \lambda}{\lambda^2} \tag{5.64}$$

which is illustrated in Figure 5.7. We note that this curve also has a limit point $dp/d\lambda = 0$ at $\lambda \approx 1.8$, followed by unstable response $dp/d\lambda < 0$ in the sense that the stretch continues to increase under decreasing pressure. According to this material model, there is no region of stable response $dp/d\lambda > 0$ past the limit point.

All reasonable material models predict the occurrence of a limit point at about the same value of $\lambda \approx 1.8$, although the peak pressure varies slightly from model to model. Therefore, we conclude that this phenomenon is essentially a geometric effect arising from incompressibility of the material and is not dependent on the details of the constitutive response. The deformed cross-sectional area decreases at a faster rate than the true membrane stress can increase, at least in the earlier stages of inflation. The post–peak slope $dp/d\lambda$ can recover and become positive (i.e., stable) for materials and material models that have a rapidly enough stiffening true stress versus stretch response. Although it could be argued that incompressibility is truly a material property, our point is that the unstable balloon response is a characteristic of a whole class of incompressible material models, not of one or a few specific models.

Although the incompressibility constraint was readily dealt with in this simple example, we briefly note here that this is not a trivial matter in general nonlinear structural analysis. It has long been known that even for linear analysis with nearly incompressible ($\nu \to 1/2$) elastic isotropic materials, significant computational difficulties (e.g., ill-conditioned equations) arise. There exists a significant body of literature on computational approaches to nonlinear analysis involving incompressible materials, involving, for example, "mixed" finite element methods that employ nodal pressure variables in addition to the usual nodal displacement unknowns. Other approaches involve separating the constitutive models into deviatoric and dilatational components.

5.4.4 Soft biological tissue

Hyperelastic material models have also found use in important biomechanical applications. We briefly discuss here one such model (Fung 1993) that was developed to model the response of arteries subjected in particular to internal pressure and longitudinal stress. Soft biological tissue is typically anisotropic and often exhibits a much more rapidly stiffening elastic response with increasing deformation (i.e., stretch) than do rubberlike materials.

5.4.4.1 Fung exponential hyperelastic model

The exponential strain energy density function originally proposed by Fung can be written in the compact form

$$W = \frac{\hat{C}}{2}(e^{Q} - 1) \tag{5.65}$$

where $\hat{C}$ is a material constant with dimensions of stress, and the scalar quantity Q is a normalized (dimensionless) strain energy variable, which incorporates additional material constants.

We may express Q in the following way:

$$Q = B_{ijkl}\varepsilon_{ij}\varepsilon_{kl} \tag{5.66}$$

where $\boldsymbol{\varepsilon}$ is the Green strain tensor and B_{ijkl} is a positive definite tensor of nondimensional elastic constants, perhaps representing transverse isotropy or other elastic symmetry appropriate for a particular application. Orthotropic or transversely isotropic elastic symmetry is often assumed for biological materials based on their underlying microstructure.

For example, human arterial vessel walls consist of three distinct layers, a thin inner layer and two outer layers, each of which remind one of a filament-wound composite pipe. In each of these layers, there are two families of collagen fibers (at angles $\pm\theta$ with respect to the longitudinal axis of the vessel) helically wound around the vessel wall. Orthotropic elastic symmetry, with homogeneous properties through the vessel wall, is probably the most reasonable simple idealization in this case. In three dimensions, nine independent elastic constants are in general required for orthotropy.

Similarly, the human cornea has a multilayered microstructure consisting of five distinct layers. The middle layer (the stroma) makes up about 90% of the thickness of the cornea (about 0.5 mm thick in an adult). It consists of perhaps up to 500 lamellae parallel to the surface of the cornea, which are composed primarily of collagen fibrils embedded in a matrix. Within a lamina, the collagen fibrils are essentially parallel, but adjacent laminae generally do not have the same fiber orientation. Therefore, transversely isotropic elastic symmetry is probably a fairly realistic idealization in this case (Kwon 2006). There are five independent elastic constants required for transverse isotropy.

We note that a power series expansion of Equation 5.65 about $Q = 0$ (i.e., $\boldsymbol{\varepsilon} = 0$) yields for the first term

$$W(0) = \frac{1}{2}\hat{C}\ B_{ijkl}\varepsilon_{ij}\varepsilon_{kl} \tag{5.67}$$

Therefore, the tensor $\hat{C}B_{ijkl}$ is essentially the initial stiffness (matrix) of the material, so that Q may also be expressed as

$$Q = \frac{1}{\hat{C}} D^0_{ijkl} \varepsilon_{ij} \varepsilon_{kl} \tag{5.68}$$

where D^0_{ijkl} is the initial elastic stiffness at $\boldsymbol{\varepsilon} = 0$.

The normalization of the elastic constants leading to the definition of the scalar Q is of course not unique. If, for example, a new scaling factor $\hat{C}_\alpha = \hat{C}/\alpha$ were to be used (whereby the tensor B_{ijkl} would also have to be multiplied by α), then Equation 5.65 would be transformed to

$$W = \frac{\hat{C}}{2\alpha}(e^{\alpha Q} - 1) \tag{5.69}$$

This leaves the initial elastic stiffness unchanged, but for $\alpha > 1$ leads to a more rapidly stiffening material response with increasing deformation, due to the magnification of the exponential term. This modification of the original Fung model was used by Kwon (2006) to allow additional modeling flexibility in his studies of the human cornea.

With the incompressibility constraint, the augmented strain energy density function is

$$W^* = \frac{\hat{C}}{2\alpha}(e^{\alpha Q} - 1) + \frac{p}{2}(I_{3c} - 1) \tag{5.70}$$

Therefore, the 2PK stress is

$$\sigma_{ij} = \frac{\partial W^*}{\partial \varepsilon_{ij}} \tag{5.71}$$

The constitutive part of Equation 5.71 gives

$$\sigma^{(1)}_{ij} = e^{\alpha Q} D^0_{ijkl} \varepsilon_{kl} \tag{5.72}$$

and the constraint gives

$$\sigma^{(2)}_{ij} = \frac{p}{2}\frac{\partial I_{c3}}{\partial \varepsilon_{ij}} = \frac{p}{2}\frac{\partial I_{c3}}{\partial C_{kl}}\frac{\partial C_{kl}}{\partial \varepsilon_{ij}} \tag{5.73}$$

where $\mathbf{C} = \mathbf{F}^T\mathbf{F}$ is the metric tensor as usual.

We use the identity

$$\frac{\partial(\det\mathbf{A})}{\partial A_{ij}} = (\det\mathbf{A})A^{-1}_{ji} \tag{5.74}$$

for the derivatives of a determinant of a matrix $\mathbf{A}$ with respect to its components in Equation 5.73 to find

$$\sigma^{(2)}_{ij} = p\, C^{-1}_{ij} \tag{5.75}$$

The 2PK stress is then

$$\sigma_{ij} = e^{\alpha Q}\ D^0_{ijkl}\varepsilon_{kl} + p\ C^{-1}_{ij} \tag{5.76}$$

Example 5.1 Fung's exponential model

We now exercise the Fung exponential model using some orthotropic material elastic properties for rabbit arteries, reported by Chuong and Fung (1983), obtained from longitudnal stretching and internal pressure tests. We will not take account here of the "cylindrical" geometry of the artery but simply explore the behavior of the model in plane stress, the "plane" of concern being defined by the circumferential and longitudinal directions, which, along with the radial direction, are the axes of material symmetry.

Chuong and Fung (1983) report their experimental results in the form of six coefficients b_i (i = 1, 6) and a coefficient c for pressure and stretching tests on four different animals. For a particular experiment (number 71), they give the following numerical values:

$$c\begin{bmatrix} b_1 & b_4 & b_6 \\ b_4 & b_2 & b_5 \\ b_6 & b_5 & b_3 \end{bmatrix} = 26.95\begin{bmatrix} 0.9925 & 0.0749 & 0.0193 \\ 0.0749 & 0.4180 & 0.0295 \\ 0.0193 & 0.0295 & 0.0089 \end{bmatrix} \text{(kPa)} \tag{5.77}$$

These coefficients can be identified with components of the initial elastic stiffness tensor as follows:

$$c\begin{bmatrix} b_1 & b_4 & b_6 \\ b_4 & b_2 & b_5 \\ b_6 & b_5 & b_3 \end{bmatrix} \Rightarrow \begin{bmatrix} D^0_{1111} & D^0_{1122} & D^0_{1133} \\ D^0_{2211} & D^0_{2222} & D^0_{2233} \\ D^0_{3311} & D^0_{3322} & D^0_{3333} \end{bmatrix} \tag{5.78}$$

where 1 = circumferential, 2 = longitudinal, and 3 = radial. Only these six components of the tensor D^0_{ijkl} are needed in this case because we consider direct stresses applied in the material axis directions.

The Cauchy principal stresses in the three material axis directions are

$$\bar{\sigma}_i = \lambda_i^2\ \sigma_i\ \text{(no sum)} \tag{5.79}$$

where we have again used the incompressibility condition $\lambda_1\lambda_2\lambda_3 = 1$.

Equation 5.76 can now be written in a compact matrix–vector format as

$$\begin{Bmatrix} \bar{\sigma}_1 - p \\ \bar{\sigma}_2 - p \\ -p \end{Bmatrix} = e^{\alpha Q}[\mathbf{C}][\mathbf{D}]\begin{Bmatrix} \varepsilon_1 \\ \varepsilon_2 \\ \varepsilon_3 \end{Bmatrix} \tag{5.80}$$

where [**D**] is the (3 × 3) matrix shown in Equation 5.78. In arriving at Equation 5.80, we have used the plane stress condition $\bar{\sigma}_3 = 0$. C is given by

$$C = \begin{bmatrix} \lambda_1^2 & 0 & 0 \\ 0 & \lambda_2^2 & 0 \\ 0 & 0 & \lambda_3^2 \end{bmatrix} \tag{5.81}$$

In this example, the Cauchy stresses are deformation driven by specified stretches $(\lambda_1, \lambda_2, \lambda_3) = (\lambda_1, \lambda_2, 1/\lambda_1\lambda_2)$, which in turn determine the Green strains $(\varepsilon_1, \varepsilon_2, \varepsilon_3)$. The flat sheet is first stretched longitudinally to $\lambda_z = 1.5$ with the perpendicular in-plane ("circumferential") dimension held constant at $\lambda_\theta = 1.0$ (Figure 5.8). This is (approximately) representative of the *in vivo* initial tension in the artery.

With the longitudinal prestretch then held constant at $\lambda_z = 1.5$, the sheet is stretched to $\lambda_z = 1.5$ in the orthogonal (cirumferential) direction (Figure 5.9) to mimic the effect of internal pressure in the artery.

Figure 5.10 shows the slight decrease in longitudinal Cauchy stress as the circumferential stretch is applied (which is not obvious in Figure 5.9).

Figure 5.11 illustrates the stiffening effect of parameter α (Equation 5.69) on the circumferential stretch response predicted by the Fung exponential model.

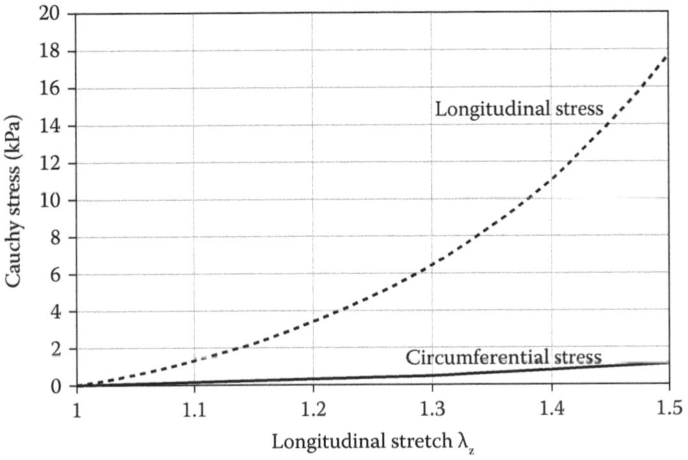

Figure 5.8 Longitudinal prestretch of flat membrane (Fung rabbit artery model).

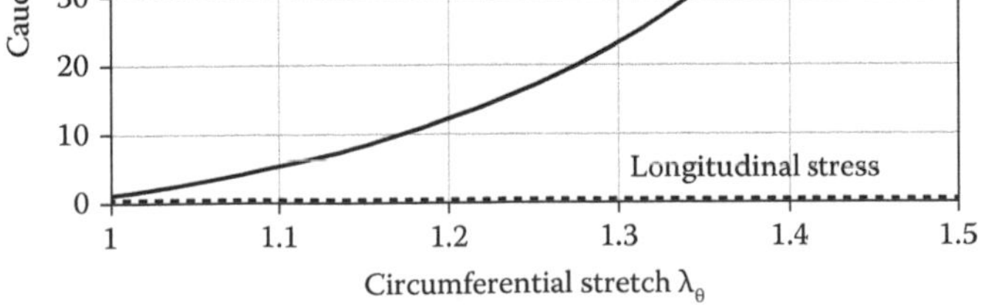

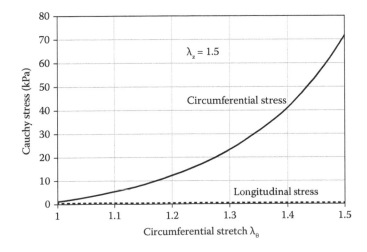

Figure 5.9 Inflation of rabbit artery following longitudinal stretch $\lambda_z = 1.5$.

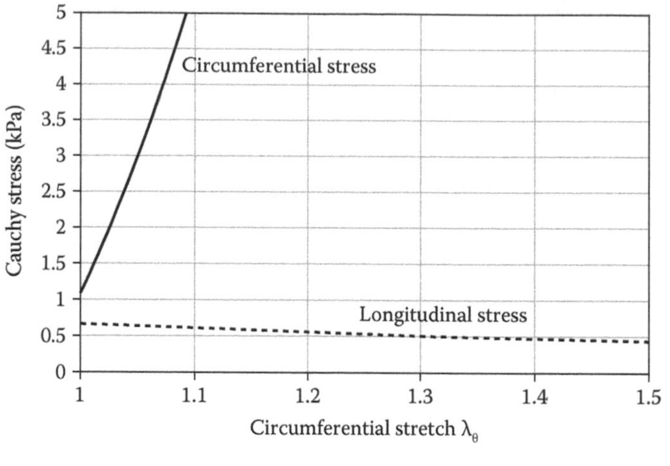

Figure 5.10 Inflation of rabbit artery (magnified vertical scale).

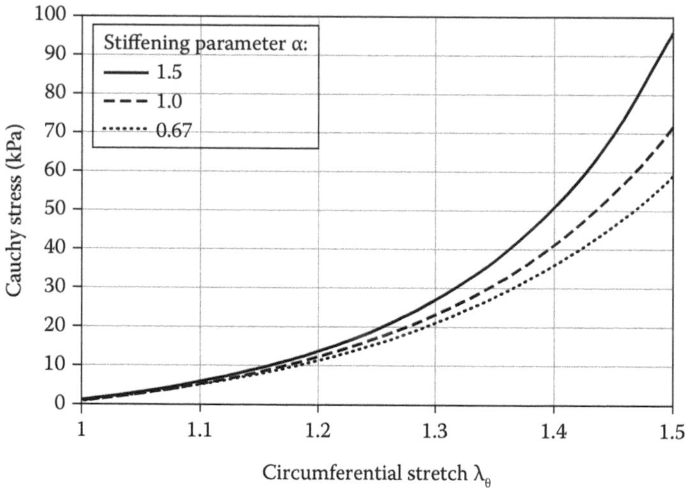

Figure 5.11 Effect of stiffening parameter α on rabbit artery inflation response.

5.5 HYPOELASTIC MATERIAL MODELS

Hypoelastic material models (originally proposed by Truesdell [1955]) are expressed directly in rate or incremental form and might be loosely characterized as the "least elastic" of the three categories of elastic material models discussed in this chapter. They are in general dissipative in nature, and are intended to be so, which may make them more realistic models for many materials. A potential drawback is that hypoelastic material models may in some cases generate (spurious) energy in closed stress or strain cycles.

For hypoelastic material models, the current stress increment (or rate) is assumed to depend on both the current strain increment (or rate) and the current stress. Therefore, in a hypoelastic material model an objective rate of Cauchy stress, say $\bar{\sigma}_{ij}^{\triangle}$, is expressed as a function of the Cauchy stress and a work-conjugate (objective) strain rate $\bar{\varepsilon}_{kl}^{\triangle}$.

$$\bar{\sigma}_{ij}^{\triangle} = f_{ij}(\bar{\sigma}_{ij})\bar{\varepsilon}_{kl}^{\triangle} \tag{5.82}$$

We remark here that there is therefore no defined "natural" state for a hypoelastic material.

We shall subsequently consider the following specific choice of stress and strain rate for time-independent behavior:

$$\dot{\bar{\sigma}}'_{ij} = \bar{D}_{ijkl}(\bar{\sigma}'_{mn})d_{kl} \tag{5.83}$$

in which the material rate of unrotated Cauchy stress $\dot{\bar{\sigma}}'$ and the unrotated deformation rate $\mathbf{d}$ are the work-conjugate stress and strain rates, and the material response tensor $\bar{\mathbf{D}}$ is a function of the unrotated Cauchy stress.

An obvious alternative choice, which we will first consider, is to relate an objective rate of Cauchy stress (e.g., Jaumann, Green–Naghdi, or Truesdell rate) to the deformation rate $\mathbf{D}$.

Integration of Equation 5.83 over different closed strain paths will in general result in different (finite) stress increments; that is, a hypoelastic material model is typically path dependent. This is of course true of most realistic inelastic material models.

5.5.1 Hypoelastic grade zero

A hypoelastic material of "grade zero" is one in which $\bar{D}_{ijkl}$ is constant. For an isotropic material, $\bar{D}_{ijkl}$ can therefore be at most a linear combination of the two isotropic fourth-order tensors

$$\begin{bmatrix} \delta_{ij}\delta_{kl} \\ \delta_{ik}\delta_{jl} + \delta_{il}\delta_{jk} \end{bmatrix} \tag{5.84}$$

That is,

$$\bar{D}_{ijkl}^{0} = \bar{\lambda}\ \delta_{ij}\delta_{kl} + \bar{\mu}\left[\delta_{ik}\delta_{jl} + \delta_{il}\delta_{jk}\right] \tag{5.85}$$

where $\bar{\lambda}$ and $\bar{\mu}$ are material constants.

5.5.2 Hypoelastic grade one

A hypoelastic material of "grade one" is one in which $\bar{D}_{ijkl}$ is a linear function of the stress. Therefore, for an isotropic material $\bar{D}_{ijkl}$ may be at most a linear combination of the five isotropic tensors in Equation 5.86, the latter three of which are themselves linear in the stress.

$$\begin{bmatrix} \delta_{ij}\delta_{kl} \\ \delta_{ik}\delta_{jl} + \delta_{il}\delta_{jk} \\ \delta_{ij}\bar{\sigma}_{kl} \\ \delta_{kl}\bar{\sigma}_{ij} \\ \delta_{ik}\bar{\sigma}_{jl} + \delta_{jl}\bar{\sigma}_{ik} + \delta_{il}\bar{\sigma}_{jk} + \delta_{jk}\bar{\sigma}_{il} \end{bmatrix} \tag{5.86}$$

The coefficients multiplying the first two of these tensors may each be linear in the first stress invariant (i.e., the mean stress).

Therefore, the most general form of $\overline{D}_{ijkl}$ for an isotropic grade one hypoelastic material is

$$\begin{aligned}\overline{D}^{1}_{ijkl} &= (c_1 + c_2\overline{\sigma}_{mm})\delta_{ij}\delta_{kl} + (c_3 + c_4\overline{\sigma}_{mm})\left[\delta_{ik}\delta_{jl} + \delta_{il}\delta_{jk}\right] \\ &+ c_5\ \delta_{ij}\overline{\sigma}_{kl} + c_6\delta_{kl}\overline{\sigma}_{ij} + c_7\left[\delta_{ik}\overline{\sigma}_{jl} + \delta_{jl}\overline{\sigma}_{ik} + \delta_{il}\overline{\sigma}_{jk} + \delta_{jk}\overline{\sigma}_{il}\right]\end{aligned} \tag{5.87}$$

involving seven material constants c_i ($i = 1\ldots7$).

The characteristics of the grade zero hypoelastic material model are important because of its widespread use in incremental-iterative solution algorithms for plasticity problems, as we will see in Chapter 7.

To begin our further exploration of the grade zero hypoelastic material model (given in Equation 5.85), we first consider the Lagrangian material tangent stiffness that corresponds to a grade zero hypoelastic model.

5.5.3 Lagrangian versus hypoelastic material tangent stiffness

In a total Lagrangian formulation, a material model is expressed as a relation between the material rate of symmetric Piola–Kirchhoff stress and the Green strain rate.

$$\dot{\boldsymbol{\sigma}} = \mathbf{D} : \dot{\boldsymbol{\varepsilon}} \tag{5.88}$$

If, for a large deformation elastic problem, we were to choose a grade zero hypoelastic material model, how would the elastic material tangent stiffness tensor $\mathbf{D}^e$ for a total Lagrangian formulation be related to $\overline{D}^{0}_{ijkl}$?

We know that the grade zero hypoelastic model can be formulated in terms of several different objective rates of Cauchy stress, and it would seem to be obvious that, for a given material, the material parameters $\overline{\lambda}$ and $\overline{\mu}$ might have to be adjusted, depending on which Eulerian stress and strain rates are used.

5.5.3.1 *Truesdell stress rate*

We first assume a linear elastic relation between the Truesdell stress rate tensor and deformation tensor,

$$\overline{\sigma}^{Tr}_{mn} = \left[\overline{\lambda}_1\ \delta_{mn}\ \delta_{kl} + \overline{\mu}_1(\delta_{mk}\ \delta_{nl} + \delta_{ml}\ \delta_{nk})\right]D_{kl} \tag{5.89}$$

with elastic constants $\overline{\lambda}_1$ and $\overline{\mu}_1$.

The following relationships between the Lagrangian and Eulerian rates were obtained earlier:

$$\begin{cases}\dot{\boldsymbol{\sigma}} = |\mathbf{F}|\overline{\mathbf{F}}\ \overline{\boldsymbol{\sigma}}^{Tr}\overline{\mathbf{F}}^{T} \\ \mathbf{D} = \overline{\mathbf{F}}^{T}\dot{\boldsymbol{\varepsilon}}\ \overline{\mathbf{F}}\end{cases} \tag{5.90}$$

which can be written in Cartesian tensor notation as

$$\begin{cases}\dot{\sigma}_{ij} = |\mathbf{F}|\overline{F}_{im}\ \overline{F}_{jn}\ \overline{\sigma}^{Tr}_{mn} \\ D_{kl} = \overline{F}_{rk}\ \overline{F}_{sl}\ \dot{\varepsilon}_{rs}\end{cases} \tag{5.91}$$

Substitution of Equations 5.91 into Equation 5.89 results in the following:

$$D^{e}_{ijkl} = |\mathbf{F}|\left[\overline{\lambda}_1\overline{C}_{ij}\ \overline{C}_{kl} + \overline{\mu}_1(\overline{C}_{ik}\overline{C}_{jl} + \overline{C}_{il}\overline{C}_{jk})\right] \tag{5.92}$$

where $\overline{C} = \overline{F}\overline{F}^T \equiv C^{-1}$, that is, the inverse of the metric tensor. We note the similar structure of Equations 5.89 and 5.92, and see that an isotropic linear hypoelastic (i.e., grade zero) relation between the Truesdell stress rate and deformation tensor leads to an isotropic nonconstant Lagrangian tangent stiffness tensor that depends on deformations (but not on stress).

Starting from other postulated grade zero hypoelastic relations (Jaumann or Green–Naghdi rates of Cauchy stress vs. deformation rate) leads to Lagrangian tangent stiffness tensors that depend on stress, as well as deformations.

5.5.3.2 *Jaumann stress rate*

If we assume instead a linear isotropic elastic relation between the Jaumann rate of Cauchy stress and the deformation rate tensor, say

$$\overline{\sigma}_{ij}^{J} = \left[\overline{\lambda}_2 \ \delta_{ij} \ \delta_{kl} + \overline{\mu}_2(\delta_{ik} \ \delta_{jl} + \delta_{il} \ \delta_{jk})\right] D_{kl} \tag{5.93}$$

with new material constants $\overline{\lambda}_2$ and $\overline{\mu}_2$, and recall the relationship between the rate of symmetric Piola–Kirchhoff stress and Jaumann stress, that is,

$$\dot{\sigma} = |F|\overline{F}\left[\ \dot{\overline{\sigma}}^{J} - D\overline{\sigma} - \overline{\sigma}D + \mathrm{tr}(D)\overline{\sigma}\right]\overline{F}^{T} \tag{5.94}$$

we obtain the resulting Lagrangian material tangent stiffness as

$$D_{ijkl}^{e} = |F|\left[\overline{\lambda}_2\overline{C}_{ij}\overline{C}_{kl} + \overline{\mu}_2(\overline{C}_{ik}\overline{C}_{jl} + \overline{C}_{il}\overline{C}_{jk})\right] + \sigma_{ij}\overline{C}_{kl} + \sigma_{ik}\overline{C}_{jl} + \sigma_{jl}\overline{C}_{ik} \tag{5.95}$$

We note that the 2PK stress appears in Equation 5.95. Of course, this equation may also be written in terms of Cauchy stress rather than 2PK stress.

Therefore, a linear hypoelastic relation between the Jaumann rate of Cauchy stress and the deformation rate implies a Lagrangian tangential stiffness that depends on the deformations (through C^{-1}) and the 2PK stress, as well as elastic constants λ, μ.

We next compare three different hypoelastic grade zero models (Truesdale rate, Jaumann rate, and Green–Naghdi rate vs. deformation rate) in the simple shear problem to illustrate their quantitative differences for large elastic (shear) strains.

5.5.4 Comparison of linear isotropic hypoelastic models in simple shear

The geometry of the simple shear problem is recalled in Figure 5.12. The kinematics that were determined previously are summarized below.

The mapping is

$$\begin{cases} \overline{x}_1 = x_1 + \kappa(t)x_2 \\ \overline{x}_2 = x_2 \\ \overline{x}_3 = x_3 \end{cases} \tag{5.96}$$

which is governed by the parameter $\kappa(t)$. The Lagrangian and Eulerian deformation gradients are

$$F = \begin{bmatrix} 1 & \kappa & 0 \\ 0 & 1 & 0 \\ 0 & 0 & 1 \end{bmatrix}; \overline{F} = \begin{bmatrix} 1 & -\kappa & 0 \\ 0 & 1 & 0 \\ 0 & 0 & 1 \end{bmatrix}; |F| = 1 \tag{5.97}$$

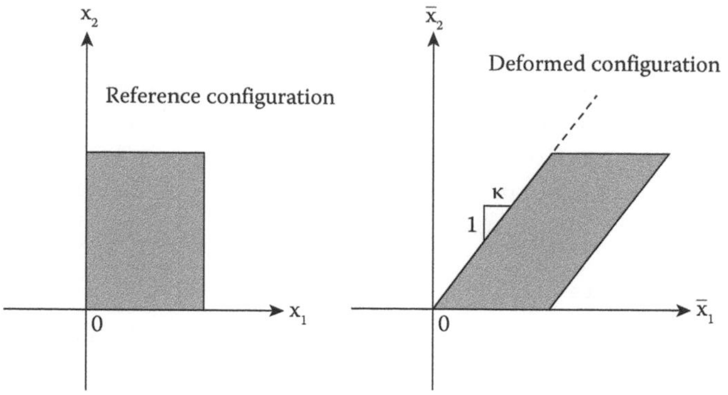

Figure 5.12 2D element under simple shear.

The velocity gradient and its decomposition into deformation rate **D** and spin **W** are

$$\mathbf{V} = \begin{bmatrix} 0 & \dot{\kappa} & 0 \\ 0 & 0 & 0 \\ 0 & 0 & 0 \end{bmatrix}; \mathbf{W} = \frac{\dot{\kappa}}{2}\begin{bmatrix} 0 & 1 & 0 \\ -1 & 0 & 0 \\ 0 & 0 & 0 \end{bmatrix}; \mathbf{D} = \frac{\dot{\kappa}}{2}\begin{bmatrix} 0 & 1 & 0 \\ 1 & 0 & 0 \\ 0 & 0 & 0 \end{bmatrix}; \mathrm{tr}(\mathbf{D}) = 0 \qquad (5.98)$$

The Green strain and Green strain rate are

$$\boldsymbol{\varepsilon} = \frac{1}{2}\begin{bmatrix} 0 & \kappa & 0 \\ \kappa & \kappa^2 & 0 \\ 0 & 0 & 0 \end{bmatrix}; \dot{\boldsymbol{\varepsilon}} = \dot{\kappa}\begin{bmatrix} 0 & 1/2 & 0 \\ 1/2 & \kappa & 0 \\ 0 & 0 & 0 \end{bmatrix} \qquad (5.99)$$

5.5.4.1 *Truesdell rate hypoelastic model*

The grade zero hypoelastic model relating Truesdell stress rate $\bar{\sigma}_{ij}^{Tr}$ and deformation rate D_{ij} is

$$\bar{\sigma}_{ij}^{Tr} = \lambda\ D_{kk}\ \delta_{ij} + 2\mu\ D_{ij} \qquad (5.100)$$

In this case, $\mathrm{tr}(\mathbf{D}) = D_{kk} = 0$, and we have

$$\bar{\boldsymbol{\sigma}}^{Tr} = \mu\ \dot{\kappa}\begin{bmatrix} 0 & 1 & 0 \\ 1 & 0 & 0 \\ 0 & 0 & 0 \end{bmatrix} \qquad (5.101)$$

We recall the definition of the Truesdell rate of Cauchy stress,

$$\bar{\boldsymbol{\sigma}}^{Tr} = \dot{\bar{\boldsymbol{\sigma}}} - \mathbf{V}\ \bar{\boldsymbol{\sigma}} - \bar{\boldsymbol{\sigma}}\ \mathbf{V}^T + \mathrm{tr}(\mathbf{D})\bar{\boldsymbol{\sigma}} \qquad (5.102)$$

Equation 5.101 then becomes

$$\begin{bmatrix} \dot{\bar{\sigma}}_{11} - 2\ \dot{\kappa}\ \bar{\sigma}_{12} & \dot{\bar{\sigma}}_{12} - \dot{\kappa}\ \bar{\sigma}_{22} & 0 \\ \dot{\bar{\sigma}}_{21} - \dot{\kappa}\ \bar{\sigma}_{22} & \dot{\bar{\sigma}}_{22} & 0 \\ 0 & 0 & 0 \end{bmatrix} = \mu\ \dot{\kappa}\begin{bmatrix} 0 & 1 & 0 \\ 1 & 0 & 0 \\ 0 & 0 & 0 \end{bmatrix} \qquad (5.103)$$

This leads to the following system of ordinary differential equations:

$$\begin{cases} \dot{\overline{\sigma}}_{11} - 2\dot{\kappa}\ \overline{\sigma}_{12} = 0 \\ \dot{\overline{\sigma}}_{22} = 0 \\ \dot{\overline{\sigma}}_{12} - \dot{\kappa}\overline{\sigma}_{22} = \mu\ \dot{\kappa} \end{cases} \tag{5.104}$$

From the second equation,

$$\overline{\sigma}_{22} = C_1 \tag{5.105}$$

and because $\overline{\sigma}_{22}\big|_{t=0} = 0$, $\overline{\sigma}_{22} \equiv 0$. Therefore, we have

$$\dot{\overline{\sigma}}_{12} = \mu\ \dot{\kappa} \tag{5.106}$$

Because $\overline{\sigma}_{12}\big|_{\kappa=0} = 0$, we obtain

$$\overline{\sigma}_{12} = \mu\ \kappa \tag{5.107}$$

and

$$\begin{cases} \overline{\sigma}_{11} = \mu\,\kappa^2 \\ \overline{\sigma}_{22} = 0 \end{cases} \tag{5.108}$$

for the Truesdell hypoelastic model.

5.5.4.2 *Jaumann rate hypoelastic model*

The grade zero hypoelastic model relating the Jaumann rate of Cauchy stress to the deformation rate is

$$\overline{\sigma}^J_{ij} = \lambda\ D_{kk}\ \delta_{ij} + 2\mu\ D_{ij} \tag{5.109}$$

where we assume the same elastic moduli λ, μ for all the hypoelastic models considered in this section. For the simple shear problem, we again have $D_{kk} = 0$, and therefore

$$\overline{\sigma}^J_{ij} = 2\mu\ \ D_{ij} \tag{5.110}$$

The Jaumann rate of Cauchy stress is defined as

$$\overline{\boldsymbol{\sigma}}^J = \dot{\overline{\boldsymbol{\sigma}}} - \mathbf{W}\ \overline{\boldsymbol{\sigma}} - \overline{\boldsymbol{\sigma}}\ \mathbf{W}^T \tag{5.111}$$

which leads to

$$\begin{bmatrix} \dot{\overline{\sigma}}_{11} - \dot{\kappa}\ \overline{\sigma}_{12} & \dot{\overline{\sigma}}_{12} - \frac{\dot{\kappa}}{2}(\overline{\sigma}_{22} - \overline{\sigma}_{11}) & 0 \\ \dot{\overline{\sigma}}_{21} - \frac{\dot{\kappa}}{2}(\overline{\sigma}_{22} - \overline{\sigma}_{11}) & \dot{\overline{\sigma}}_{22} + \dot{\kappa}\overline{\sigma}_{12} & 0 \\ 0 & 0 & 0 \end{bmatrix} = \mu\dot{\kappa}\begin{bmatrix} 0 & 1 & 0 \\ 1 & 0 & 0 \\ 0 & 0 & 0 \end{bmatrix} \tag{5.112}$$

Therefore, we have the following equations:

$$\begin{cases} \dot{\bar{\sigma}}_{11} - \dot{\kappa}\bar{\sigma}_{12} = 0 \\ \dot{\bar{\sigma}}_{22} + \dot{\kappa}\bar{\sigma}_{12} = 0 \\ \dot{\bar{\sigma}}_{12} - \frac{\dot{\kappa}}{2}(\bar{\sigma}_{22} - \bar{\sigma}_{11}) = \mu\dot{\kappa} \end{cases} \tag{5.113}$$

From the first two equations above,

$$\begin{cases} \bar{\sigma}_{11} = \int_0^{\kappa} \bar{\sigma}_{12}\ d\kappa \\ \bar{\sigma}_{22} = -\int_0^{\kappa} \bar{\sigma}_{12}\ d\kappa \end{cases} \tag{5.114}$$

and therefore

$$\dot{\bar{\sigma}}_{12} + \dot{\kappa}\int_0^{\kappa} \bar{\sigma}_{12}\ d\kappa = \mu\dot{\kappa} \tag{5.115}$$

which is transformed to the following ordinary differential equation:

$$\frac{d^2\bar{\sigma}_{12}}{d\kappa^2} + \bar{\sigma}_{12} = 0 \tag{5.116}$$

whose solution is

$$\bar{\sigma}_{12} = C_1\ \cos\kappa + C_2\ \sin\kappa \tag{5.117}$$

From the initial condition $\bar{\sigma}_{12}|_{\kappa=0} = 0$, $C_1 = 0$, giving

$$\bar{\sigma}_{12} = C_2\ \sin\kappa \tag{5.118}$$

Substitution of Equation 5.118 into Equation 5.115 provides

$$C_2\ \dot{\kappa}\cos\kappa + \dot{\kappa}\int_0^{\kappa} C_2\ \sin\kappa\ d\kappa = \mu\ \dot{\kappa} \tag{5.119}$$

from which $C_2 = \mu$. Therefore,

$$\bar{\sigma}_{12} = \mu\ \sin\kappa \tag{5.120}$$

for the Jaumann hypoelastic model.

This well-known result has been discussed in many papers (e.g., Dienes 1979; Johnson and Bammann 1984).

It demonstrates the disturbing fact that the Jaumann rate hypoelastic model, which was incorporated in many early computer codes for large deformation inelastic (static and dynamic) analysis, predicts an oscillating (sinusoidal) true (Cauchy) stress at very large shear strains.

Therefore, we conclude that despite the fact that the Jaumann rate of Cauchy stress is objective, it is not suitable for modeling material behavior involving large shear deformations or significant rotation of the material principal axes.

5.5.4.3 *Green–Naghdi rate hypoelastic model*

The grade zero hypoelastic model relating the Green–Naghdi rate of Cauchy stress to the deformation rate is

$$\overline{\sigma}_{ij}^{GN} = \lambda\ D_{kk}\ \delta_{ij} + 2\mu\ D_{ij} \tag{5.121}$$

which is equivalent to the identical relation between the rate of unrotated Cauchy stress and the unrotated deformation rate, that is,

$$\dot{\overline{\sigma}}'_{ij} = \lambda\ d_{kk}\ \delta_{ij} + 2\mu\ d_{ij} \tag{5.122}$$

With tr(**D**) = 0 (and tr(**d**) = 0), we have

$$\begin{cases} \overline{\sigma}_{ij}^{GN} = 2\mu D_{ij} \\ \dot{\overline{\sigma}}'_{ij} = 2\mu d_{ij} \end{cases} \tag{5.123}$$

Proceeding with the second of Equations 5.123, we first "unrotate" the deformation rate via

$$\mathbf{d} = \mathbf{R}^T \mathbf{D}\ \mathbf{R} \tag{5.124}$$

where

$$\mathbf{R} = \begin{bmatrix} \cos\theta & \sin\theta & 0 \\ -\sin\theta & \cos\theta & 0 \\ 0 & 0 & 1 \end{bmatrix} \tag{5.125}$$

and $\tan\theta = \kappa/2$ (Dienes 1979) to obtain

$$\mathbf{d} = \frac{\dot{\kappa}}{2} \begin{bmatrix} -\sin 2\theta & \cos 2\theta & 0 \\ \cos 2\theta & \sin 2\theta & 0 \\ 0 & 0 & 0 \end{bmatrix} \tag{5.126}$$

Therefore,

$$\frac{d}{d\theta}\overline{\boldsymbol{\sigma}}' = \frac{2\mu}{\cos^2\theta} \begin{bmatrix} -\sin 2\theta & \cos 2\theta & 0 \\ \cos 2\theta & \sin 2\theta & 0 \\ 0 & 0 & 0 \end{bmatrix} \tag{5.127}$$

where we have used $d\kappa/d\theta = 1/\cos^2\theta$. Integration of these relations (with the initial condition that all stresses are zero at $\theta = 0$) yields the unrotated Cauchy stresses as

$$\begin{cases} \overline{\sigma}'_{11} = -\overline{\sigma}'_{22} = 4\mu\ \ln(\cos\ \theta) \\ \overline{\sigma}'_{12} = 2\mu(2\theta - \tan\theta) \end{cases} \tag{5.128}$$

Finally, rotation back to the fixed Cartesian system provides the expressions for the Cauchy stresses:

$$\begin{cases} \overline{\sigma}_{11} = -\overline{\sigma}_{22} = 4\mu\left[\cos 2\theta\ \ln(\cos\theta) + \theta\ \sin\theta - \sin^2\theta\right] \\ \overline{\sigma}_{12} = 2\mu\ \cos 2\theta\left[2\theta - 2\ \tan 2\theta\ \ln(\cos\theta) - \tan\theta\right] \end{cases} \tag{5.129}$$

Equations 5.129 were given by Dienes (1979).

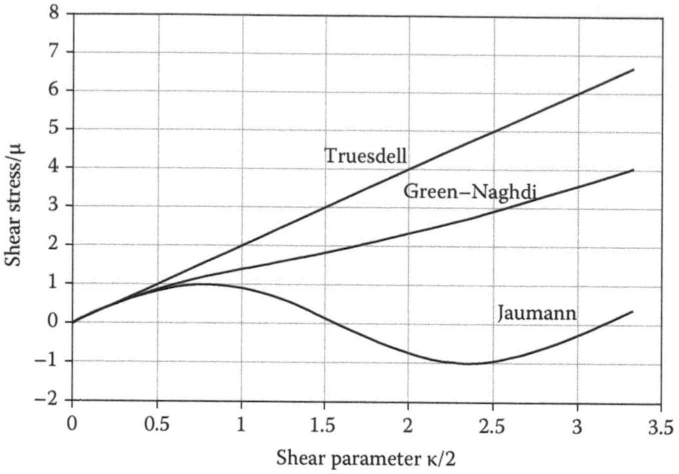

Figure 5.13 Behavior of hypoelastic constitutive models in simple shear.

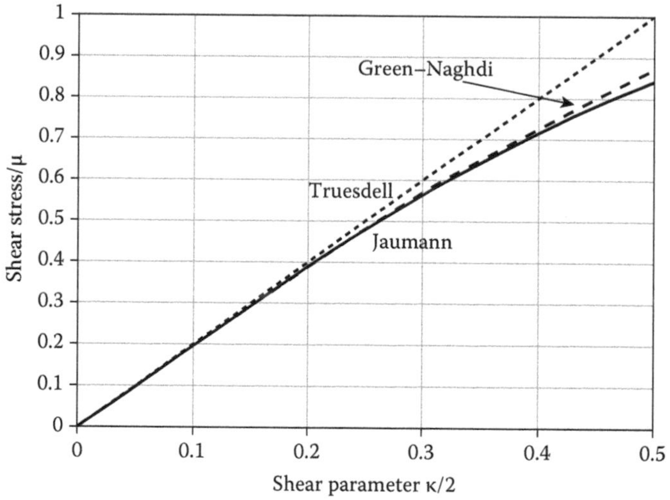

Figure 5.14 Hypoelastic material models compared at smaller shear deformations.

The three grade zero hypoelastic material models are compared in Figure 5.13.

We note the physically unrealistic behavior of the Jaumann rate of Cauchy stress that is illustrated in Figure 5.13. Even though the elastic shear deformations represented in Figure 5.13 are far beyond the range of practical interest, this striking result has nevertheless led to the use of other objective rates of Cauchy stress in preference to the Jaumann rate.

At the more realistic, although still very large, shear deformations shown in Figure 5.14, we see that there is little difference between the three hypoelastic models we have considered.

5.6 NUMERICAL EVALUATION OF A LINEAR ISOTROPIC HYPOELASTIC MODEL

Simo and Pister (1984) showed that in general, a linear isotropic hypoelastic material model cannot be energy conserving over closed strain cycles. Kojic and Bathe (1987) provided a closed-form analytical verification of this fact for the specific case of an updated Lagrangian formulation employing the Jaumann rate of Cauchy stress.

Despite this undesirable feature, the linear isotropic hypoelastic model is still widely used for stress updating in numerical analyses of elastoplastic problems. This is usually seen as an acceptable approximation, and justified on the basis that elastic strain increments (especially for metal plasticity) are typically very small compared with plastic strain increments.

In this section, we quantitatively evaluate this hypoelastic model over representative closed strain cycles. A convenient method to generate a variety of closed strain cycles and to calculate the net work done by the stresses over the strain cycle is to employ the simple isoparametric finite element shown in Figure 5.15.

The triangular three-noded element shown in Figure 5.15 can be viewed as the projection on the $(\bar{x}, \bar{y})$ plane of a prismatic three-dimensional (3D) solid that extends indefinitely in the $\bar{z}$ direction, but we restrict ourselves here to cycles of two-dimensional (2D) plane strain and therefore need only consider deformations in the $(\bar{x}, \bar{y})$ plane. We note that a 3D simplex element (i.e., a tetrahedron) could be employed in exactly the same way to efficiently generate families of 3D strain cycles.

5.6.1 Natural coordinate system for triangle

The location of a generic point P(x, y) in Figure 5.16 is described using the triangular (area) coordinates (L_1, L_2, L_3), that is,

$$\begin{cases} L_i = \dfrac{A_i}{A}, \quad i = 1, 2, 3 \\ A \equiv A_1 + A_2 + A_3 \end{cases} \tag{5.130}$$

which are natural coordinates for the trianglular shape and are independent of the orientation of the triangle in the (x, y) coordinate system (Zienkiewicz 1977).

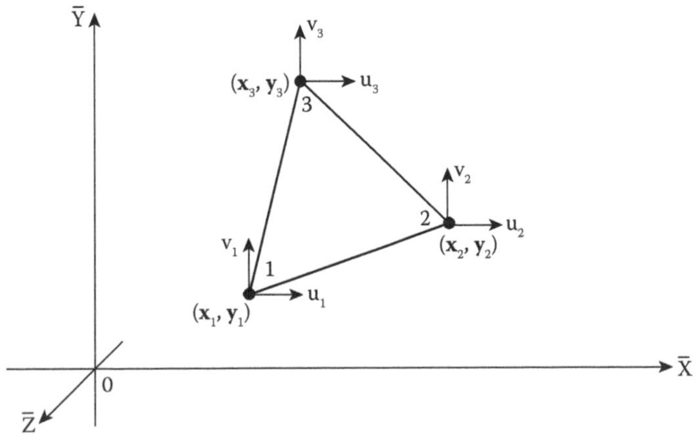

Figure 5.15 Isoparametric (simplex) finite element.

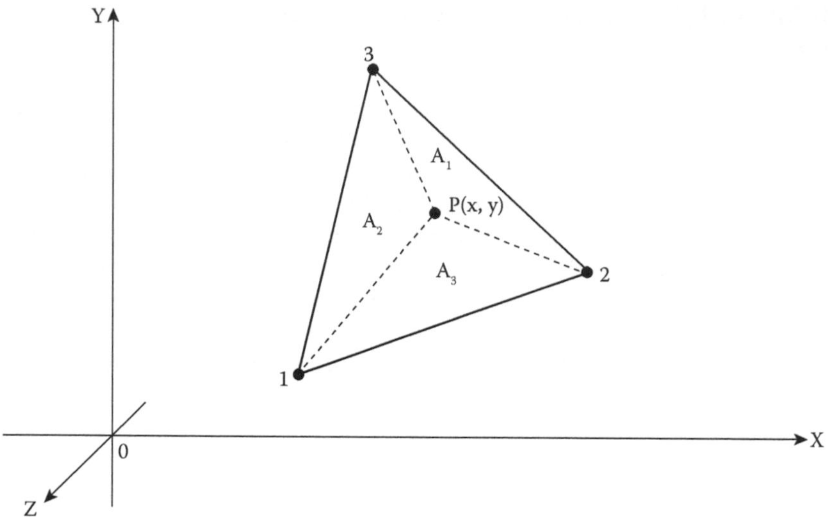

Figure 5.16 Triangular (area) coordinates.

The area A of the triangle can be expressed in terms of the Cartesian coordinates of the nodes as the determinant

$$2A = \begin{vmatrix} 1 & x_1 & y_1 \\ 1 & x_2 & y_2 \\ 1 & x_3 & y_3 \end{vmatrix} \tag{5.131}$$

which can be expanded along the second column to yield the formula

$$\begin{cases} 2A = \sum_{i=1,2,3} x_i y_{jk}, (i, j, k) \text{cyclic} \\ \text{and } y_{jk} \equiv y_j - y_k \end{cases} \tag{5.132}$$

The area coordinates of a generic point P(x, y) in the triangle are related to its Cartesian coordinates via the following:

$$\begin{aligned} &L_i(x, y) = \frac{1}{2A}\left[2A_{jk} + x\ y_{jk} + y\ x_{kj}\right] \\ &i = 1, 2, 3 \text{ and } (i, j, k) \text{ cyclic} \end{aligned} \tag{5.133}$$

where A_{jk} is the area of the triangle with vertices at (0, 0), (x_j, y_j), (x_k, y_k). The inverse relations are

$$\begin{cases} x = x_1 L_1 + x_2 L_2 + x_3 L_3 \\ y = y_1 L_1 + y_2 L_2 + y_3 L_3 \end{cases} \tag{5.134}$$

5.6.2 Kinematics

The initial nodal coordinates are (x_i, y_i), i = 1, 2, 3. The nodal displacements are (u_i, v_i), i = 1, 2, 3. The current nodal coordinates in the deformed configuration are therefore

$$\left.\begin{aligned} \bar{x}_i &= x_i + u_i \\ \bar{y}_i &= y_i + v_i \end{aligned}\right\}, \quad i = 1, 2, 3 \tag{5.135}$$

where (u_i, v_i) are understood to be functions of a time-like loading parameter t, that is, $u_i = u_i(t)$, $v_i = v_i(t)$. The coordinates $(\bar{x}, \bar{y})$ of a generic point in the deformed configuration are interpolated in terms of the nodal coordinates in the deformed configuration as in Equation 5.134, that is,

$$\begin{cases} \bar{x} = \sum\limits_{i=1,3} \bar{x}_i L_i \\ \bar{y} = \sum\limits_{i=1,3} \bar{y}_i L_i \end{cases} \tag{5.136}$$

and therefore the displacements (u, v) of a generic point within the triangle in terms of the nodal displacements are

$$\begin{cases} u = \sum\limits_{i=1,3} u_i L_i \\ v = \sum\limits_{i=1,3} v_i L_i \end{cases} \tag{5.137}$$

Since the displacements vary linearly within the triangular element, the displacement gradients are constant.

If the displacements of one node, say node 1, are set to zero, thus eliminating rigid body translation, the remaining four degrees of freedom at nodes 2 and 3 can be selected to generate a rigid body rotation (i.e., representing a rotation of the principal material axes) and three independent displacement gradients from which, for example, the Green strains or the deformation rate can be determined.

In principle, any desired (2D) complete strain cycle can be generated by specifying closed displacement cycles for the two "free" nodes (nodes 2 and 3 here). It will suffice for our purposes to choose simple cyclic displacement histories that can be easily interpreted.

The deformation gradient is determined from

$$\mathbf{F}^T \equiv \begin{bmatrix} \dfrac{\partial \bar{x}}{\partial x} & \dfrac{\partial \bar{y}}{\partial x} \\ \dfrac{\partial \bar{x}}{\partial y} & \dfrac{\partial \bar{y}}{\partial y} \end{bmatrix} = \begin{bmatrix} \dfrac{\partial L_1}{\partial x} & \dfrac{\partial L_2}{\partial x} & \dfrac{\partial L_3}{\partial x} \\ \dfrac{\partial L_1}{\partial y} & \dfrac{\partial L_2}{\partial y} & \dfrac{\partial L_3}{\partial y} \end{bmatrix} \begin{bmatrix} \bar{x}_1 & \bar{y}_1 \\ \bar{x}_2 & \bar{y}_2 \\ \bar{x}_3 & \bar{y}_3 \end{bmatrix} \tag{5.138}$$

which can be expressed, using Equation 5.133, as

$$\mathbf{F}^T = \frac{1}{2A} \begin{bmatrix} y_{23} & y_{31} & y_{12} \\ x_{32} & x_{13} & x_{21} \end{bmatrix} \begin{bmatrix} \bar{x}_1 & \bar{y}_1 \\ \bar{x}_2 & \bar{y}_2 \\ \bar{x}_3 & \bar{y}_3 \end{bmatrix} \tag{5.139}$$

Define the (2×3) matrix $\mathbf{T}$ as

$$\mathbf{T} = \frac{1}{2A} \begin{bmatrix} y_{23} & y_{31} & y_{12} \\ x_{32} & x_{13} & x_{21} \end{bmatrix} \tag{5.140}$$

so that

$$\mathbf{F}^{\mathrm{T}} = \mathbf{T}\begin{bmatrix} \overline{x}_1 & \overline{y}_1 \\ \overline{x}_2 & \overline{y}_2 \\ \overline{x}_3 & \overline{y}_3 \end{bmatrix} \tag{5.141}$$

The corresponding Lagrangian displacement gradient is

$$\mathbf{G}^{\mathrm{T}} \equiv \begin{bmatrix} \dfrac{\partial u}{\partial x} & \dfrac{\partial v}{\partial x} \\ \dfrac{\partial u}{\partial y} & \dfrac{\partial v}{\partial y} \end{bmatrix} = \mathbf{T}\begin{bmatrix} u_1 & v_1 \\ u_2 & v_2 \\ u_3 & v_3 \end{bmatrix} \tag{5.142}$$

5.6.3 Nodal displacement patterns

Without loss of generality, we choose the nodal displacements of node 1 to be zero, that is, $u_1 = v_1 = 0$, and therefore $\overline{x}_1 = x_1$ and $\overline{y}_1 = y_1$.

A set of nodal displacement patterns are chosen to produce displacement gradients that are independent of the size, shape, and orientation of the triangular element. For example, the nodal displacement pattern

$$\begin{bmatrix} u_1 & v_1 \\ u_2 & v_2 \\ u_3 & v_3 \end{bmatrix}_{\gamma} = \begin{bmatrix} 0 & 0 \\ y_{21} & x_{21} \\ y_{31} & x_{31} \end{bmatrix} \tag{5.143}$$

produces a Lagrangian displacement gradient

$$\mathbf{G}^{\mathrm{T}} = \begin{bmatrix} 0 & 1 \\ 1 & 0 \end{bmatrix} \tag{5.144}$$

which for small displacements corresponds to a pure shear strain. The nodal displacement pattern

$$\begin{bmatrix} u_1 & v_1 \\ u_2 & v_2 \\ u_3 & v_3 \end{bmatrix}_{\theta} = \begin{bmatrix} 0 & 0 \\ -y_{21} & x_{21} \\ -y_{31} & x_{31} \end{bmatrix} \tag{5.145}$$

produces a Lagrangian displacement gradient

$$\mathbf{G}^{\mathrm{T}} = \begin{bmatrix} 0 & 1 \\ -1 & 0 \end{bmatrix} \tag{5.146}$$

which corresponds to an infinitesimal rotation.

Similarly,

$$\begin{aligned} &\begin{bmatrix} 0 & 0 \\ x_{21} & 0 \\ x_{31} & 0 \end{bmatrix} \rightarrow \begin{bmatrix} 1 & 0 \\ 0 & 0 \end{bmatrix} \\ &\begin{bmatrix} 0 & 0 \\ 0 & y_{21} \\ 0 & y_{31} \end{bmatrix} \rightarrow \begin{bmatrix} 0 & 0 \\ 0 & 1 \end{bmatrix} \end{aligned} \tag{5.147}$$

5.6.4 Strain cycles

Closed strain cycles are generated by displacing nodes 2 and 3 in closed displacement loops, each consisting of four straight-line branches, as illustrated schematically in Figure 5.17.

The nodal displacements of nodes 2 and 3 are linked via patterns of the type illustrated in Equation 5.143, and scaled appropriately to produce strain cycles of desired magnitudes. The straight-line displacement branches shown schematically in Figure 5.17 trace out a parallelogram.

Much more complex nodal displacement histories could of course be generated, but they are not needed for the following reason: we think of each displacement branch as a straight-line increment in strain space, which corresponds to the way in which the hypoelastic model is typically used as an elastic predictor and radial corrector for stress updating in numerical elastoplastic analyses.

Numerical calculations are carried out for two types of closed strain cycles, as follows:

- *Strain cycle A*: Shear followed by uniaxial tension
- *Strain cycle B*: Shear followed by equibiaxial tension

5.6.5 Calculation procedure

The final results are presented in nondimensional form, but for numerical computation we use the following values: E = 30,000 ksi, Poisson's ratio $\nu = 0.3$, and uniaxial yield stress $\sigma_Y = 60$ ksi. The corresponding uniaxial yield strain $e_Y = \sigma_Y/E = 0.002$.

Each branch of the nodal displacement cycle is divided into a large number N of equal-sized increments **du**, where **u** represents the (3, 2) matrix of nodal displacements, for example, Equation 5.143.

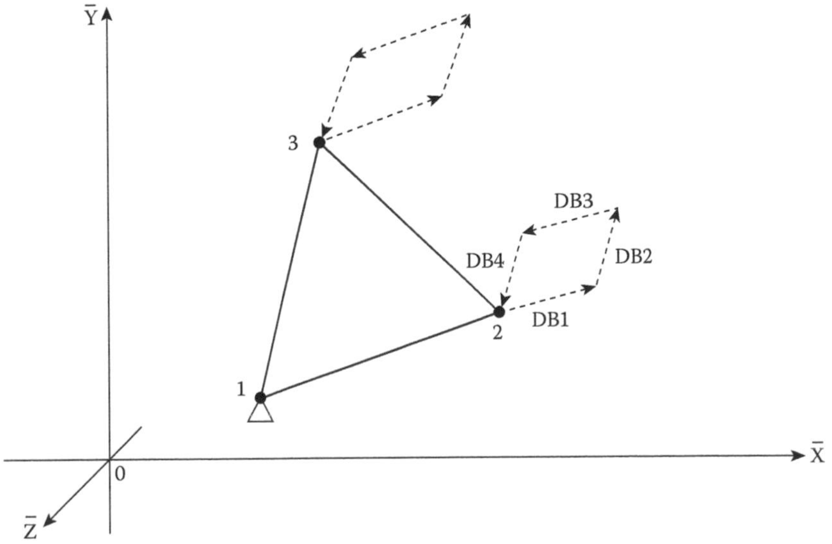

Figure 5.17 Nodal displacement cycles (schematic).

The calculations proceed as follows:

1. Nodal coordinate update

$$\mathbf{u}_{n+1} = \mathbf{u}_n + \mathbf{du} \tag{5.148}$$

2. Deformation gradient and increment of deformation gradient

$$\begin{aligned} \mathbf{F}_{n+1}^{T} &= \mathbf{T}\ \mathbf{u}_{n+1} \\ \mathbf{dF}_{n+1}^{T} &= \mathbf{T}\ \mathbf{du}_{n+1} \end{aligned} \tag{5.149}$$

3. Velocity gradient and deformation rate

$$\begin{aligned} \mathbf{V}_{n+1} &= \mathbf{dF}_{n+1}\ \mathbf{F}_{n+1}^{-1} \\ \mathbf{D}_{n+1} &= \frac{1}{2}(\mathbf{V}_{n+1} + \mathbf{V}_{n+1}^{T}) \end{aligned} \tag{5.150}$$

4. Green strain

$$\boldsymbol{\varepsilon}_{n+1} = \frac{1}{2}(\mathbf{F}_{n+1}^{T}\mathbf{F}_{n+1} - \mathbf{I}) \tag{5.151}$$

5. Green strain rate

$$\mathbf{d}\boldsymbol{\varepsilon}_{n+1} = \mathbf{F}_{n+1}^{T}\mathbf{D}_{n+1}\mathbf{F}_{n+1} \tag{5.152}$$

6. Right stretch **U** and material rotation **R** (spectral decomposition)

$$\begin{aligned} \mathbf{U}_{n+1} &= \sqrt{\mathbf{F}_{n+1}^{T}\mathbf{F}_{n+1}} \\ \mathbf{R}_{n+1} &= \mathbf{F}_{n+1}\mathbf{U}_{n+1}^{-1} \end{aligned} \tag{5.153}$$

7. Unrotated deformation rate

$$\mathbf{d}_{n+1} = \mathbf{R}_{n+1}^{T}\mathbf{D}_{n+1}\mathbf{R}_{n+1} \tag{5.154}$$

8. Unrotated Cauchy stress update

$$\bar{\boldsymbol{\sigma}}'_{n+1} = \bar{\boldsymbol{\sigma}}'_{n} + \lambda\ \mathbf{I}\ \mathrm{tr}(\mathbf{d}_{n+1}) + 2\mu\ \mathbf{d}_{n+1} \tag{5.155}$$

9. Cauchy stress update

$$\bar{\boldsymbol{\sigma}}_{n+1} = \mathbf{R}_{n+1}\bar{\boldsymbol{\sigma}}'_{n+1}\mathbf{R}_{n+1}^{T} \tag{5.156}$$

10. 2PK stress update

$$\boldsymbol{\sigma}_{n+1} = \frac{1}{|\mathbf{F}|_{n+1}}\mathbf{F}_{n+1}\bar{\boldsymbol{\sigma}}_{n+1}\mathbf{F}_{n+1}^{T} \tag{5.157}$$

11. Work increments

$$\begin{cases} \mathrm{d}\overline{\mathrm{W}}_{n+1} = (\bar{\boldsymbol{\sigma}}_{n+1} : \mathbf{D}_{n+1})\overline{\mathrm{A}}_{n+1} \\ \mathrm{dW}_{n+1} = (\boldsymbol{\sigma}_{n+1} : \mathbf{d}\boldsymbol{\varepsilon}_{n+1})\mathrm{A}_{n+1} \end{cases} \tag{5.158}$$

12. Total work update

$$\begin{cases} \overline{W}_{n+1} = \overline{W}_n + d\overline{W}_{n+1} \\ W_{n+1} = W_n + dW_{n+1} \end{cases} \tag{5.159}$$

Calculation of the Lagrangian variables is carried out as a check on the computations.

5.6.6 Numerical results

Figure 5.18 shows the principal extensions $e_i \equiv \lambda_i - 1$, $(i = 1, 2)$, where λ_i are the principal stretches as usual, normalized by uniaxial yield strain $e_Y = \sigma_Y/E = 0.002$. The (normalized) principal extensions shown in Figure 5.18 are, of course, not necessarily in fixed Cartesian directions. We remark that the strain increments corresponding to each straight-line nodal displacement branch would be considered fairly large in a typical elastoplastic numerical analysis.

In the figure, the first displacement branch terminating at (–0.10, 0.10) is the same for cycles A and B. The third and fourth branches of cycle B together coincide with the fourth branch of cycle A (from [0.20, 0] to [0, 0]).

A scalar normalized Cauchy stress $\overline{S}_N$ is defined as

$$\overline{S}_N = \sqrt{\frac{3}{2}\bar{s}_{ij}\bar{s}_{ij}}/\sigma_Y \tag{5.160}$$

where σ_Y is the uniaxial yield stress. $\overline{S}_N$ is the ratio of the effective Cauchy stress to that required to cause (von Mises) yield.

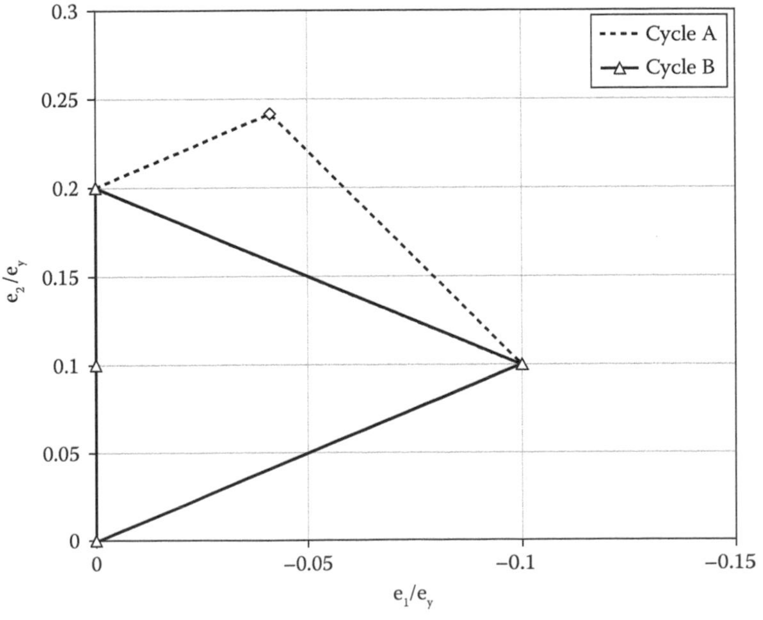

Figure 5.18 Normalized principal extensions for strain cycles.

Table 5.1 (Normalized) dissipated energy and residual stress in closed strain cycles

	W_N	S_N
Cycle A × 10	3.195×10^{-4}	4.469×10^{-7}
	3.196×10^{-2}	4.456×10^{-5}
Cycle B × 10	2.485×10^{-4}	2.894×10^{-7}
	2.487×10^{-2}	2.888×10^{-5}

A scalar normalized work quantity $\overline{W}_N$ is defined as

$$\overline{W}_N = \overline{W}/\overline{W}_{sh} \tag{5.161}$$

where $\overline{W}$ is the calculated net work done by the Cauchy stresses over a complete strain cycle. The normalizing factor $\overline{W}_{sh}$ is defined as

$$\overline{W}_{sh} = \overline{A}\sigma_Y^2/6\mu \tag{5.162}$$

which is equal to the stored elastic shear strain energy at von Mises yield.

For each of cycles A and B, results were also calculated for nodal displacements 10× those illustrated in Figure 5.18. These results are shown in Table 5.1.

We note that the normalized net work (i.e., dissipated elastic strain energy) and the normalized residual stress for both of the smaller cycles A and B are indeed very small. As noted above, these "smaller" cycles are still reasonably large (a strain increment on the order of 1/10 of the uniaxial yield strain) when viewed in the context of a predictor and corrector stress update application.

We observe also that the net work and residual stress quantities for the (10×) larger cycles are essentially 100× the corresponding values for the smaller cycles, as expected.

These numerical results are admittedly limited in scope. However, they do seem to support the view that use of the linear isotropic hypoelastic model (in standard predictor and corrector stress update algorithms) is a reasonable approximation, despite objections that can be raised on theoretical grounds.

Chapter 6

Stress invariants and material tests

6.1 INTRODUCTION

In Chapter 5, we discussed constitutive laws governing the elastic behavior of materials, including Cauchy elasticity, hyperelasticity, and hypoelasticity. Elastic material behavior in general may be nonlinear but is path independent (nondissipative). However, when we talk about material nonlinearity, very often we refer to inelastic or plastic deformations of materials. Plasticity theory (discussed in Chapter 7) deals with the modeling of inelastic behavior of materials. In preparation for the discussion of plasticity theory, we first discuss various ways that stresses can be represented.

In this chapter, we first look at representing stresses and strains in volumetric and deviatoric form. Deviatoric stresses play an important role in metal plasticity. Stresses and deviatoric stresses are each characterized by principal stresses and three invariants that are independent of the frame of reference. We also discuss an alternative form of stress invariants that is useful in the modeling of frictional materials.

All material models are based partly on the observation of material behavior in controlled laboratory material tests. There are a number of standard material tests that are routinely used, not only to observe behavioral characteristics, but also to determine the parameters of specific material models. We discuss some of the most widely used material tests and examine various representations of the stress paths that the test specimen follows in these tests.

6.2 VOLUMETRIC AND DEVIATORIC STRESSES AND STRAINS

Most constitutive models of material behavior, such as plasticity models, are defined in terms of stress invariants. In this way, invariance under coordinate rotation of a constitutive model of material behavior is assured. The modern theory of metal plasticity is partially based on the experimental observation that changes in *compressive* mean stress (i.e., hydrostatic compression) have no effect on yield or irreversible deformation (Bridgman 1923, 1952). This observation has led to a number of well-known metal plasticity models, which will be described in Chapter 7. As a result, it is convenient to decompose stress and strain into their volumetric and deviatoric components.

The hydrostatic stress p, also known as the volumetric stress, is defined as the average of the three normal stresses (σ_{11}, σ_{22}, σ_{33}). The volumetric strain ε_v is defined as the sum of the three normal strains.

$$\begin{cases} p = \dfrac{1}{3}\sigma_{kk} \\ \varepsilon_v = \varepsilon_{kk} \end{cases} \tag{6.1}$$

Deviatoric stresses s_{ij} and deviatoric strains e_{ij} are then defined as

$$\begin{cases} s_{ij} = \sigma_{ij} - p\ \delta_{ij} \\ e_{ij} = \varepsilon_{ij} - \dfrac{\varepsilon_v}{3}\ \delta_{ij} \end{cases} \tag{6.2}$$

so that their first invariant is zero, that is, $s_{kk} = 0$ and $e_{kk} = 0$.

With these definitions, the stresses and strains can be expressed as the sum of their deviatoric and volumetric components. Linear elastic stress–strain relations were discussed in detail in Chapter 5. We saw there that for an *isotropic* elastic material, only two independent material constants are required: for example, the Lamé constants (λ, μ), the modulus of elasticity and Poison's ratio (E,ν), or the bulk and shear moduli (K, G). The volumetric and deviatoric stress–strain relations can then be expressed as

$$\begin{cases} p = K\varepsilon_v \\ s_{ij} = 2G\ e_{ij} \end{cases} \tag{6.3}$$

That is, the volumetric and deviatoric stress–strain relations are uncoupled for *isotropic* elastic materials. Moreover, for isotropic materials it can be verified that the principal directions (axes) of the stress and strain tensors are the same. This is an important property for material modeling.

6.3 PRINCIPAL STRESSES AND STRESS INVARIANTS

For any state of stress, there exists a set of three mutually orthogonal directions such that the stress tensor referred to these directions is diagonal (i.e., the shear stresses are zero). These special directions, or principal directions, are the eigenvectors of the stress tensor. The normal stresses in the principal directions (i.e., the principal stresses) are the three roots ($\sigma_1 \geq \sigma_2 \geq \sigma_3$) of the characteristic equation

$$\det(\sigma_{ij} - \sigma\ \delta_{ij}) = 0 \tag{6.4}$$

In Chapter 5, we discussed the characteristic equation for the strain tensor. The characteristic equation for the stress tensor is similar, that is,

$$\sigma^3 - I_1\,\sigma^2 + I_2\,\sigma - I_3 = 0 \tag{6.5}$$

The coefficients in this last equation are the three invariants (I_1, I_2, I_3) of the stress tensor:

$$\begin{cases} I_1 = \operatorname{tr}(\sigma) \\ I_2 = \begin{vmatrix} \sigma_{11} & \sigma_{12} \\ \sigma_{21} & \sigma_{22} \end{vmatrix} + \begin{vmatrix} \sigma_{22} & \sigma_{23} \\ \sigma_{32} & \sigma_{33} \end{vmatrix} + \begin{vmatrix} \sigma_{11} & \sigma_{13} \\ \sigma_{31} & \sigma_{33} \end{vmatrix} \\ I_3 = \det(\sigma) \end{cases} \tag{6.6}$$

or

$$\begin{cases} I_1 = \sigma_{kk} \\ I_2 = \dfrac{1}{2}[(\sigma_{kk})^2 - \sigma_{ij}\,\sigma_{ij}] \\ I_3 = \dfrac{1}{6}(2\sigma_{ij}\,\sigma_{jk}\,\sigma_{ki} - 3\,\sigma_{ij}\,\sigma_{ij}\,\sigma_{kk} + \sigma_{kk}^3) \end{cases} \tag{6.7}$$

The stress invariants can also be expressed in terms of principal stresses, that is,

$$\begin{cases} I_1 = \sigma_1 + \sigma_2 + \sigma_3 \\ I_2 = \sigma_1 \sigma_2 + \sigma_2 \sigma_3 + \sigma_3 \sigma_1 \\ I_3 = \sigma_1 \sigma_2 \sigma_3 \end{cases} \tag{6.8}$$

The invariants and principal values of the deviatoric stress tensor can be similarly determined from the corresponding characteristic equation of the deviatoric stress tensor:

$$\begin{cases} \det(s_{ij} - s\,\delta_{ij}) = 0 \\ s^3 - J_1 s^2 + J_2 s - J_3 = 0 \end{cases} \tag{6.9}$$

The roots of this equation are the principal deviatoric stresses $s_1 \geq s_2 \geq s_3$. We note that the principal directions of the deviatoric stress tensor are the same as the principal directions of the stress tensor. The invariants of the deviatoric stress tensor are expressed as follows:

$$\begin{cases} J_1 = \mathrm{tr}(s) = s_{kk} = s_1 + s_2 + s_3 = 0 \\ J_2 = \dfrac{1}{2} s_{ij} s_{ij} = \dfrac{1}{2}(s_1^2 + s_2^2 + s_3^2) \\ J_3 = \det(s) = \dfrac{1}{3}(s_{ij} s_{jk} s_{ki}) = s_1 s_2 s_3 \end{cases} \tag{6.10}$$

6.4 ALTERNATIVE FORMS OF STRESS INVARIANTS

It is observed that these invariants are often difficult to work with because, except for the first invariant p, it is difficult to assign to them physical and geometric interpretations. With this in mind, we define a new set of invariants (p, q, θ). The hydrostatic or mean stress p has already been defined as

$$p = \frac{1}{3}\sigma_{kk} = \frac{1}{3} I_1 \tag{6.11}$$

The new stress invariant q is defined as

$$q = \sqrt{\frac{3}{2} s_{ij} s_{ij}} = \sqrt{3 J_2} \tag{6.12}$$

The stress invariant q is used in place of the second invariant of deviatoric stress J_2 since it has a simple meaning for triaxial stress test paths, as will be seen later in this chapter. The third invariant, called the Lode angle, is defined as follows:

$$\theta = \frac{1}{3}\sin^{-1}\left(-\frac{27 J_3}{2 q^3}\right) \tag{6.13}$$

The arcsine in the equation above ranges between $-\pi/2$ and $\pi/2$. Therefore, the range of the Lode angle is

$$-\frac{\pi}{6} \leq \theta \leq \frac{\pi}{6} \tag{6.14}$$

The principal stresses can be represented in terms of these stress invariants as follows (Nayak and Zienkiewicz 1972):

$$\begin{Bmatrix} \sigma_1 \\ \sigma_2 \\ \sigma_3 \end{Bmatrix} = \frac{2q}{3} \begin{Bmatrix} \sin\left(\theta + \frac{2\pi}{3}\right) \\ \sin(\theta) \\ \sin\left(\theta + \frac{4\pi}{3}\right) \end{Bmatrix} + p \begin{Bmatrix} 1 \\ 1 \\ 1 \end{Bmatrix} \tag{6.15}$$

6.5 OCTAHEDRAL STRESSES

We consider a Cartesian coordinate system oriented in the principal stress directions (1, 2, 3) shown in Figure 6.1.

The three coordinate planes, together with an oblique plane with unit normal **n**, form a (differential) tetrahedron. The particular oblique plane with unit normal

$$\mathbf{n} = \frac{1}{\sqrt{3}}(1, 1, 1)^T \tag{6.16}$$

is termed the *octahedral plane*. It is so called because a similar oblique plane could be formed in each octant of this particular three-dimensional Cartesian space, each one of the eight planes having one of the eight normal vectors,

$$\mathbf{n} = \frac{1}{\sqrt{3}}(\pm 1, \pm 1, \pm 1)^T \tag{6.17}$$

The normal stresses on the coordinate planes are σ_1, σ_2, and σ_3, and the shear stresses are zero, as shown in Figure 6.2. The traction vector **t** on the oblique (octahedral) plane is determined in the usual way and is given by

$$\mathbf{t} = [\boldsymbol{\sigma}]\mathbf{n} = \begin{bmatrix} \sigma_1 & & \\ & \sigma_2 & \\ & & \sigma_3 \end{bmatrix} \frac{1}{\sqrt{3}} \begin{Bmatrix} 1 \\ 1 \\ 1 \end{Bmatrix} = \frac{1}{\sqrt{3}} \begin{Bmatrix} \sigma_1 \\ \sigma_2 \\ \sigma_3 \end{Bmatrix} \tag{6.18}$$

The magnitude of the normal stress σ_{oct} on the octahedral plane is $\mathbf{n}^T\mathbf{t}$ or

$$\sigma_{oct} = \frac{1}{3}(\sigma_1 + \sigma_2 + \sigma_3) = p \tag{6.19}$$

The magnitude of *resultant* shear stress on the oblique plane is defined as the *octahedral shear stress* τ_{oct}, which is given by

$$\tau_{oct}^2 = \mathbf{t}^T\mathbf{t} - \sigma_{oct}^2 = \frac{1}{3}(\sigma_1^2 + \sigma_2^2 + \sigma_3^2) - p^2 \tag{6.20}$$

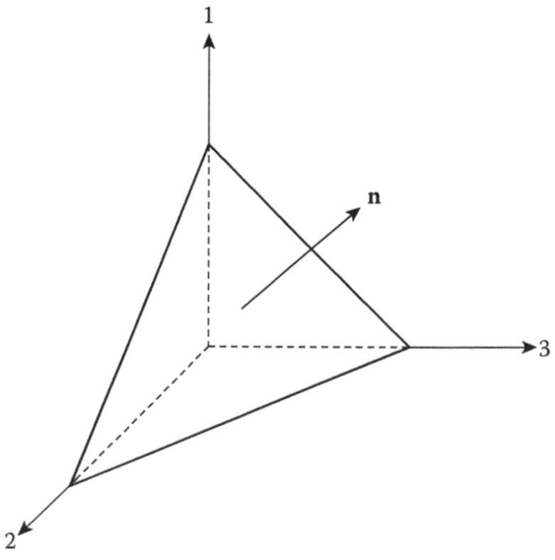

Notation: (1, 2, 3) principal stress axes

Figure 6.1 Octahedral plane in principal stress coordinate system.

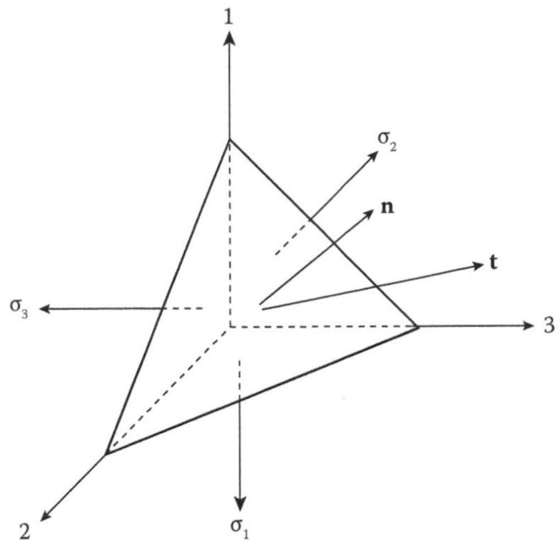

Figure 6.2 Traction vector on octahedral plane.

Putting this last expression in terms of the principal deviator stresses (s_1, s_2, s_3), we find

$$\tau_{oct}^2 = \frac{1}{3}(s_1^2 + s_2^2 + s_3^2) = \frac{2}{3}J_2 \qquad (6.21)$$

where J_2 is the second invariant of the deviator stress that was given in Equation 6.10. The octahedral shear stress is related to the invariant $q(=\sqrt{3J_2})$ introduced in the previous section, by

$$q = \frac{3}{\sqrt{2}}\tau_{oct} \qquad (6.22)$$

6.6 PRINCIPAL STRESS SPACE

A general three-dimensional state of stress can be represented as a point with coordinates (σ_1, σ_2, σ_3) in a three-dimensional Cartesian space known as *principal stress space* (Figure 6.3), in which the coordinate axes are aligned with the principal stress directions. The principal stress space is also sometimes called the Haigh–Westergaard stress space (Mendelson 1968).

This representation of general three-dimensional stress states is very convenient for visualization of material yield and failure criteria, as well as the stress paths followed by material points in different laboratory tests.

We define a principal stress vector $\boldsymbol{\sigma}$ (a true vector) in this space as

$$\boldsymbol{\sigma} = (\sigma_1, \sigma_2, \sigma_3)^T \qquad (6.23)$$

The *hydrostatic axis* on which $\sigma_1 = \sigma_2 = \sigma_3$ is the space diagonal defined by the unit vector **n**:

$$\mathbf{n} = \frac{1}{\sqrt{3}}(1, 1, 1)^T \qquad (6.24)$$

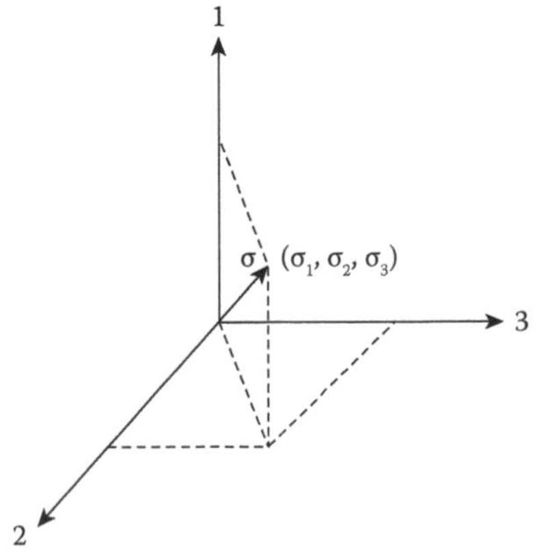

Figure 6.3 Principal stress vector $\boldsymbol{\sigma}$.

Any stress point lying on the hydrostatic axis corresponds to a hydrostatic pressure and zero deviator stresses. We may also define a principal deviator stress vector s as

$$s = (s_1, s_2, s_3)^T \qquad (6.25)$$

The *π-plane* is orthogonal to the space diagonal and contains the origin ($\sigma_1 = \sigma_2 = \sigma_3 = 0$) of principal stress space. The principal deviator stress vector s lies in the *π-plane* since $\mathbf{n}^T s \equiv 0$.

Looking down the space diagonal **n**, we see the π-plane in true perspective, coinciding with the plane of the paper. In this view, the *projections* of the principal stress axes will appear as shown in Figure 6.4, lying at 120° angles to one another. In this projection, states of stress with identical principal deviator stresses but different hydrostatic stresses will appear to coincide.

The location of a specific state of stress $(\sigma_1, \sigma_2, \sigma_2)$ can be specified as a distance $\sqrt{3}p$ measured along the space diagonal, and its "polar" coordinates in the π-plane: a scalar radial distance r (say) and the Lode angle θ. The true radius r is equal to the length of the principal deviator stress vector, that is,

$$r = |s| = \sqrt{2J_2} = \sqrt{\frac{2}{3}}q \tag{6.26}$$

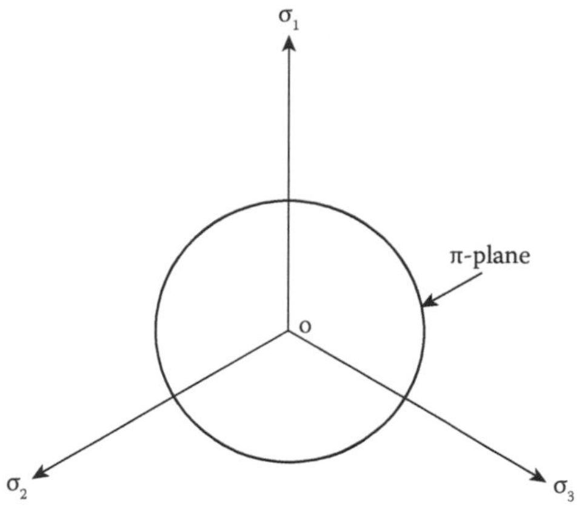

Figure 6.4 View of the π-plane looking down the space diagonal **n**.

If we take $\sigma_1 \geq \sigma_2 \geq \sigma_3$ as is conventional, then a pure shear stress path is defined by $\sigma_1 = \tau$, $\sigma_2 = 0$, $\sigma_3 = -\tau$. Since $p = 0$, $s_1 = \tau$, $s_2 = 0$, $s_3 = -\tau$ on the π-plane. The Lode angle (Figure 6.5) defined in Equation 6.13 is zero on the pure shear stress path $s_2 = 0$ because $J_3(\equiv s_1 s_2 s_3)$ is zero there, and is counterclockwise positive.

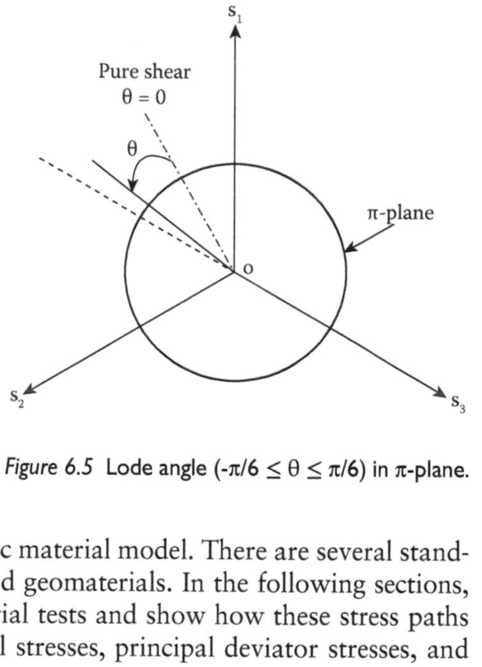

Figure 6.5 Lode angle ($-\pi/6 \leq \theta \leq \pi/6$) in π-plane.

6.7 STANDARD MATERIAL TESTS AND STRESS PATHS

In a material test, a specimen of the material is subjected to a predetermined stress path or a strain path, or a combination thereof. The objective is often to observe the material stress–strain response or to determine the parameters for a specific material model. There are several standard material tests that are performed for metals and geomaterials. In the following sections, we discuss the stress paths for some of these material tests and show how these stress paths can be conveniently described in terms of principal stresses, principal deviator stresses, and their invariants.

6.7.1 Representations of stress paths

Stress paths can be shown in a number of ways, each displaying some particular useful information.

- *Principal stress space* provides the most general means of displaying stress paths and yield and failure criteria.
- The *π-plane* is often used for displaying stress paths (and yield criteria) for metals because (compressive) mean stress essentially does not affect yield stress.

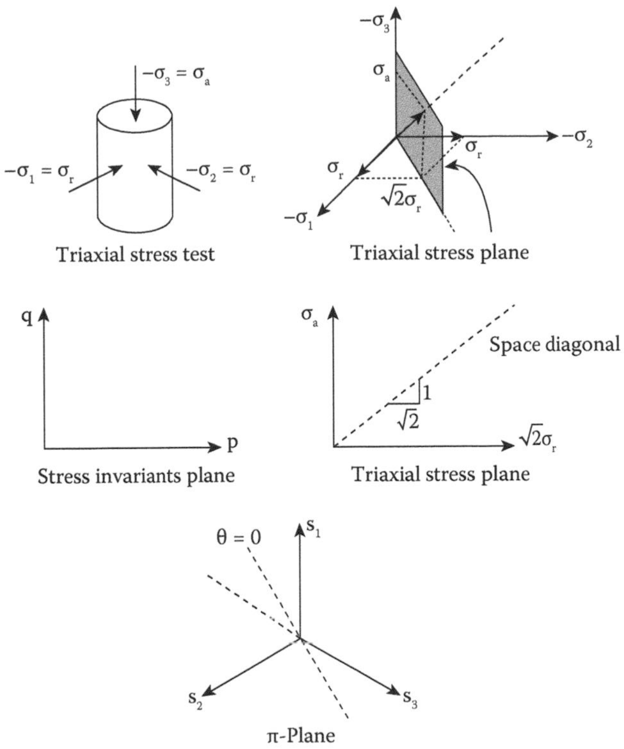

Figure 6.6 Representation of triaxial stress state.

- The *stress invariants plane* (q vs. |p|) can be used to show the effect of volumetric compressive stress on the deviatoric stress for a particular stress path, which provides very useful information for frictional materials.
- The *triaxial stress plane*, which contains the σ_3 axis and bisects the σ_1 and σ_2 axes in principal stress space, is often used to display the results of triaxial compression tests on geomaterials. We illustrate in Figure 6.6 a particular type of triaxial test in which a cylindrical specimen is subjected to an axisymmetric state of stress consisting of a radial compression σ_r and an axial compression σ_a, which are increased proportionally (other types of triaxial tests are described later in this chapter). Therefore, in principal stress space, triaxial compression stress paths are contained in the triaxial stress plane with axes σ_a and $\sqrt{2}\sigma_r$, as shown in Figure 6.6.

6.7.2 Uniaxial tests

6.7.2.1 Uniaxial tension

In a uniaxial test, a sample of the material is subjected to a single component of stress along the axis of the sample. In metals, this is the most common type of test. The tensile test specimen often is cylindrical in shape over a central gage length (L in Figure 6.7) and has enlarged "shoulders" at its ends. The sample is subjected to axial tension until it yields, and the test is usually continued to failure. Cyclic (loading and unloading) tests are sometimes also carried out. The results of the test are usually represented in the form of a (nominal) stress versus (engineering) strain curve. Typical results of uniaxial tension tests of metals are shown schematically in Figure 6.7.

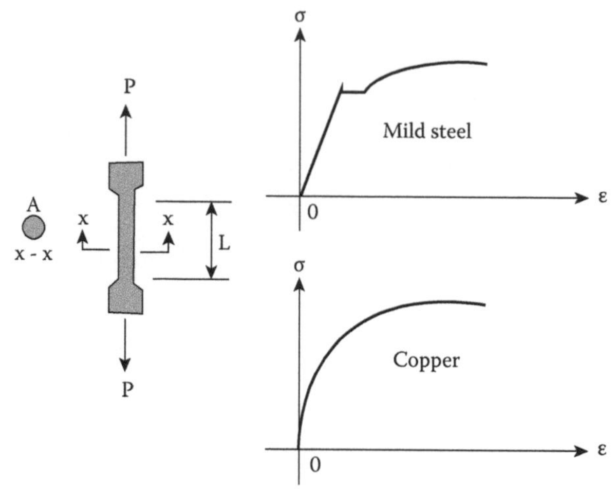

Figure 6.7 Typical behavior of metals in uniaxial tension tests (schematic).

If, in a Cartesian coordinate system, the z axis coincides with the axis of the specimen, then we have the following state of stress in a uniaxial tension test:

$$\begin{cases} \sigma_{xx} = \sigma_{yy} = 0 \\ \sigma_{zz} = \sigma \end{cases} \tag{6.27}$$

in which σ denotes the tensile axial stress. Since there is no shear stress acting on the specimen, the principal stresses are

$$\begin{cases} \sigma_1 = \sigma \\ \sigma_2 = 0 \\ \sigma_3 = 0 \end{cases} \tag{6.28}$$

Deviatoric stresses, invariants, and alternative invariants are as follows:

$$p = \frac{1}{3}(\sigma_1 + \sigma_2 + \sigma_3) = \frac{1}{3}\sigma \tag{6.29}$$

$$\begin{cases} s_1 = \sigma_1 - p = \dfrac{2}{3}\sigma \\ s_2 = \sigma_2 - p = -\dfrac{1}{3}\sigma \\ s_3 = \sigma_3 - p = -\dfrac{1}{3}\sigma \end{cases} \tag{6.30}$$

$$\begin{cases} J_2 = \dfrac{1}{2}(s_1^2 + s_1^2 + s_1^2) = \dfrac{1}{3}\sigma^2 \\ J_3 = s_1\, s_2\, s_3 = \dfrac{2}{27}\sigma^3 \end{cases} \tag{6.31}$$

$$q = \sqrt{3 J_2} = \sigma \tag{6.32}$$

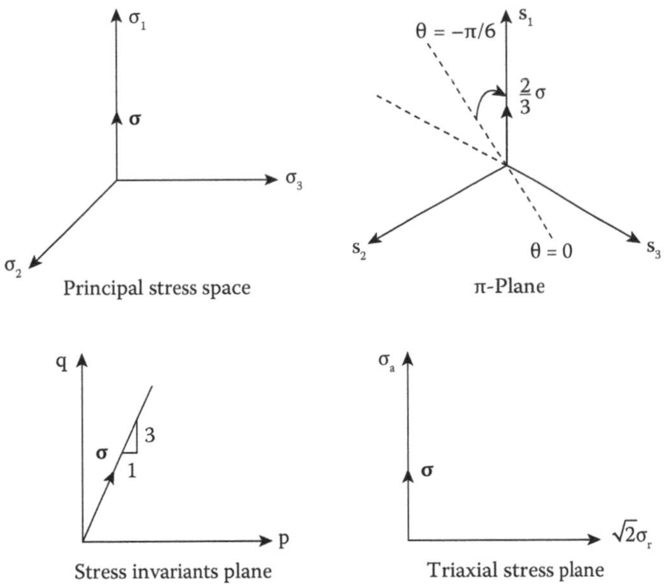

Figure 6.8 Representations of stress path in a uniaxial tension test.

The alternative set of invariants is

$$p = \frac{1}{3}\sigma; \quad q = \sigma; \quad \theta = -\frac{\pi}{6} \tag{6.33}$$

The stress path is shown in Figure 6.8. We note that for the uniaxial tension test, the stress path is a line along the positive s_1 axis in the π-plane.

6.7.2.2 *Uniaxial compression*

For geomaterials, such as soil, rock, and concrete, a cylindrical specimen is subjected to uniaxial compression. The axial compressive stress may be increased to failure or it may be unloaded and reloaded cyclically. In geomaterials, this type of test is often called an *unconfined compression test.* The result of a typical cyclic uniaxial test on plain concrete (Karsan and Jirsa 1969) is shown schematically in Figure 6.9.

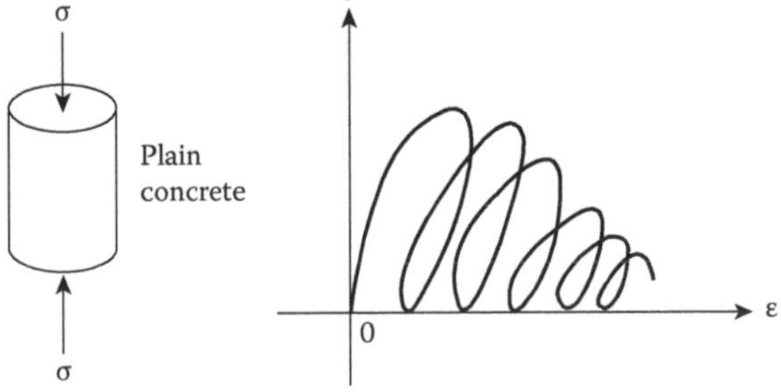

Figure 6.9 Cyclic uniaxial compression test on plain concrete.

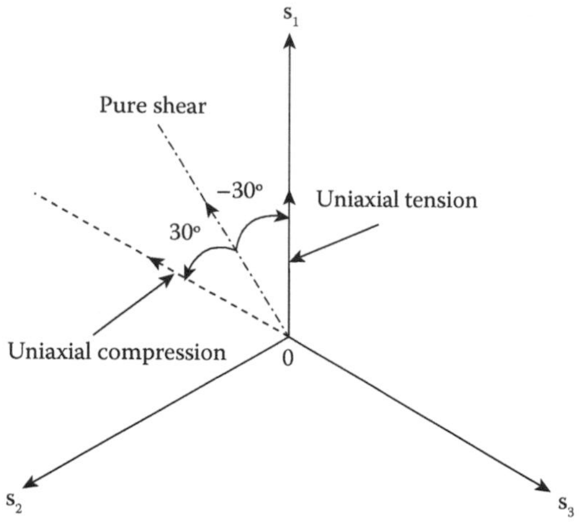

Figure 6.10 Stress paths in uniaxial tension, uniaxial compression, and pure shear.

For the uniaxial compression test,

$$\begin{cases} \sigma_1 = 0 \\ \sigma_2 = 0 \\ \sigma_3 = -\sigma \end{cases} \tag{6.34}$$

where σ is the compressive axial stress. We also have

$$\begin{cases} s_1 = \dfrac{1}{3}\sigma \\ s_2 = \dfrac{1}{3}\sigma \\ s_3 = -\dfrac{2}{3}\sigma \end{cases} \tag{6.35}$$

$$\begin{cases} J_2 = \dfrac{1}{3}\sigma^2 \\ J_3 = -\dfrac{2}{27}\sigma^3 \end{cases} \qquad \theta = \frac{\pi}{6} \tag{6.36}$$

The stress path in the π-plane for uniaxial compression is along the negative s_3 axis (i.e., $\theta = \pi/6$), as shown in Figure 6.10. The figure also shows for comparison the stress paths for uniaxial tension ($\theta = -\pi/6$) and pure shear ($\theta = 0$).

6.7.3 Biaxial tests

Kupfer et al. (1969) performed a well-known series of tests on two-dimensional plain concrete specimens under combinations of in-plane loads (i.e., principal stresses) using brush loading heads to ensure that lateral expansion of the specimen could take place without spurious

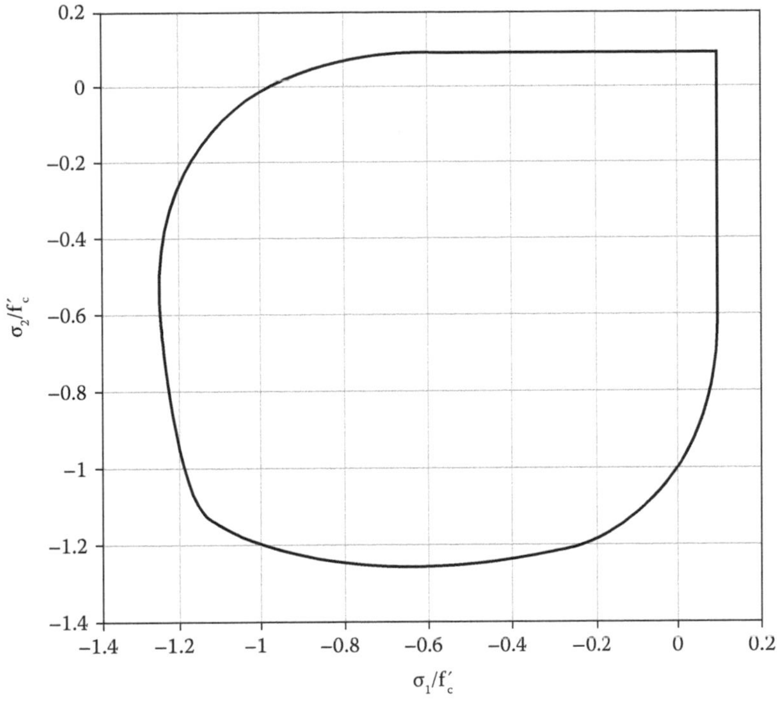

Figure 6.11 Biaxial yield and failure envelope for plain concrete (schematic).

shear stresses being induced at the specimen boundaries. Figure 6.11 shows the peak values of principal stresses σ_1 and σ_2 normalized by the uniaxial compressive strength f'_c.

6.7.4 Isotropic compression tests

In an isotropic compression test, usually done on geomaterials, the stresses applied on the sample are the same in all directions. The test is conducted either as a consolidation test or an isotropic compression prior to triaxial testing. In either case, if the porous sample is saturated, the test is done in a drained condition. The sample is subjected to equal all-around compressive stress σ. Therefore,

$$\sigma_{xx} = \sigma_{yy} = \sigma_{zz} = -\sigma \tag{6.37}$$

$$\sigma_1 = \sigma_2 = \sigma_3 = -\sigma \tag{6.38}$$

$$p = \frac{1}{3}(\sigma_1 + \sigma_2 + \sigma_3) = -\sigma \tag{6.39}$$

The principal deviatoric stresses, their invariants, and q are all zero, and the Lode angle is undefined. The stress paths are shown in Figure 6.12. Since the principal stress vector lies in the (negative) direction of the space diagonal, its projection on the π-plane is a point at the origin of the coordinate system.

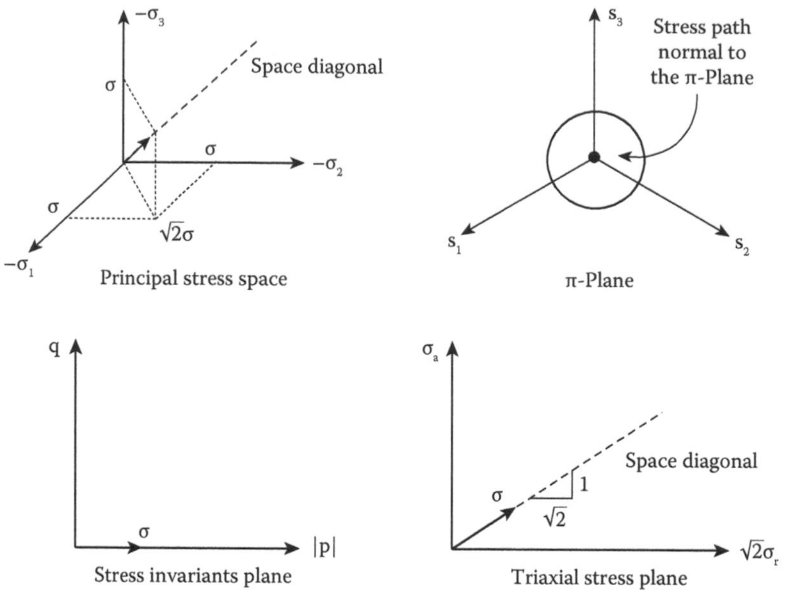

Figure 6.12 Stress path in an isotropic compression test.

6.7.5 Triaxial tests

Triaxial tests are usually performed on geomaterials such as soils, rocks, and concrete. A triaxial testing apparatus generally consists of a pressure chamber, together with an axial loading device. The cylindrical test specimen is first subjected to an all-around confining pressure σ_c, and an additional axial compressive stress σ can then be applied. Both the confining pressure σ_c and the axial stress σ can be varied. Several types of triaxial tests are possible, and each follows a specific stress path. Five well-known types of triaxial tests and their stress paths are discussed here.

1. **Type A_c.** The triaxial compression test with constant confining pressure, shown in Figure 6.13a. In this test, the all-around pressure is kept constant, and the axial compression is increased. That is, $\sigma_r = \sigma_c$ (= constant), $\sigma_a = \sigma_c + \sigma$.
2. **Type A_e.** The triaxial extension test with constant confining pressure, shown in Figure 6.13b. In this test, the all-around confining pressure σ_c is kept constant, and an increasing axial tensile stress σ is applied. That is, $\sigma_r = \sigma_c$ (= constant), $\sigma_a = \sigma_c - \sigma$.
3. **Type B_c.** The triaxial compression test with constant axial stress, shown in Figure 6.13c. In this test, the initial all-around confining pressure σ_{c0} is reduced by $\Delta\sigma_c$, so that on the loading path the confining pressure is $\sigma_c = \sigma_{c0} - \Delta\sigma_c$. The axial stress on the loading path is kept constant by applying a compressive stress σ equal to $\Delta\sigma_c$, so that the axial stress remains equal to the initial confining pressure σ_{c0}. The loading path can be described as $\sigma_r = \sigma_{c0} - \Delta\sigma_c$, $\sigma_a = \sigma_{c0} - \Delta\sigma_c + \sigma = \sigma_{c0}$ (constant).
4. **Type B_e.** The triaxial extension test with constant axial stress, shown in Figure 6.13d. In this test, the initial all-around confining pressure σ_{c0} is increased by $\Delta\sigma_c$. The axial stress (loading) is kept constant by applying $\sigma = -\Delta\sigma_c$, so that the axial stress remains equal to the initial confining pressure σ_{c0}. The loading path can be described as $\sigma_r = \sigma_{c0} + \Delta\sigma_c$, $\sigma_a = \sigma_{c0} + \Delta\sigma_c + \sigma = \sigma_{c0}$ (constant).

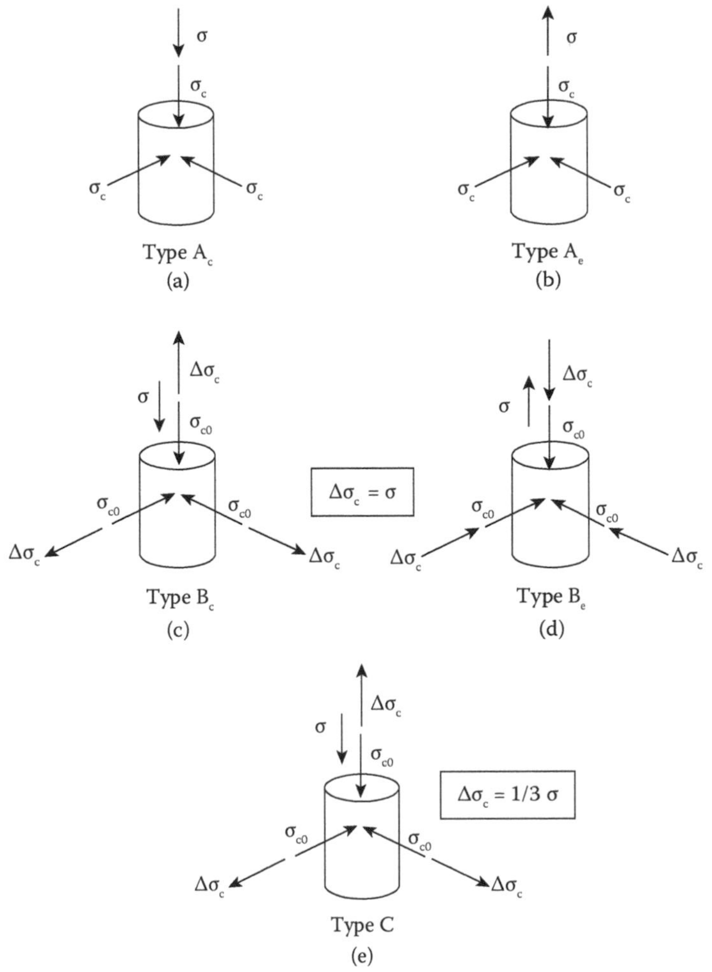

Figure 6.13 (a–e) Five different types of triaxial test.

5. **Type C.** The triaxial test with constant mean pressure, shown in Figure 6.13e. In this test, the initial all-around confining pressure σ_{c0} is reduced by $\Delta\sigma_c$, and the mean pressure p is kept constant by applying an additional axial compression σ to compensate for the reduction in radial pressure, such that $\Delta\sigma_c = \sigma/3$. The loading path can be described as $\sigma_r = \sigma_{c0} - \sigma/3$, $\sigma_a = \sigma_{c0} - \Delta\sigma_c + \sigma = \sigma_{c0} + 2\sigma/3$ (i.e., p = constant).

These five triaxial tests are shown schematically in Figure 6.13. The stresses, principal stresses, and stress invariants for the five types of triaxial tests discussed above are summarized in Table 6.1, and the representations of their stress paths are shown in Figure 6.14.

For illustrative purposes, we summarize the calculations of the stress invariants for the triaxial compression test with constant confining pressure (Type A_c) in Table 6.1. In this test, the stress components at a point in the sample are as follows:

$$\begin{cases} \sigma_{xx} = -\sigma_c \\ \sigma_{yy} = -\sigma_c \\ \sigma_{zz} = -(\sigma_c + \sigma) \end{cases} \tag{6.40}$$

Table 6.1 Stresses, principal stresses, and stress invariants for five triaxial tests

	Type A_c	*Type* A_e	*Type* B_c	*Type* B_e	*Type C*
σ_{xx}	σ_c	σ_c	$\sigma_{c0} - \sigma$	$\sigma_{c0} + \sigma$	$\sigma_{c0} - 1/3\sigma$
σ_{yy}	σ_c	σ_c	$\sigma_{c0} - \sigma$	$\sigma_{c0} + \sigma$	$\sigma_{c0} - 1/3\sigma$
σ_{zz}	$\sigma_c + \sigma$	$\sigma_c - \sigma$	σ_{c0}	σ_{c0}	$\sigma_{c0} + 2/3\sigma$
σ_1	$\sigma_c + \sigma$	σ_c	σ_{c0}	$\sigma_{c0} + \sigma$	$\sigma_{c0} + 2/3\sigma$
σ_2	σ_c	σ_c	$\sigma_c - \sigma$	$\sigma_{c0} + \sigma$	$\sigma_{c0} - 1/3\sigma$
σ_3	σ_c	$\sigma_c - \sigma$	$\sigma_c - \sigma$	σ_{c0}	$\sigma_{c0} - 1/3\sigma$
s_1	$2/3\sigma$	$1/3\sigma$	$2/3\sigma$	$1/3\sigma$	$2/3\sigma$
s_2	$-1/3\sigma$	$1/3\sigma$	$-1/3\sigma$	$1/3\sigma$	$-1/3\sigma$
s_3	$-1/3\sigma$	$-2/3\sigma$	$-1/3\sigma$	$-2/3\sigma$	$-1/3\sigma$
p	$\sigma_c + 1/3\sigma$	$\sigma_c - 1/3\sigma$	$\sigma_{c0} - 2/3\sigma$	$\sigma_{c0} + 2/3\sigma$	σ_{c0}
q	σ	σ	σ	σ	σ
θ	$-\pi/6$	$\pi/6$	$-\pi/6$	$\pi/6$	$-\pi/6$

Note: This table is valid for compression positive only.

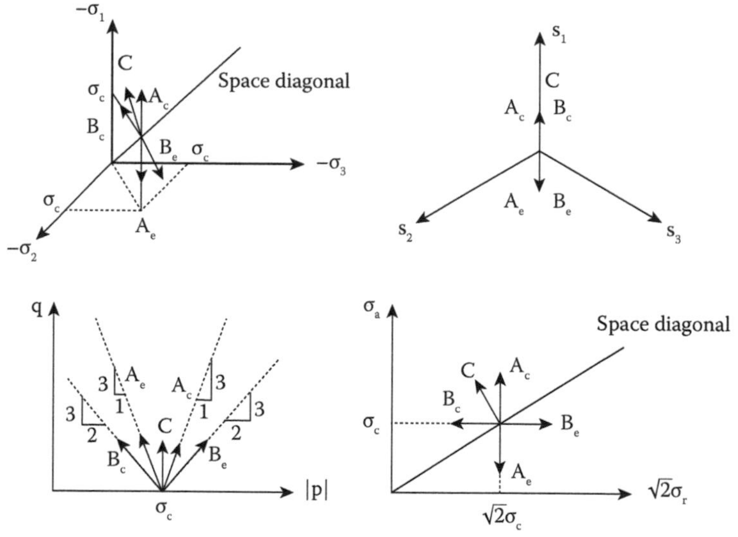

Figure 6.14 Triaxial stress paths corresponding to five types of triaxial test.

We have the following stresses and invariants:

$$\begin{cases} \sigma_1 = -\sigma_c \\ \sigma_2 = -\sigma_c \\ \sigma_3 = -(\sigma_c + \sigma) \end{cases} \tag{6.41}$$

$$p = -\sigma_c - \frac{1}{3}\sigma \tag{6.42}$$

$$\begin{cases} s_1 = \sigma_1 - p = \dfrac{1}{3}\sigma \\ s_2 = \sigma_2 - p = \dfrac{1}{3}\sigma \\ s_3 = \sigma_3 - p = -\dfrac{2}{3}\sigma \end{cases} \tag{6.43}$$

We note that the only difference between the uniaxial test and triaxial compression test is the existence of all-around compressive stress in the triaxial test. Consequently, the hydrostatic stress differs in these two tests. In the triaxial compression test deviatoric stresses, their invariants, q and θ, are the same as in the uniaxial test given in Equations 6.31 through 6.34. The stress path corresponding to the test is illustrated in Figure 6.14.

6.7.6 True triaxial tests

A number of different experimental methods have been developed that are capable of imposing three independent stresses or strains on geomaterial samples (Figure 6.15). A sample is normally subjected to an isotropic state of stress first and then subjected to independently varying individual stresses.

Special types of tests can be performed in a true triaxial test, for example, a plane strain test in which the strain in one direction is constrained to zero, and the stresses in the other two directions are varied; a pure shear test is another, as described next.

6.7.6.1 Pure shear test

The sample is first subjected to a hydrostatic all-around state of stress σ_c. Then the stress in one direction is increased by σ (e.g., the z direction), and in a second direction (e.g., the x direction) it is decreased by the same magnitude. The stress in the third direction is kept constant.

The stresses in the x-y-z directions are given by the following equations. Since there is no applied shear stress, these are principal stresses.

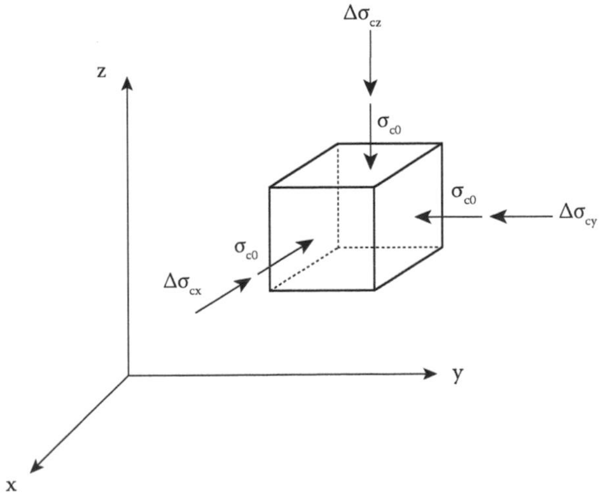

Figure 6.15 Typical true triaxial test conducted on a cubic specimen.

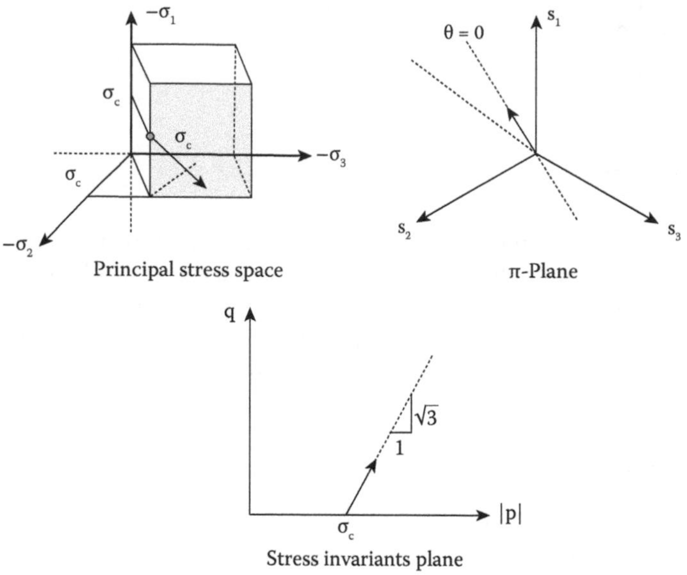

Figure 6.16 Stress path for pure shear test.

$$\begin{cases} \sigma_{xx} = \sigma_1 = -\sigma_c + \sigma \\ \sigma_{yy} = \sigma_2 = -\sigma_c \\ \sigma_{zz} = \sigma_3 = -\sigma_c - \sigma \end{cases} \tag{6.44}$$

The hydrostatic pressure $p = -\sigma_c$. Deviatoric stresses, their invariants, and the alternative invariants are given in the following equations:

$$\begin{cases} s_1 = +\sigma \\ s_2 = 0 \\ s_3 = -\sigma \end{cases} \tag{6.45}$$

$$\begin{cases} J_2 = \frac{1}{2}\left(s_1^2 + s_1^2 + s_1^2\right) = \sigma^2 \\ J_3 = s_1\, s_2\, s_3 = 0 \end{cases} \tag{6.46}$$

$$\begin{cases} p = -\sigma_c \\ q = \sqrt{3}\sigma \\ \theta = 0 \end{cases} \tag{6.47}$$

The stress path for the pure shear test with $\sigma_c = 0$ is shown in Figure 6.16.

Chapter 7

Elastoplastic material models

7.1 INTRODUCTION

The most salient feature of elastic material models is that there exists a unique one-to-one relationship between stress and strain; that is, the current state of stress depends only on the current state of deformation. This is clearly a far-reaching idealization of material behavior.

Many materials, such as metals, composites, and geomaterials, exhibit inelastic characteristics even for small deformations. Significant path-dependent material response may include plastic deformation, which is time independent, as well as viscoelastic or viscoplastic deformation, which are time dependent. We here use the term *plastic deformation* to refer to (rate-independent) permanent deformation that remains after the removal of loading. In conventional plasticity material models, time appears as a pseudoloading parameter. Creep and stress relaxation processes, on the other hand, are inherently time dependent.

Sometimes, for convenience, we tend to classify material behavior as brittle or ductile, depending on the amount of deformation the material can sustain prior to failure, which is obviously an extremely important attribute for materials of construction.

If a material has little capacity to deform before reaching failure, it is termed brittle. In this case, since very little plastic deformation can occur, we can with reasonable accuracy employ an elastic material model (linear or nonlinear). Some materials occurring in nature, such as rock and ice, tend to exhibit a brittle type of behavior.

A material is said to be ductile if it is capable of undergoing large deformations before reaching failure. An important benefit in an engineered structure is the potential for stress redistribution prior to structural failure, rather than a sudden "weakest-link" type of failure.

Structural steel is an example of the kind of engineering material that exhibits ductile behavior. Of course, the boundary between a brittle deformation and a plastic deformation is not clearly defined in an absolute sense.

There are two established viewpoints on studying the plastic behavior of materials. The phenomenological view considers material behavior on a macroscale, and the plastically deformable body is viewed as a continuum. This is basically the realm of the classical mathematical theory of plasticity that deals with the development of idealized mathematical models of the observed behavior of materials. It has its origin in early twentieth-century work in generalizing the observed uniaxial behavior of materials to multiaxial stress states (Hill 1950; Drucker and Prager 1952; Martin 1975).

The physical viewpoint considers material behavior on a microscale, which forms the basis of disciplines such as dislocation theory and particle mechanics. Ideally, we strive to use physical understanding of the material behavior on the microscale as the basis for mathematical modeling of inelastic material behavior.

In this chapter, we present an introduction to the mathematical theory of plasticity. In developing plasticity-based constitutive modeling of material behavior and generalizing from observed uniaxial behavior to multiaxial stress states, we have to ensure that certain basic laws of mechanics are satisfied. We have already discussed the requirements of material frame indifference relative to deformation and stress and their rates. Additionally, constitutive models must obey certain work inequalities that ensure that, for example, spurious energy creation cannot occur over an arbitrary closed stress cycle.

In the following sections, we present an introduction to the rate-independent, isothermal theory of plasticity.

7.2 BEHAVIOR OF METALS UNDER UNIAXIAL STRESS

As discussed earlier, uniaxial stress tests on metals are probably the simplest and most commonly performed material tests. A typical stress–strain curve for mild steel subjected to a stress cycle of loading and unloading is shown in Figure 7.1. Also, shown in the same figure is an idealized uniaxial stress–strain curve that is representative of the behavior of many materials.

It can be observed that there is a well-defined yield stress σ_Y and that the material behavior is different below and above the yield stress. Metals seem to behave elastically below the yield stress. When the stress is increased beyond the yield stress, a permanent irrecoverable strain develops. After exceeding the yield stress, if the stress is reduced to zero, the strain does not disappear. The remaining (residual) strain is called the inelastic strain or plastic strain.

It can be observed from the uniaxial stress–strain curves that unloading is essentially elastic. Reloading also seems to be elastic up to the previous maximum stress from which the unloading started. This indicates that the yield stress is not constant, but that it changes when the material is stressed beyond the initial yield stress. In the case of a uniaxial and unidirectional stress state, the yield stress is approximately equal to the maximum stress reached up to that point. Another important consequence of this observation is that unlike elastic behavior, the inelastic behavior of materials is path dependent. The current state of stress, by itself, does not determine the subsequent material behavior. The past history of loading and unloading has a permanent influence on the material behavior. It is for this reason that materials exhibiting elastoplastic behavior are sometimes called materials with memory. Materials exhibiting time-dependent elastoplastic behavior are sometimes called materials with fading memory.

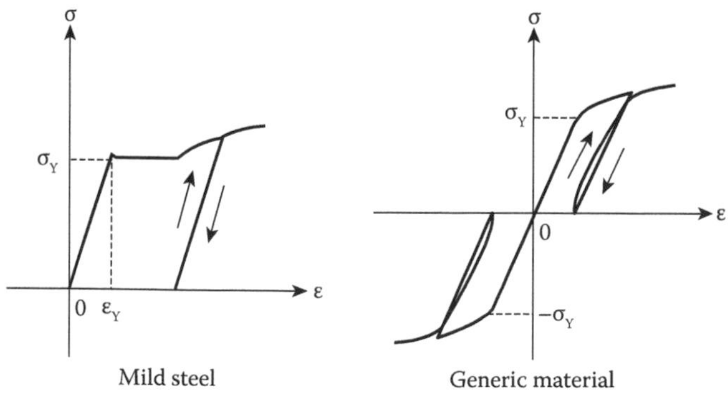

Figure 7.1 Typical uniaxial stress–strain curves.

7.2.1 Fundamental assumptions of classical plasticity theory

These observations of uniaxial tests of metals lead to two of the basic assumptions of the classical theory of plasticity.

The first is that the strain rates can be additively decomposed into an elastic part and a plastic part.

$$d\varepsilon = d\varepsilon^e + d\varepsilon^p \tag{7.1}$$

Strictly speaking, this additive decomposition of *incremental* strain implies that the elastic strains are "small" in some sense. Elastic strains will truly be small in most cases where a plasticity material model is appropriate; however, large plastic strains can occur in practically important stress analysis situations. We note that multiplicative decomposition of the deformation gradient, that is, $\mathbf{F} = \mathbf{F}_e\ \mathbf{F}_p$, has been suggested (e.g., Lee 1981) as a basis for large strain plasticity formulations.

The second observation is that *stress rates* are related to the *elastic* part of the strain rate, through *elastic* stress–strain relations:

$$d\boldsymbol{\sigma} = \mathbf{D}^e : d\boldsymbol{\varepsilon}^e = \mathbf{D}^e : (d\boldsymbol{\varepsilon} - d\boldsymbol{\varepsilon}^p) \tag{7.2}$$

We remark that Equation 7.2 is actually a *hypoelastic* material model, which again will require that the elastic strains be small. In the formulation that follows, we will think of the stress as unrotated Cauchy stress $\bar{\boldsymbol{\sigma}}'$ and the strain rates as the unrotated deformation rate $\mathbf{d}$ (or Cauchy stress and deformation rate, as appropriate), although we shall not explicitly indicate that in order to simplify the notation.

7.3 INELASTIC BEHAVIOR UNDER MULTIAXIAL STATES OF STRESS

In extending the concept of the yield stress to multiaxial states of stress, we must first determine what combination of stresses will cause yielding. This combination of stresses is characterized via a yield surface, as shown in Figure 7.2.

There are two related concepts involved: the initial yield surface and the subsequent evolution of the yield surface as a result of plastic deformation. The initial yield surface determines the combination of stresses that will cause first yielding. The subsequent evolution of the yield

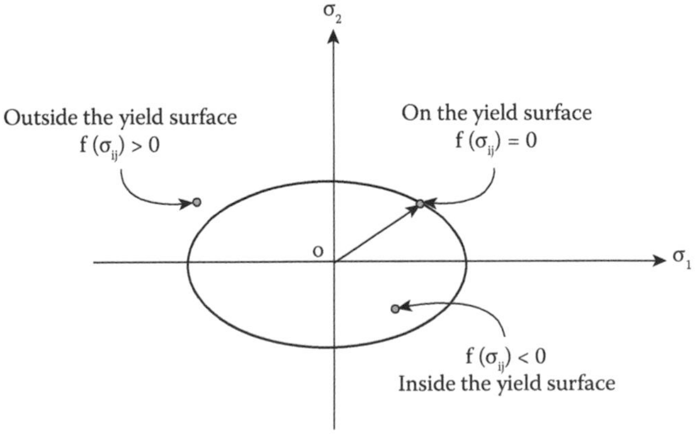

Figure 7.2 Representation of a typical yield surface.

surface after the onset of inelasticity is described by hardening rules. The so-called flow rules determine the direction and magnitude of changes in the plastic strain. These rules, along with some basic principles, which will be described later, enable the development of incremental, elastoplastic stress–strain relations.

7.3.1 Yield surface

Consider a virgin material that is probed along radial (proportional loading) stress paths, starting from the origin in stress space. Along each stress path, the point where first yielding occurs is observed. These points form a surface in the stress space that is called the yield surface. A generic yield surface is shown schematically in Figure 7.2 in a two-dimensional stress space.

The yield surface can be expressed as

$$f(\sigma_{ij}) = 0 \tag{7.3}$$

such that for stresses inside the yield surface, $f(\sigma_{ij}) < 0$, and for stresses outside the yield surface, $f(\sigma_{ij}) > 0$.

For stress changes inside the yield surface, either loading or unloading, the behavior is elastic. Plastic deformations start occurring when the stress point reaches the yield surface. Points outside the yield surface are inaccessible unless the material hardens. In that case, with increasing inelastic deformation the yield surface evolves and its position, shape, and size can change, depending on the history of stresses and strains.

At this point, we introduce some history parameters $\mathbf{H}(\varepsilon^p)$, which are related to the plastic strain, and which provide the rules for the evolution of the yield surface. Therefore, the general form of the subsequent yield surface is given by

$$f(\sigma_{ij}, \mathbf{H}) = 0 \tag{7.4}$$

The evolution of the yield surface is shown schematically in Figure 7.3 in a two-dimensional stress space.

As shown in the figure, the stress vector reaches the initial yield surface at point A. From that point on, the plastic strains change, which results in changes in the history parameters **H**. The shape, size, and position of the yield surface change in such a way that the stress point

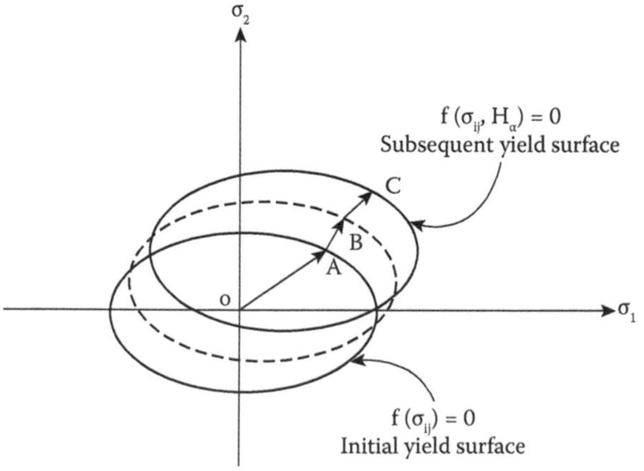

Figure 7.3 Evolution of yield surface.

always remains on the yield surface. This is referred to as the *consistency condition*. Therefore, on the subsequent yield surface, we have

$$f(\boldsymbol{\sigma} + d\boldsymbol{\sigma}, \mathbf{H} + d\mathbf{H}) = f(\boldsymbol{\sigma}, \mathbf{H}) + df = 0 \tag{7.5}$$

for *incremental* $d\boldsymbol{\sigma}$ and $d\mathbf{H}$. The consistency condition therefore requires $df = 0$. That is,

$$df = \frac{\partial f}{\partial \sigma_{ij}} d\sigma_{ij} + \frac{\partial f}{\partial H_k} dH_k = 0 \tag{7.6}$$

and the evolution of the yield surface is described by a *hardening rule*. Specific hardening rules will be discussed later.

Since the yield surface is a *generalization* of the yield point in uniaxial tests to multiaxial stress space, the shape of the yield surface must be consistent with observed simple material tests, including the uniaxial stress test. The shape of the yield surface is further constrained by certain symmetry requirements, invariance under coordinate rotation, conservation laws, and material stability conditions.

7.4 WORK AND STABILITY CONSTRAINTS

Several work–energy and stability constraints, that is, inequalities, have been proposed that provide a general framework for multiaxial generalization of uniaxial behavior. As a result, certain behaviors are excluded from the idealized elastoplastic material models, some of which may have physically impossible or undesired consequences. A requirement that has already been mentioned is that in a closed stress cycle (Figure 7.4), the material model must not create (spurious) energy.

7.4.1 Work and energy

The work done by stresses acting on the material in a closed stress cycle (Figure 7.4) must be nonnegative; that is, work is required to produce inelastic deformation. Therefore, if inelastic deformations are produced, the work done by the stresses must be positive, meaning that there is a net energy loss over the stress cycle.

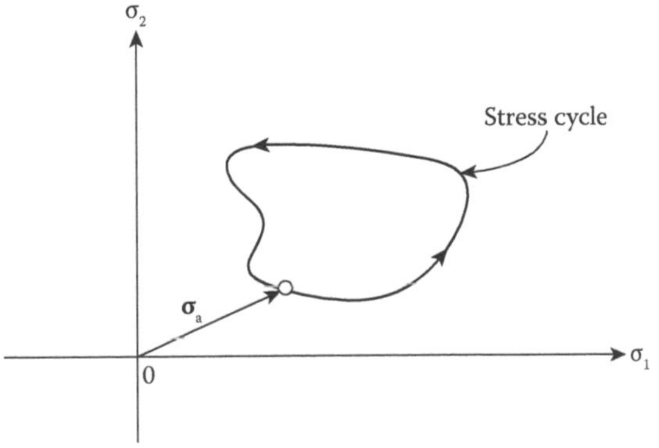

Figure 7.4 Closed multiaxial stress cycle (schematic).

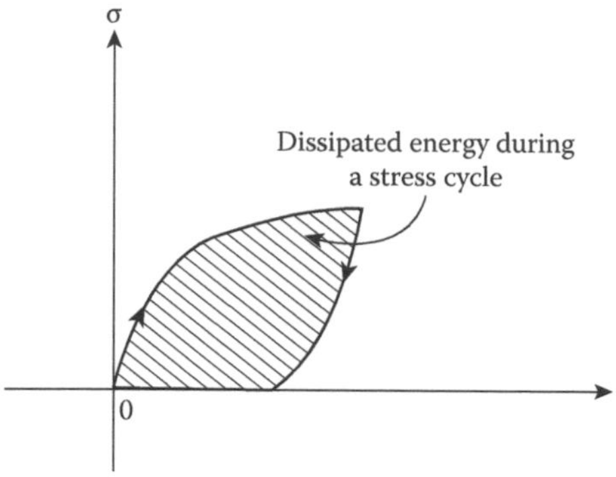

Figure 7.5 Stress–strain curve for a closed uniaxial stress cycle.

If no inelastic deformation is produced during the stress cycle, the work is stored as elastic strain energy, which is recovered on completion of the stress cycle, and so in this case, the net work done by the stresses over the cycle is zero.

Figure 7.5 illustrates a hypothetical uniaxial stress cycle in which positive work is done by the stress, accompanied by residual inelastic deformation, over the complete stress cycle.

Figure 7.6 shows the work done by the stress in separate stages of the uniaxial stress cycle. In the loading stage, both elastic and inelastic strains are produced. In the unloading portion of the cycle, elastic strain energy is recovered.

The work inequality is expressed in multiaxial form as

$$\oint_{\sigma} \sigma_{ij} d\varepsilon_{ij} \geq 0 \tag{7.7}$$

where the integral is over an arbitrary closed cycle in stress space.

A second postulate, termed "stability in the small" by Drucker, defines material stability in a specific way. When generalized to the multiaxial stress case, the concept provides an additional fundamental inequality that is used in the formulation of the elastoplastic constitutive relations.

7.4.2 Stability in the small

The uniaxial "material stiffness," that is, the slope of the tangent to the stress–strain curve, is a measure of the "stability" of the material at that point. If the slope becomes negative at some point, it indicates that the material response is no longer stable, in the sense that the current stress cannot be sustained. Some examples of "unstable" behavior in this sense are shown in Figure 7.7.

The unstable behaviors illustrated in Figure 7.7 are both characterized by the condition $d\sigma\ d\varepsilon < 0$. Therefore, the requirement that $d\sigma\ d\varepsilon \geq 0$ precludes both unstable portions of the uniaxial responses shown there. This requirement is generalized to the multiaxial case in the form

$$d\sigma_{ij} d\varepsilon_{ij} \geq 0 \tag{7.8}$$

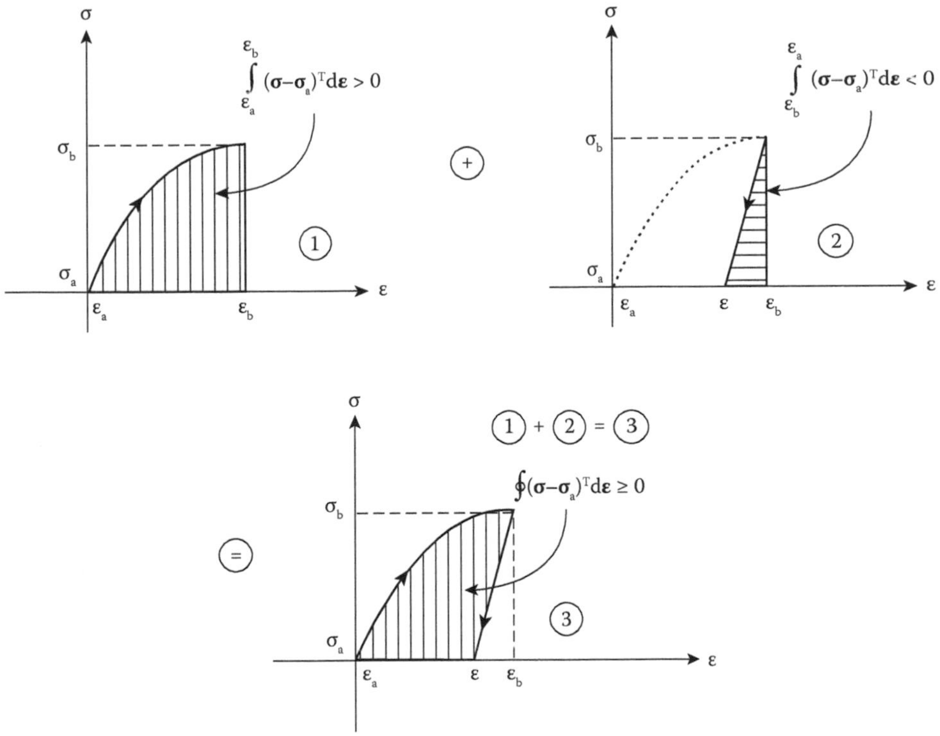

Figure 7.6 Examples of uniaxial closed stress cycles.

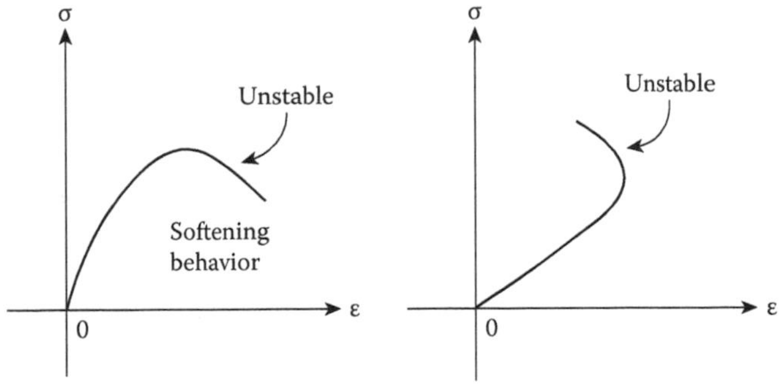

Figure 7.7 Examples of unstable behavior that violate the condition of stability in the small.

If we recast Equation 7.8 in the form

$$d\varepsilon_{ij} D^t_{ijkl} d\varepsilon_{kl} \geq 0 \tag{7.9}$$

where $\mathbf{D}^t$ is the (instantaneous) material tangent stiffness, we see that Equation 7.9 is the requirement that the material tangent stiffness be positive semidefinite.

Equation 7.9 of course applies in the neighborhood of a material point, that is, for a condition of homogeneous stress and strain. However, material models that violate Equation 7.9, that is, that soften, can be used in structural-level simulations because of the possibility of stress redistribution.

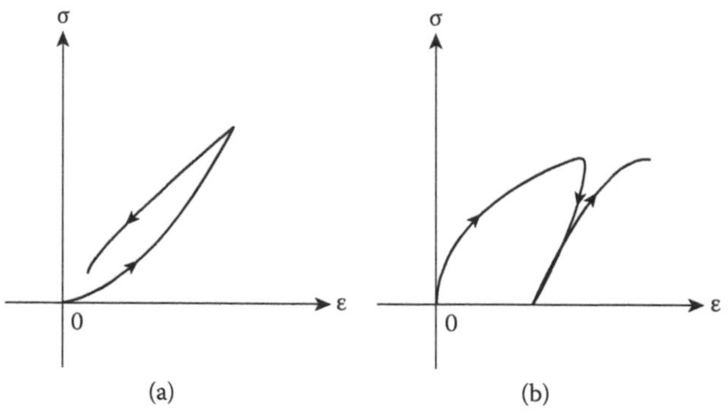

Figure 7.8 (a, b) Types of uniaxial material behavior excluded by the complementary work inequality.

There are of course many important physical processes in which softening occurs, for example, in the realm of ductile fracture, due to the nucleation, growth, and coalescence of microvoids. Phenomenological plasticity models incorporating a shrinking yield surface based on the early work by Gurson (1975, 1977) are but one example.

The inequalities shown in Equations 7.7 and 7.8 may be replaced by alternative expressions that are essentially equivalent and that may be more useful for some purposes.

7.4.3 Complementary work

The restriction of Equation 7.7 can be replaced by an essentially equivalent inequality involving complementary work for an arbitrary closed stress cycle:

$$\oint_{\sigma} \varepsilon_{ij} d\sigma_{ij} \leq 0 \tag{7.10}$$

However, Equation 7.10 additionally precludes the uniaxial behavior illustrated in Figure 7.8b. For the stress cycle starting at the peak stress, followed by unloading and reloading back to the same previous maximum value, the complementary work is positive, violating Equation 7.10. Of course, in the framework of classical plasticity, unloading and reloading are assumed to be elastic inside the yield surface.

The behavior shown in Figure 7.8a is clearly unrealistic, and it obviously violates the complementary work inequality. The fact that the behavior shown in Figure 7.8b also violates the complementary work inequality may not be so obvious; it is illustrated in Figure 7.9. This type of behavior is sometimes encountered in poorly formulated material models.

Figure 7.10 illustrates the complementary work inequality in a uniaxial stress–strain plot.

7.4.4 Net work

We now consider work and complementary work done for an arbitrary path in stress space from point a, at which the stress is σ_{ij}^{a} and the initial strain is ε_{ij}^{a} to point b, at which the stress is σ_{ij}^{b} and the strain is ε_{ij}^{b}. For this arbitrary path,

$$\int_{a}^{b} d(\sigma_{ij}\varepsilon_{ij}) = \int_{a}^{b} \varepsilon_{ij} d\sigma_{ij} + \int_{a}^{b} \sigma_{ij} d\varepsilon_{ij} \equiv \sigma_{ij}^{b}\varepsilon_{ij}^{b} - \sigma_{ij}^{a}\varepsilon_{ij}^{a} \tag{7.11}$$

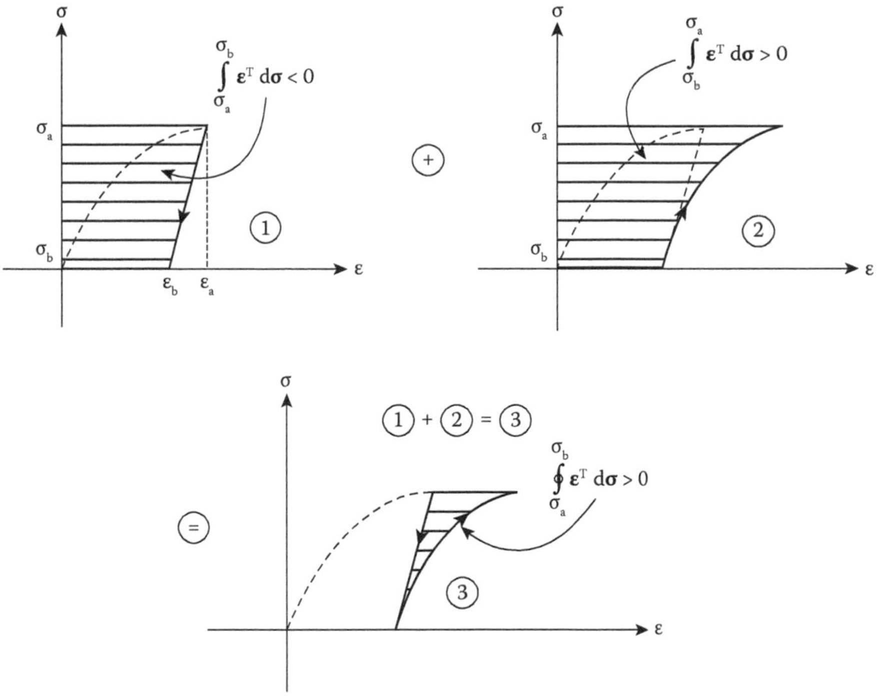

Figure 7.9 Violation of the complementary work inequality in Figure 7.8b.

If we consider the case in which point b is the original starting point a, that is, a complete stress cycle in stress space, then the right-hand side of Equation 7.11 can be written as

$$\sigma_{ij}^{a}(\varepsilon_{ij}^{A} - \varepsilon_{ij}^{a}) = \sigma_{ij}^{a} \int_{\varepsilon^{a}}^{\varepsilon^{A}} d\varepsilon_{ij} \tag{7.12}$$

in which ε_{ij}^{A} is the final strain at point a upon completion of the stress cycle. Note that the integral in Equation 7.12 is evaluated along the path in stress space. Equation 7.11 can now be written as

$$\oint \varepsilon_{ij} d\sigma_{ij} + \oint (\sigma_{ij} - \sigma_{ij}^{a}) d\varepsilon_{ij} = 0 \tag{7.13}$$

Therefore, in view of Equation 7.10,

$$\oint (\sigma_{ij} - \sigma_{ij}^{a}) d\varepsilon_{ij} \geq 0 \tag{7.14}$$

By considering the stress cycle in Equation 7.14 to consist of an arbitrary path from point a to point b, followed by a straight-line return (in stress space), that is, proportional (un)loading, to point a, it can be shown that for *any arbitrary path in stress space* (not necessarily a stress cycle) beginning at point a with stress σ_{ij}^{a},

$$\int_{\sigma^{a}}^{\sigma^{b}} (\sigma_{ij} - \sigma_{ij}^{a}) d\varepsilon_{ij} \geq 0 \tag{7.15}$$

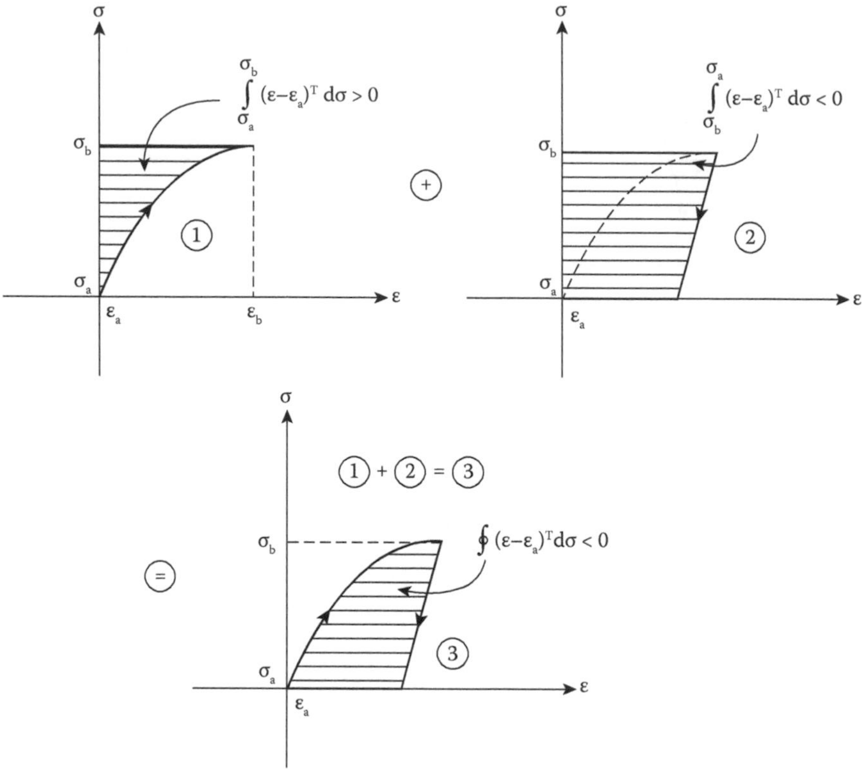

Figure 7.10 Complementary work inequality over a closed stress cycle.

The work integral in Equation 7.15 is termed the *net work* (Martin 1975). It provides the basis for important results regarding the shape of the yield surface and the direction of plastic flow.

The behavior of real materials, especially geomaterials, over a closed stress cycle is typically of the form shown in Figure 7.11. In this figure, the area A represents the energy dissipated by the material and the area B represents the energy generated by the material. The area A is always larger than the area B, so that the net value of the complementary energy over a closed cycle is always negative, in compliance with the complementary work inequality, Equation 7.10.

7.5 ASSOCIATED PLASTICITY MODELS

We now apply the work and stability postulates to develop important results related to the shape of the yield surface and the direction of plastic flow.

7.5.1 Drucker's postulate

We start with the net work inequality over a closed stress cycle, Equation 7.15. Consider the stress cycle shown in Figure 7.12, which consists of three segments.

The first segment is from a stress $\boldsymbol{\sigma}_a$ inside the yield surface to a stress $\overline{\boldsymbol{\sigma}}$ on the current yield surface. On this segment, the elastic strain $\boldsymbol{\varepsilon}^e$ changes, but the plastic strain remains equal to $\boldsymbol{\varepsilon}_a^p$,

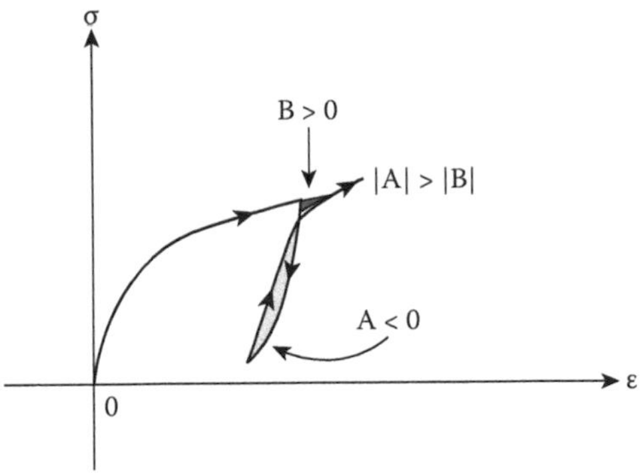

Figure 7.11 Typical material behavior over a closed stress cycle.

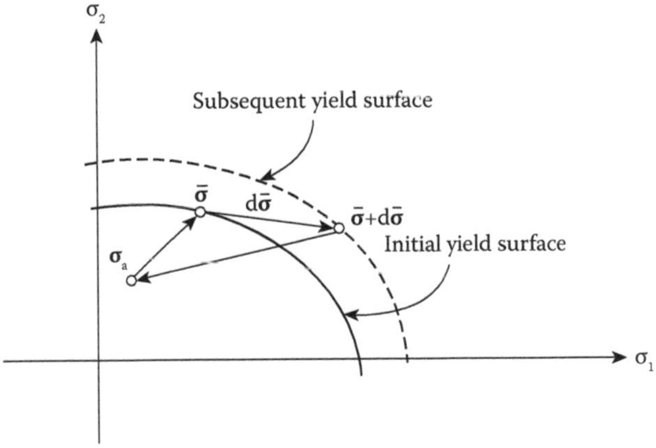

Figure 7.12 Net work inequality over a closed stress cycle.

its value at point a, since the segment is entirely within the yield surface. The second segment is an infinitesimal increment of stress from $\bar{\boldsymbol{\sigma}}$ to $\bar{\boldsymbol{\sigma}} + d\bar{\boldsymbol{\sigma}}$, during which additional plastic strains $d\boldsymbol{\varepsilon}^p$, as well as elastic strains $d\boldsymbol{\varepsilon}^e$, are generated. The stress $\bar{\boldsymbol{\sigma}} + d\bar{\boldsymbol{\sigma}}$ is on the subsequent yield surface, which evolves due to the plastic strain $d\boldsymbol{\varepsilon}^p$. The third segment is from $\bar{\boldsymbol{\sigma}} + d\bar{\boldsymbol{\sigma}}$ back to the initial stress $\boldsymbol{\sigma}_a$ at point a, during which the behavior is elastic. The net work over the complete stress cycle can be written as

$$\begin{aligned}\oint (\sigma_{ij} - \sigma_{ija}) d\varepsilon_{ij} &= \int_{\boldsymbol{\sigma}_a}^{\bar{\boldsymbol{\sigma}}} (\sigma_{ij} - \sigma_{ija}) d\varepsilon_{ij}^e + \int_{\bar{\boldsymbol{\sigma}}}^{\bar{\boldsymbol{\sigma}} + d\bar{\boldsymbol{\sigma}}} (\sigma_{ij} - \sigma_{ija})(d\varepsilon_{ij}^e + d\varepsilon_{ij}^p) \\ &+ \int_{\bar{\boldsymbol{\sigma}} + d\bar{\boldsymbol{\sigma}}}^{\boldsymbol{\sigma}_a} (\sigma_{ij} - \sigma_{ija}) d\varepsilon_{ij}^e \\ &= \oint (\sigma_{ij} - \sigma_{ija}) d\varepsilon_{ij}^e + \int_{\bar{\boldsymbol{\sigma}}}^{\bar{\boldsymbol{\sigma}} + d\bar{\boldsymbol{\sigma}}} (\sigma_{ij} - \sigma_{ija}) d\varepsilon_{ij}^p \geq 0\end{aligned} \tag{7.16}$$

We note that strains that remain constant on a segment of the stress cycle do not enter into the net work evaluation for that particular segment. Equation 7.16 reduces to

$$\int_{\bar{\sigma}}^{\bar{\sigma}+d\bar{\sigma}} (\sigma_{ij} - \sigma_{ija}) d\varepsilon_{ij}^{p} \geq 0 \tag{7.17}$$

and in the limit as $d\bar{\sigma}$ is made vanishingly small,

$$(\bar{\sigma}_{ij} - \sigma_{ija}) d\bar{\varepsilon}_{ij}^{p} \geq 0 \tag{7.18}$$

in which $d\bar{\varepsilon}^{p}$ is the plastic strain increment resulting from the stress increment $d\bar{\boldsymbol{\sigma}}$. We note that this inequality applies for any stress $\boldsymbol{\sigma}_a$ within or on the yield surface and any stress $\bar{\boldsymbol{\sigma}}$ on the yield surface.

We note that Equation 7.18 can also be obtained via the application of the complementary work inequality to the same stress cycle. Equation 7.18 has very significant consequences, which we consider next.

7.5.2 Convexity and normality

Satisfaction of Drucker's postulate without any further restrictions places constraints on the direction of plastic flow and on the shape of the yield surface.

We remark first that in the absence of hardening (i.e., ideal plasticity), $d\sigma_{ij}$ is tangent to the yield surface. Since stability in the small requires that during plastic flow, $d\sigma_{ij} d\varepsilon_{ij}^{p} \geq 0$, the direction of the plastic strain increment must be such that it has a nonnegative projection on the normal to the yield surface.

We now consider a smooth yield surface in stress space, as shown schematically in Figure 7.13.

In stress space, $\bar{\boldsymbol{\sigma}} - \boldsymbol{\sigma}_a$ is a vector from any point $\boldsymbol{\sigma}_a$ within, or possibly on, the yield surface to a generic point $\bar{\boldsymbol{\sigma}}$ on the yield surface. Two of the infinitely many possible positions of $\boldsymbol{\sigma}_a$ are shown. Also shown in the same figure is a plastic strain increment vector $d\bar{\boldsymbol{\varepsilon}}^{P}$ (in an arbitrary outward direction) superimposed on the yield surface. We note here that the elastic constitutive tensor $\mathbf{D}^{e}$ maps strains from strain space into stress space.

In order for the inner product of these two vectors to *always* be nonnegative, the angle between the two vectors must be less than or equal to $\pi/2$, and this in turn requires that

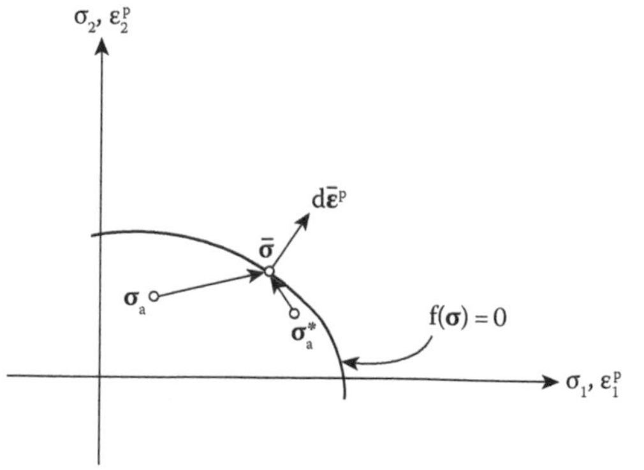

Figure 7.13 Convexity and normality from Equation 7.18.

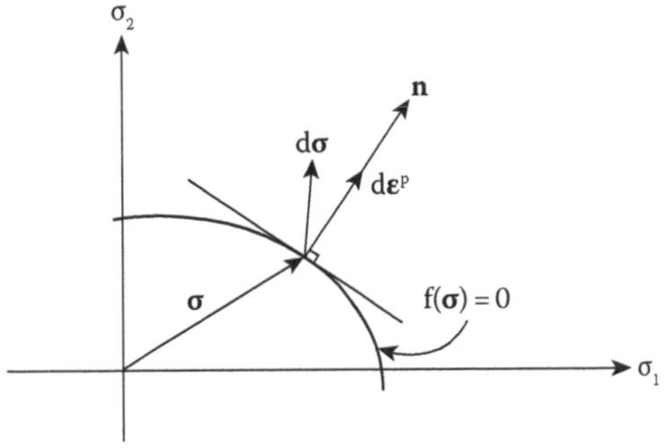

Figure 7.14 Illustration of the normality rule in Drucker's postulate.

the yield surface be convex and that the plastic strain increment be normal to the yield surface (Figure 7.14), as elaborated next.

7.5.2.1 *Normality of the plastic strain increment*

It is evident that if the plastic strain increment is in the direction normal to the yield surface, that is,

$$d\varepsilon_{ij}^{p} = d\lambda \frac{\partial f}{\partial \sigma_{ij}} \tag{7.19}$$

where $d\lambda \geq 0$, then the requirement of stability in the small is met.

7.5.2.2 *Convexity of the yield surface*

If Equation 7.19 is adopted, then Equation 7.18 becomes

$$\frac{\partial f}{\partial \overline{\sigma}_{ij}}(\overline{\sigma}_{ij} - \sigma_{ija}) \geq 0 \tag{7.20}$$

that is, the yield surface is *convex*. A convex set in Euclidean space is defined as one in which a straight line connecting any two members of the set (e.g., $\boldsymbol{\sigma}_a$ and $\boldsymbol{\sigma}_b$) remains in the set. In other words, if $\boldsymbol{\sigma}_a$ and $\boldsymbol{\sigma}_b$ are members of a convex set $f(\boldsymbol{\sigma}) \leq 0$, then $\boldsymbol{\sigma}_c = (1 - \alpha)\ \boldsymbol{\sigma}_a + \alpha\ \boldsymbol{\sigma}_b$, where $0 \leq \alpha \leq 1$ is also a member of that convex set. We have essentially used this property in some of the previous work and complementary work developments.

Therefore, normality of the plastic strain increment and the consequent convexity of the yield surface are *sufficient* to satisfy the work–energy and material stability criteria that have been discussed. We have not, however, shown that they are *necessary*.

The formulation that has been presented here is known as *associated plasticity*.

Subsequent studies have shown that Drucker's postulate can be satisfied without invoking the normality rule (Mroz 1963). In fact, experiments have shown that many real materials do not obey the normality rule at all. This point will be further discussed in a later section.

A more general *nonassociated plasticity* theory introduces, in addition to a yield surface, a plastic potential $g(\boldsymbol{\sigma})$, whose outward normal defines the direction of plastic flow; that is, $d\varepsilon_{ij}^{p} = d\lambda\, \partial g / \partial \sigma_{ij}$ replaces the associated flow rule, Equation 7.19.

7.6 INCREMENTAL STRESS–STRAIN RELATIONS

Incremental stress–strain relations are needed for elastoplastic finite element analysis. Our objective in this section is to develop an expression for the incremental elastoplastic constitutive matrix.

7.6.1 Equivalent uniaxial stress and plastic strain

We will find it convenient to develop the necessary relations in tensor form, and then to convert the results into matrix–vector form for computational purposes.

The following notation for the outward normal to the yield surface is introduced:

$$n_{ij} \equiv \frac{\partial f}{\partial \sigma_{ij}} \tag{7.21}$$

We note that n_{ij} is not necessarily a *unit* normal.

The normality rule is then written in the form

$$d\varepsilon_{ij}^{p} = d\lambda\ n_{ij}, \quad d\lambda \geq 0 \tag{7.22}$$

Equation 7.22 is often called the *associated flow rule* to distinguish it from the more general plasticity theory mentioned previously, which defines the direction of plastic flow via a plastic potential.

Hardening can be conveniently incorporated in the incremental stress–strain relations by defining a (scalar) equivalent stress σ_e and equivalent plastic strain ε_e^p in terms of the corresponding multiaxial quantities. Uniaxial stress–strain test data can then be used to define a "plastic modulus" $h(\varepsilon_e^p)$, which is considered to be a function of plastic strain, in the form

$$d\sigma_e = h(\varepsilon_e^p)d\varepsilon_e^p \tag{7.23}$$

We define the increment of equivalent stress as

$$d\sigma_e = n_{ij}d\sigma_{ij} \tag{7.24}$$

which is a measure of the magnitude of the stress increment normal to the yield surface. It can be nonzero only due to hardening, as is evident from Equation 7.23.

The increment of equivalent plastic strain is defined as

$$d\varepsilon_e^p = d\lambda \tag{7.25}$$

We note that with these definitions,

$$d\varepsilon_{ij}^p = d\varepsilon_e^p\ n_{ij} \tag{7.26}$$

and

$$d\sigma_{ij}d\varepsilon_{ij}^p = d\sigma_e d\varepsilon_e^p \tag{7.27}$$

If the yield surface is expressed in the form

$$f(\boldsymbol{\sigma}, H) = \sigma_e - Y(\varepsilon_e^p) = 0 \tag{7.28}$$

then the consistency condition df = 0 requires

$$df = n_{ij} d\sigma_{ij} - Y'(\varepsilon_e^p) d\varepsilon_e^p = 0 \tag{7.29}$$

in which

$$Y' \equiv \frac{dY}{d\varepsilon_e^p} \tag{7.30}$$

Equation 7.29 leads to the definition of the plastic modulus as a function of equivalent plastic strain as

$$h(\varepsilon_e^p) \equiv Y'(\varepsilon_e^p) \tag{7.31}$$

The plastic modulus is determined from test data, often a uniaxial test, and is then used in conjunction with a specific hardening rule, to be discussed subsequently, which determines how the yield surface evolves as a function of plastic strain.

The equivalent plastic strain increment is related to the plastic strain increment via

$$d\varepsilon_e^p = \frac{\sqrt{d\varepsilon_{ij}^p d\varepsilon_{ij}^p}}{\sqrt{n_{ij} n_{ij}}} \tag{7.32}$$

7.6.2 Loading and unloading criteria

We require the elastoplastic constitutive relations in rate form, that is,

$$d\sigma_{ij} = D_{ijkl}^{ep} d\varepsilon_{kl} \tag{7.33}$$

The stress increment is related to the elastic strain increment as follows:

$$d\sigma_{ij} = D_{ijkl}^{e} \ (d\varepsilon_{kl} - d\varepsilon_{kl}^p) \tag{7.34}$$

We substitute the expression for the plastic strain increment, Equation 7.22 in Equation 7.34, to obtain

$$d\sigma_{ij} = D_{ijkl}^{e} (d\varepsilon_{kl} - d\lambda \, n_{kl}) \tag{7.35}$$

and then contract this result with n_{ij} to obtain the scalar relation,

$$d\sigma_e = n_{ij} D_{ijkl}^{e} (d\varepsilon_{kl} - d\lambda \, n_{kl}) \tag{7.36}$$

We next use Equations 7.23 and 7.25 in Equation 7.36 to obtain an expression for $d\lambda$.

$$d\lambda = \frac{n_{ij} D_{ijkl}^{e} d\varepsilon_{kl}}{h + n_{ij} D_{ijkl}^{e} n_{kl}} \tag{7.37}$$

We note first that the denominator in Equation 7.37 is always positive. In fact, the two terms in the denominator represent, respectively, the plastic and the effective elastic stiffness in the direction (n_{ij}) of the incremental plastic strain normal to the yield surface. Accordingly, we introduce the notation

$$k_{nn}^{e} \equiv n_{ij} D_{ijkl}^{e} n_{kl} \tag{7.38}$$

for this scalar effective elastic stiffness. The stress appearing in the numerator of Equation 7.37, that is,

$$d\sigma_{ij}^{e} \equiv D_{ijkl}^{e}\, d\varepsilon_{kl} \tag{7.39}$$

is termed the *trial elastic stress increment*, computed from the total strain increment, and $n_{ij} d\sigma_{ij}^{e}$ is its (scalar) projection on the normal to the yield surface, which therefore determines the sign of $d\lambda$. With this more compact notation, Equations 7.35 through 7.37 can be put in the equivalent forms,

$$\begin{cases} d\sigma_{ij} = d\sigma_{ij}^{e} - D_{ijkl}^{e} n_{kl} d\lambda \\ d\sigma_{e} = n_{ij} d\sigma_{ij}^{e} - k_{nn}^{e} d\lambda \\ d\lambda = \dfrac{n_{ij} d\sigma_{ij}^{e}}{h + k_{nn}^{e}} \end{cases} \tag{7.40}$$

which will be useful for interpretation purposes later.

In a hardening material, if the current stress is on the yield surface, then $d\lambda > 0$ indicates plastic loading and $d\lambda < 0$ indicates elastic unloading, that is,

$$f(\boldsymbol{\sigma}, H) = 0 \begin{cases} d\lambda > 0 \Rightarrow \text{Plastic Loading} \\ d\lambda < 0 \Rightarrow \text{Elastic Unloading} \end{cases} \tag{7.41}$$

If the current stress is inside the yield surface, the updated trial stress, that is, $\sigma_{ij} + d\sigma_{ij}^{e}$, is evaluated:

$$\begin{cases} f(\boldsymbol{\sigma} + d\boldsymbol{\sigma}^{e}, H) < 0 \Rightarrow \text{Elastic} \\ f(\boldsymbol{\sigma} + d\boldsymbol{\sigma}^{e}, H) > 0 \Rightarrow \text{Elastic/Plastic Transition} \end{cases} \tag{7.42}$$

7.6.3 Continuum tangent stiffness

If $d\lambda \geq 0$, that is, we have (plastic) loading, $d\lambda$ from Equation 7.37 is inserted into Equation 7.35 to give

$$d\sigma_{ij} = \left[D_{ijkl}^{e} - \frac{D_{ijrs}^{e} n_{rs} n_{mn} D_{mnkl}^{e}}{[h + k_{nn}^{e}]} \right] d\varepsilon_{kl} \tag{7.43}$$

The elastoplastic tangent stiffness D_{ijkl}^{ep} (often termed the *continuum tangent stiffness*) is given by

$$D_{ijkl}^{ep} = \left[D_{ijkl}^{e} - \frac{D_{ijrs}^{e} n_{rs} n_{mn} D_{mnkl}^{e}}{[h + k_{nn}^{e}]} \right] \tag{7.44}$$

Equation 7.44 may be written in direct tensor notation as

$$\mathbf{D}^{ep} = \left[\mathbf{D}^{e} - \frac{\mathbf{D}^{e} : \mathbf{n} \otimes \mathbf{n} : \mathbf{D}^{e}}{[h + k_{nn}^{e}]} \right] \tag{7.45}$$

where the tensor outer product $\mathbf{a} \otimes \mathbf{b}$ is defined by

$$(\mathbf{a} \otimes \mathbf{b})_{ijkl} = a_{ij} b_{kl} \tag{7.46}$$

and the tensor inner product $\mathbf{a} : \mathbf{b}$ is defined by

$$(\mathbf{a} : \mathbf{b}) = a_{ij} b_{ij} \tag{7.47}$$

These various relationships can be more easily understood and interpreted in a (three-dimensional) principal stress space in which the stresses (and strains) are (3×1) vectors, the elastic and elastoplastic tangent stiffnesses are (3×3) matrices, and the various yield surfaces can be visualized. We will utilize that representation later in this chapter.

7.6.4 Elastic–perfectly plastic behavior

An example of elastic–perfectly plastic material behavior is the behavior of mild steel in uniaxial tension, right after yielding and prior to the onset of hardening. An idealized stress–strain relation for mild steel is shown in Figure 7.15, where the elastic–perfectly plastic portion is shown as solid lines. The hardening portion is shown as a dashed line.

In a uniaxial tension test, when the stress reaches the yield stress σ_Y, the material flows. It appears that when the material yields, the plastic strain keeps increasing at a constant stress. In other words, the yield stress is constant and does not change with increasing plastic deformation.

In a multiaxial state of stress, elastic–perfectly plastic behavior (no hardening, $h = 0$) implies that the yield surface remains unchanged as the plastic deformations increase. If the plastic modulus $h = 0$, Equation 7.43 becomes

$$d\sigma_{ij} = \left[D^e_{ijkl} - \frac{D^e_{ijrs} n_{rs} n_{mn} D^e_{mnkl}}{k^e_{nn}} \right] d\varepsilon_{kl} \tag{7.48}$$

It may not be obvious from this incremental stress–strain relation how plastic flow can occur in this case. Under a multiaxial state of stress, unlimited plastic flow, similar to the plastic flow in uniaxial tension, need not occur except under special conditions.

First, we observe that the tangent stiffness D^{ep}_{ijkl} shown above has a zero eigenvalue with corresponding eigenvector n_{ij}; that is, it is positive semidefinite. This can be easily verified

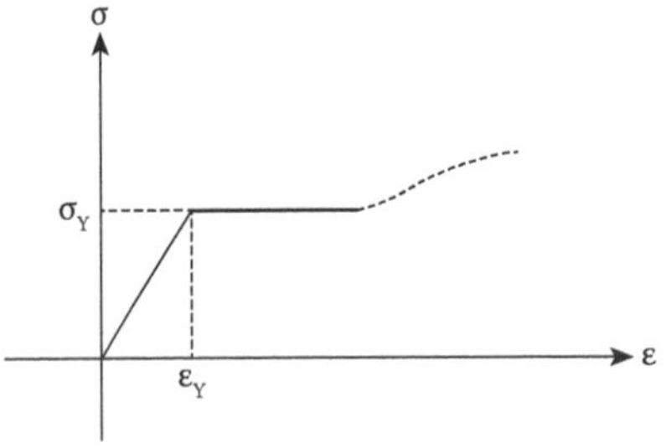

Figure 7.15 Typical stress–strain relation for mild steel.

by specifying a strain increment $d\varepsilon_{ij} = d\alpha\, n_{ij}$ and determining the corresponding stress increment $d\sigma_{ij}$ as follows:

$$\begin{aligned} d\sigma_{ij} &= \left[D^e_{ijkl} - \frac{D^e_{ijrs} n_{rs} n_{mn} D^e_{mnkl}}{k^e_{nn}} \right] n_{kl}\; d\alpha \\ &= \left[D^e_{ijkl} n_{kl} - D^e_{ijrs} n_{rs} \left(\frac{n_{mn} D^e_{mnkl} n_{kl}}{k^e_{nn}} \right) \right] d\alpha = [0] \end{aligned} \tag{7.49}$$

Therefore, any strain increment in the direction of **n**, for example, $d\boldsymbol{\varepsilon} = d\alpha\,\mathbf{n}$, will produce a stress increment $d\boldsymbol{\sigma}$ equal to zero. In direct tensor notation,

$$d\boldsymbol{\sigma} = \mathbf{D}^{ep} : (d\alpha\, \mathbf{n}) = d\alpha\, \mathbf{D}^{ep} : \mathbf{n} = 0 \tag{7.50}$$

This demonstrates that for elastic–perfectly plastic behavior, unlimited plastic flow similar to that in a uniaxial tension test can occur only if the material is subjected to strain increments in the direction of the outward normal to the yield surface, which is shown schematically in Figure 7.16a.

If the strain increment is not orthogonal to the yield surface, then the stress point can move *along* the yield surface. In this case, we choose a strain increment that has components in both **n** and **m** directions, where **m** is tangent to the yield surface, as shown in Figure 7.16b, that is,

$$d\varepsilon_{ij} = d\alpha\, n_{ij} + d\beta\, m_{ij}, \quad n_{ij} m_{ij} \equiv 0, \quad m_{ij} m_{ij} = 1 \tag{7.51}$$

The corresponding stress increment is

$$\begin{aligned} d\sigma_{ij} &= d\alpha\, D^{ep}_{ijkl} n_{kl} + d\beta\, D^{ep}_{ijkl} m_{kl} \\ &= d\beta\, D^{ep}_{ijkl} m_{kl} \end{aligned} \tag{7.52}$$

and its scalar tangential component is

$$\begin{aligned} m_{ij} d\sigma_{ij} &= \left[(m_{ij} D^e_{ijkl} m_{kl}) - \frac{(m_{ij} D^e_{ijrs} n_{rs})(n_{mn} D^e_{mnkl} m_{kl})}{k^e_{nn}} \right] d\beta \\ &= \left[k^e_{mm} - \frac{k^e_{mn} k^e_{nm}}{k^e_{nn}} \right] d\beta \end{aligned} \tag{7.53}$$

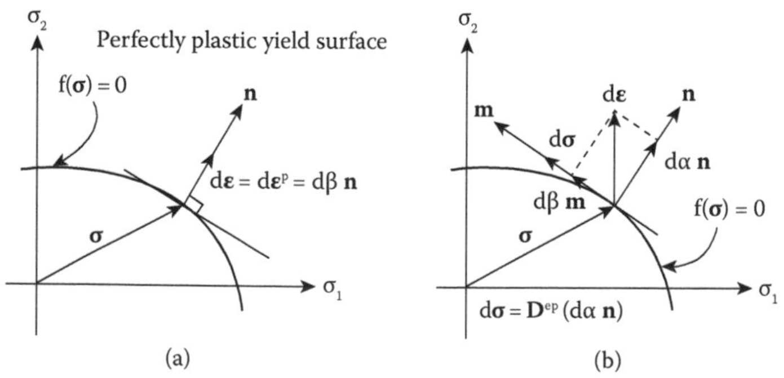

Figure 7.16 Direction of stress increment for conditions: (a) strain increment orthogonal to yield surface and (b) strain increment not orthogonal to yield surface.

We see that the tangential stress increment is an elastic response. The effective elastic stiffness coefficients in Equation 7.53 are defined in an obvious way. They are summarized below.

$$\begin{cases} k^e_{nn} \equiv n_{ij} D^e_{ijkl} n_{kl} \\ k^e_{mn} \equiv m_{ij} D^e_{ijkl} n_{kl} \\ k^e_{nm} \equiv n_{ij} D^e_{ijkl} m_{kl} \\ k^e_{mm} \equiv m_{ij} D^e_{ijkl} m_{kl} \end{cases} \tag{7.54}$$

As a result of the symmetries of the elastic material stiffness D^e_{ijkl}, $k^e_{mn} = k^e_{nm}$.

7.6.5 Interpretation of incremental stresses

In this section, we present additional geometric interpretations of the elastoplastic stress–strain relations that have been developed to this point. The geometric interpretations will be helpful in understanding the numerical algorithms that will be discussed in due course.

7.6.5.1 *Elastic–perfectly plastic*

When there is no hardening ($h=0$), the behavior is elastic–perfectly plastic and the increment of equivalent stress $d\sigma_e=0$. Equations 7.40 then become

$$\begin{cases} d\sigma_e = 0 \\ d\lambda = n_{rs} d\sigma^e_{rs} / k^e_{nn} \\ d\sigma_{ij} = d\sigma^e_{ij} - D^e_{ijkl} d\varepsilon^p_{ij} \equiv d\sigma^e_{ij} + d\sigma^c_{ij} \end{cases} \tag{7.55}$$

where $d\boldsymbol{\sigma}^c$ is the incremental "correction" to the trial elastic stress increment to yield the elastoplastic stress increment $d\boldsymbol{\sigma}$, that is, $d\boldsymbol{\sigma} = d\boldsymbol{\sigma}^e + d\boldsymbol{\sigma}^c$, where $d\boldsymbol{\sigma}^c \equiv -\mathbf{D}^e : d\boldsymbol{\varepsilon}^p$ (Figure 7.17). The components of the incremental stress correction are

$$d\sigma^c_{ij} = \underbrace{-D^e_{ijkl}\, n_{kl}\, \underbrace{n_{rs} d\sigma^e_{rs}/k^e_{nn}}_{d\lambda}}_{d\varepsilon^p_{ij}} \tag{7.56}$$

We remark again that $\mathbf{D}^e$ maps the incremental plastic strain vector into stress space so that the direction of the required incremental stress correction is not normal to the yield surface.

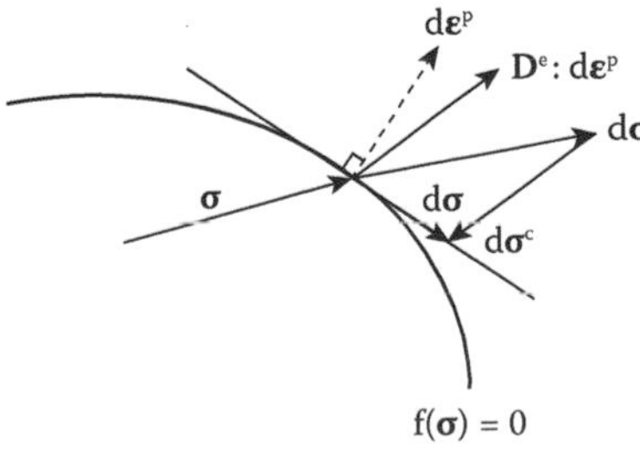

Figure 7.17 Incremental stress vectors—elastic–perfectly plastic.

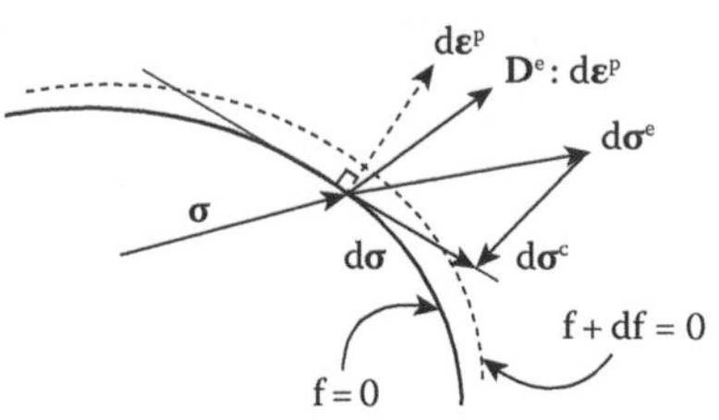

Figure 7.18 Incremental stress vectors—elastic–hardening plasticity.

7.6.5.2 *Hardening plasticity*

If the material hardens with plastic strain ($h > 0$), the incremental stress is no longer tangent to the yield surface (Figure 7.18). The incremental stress correction is still defined by

$$d\sigma_{ij}^{c} \equiv -D_{ijkl}^{e} d\varepsilon_{kl}^{p} \tag{7.57}$$

but the plastic strain increment $d\varepsilon^p$ differs from that for elastic–perfectly plastic behavior.

In the hardening case, we have

$$d\varepsilon_{kl}^{p} = \frac{n_{kl} n_{rs} d\sigma_{rs}^{e}}{h + k_{nn}^{e}} \tag{7.58}$$

Compared with the elastic–perfectly plastic case (for the same strain increment $d\boldsymbol{\varepsilon}$ and current stress point $\boldsymbol{\sigma}$), we see that the plastic strain increment is simply scaled (down) by a factor of $k_{nn}^{e}/(h + k_{nn}^{e})$. Therefore, the incremental stress correction is also scaled down by the same factor, but its direction is unchanged. As a result of the hardening, the yield surface is updated from $f = 0$ to $f + df = 0$.

Of course, in neither case (perfectly plastic nor hardening plastic) does the stress $\boldsymbol{\sigma} + d\boldsymbol{\sigma}$ lie precisely on the current or updated yield surface for finite-sized strain increments. An efficient and accurate stress update algorithm that does ensure this is an essential component of any incremental-iterative finite element analysis.

7.7 YIELD SURFACES IN PRINCIPAL STRESS SPACE

One of the important early observations in metal plasticity was the fact that the change in the mean (compressive) stress p has no effect on the state of the permanent deformation even at very high mean compressive stress (Bridgman 1923, 1952).

The primary implication of this observation is that the yield surfaces for metals are of cylindrical shape, with the axis of the cylinder parallel to the space diagonal in principal stress space. Therefore, in studying the shape of metal yield surfaces, we may confine our attention to the shape of the cross section of the yield surface on the octahedral plane (π-plane).

7.7.1 Material isotropy and symmetry requirements

We first assume that the yield stress has been determined from a uniaxial stress test. These data provide three points on the yield surface in the positive direction (tensile stress), along the principal stress axes.

$$\begin{Bmatrix} \sigma_1 \\ \sigma_2 \\ \sigma_3 \end{Bmatrix} = \begin{Bmatrix} \sigma_Y \\ 0 \\ 0 \end{Bmatrix}; \quad \begin{Bmatrix} \sigma_1 \\ \sigma_2 \\ \sigma_3 \end{Bmatrix} = \begin{Bmatrix} 0 \\ \sigma_Y \\ 0 \end{Bmatrix}; \quad \begin{Bmatrix} \sigma_1 \\ \sigma_2 \\ \sigma_3 \end{Bmatrix} = \begin{Bmatrix} 0 \\ 0 \\ \sigma_Y \end{Bmatrix} \tag{7.59}$$

Three additional points on the yield surface are obtained by assuming that the yield stress in uniaxial compression is the same as in uniaxial tension.

$$\begin{Bmatrix} \sigma_1 \\ \sigma_2 \\ \sigma_3 \end{Bmatrix} = \begin{Bmatrix} -\sigma_Y \\ 0 \\ 0 \end{Bmatrix}; \quad \begin{Bmatrix} \sigma_1 \\ \sigma_2 \\ \sigma_3 \end{Bmatrix} = \begin{Bmatrix} 0 \\ -\sigma_Y \\ 0 \end{Bmatrix}; \quad \begin{Bmatrix} \sigma_1 \\ \sigma_2 \\ \sigma_3 \end{Bmatrix} = \begin{Bmatrix} 0 \\ 0 \\ -\sigma_Y \end{Bmatrix} \tag{7.60}$$

Looking down the space diagonal (i.e., the normal to the octahedral plane), we see the true shape of the octahedral plane. These six points plot on the octahedral plane at equal distances $Y = \sqrt{2/3}\,\sigma_Y$ from the hydrostatic axis, as shown in Figure 7.19. Figure 7.20 shows the location of these six points on the octahedral plane.

The shape of the yield surface on the octahedral plane is constrained by two sets of symmetry requirements. The first set of symmetry requirements is based on the assumption that the material is initially isotropic.

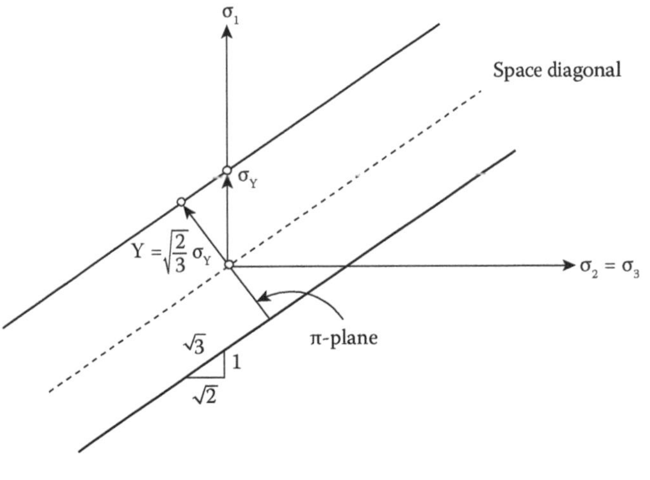

Figure 7.19 Uniaxial tension yield stress plotted on octahedral plane in principal stress space.

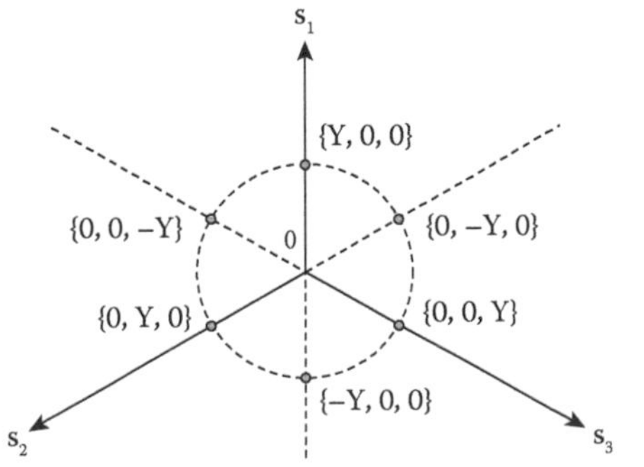

Figure 7.20 Six points on the yield surface determined from uniaxial stress test.

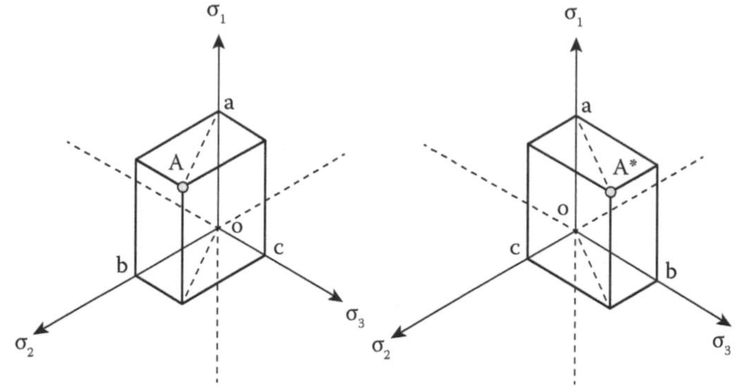

Figure 7.21 Symmetry observation leading to symmetry of yield surface with respect to the σ_1 axis on the octahedral plane.

For an isotropic material, there are no preferential material directions, and therefore relabeling the principal axis directions can have no effect on the initial yield surface. Consequently, if a stress point with principal values {a, b, c} is on the yield surface, three additional points with coordinates {a, c, b}, {c, b, a}, and {b, a, c} will also be on the yield surface.

This shows that there are three planes of symmetry; each plane contains one coordinate axis and bisects the remaining two coordinate axes. These three planes of symmetry contain the space diagonal in principal stress space, and their projections on the octahedral plane therefore coincide with those of the principal stress axes.

Figure 7.21 shows the effect of interchanging the last two principal stresses. It is apparent that points A (a, b, c) and A* (a, c, b) are symmetric with respect to the σ_1 axis on the π-plane.

The assumption that the material behavior in tension and compression is the same requires that the yield surface have three additional planes of symmetry. Therefore, changing the *sign* of the principal stresses should not affect the yield state of the material. If the stress point {a, b, c} is on the yield surface, then three other points, with coordinates {–a, –c, –b}, {–b, –a, –c}, and {–c, –b, –a}, are also on the yield surface. This is demonstrated in Figure 7.22, which shows that if point {a, b, c} is on the yield surface, so is {–a, –c, –b}. This implies the existence of a plane of symmetry that contains the space diagonal and is *orthogonal* to the projection of the σ_1 axis on the octahedral plane (i.e., orthogonal to the s_1 axis), as shown in Figure 7.22. Using similar arguments, we find two additional planes of symmetry, as shown in Figure 7.23.

For isotropic behavior, the axes of symmetry divide the octahedral plane into 12 equal segments (each segment subtending an angle $\pi/6$). The shape of the complete yield surface in the octahedral plane can therefore be generated by rotating any one of these segments (about the space diagonal) through an integer multiple of $\pi/6$. This also means that the shape of the yield surface can be determined via experiments following stress paths lying in any one of the $\pi/6$ segments.

In summary, the following idealizations based on observations are used in constraining the shape of the initial yield surface for metals.

1. The mean stress has no effect on yielding. Therefore, the initial yield surfaces are cylinders parallel to the principal stress space diagonal.
2. The yield stress determined from a uniaxial tension test provides three points on the initial yield surface in the octahedral plane.

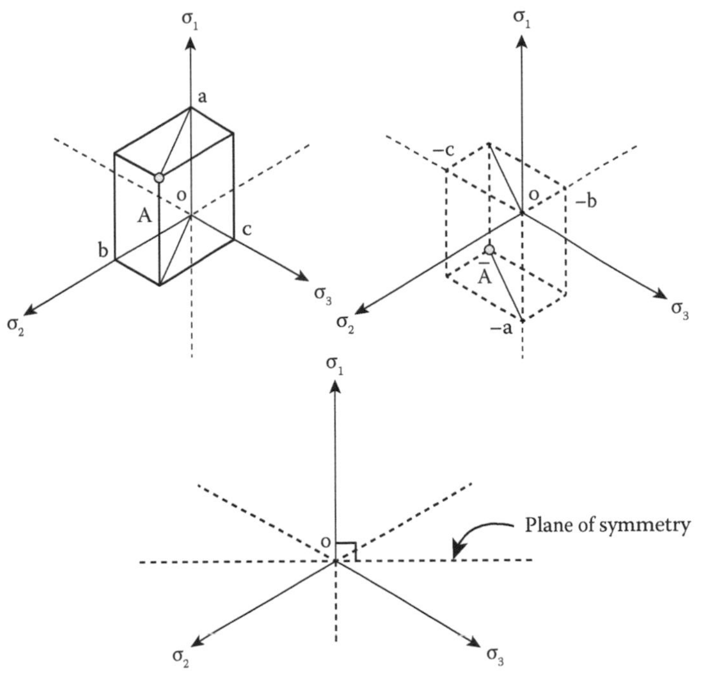

Figure 7.22 Demonstration of the fact that the two stress points {a, b, c} and {−a, −c, −b} are on the yield surface and the resulting plane of symmetry.

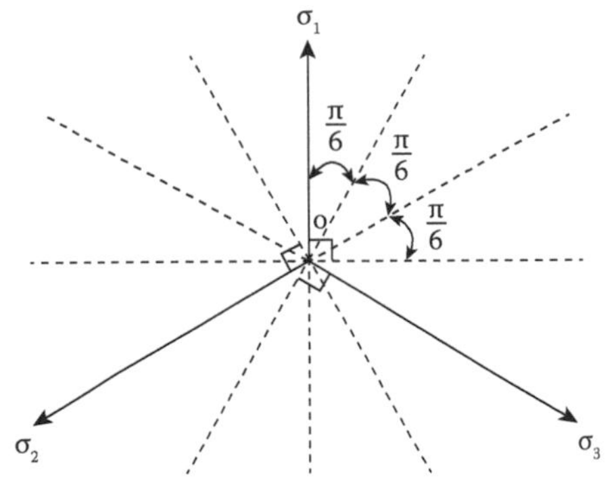

Figure 7.23 Three planes of symmetry whose intersections with the octahedral plane are orthogonal to the projections of the principal stress axes on the octahedral plane. The three planes of symmetry are also orthogonal to the octahedral plane.

3. The assumption that the initial yield stress is the same in tension and compression provides three additional points on the initial yield surface in the octahedral plane.
4. The assumption of initial material isotropy requires that the shape of the yield surface in the octahedral plane have sixfold rotational symmetry (rotations about the space diagonal). Drucker's postulate additionally requires that the yield surface be convex.

These considerations restrict the projection of the yield surface on the octahedral plane to lie within the narrow shaded area bounded by the two regular hexagons shown in Figure 7.24. We observe that each hexagon passes through the six data points determined from a uniaxial

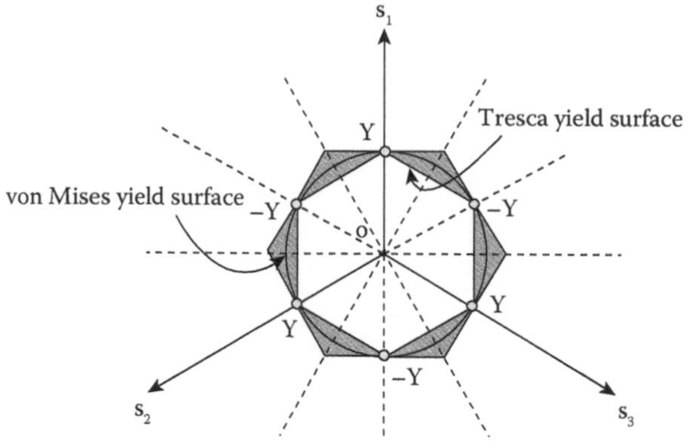

Figure 7.24 Zone of possible yield surfaces in metal plasticity.

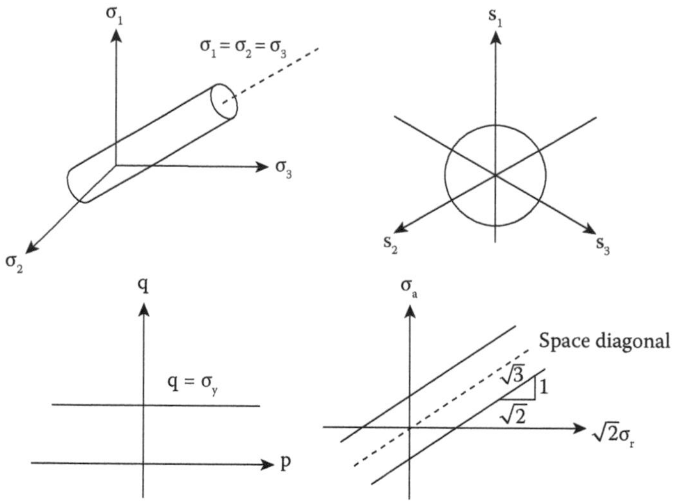

Figure 7.25 Representations of von Mises yield surface.

tension test and is convex. Therefore, either hexagon, as well as any convex shape lying between them, is an admissible yield surface.

We recall that mathematically, convexity means that if we connect any two points in a set with a straight line, all the points on the line between the two points are in the same set. If a yield surface were to lie outside of the shaded zone, it would not be convex even though it may pass through the six uniaxial yield points and satisfy the symmetry requirements. The hexagon on the inside of the shaded zone is known as the *Tresca* yield surface. Another well-known yield surface, the *von Mises* yield surface, is a circle in the shaded zone.

7.7.2 von Mises yield surface

The von Mises yield surface has a *circular* cylindrical shape in principal stress space with axis lying along the space diagonal, as shown in Figure 7.25. The radius of the circle is related to the uniaxial yield stress σ_Y, as described below.

The von Mises yield surface is often defined by the following equation:

$$f(\boldsymbol{\sigma}) = q - \sigma_Y = 0 \qquad (7.61)$$

In this equation,

$$q \equiv \sqrt{3J_2} = \sqrt{\frac{3}{2} s_{ij} s_{ij}}$$

from Equation 6.12 in Chapter 6.

The equivalent stress q in Equation 7.61 is called the *von Mises stress*. It is convenient to use because of the fact that $q = \sigma_1$ on a uniaxial stress path. The von Mises yield function can also be written in slightly different forms, one of which we have introduced previously for yield surfaces in general, that is,

$$f(\boldsymbol{\sigma}) = \sigma_e - Y \tag{7.62}$$

Here, $\sigma_e \equiv \sqrt{2J_2} = \sqrt{s_{ij}s_{ij}}$ and $Y = \sqrt{\frac{2}{3}}\sigma_Y$. Equation 7.62 expresses the geometric fact that the radius of the yield surface r $(=\sigma_e)$ in the octahedral plane is equal to $Y\left(=\sqrt{\frac{2}{3}}\sigma_Y\right)$.

Using the definitions of Equation 7.62, we find the outward normal to the yield surface as

$$n_{ij} \equiv \frac{\partial f}{\partial \sigma_{ij}} = \frac{\partial f}{\partial s_{kl}}\frac{\partial s_{kl}}{\partial \sigma_{ij}} = \frac{s_{ij}}{\sigma_e} \tag{7.63}$$

which happens, for the von Mises yield surface, to be a unit normal, that is,

$$n_{ij}n_{ij} = 1 \tag{7.64}$$

In principal stress space, the unit outward normal vector {n} to the von Mises yield surface obviously lies in the radial direction in the octahedral plane.

$$\{n\} = \frac{1}{Y}\begin{Bmatrix} s_1 \\ s_2 \\ s_3 \end{Bmatrix} = \frac{\sqrt{3}}{\sqrt{2}\,\sigma_Y}\begin{Bmatrix} s_1 \\ s_2 \\ s_3 \end{Bmatrix} \tag{7.65}$$

For isotropic materials, the elastic constitutive tensor takes a particularly simple form involving only two independent material parameters, which we take here as the shear modulus G and Poisson's ratio ν.

$$D^e_{ijkl} = 2G\left[\frac{1}{2}(\delta_{ik}\delta_{jl} + \delta_{il}\delta_{jk}) + \frac{\nu}{1-2\nu}(\delta_{ij}\delta_{kl})\right] \tag{7.66}$$

Using Equation 7.64 and the fact that $s_{kk} = 0$ in the last equation, we find the scalar effective elastic stiffness k^e_{nn} that was defined previously:

$$k^e_{nn} = n_{ij}D^e_{ijkl}n_{kl} = 2G \tag{7.67}$$

The incremental elastoplastic stress–strain relation is then found from Equation 7.48 as

$$d\sigma_{ij} = \left[D^e_{ijkl} - \frac{2G\, s_{ij}s_{kl}}{Y^2}\right] d\varepsilon_{kl} \tag{7.68}$$

We now briefly explore some of the implications of the incremental stress–strain relations based on the von Mises yield function.

First, we write the plastic strain increment $d\varepsilon^p_{ij}$ as

$$d\varepsilon^p_{ij} = d\lambda\ n_{ij} = d\lambda \frac{s_{ij}}{Y} \tag{7.69}$$

and we see that the *plastic* volume change $d\varepsilon^p_{kk} = 0$ since $s_{kk} = 0$ by definition. This conclusion is true for any yield surface function that is independent of mean stress, not just the von Mises yield surface, and holds whether or not the material is hardening.

Next, we rewrite Equation 7.68 in the form of an incremental elastic trial stress and an incremental stress corrector as

$$d\sigma_{ij} = d\sigma^e_{ij} + d\sigma^c_{ij} \tag{7.70}$$

For the von Mises yield function, the incremental stress corrector is

$$d\sigma^c_{ij} = -\frac{2G}{Y^2} s_{ij} s_{kl}\, d\varepsilon_{kl} \tag{7.71}$$

Its trace is

$$d\sigma^c_{mm} = -\frac{2G}{Y^2} s_{mm} s_{kl}\, d\varepsilon_{kl} = 0 \tag{7.72}$$

Therefore, the incremental stress corrector lies in the octahedral plane, even though the trial elastic stress increment has in general a mean stress component. This is, of course, also a manifestation of the fact that the von Mises yield function is independent of mean stress.

If we wish, we can separate the purely elastic mean stress versus volume change response to obtain incremental deviator stress versus deviator strain relations. Accordingly, we write

$$d\varepsilon_{kl} = de_{kl} + \frac{1}{3} d\varepsilon_{mm}\delta_{kl} \tag{7.73}$$

and evaluate the right-hand side of Equation 7.68. We find

$$d\sigma^e_{ij} = D^e_{ijkl} d\varepsilon_{kl} = \left(2G\ de_{ij} + \frac{E}{3(1-2\nu)} d\varepsilon_{mm}\delta_{ij}\right) \tag{7.74}$$

which leads to

$$\begin{cases} ds_{ij} = 2G\left[\delta_{ij}\delta_{kl} - \dfrac{s_{ij}s_{kl}}{Y^2}\right] de_{kl} \\ dp = K\ d\varepsilon_{kk} \end{cases} \tag{7.75}$$

where K is the elastic bulk modulus defined by

$$K \equiv \lambda + \frac{2G}{3} \equiv \frac{E}{3(1-2\nu)} \tag{7.76}$$

The first of Equations 7.75 referred to the current principal stress axes becomes

$$\begin{Bmatrix} ds_1 \\ ds_2 \\ ds_3 \end{Bmatrix} = 2G \begin{bmatrix} 1 - \frac{s_1^2}{Y^2} & \frac{-s_1 s_2}{Y^2} & \frac{-s_1 s_3}{Y^2} \\ & 1 - \frac{s_2^2}{Y^2} & \frac{-s_2 s_3}{Y^2} \\ \text{Symm} & & 1 - \frac{s_3^2}{Y^2} \end{bmatrix} \begin{Bmatrix} de_1 \\ de_2 \\ de_3 \end{Bmatrix} \quad (7.77)$$

These incremental stress–strain relations for perfectly plastic von Mises plasticity are known as the *Prandtl–Reuss equations*. We see, perhaps as expected, that plastic flow may often result in constitutive properties with stress-induced anisotropy.

It can be easily verified that the matrix in Equation 7.77 is singular, as we expect for ideal plasticity.

In the general three-dimensional case, the tensor form of the incremental stress–strain relations can also be converted to matrix–vector form for numerical computations. We will address that conversion in a subsequent section.

7.7.2.1 Biaxial (plane) stress

Assume a biaxial state of stress in the 1–2 (xy) plane, with the principal stress vector $\boldsymbol{\sigma} = \{\sigma_1, \sigma_2, 0\}$. The deviator stress vector and the stress invariants are defined in terms of the two nonzero principal stresses.

$$p = \frac{1}{3}\sigma_{kk} = \frac{1}{3}(\sigma_1 + \sigma_2) \quad (7.78)$$

$$s = \begin{Bmatrix} \sigma_1 \\ \sigma_2 \\ 0 \end{Bmatrix} - p \begin{Bmatrix} 1 \\ 1 \\ 1 \end{Bmatrix} = \frac{1}{3} \begin{Bmatrix} 2\sigma_1 - \sigma_2 \\ -\sigma_1 + 2\sigma_2 \\ -\sigma_1 - \sigma_2 \end{Bmatrix} \quad (7.79)$$

$$\begin{cases} J_2 = \frac{1}{2}(s_1^2 + s_2^2 + s_3^2) = \frac{1}{3}(\sigma_1^2 + \sigma_2^2 - \sigma_1\sigma_2) \\ J_3 = s_1 s_2 s_3 = \frac{1}{27}(3\sigma_2\sigma_1^2 + 3\sigma_1\sigma_2^2 - 2\sigma_1^3 - 2\sigma_2^3) \end{cases} \quad (7.80)$$

$$q^2 = 3J_2 = (\sigma_1^2 + \sigma_2^2 - \sigma_1\sigma_2) \quad (7.81)$$

The equation for the von Mises yield surface is written in the following convenient form:

$$f(\boldsymbol{\sigma}) = q^2 - \sigma_Y^2 = 0 \quad (7.82)$$

or

$$f(\boldsymbol{\sigma}) = \sigma_1^2 + \sigma_2^2 - \sigma_1\sigma_2 - \sigma_Y^2 = 0 \quad (7.83)$$

Therefore, the circular cylindrical von Mises yield surface projects onto an ellipse in the $\{\sigma_1, \sigma_2\}$ plane, which is shown in Figure 7.26. We note that the effective yield stress (i.e., the "mean radius" of the yield surface) is smaller in the two tension–compression quadrants.

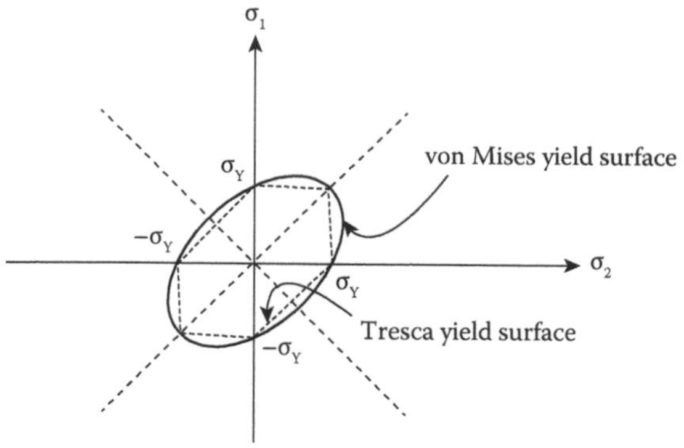

Figure 7.26 von Mises yield surface in biaxial state of stress.

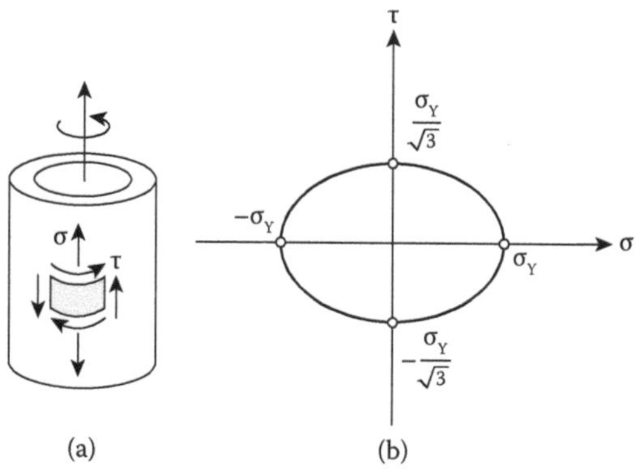

Figure 7.27 (a, b) von Mises yield surface in the tension–torsion test.

(By "radius," we mean in this case the distance of a stress point on the yield surface from the origin of the principal stress space.)

The (octahedral) shear stress is smaller in the two tension–compression quadrants than in either the tension–tension or compression–compression quadrant. We recall that a pure shear test is equivalent to an equibiaxial compression–tension test ($\sigma_1 = -\sigma_2$), and we note that for such a stress path, the radius of the projected von Mises yield surface is minimum. For metals, the observed yield point in pure shear is smaller than that observed in uniaxial tension. This general property is therefore properly reflected by the von Mises yield surface.

7.7.2.2 *Tension–torsion test*

A tension–torsion test on a thin-walled cylindrical sample results in normal stress σ and shear stress τ, as shown in Figure 7.27a. This test generates a plane stress-like state in the material. It was originally designed for testing of metals in order to understand their behavior in multiaxial stress states. (The test has also been used in testing plain concrete samples, in which an axial compression force is applied instead of the tension force.) In fact, without the action of the tension force, the test becomes a pure torsion test, from which a pure shear stress state is generated, provided that the cylindrical sample is sufficiently thin walled. If we consider a small element in the sample, remote from any significant end effects, the stresses and stress invariants are as follows:

$$\boldsymbol{\sigma} = \begin{bmatrix} \sigma & \tau & 0 \\ \tau & 0 & 0 \\ 0 & 0 & 0 \end{bmatrix}; \quad s = \begin{bmatrix} 2\sigma/3 & \tau & 0 \\ \tau & -\sigma/3 & 0 \\ 0 & 0 & -\sigma/3 \end{bmatrix} \tag{7.84}$$

$$J_2 = \frac{1}{2} s_{ij} s_{ij} = \frac{1}{3}\sigma^2 + \tau^2 \tag{7.85}$$

$$q^2 = 3J_2 = \sigma^2 + 3\tau^2 \tag{7.86}$$

The equation for the von Mises yield surface in terms of (σ, τ) is

$$f(\boldsymbol{\sigma}) = \sigma^2 + 3\tau^2 - \sigma_Y^2 = 0 \tag{7.87}$$

The yield surface is an ellipse, as shown in Figure 7.27b. We observe again in this view the smaller effective yield point $(\tau_Y = \sigma_Y/\sqrt{3})$ in pure shear.

7.7.3 Tresca yield surface

The Tresca yield criterion assumes that yielding occurs when the maximum shear stress reaches a critical value. The Tresca initial yield surface can be expressed as

$$f(\boldsymbol{\sigma}) = \tau_{max} - \tau_Y = 0 \tag{7.88}$$

In this equation,τ_Y is the shear stress at initial yield. If the principal stresses are ordered $\sigma_1 \geq \sigma_2 \geq \sigma_3$, then the maximum shear is

$$\tau_{max} = \frac{1}{2}(\sigma_1 - \sigma_3) \tag{7.89}$$

and the intermediate principal stress has no effect on yield according to this criterion. The limiting value of the maximum shear can again be determined from the yield stress in a uniaxial stress test, where $\sigma_1 = \sigma_Y$ and $\sigma_3 = 0$. Thus, $\tau_Y = \sigma_Y/2$. Therefore, we have the following expression for the Tresca yield surface:

$$f(\boldsymbol{\sigma}) = \frac{1}{2}(\sigma_1 - \sigma_3) - \frac{1}{2}\sigma_Y = 0 \tag{7.90}$$

This equation applies just in the $(\sigma_1, -\sigma_3)$ quadrant on the π-plane. In order to determine the shape of the yield surface on the octahedral plane, it is useful to write the equation for the yield surface in terms of the invariants p, q, θ, or equivalently, p, r, θ. First, it can be observed that the yield surface must be independent of p, since it has no effect on the maximum shear. Next, we can substitute for (σ_1, σ_3) in terms of the invariants p, q, θ from the relations obtained earlier.

$$\begin{cases} \sigma_1 = p + \frac{2}{3} q \sin\left(\theta + \frac{2\pi}{3}\right) \\ \sigma_3 = p + \frac{2}{3} q \sin\left(\theta + \frac{4\pi}{3}\right) \end{cases} \tag{7.91}$$

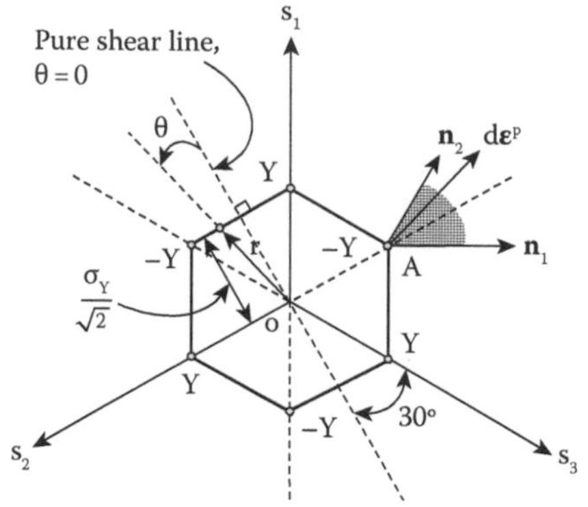

Figure 7.28 Tresca yield surface on the octahedral plane.

Using these equations and substituting in Equation 7.90, we arrive at the following expression for the Tresca yield surface:

$$f(\boldsymbol{\sigma}) = q\ \cos\theta - \frac{\sqrt{3}}{2}\sigma_Y = 0 \tag{7.92}$$

Equation 7.92 may also be written in the equivalent form:

$$f(\boldsymbol{\sigma}) = r\ \cos\theta - \frac{\sqrt{3}}{2}Y = 0 \tag{7.93}$$

where $r \equiv \sqrt{2J_2}$ is the true radial distance from the space diagonal in the octahedral plane, and by definition, $r \equiv \sqrt{2/3}q$.

In a polar coordinate system in the octahedral plane, the above two equations, which apply in the range $-\pi/6 \leq \theta \leq \pi/6$, represent a straight line.

Therefore, the shape of the Tresca yield surface on the octahedral plane is a regular hexagon that intersects the principal stress axes at ±Y, as shown in Figure 7.28. It has a (hexagonal) cylindrical shape with axis coinciding with the principal stress space diagonal.

The pure shear line (θ = 0) shown in Figure 7.28 is normal to the Tresca yield surface. On the pure shear line, $r = \sqrt{2}\tau$, and therefore yield in pure shear, according to the Tresca yield criterion occurs at $\tau_Y = \sigma_Y/2$ compared with $\tau_Y = \sigma_Y/\sqrt{3}$ according to the von Mises yield criterion.

We observe that corners are formed on the Tresca yield surface. Nevertheless, it is convex. At all points other than the corner points on the yield surface, the normality rule is also satisfied. However, the corner points warrant some special consideration in order to not violate the normality rule. For instance, we consider a corner point A as shown in Figure 7.28. According to Drucker's postulate, at any point on the yield surface, the following relationship should be satisfied:

$$(\sigma_{ij} - \sigma_{ij}^{a})\varepsilon_{ij}^{P} \geq 0 \tag{7.94}$$

This requires that at the corner, the direction of the plastic strain increment must lie within the shaded zone defined by the normal vectors $\mathbf{n}_1$ and $\mathbf{n}_2$ on the adjacent branches of the yield surface, as illustrated in Figure 7.28. For frictional materials, sharp corners on the yield (failure) surface have been observed from material tests conducted in the laboratory.

7.8 HARDENING PLASTICITY MODELS

Earlier in this chapter, we discussed the effect of hardening in general on the direction of the elastoplastic stress increment $d\boldsymbol{\sigma}$ and on the magnitude of the plastic strain increment $d\varepsilon^p$. We now describe two specific simple hardening models (*isotropic* hardening and *kinematic* hardening) that can be used to describe the evolution of the yield surface with accumulated plastic strain.

The important features of the elastoplastic hardening behavior of metals that is typically observed in a uniaxial tension test are illustrated in Figure 7.29. As the material is stressed, the first yield occurs at the initial yield stress $\sigma_Y(0)$. After first yielding, the stress continues to increase with the strain, but at a (perhaps much) reduced rate. Partial or full unloading of the stress results in a partial strain recovery along a path that is approximately parallel to the original loading path, that is, an elastic path. Since the unloading–reloading is therefore idealized as elastic, the stress at any point on the hardening portion of the stress–strain curve must be considered the current yield stress. The accumulated plastic strain at any level of stress is the residual (unrecovered) strain after complete unloading from that particular stress. It appears that the current yield stress is a function of the accumulated plastic strain, as a good approximation. Generalizations of these basic observations to multiaxial stress states form the basis for simple models of hardening plasticity.

In a multiaxial state of stress, the initial yielding of course occurs as the stress point reaches the initial yield surface. It was shown in an earlier section that in perfect plasticity, the yield surface remains unchanged. However, with hardening material behavior, the yield surface changes as the material is loaded beyond the first yield. The hardening rule specifies how the yield surface is assumed to evolve.

Yield surfaces can be characterized by their orientation, size, and shape. For metals, there is experimental evidence that all three are affected after the initial yield. In this section, we consider several simple hardening rules and assume that the changes in the yield surface are described by a single scalar hardening parameter h, which can be related to test results. Therefore, we assume

$$f(\boldsymbol{\sigma}, h) = 0 \tag{7.95}$$

As was shown earlier, the hardening parameter enters the incremental elastoplastic stress–strain relations, that is,

$$d\sigma_{ij} = \left(D^e_{ijkl} - \frac{D^e_{ijrs} n_{rs} n_{mn} D^e_{mnkl}}{h + k^e_{nn}} \right) d\varepsilon_{kl} \tag{7.96}$$

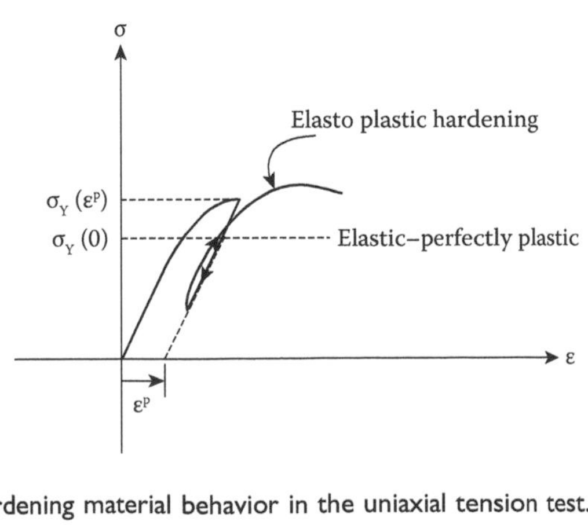

Figure 7.29 Elastoplastic hardening material behavior in the uniaxial tension test.

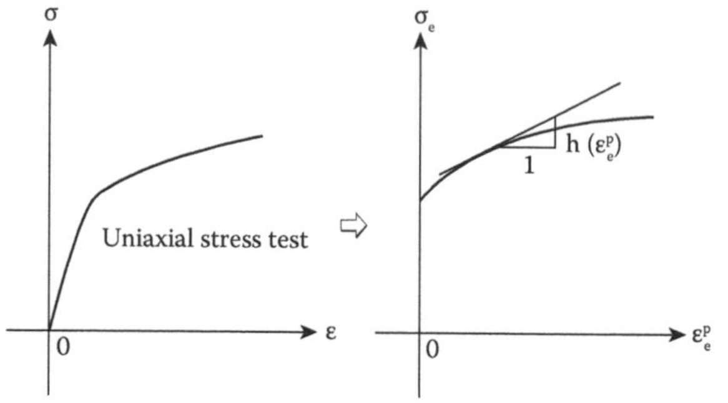

Figure 7.30 Determination of the effective stress–plastic strain curve from the uniaxial stress–strain curve.

We recall that the hardening parameter h was defined by the relation $d\sigma_e = h\ d\varepsilon_e^p$, where $d\sigma_e \equiv n_{ij} d\sigma_{ij}$ and $d\varepsilon_e^p \equiv \sqrt{d\varepsilon_{ij}^p d\varepsilon_{ij}^p} / \sqrt{n_{ij} n_{ij}}$.

Before considering specific hardening rules (i.e., how the initial yield surface evolves), we first indicate how the hardening parameter h can be determined from uniaxial test data.

As an example, we consider an idealized uniaxial stress–strain curve that may resemble that shown in Figure 7.30.

7.8.1 Determination of hardening parameter from uniaxial test

A simplified bilinear version is shown in Figure 7.31. The first segment has a tangent stiffness or modulus of E, and the second segment has a tangent stiffness of E_p. After yielding, on the second segment, we have the following stress–strain relation:

$$d\sigma_1 = E_p d\varepsilon_1 = E_p (d\varepsilon_1^e + d\varepsilon_1^p) \tag{7.97}$$

Using the elastic stress–strain relation $d\sigma_1 = E\ d\varepsilon_1^e$, we obtain the following relation:

$$d\sigma_1 = \frac{E\ E_p}{E - E_p} d\varepsilon_1^p \tag{7.98}$$

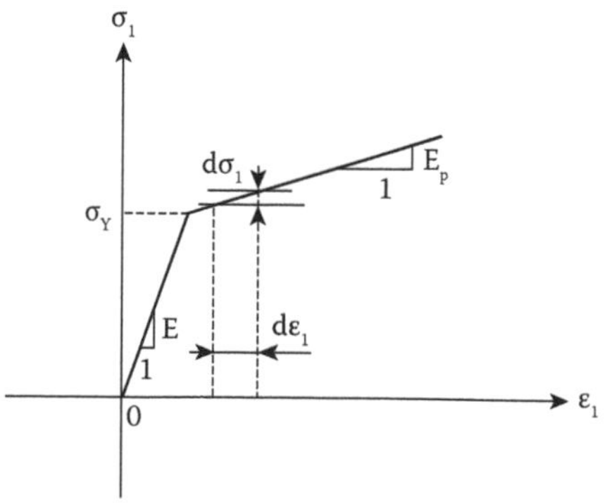

Figure 7.31 Idealized stress–strain curve consisting of two linear segments from a uniaxial tension test.

In order to determine the plastic modulus h, it is necessary to specify the yield surface with which it is to be used. We assume a von Mises yield surface, for which the normal **n** was determined earlier as

$$n_{ij} = \frac{s_{ij}}{\sigma_e} \tag{7.99}$$

For a uniaxial stress path, the incremental equivalent stress is

$$d\sigma_e = \sqrt{\frac{2}{3}} d\sigma_1 \tag{7.100}$$

To determine the incremental equivalent plastic strain for the uniaxial stress path, we first recall that $d\varepsilon^P_{kk} = 0$ for yield surfaces that are independent of mean stress. Therefore,

$$\sqrt{d\varepsilon^P_{ij} d\varepsilon^P_{ij}} = \sqrt{\frac{2}{3}} d\varepsilon^P_1 \tag{7.101}$$

and since $n_{ij}n_{ij}=1$,

$$d\varepsilon^P_e = \sqrt{\frac{2}{3}} d\varepsilon^P_1 \tag{7.102}$$

Therefore,

$$h \equiv \frac{d\sigma_e}{d\varepsilon^P_c} = \frac{d\sigma_1}{d\varepsilon^P_1} \tag{7.103}$$

and using Equation 7.98,

$$h = \frac{E\ E_p}{E - E_p} \tag{7.104}$$

We remark that the plastic modulus h is determined from *one* experiment only, and that the same uniaxial test data may lead to different values of plastic modulus for other yield surfaces.

A variable hardening modulus can be obtained from uniaxial test curves like Figure 7.30 by, for example, using a multilinear approximation, or by fitting a smooth curve to the test data.

7.8.2 Isotropic hardening

The isotropic hardening rule is based on the assumption that only the size of the yield surface is affected. Its position and shape remain unchanged, and hardening causes the yield surface to expand isotropically (i.e., equally in all directions), as shown in Figure 7.32 for the von Mises yield surface on the octahedral plane.

As we have seen, the "radius" of the yield surface is specified by the parameter Y in the following equation:

$$f(\boldsymbol{\sigma}, h) = \sigma_e - Y = 0 \tag{7.105}$$

in which Y evolves (increases) as the material hardens.

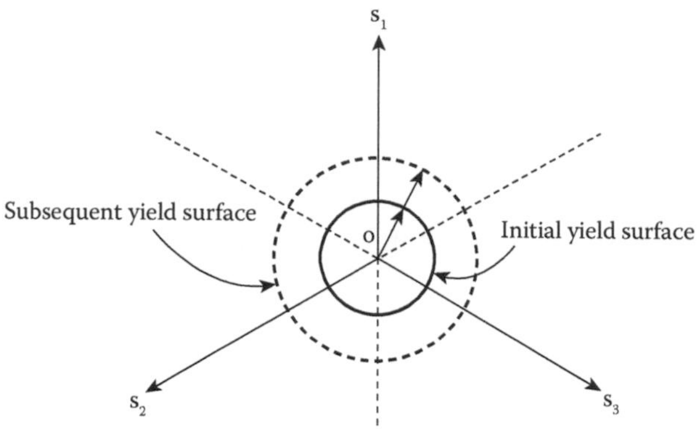

Figure 7.32 Isotropic hardening for von Mises yield surface.

Up to this point, we have assumed that the material *strain hardens* (i.e., the evolution of Y is due to the accumulation of plastic strain).

Alternatively, it can be assumed that the material *work hardens* (i.e., the evolution of Y is due to plastic work done by the stresses on the material).

In either case, the definition of the hardening parameter h remains the same (i.e., $d\sigma_e = h\ d\varepsilon_e^p$), as does the determination of h from a uniaxial test.

The consistency condition that ensures that the stress point always remains on the yield surface during hardening is

$$df = d\sigma_e - dY = 0 \tag{7.106}$$

For strain hardening models,

$$dY = d\sigma_e = h\ d\varepsilon_e^p \tag{7.107}$$

and for work hardening models,

$$dY = \frac{h}{\sigma_e} dW^p \tag{7.108}$$

where the plastic work is defined as

$$dW^p \equiv \sigma_{ij} d\varepsilon_{ij}^p = \sigma_{ij} d\varepsilon_e^p n_{ij} = \sigma_e d\varepsilon_e^p \tag{7.109}$$

Therefore, we have the evolutionary equations

$$\begin{cases} Y(\varepsilon_e^p + d\varepsilon_e^p) = Y(\varepsilon_e^p) + h\ d\varepsilon_e^p \\ Y(W^p + dW^p) = Y(W^p) + \dfrac{h}{\sigma_e} dW^p \end{cases} \tag{7.110}$$

The strain hardening and work hardening evolutionary equations are general and can be applied to any yield surface.

Isotropic hardening is a reasonable model for monotonic proportional loading paths. However, it is not an accurate representation of material behavior under cyclic stresses (Figure 7.33).

Consider how an isotropic hardening model behaves in a uniaxial loading and unloading cycle in which the initial yield stress is exceeded in tension. If the stress is reversed, yielding in compression does not occur until the previous maximum tension yield stress is exceeded. This does not agree with observations from uniaxial cyclic tests on metals, in which the initial yield

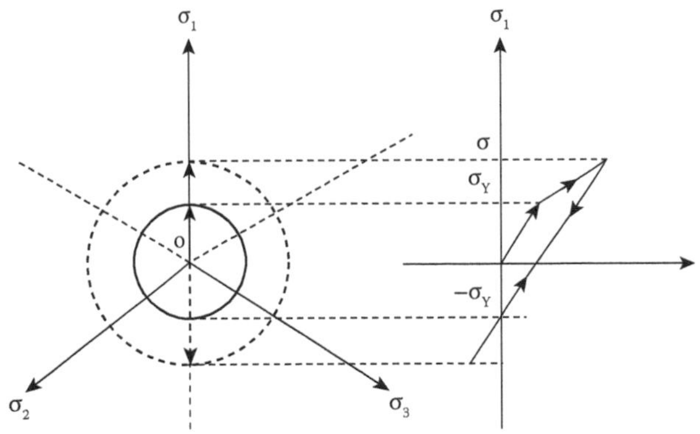

Figure 7.33 Isotropic hardening model for cyclic stress.

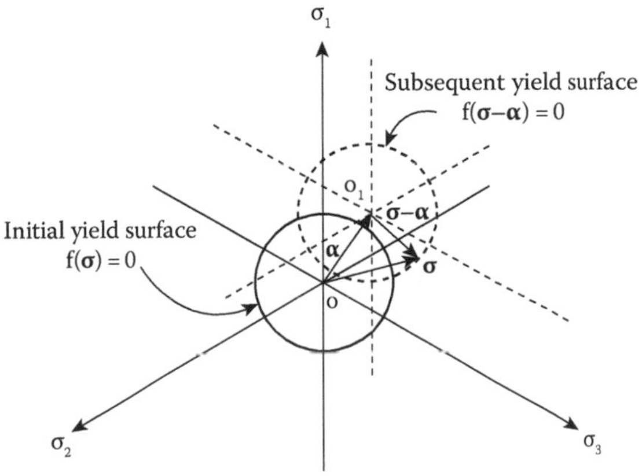

Figure 7.34 Illustration of kinematic hardening.

stress in compression is *reduced*, not increased, by prior yielding in tension. This phenomenon is known as the *Bauschinger effect.*

In addition, if after first yielding in tension a series of tension and compression cycles are performed at stresses below the first tension yield, no hysteresis occurs in the isotropic hardening model. This again does not comport with observations from conducting cyclic testing on metals.

A more reasonable hardening rule will allow at least a simplified representation of hysteresis under cyclic stress. The kinematic hardening rule, which will be described next, is such a hardening rule.

7.8.3 Kinematic hardening and back stress

Kinematic hardening is based on the assumption that hardening causes the yield surface to translate in stress space while its size and shape remain unchanged. To describe the movement of the yield surface, a tensor $\boldsymbol{\alpha}$, called the *back stress*, is introduced that defines the current origin of the shifted yield surface, that is, the elastic domain.

As shown in Figure 7.34, the yield surface is defined by

$$f(\boldsymbol{\sigma} - \boldsymbol{\alpha}) = 0 \tag{7.111}$$

The consistency condition for kinematic hardening is

$$\frac{\partial f}{\partial(\sigma_{ij}-\alpha_{ij})}(d\sigma_{ij}-d\alpha_{ij}) = n_{ij}d\sigma_{ij} - n_{ij}d\alpha_{ij} = 0 \tag{7.112}$$

where

$$n_{ij} \equiv \frac{\partial f}{\partial(\sigma_{ij}-\alpha_{ij})} \tag{7.113}$$

and therefore

$$n_{ij}d\alpha_{ij} = d\sigma_e = h\ d\varepsilon_e^p \tag{7.114}$$

The scalar Equation 7.114 does not completely determine the increment of back stress; we also need to define (choose) its direction.

According to *Prager's rule*, the increment of back stress is taken to be in the direction of the normal to the shifted yield surface, that is, $d\boldsymbol{\alpha} = d\mu\mathbf{n}$, which leads to

$$d\alpha_{ij} = \frac{h\ d\varepsilon_e^p}{n_{kl}n_{kl}} n_{ij} \tag{7.115}$$

Prager's rule for kinematic hardening is illustrated in Figure 7.35a for a yield surface in tension–torsion stress space.

Another well-known kinematic hardening model is *Ziegler's rule*, which assumes the increment of back stress is in the direction of the shifted stress, that is, $d\boldsymbol{\alpha} = d\eta(\boldsymbol{\sigma} - \boldsymbol{\alpha})$, which leads to

$$d\alpha_{ij} = \frac{h d\varepsilon_e^p}{n_{kl}(\sigma_{kl}-\alpha_{kl})}(\sigma_{ij}-\alpha_{ij}) \tag{7.116}$$

Ziegler's rule is illustrated in Figure 7.35b for a yield surface in tension–torsion stress space.

Figure 7.36 shows how a kinematic hardening model behaves in a uniaxial loading and unloading cycle.

We see that the initial yield stress in compression is *reduced*, rather than increased, by the prior plastic excursion in tension. In fact, the total elastic range (from current compression yield to current tension yield) remains constant in kinematic hardening models, whereas in isotropic hardening models, the total elastic range continually expands with additional hardening.

While kinematic hardening may be more realistic than isotropic hardening, it does not accurately replicate the Bauschinger effect either. It appears that for most metals, the reduction in compression yield stress lies somewhere between that predicted by isotropic hardening and that predicted by kinematic hardening, which naturally suggests the use of a combined isotropic–kinematic hardening model.

7.8.4 Combined isotropic and kinematic hardening

Uniaxial tests on ductile metals indicate that hardening behavior after yielding contains some of the elements of both simple isotropic and kinematic hardening. A combination of these two hardening models allows both the position and the size of the yield surface to change, while preserving its shape, as shown in Figure 7.37.

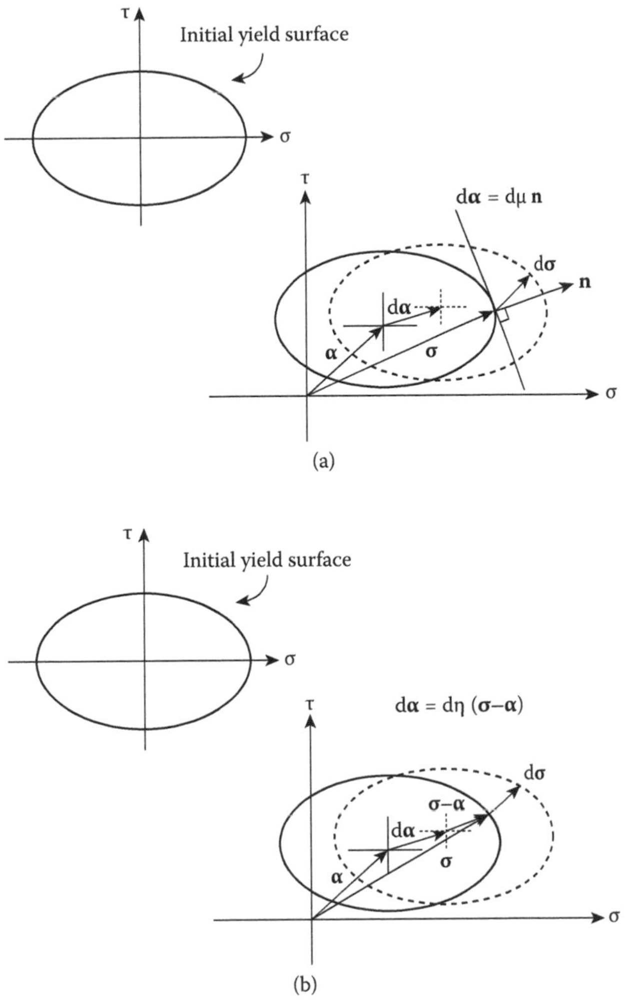

Figure 7.35 Two kinematic hardening models for von Mises yield surface in tension–torsion test: (a) Prager's rule and (b) Ziegler's rule.

The yield surface for combined isotropic–kinematic hardening can be written in the following general form:

$$f(\boldsymbol{\sigma} - \boldsymbol{\alpha}, Y(h)) = 0 \tag{7.117}$$

The consistency condition follows as

$$\frac{\partial f}{\partial(\sigma_{ij} - \alpha_{ij})}(d\sigma_{ij} - d\alpha_{ij}) - dY = 0 \tag{7.118}$$

which leads to

$$n_{ij} d\alpha_{ij} + dY = h\ d\varepsilon_e^p \tag{7.119}$$

The total hardening is therefore split into two components, which can be apportioned as desired.

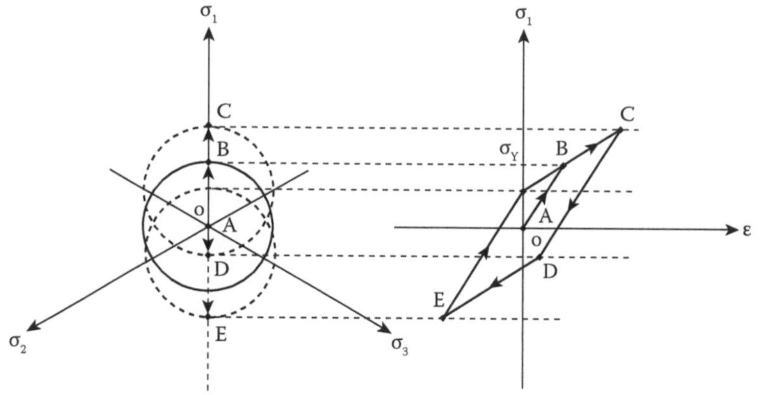

Figure 7.36 Kinematic hardening and cyclic stresses.

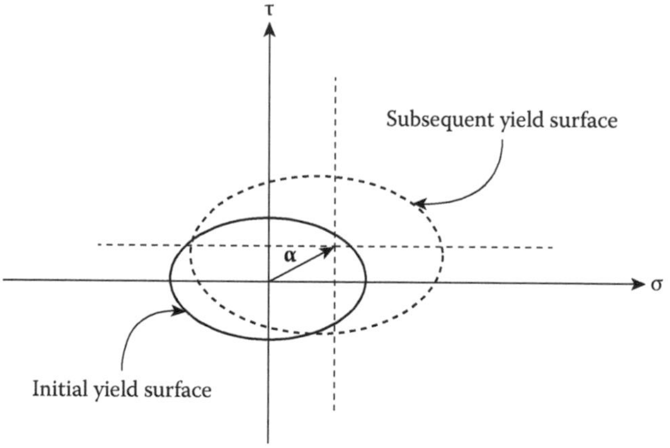

Figure 7.37 Combined isotropic and kinematic hardening.

We take, say, a proportion β ($0 \leq \beta \leq 1$) of the total hardening as kinematic hardening (with Prager's rule) and a proportion ($1 - \beta$) as isotropic hardening, leading to

$$d\alpha_{ij} = \beta \frac{h\ d\varepsilon_e^p}{n_{kl} n_{kl}}\ n_{ij} \tag{7.120}$$

$$dY = (1 - \beta)\ h\ d\varepsilon_e^p \tag{7.121}$$

With the combined isotropic–kinematic hardening model, different "degrees" of Bauschinger effect can be modeled by adjusting the parameter β.

With these three simple hardening models, the *shape* of the yield surface remains unchanged. This should not be taken to mean that this is an observed material behavior, although it is difficult experimentally to change the shape of the yield surface.

A change of shape usually consists of the formation of corners in the loading direction. This is due to dislocations in the crystalline structure of the materials, especially metals and geomaterials. Many researchers have observed the formation of corners in the yield surface during the hardening process in tests on metals.

This shows again that real material behavior, even for ductile metals, is much more complex than that represented by the simple isotropic and kinematic hardening models discussed here.

7.9 STRESS UPDATE

In a total Lagrangian formulation, as the nonlinear numerical solution is advanced step-by-step, the state variables are stored and referenced to the original undeformed configuration.

Stresses are updated at each iteration after an updated strain increment at each material point has been determined from a structural-level iteration.

The stress update at a material point is determined by numerically integrating a rate equation over the finite strain increment. The (nonobjective) material rate of Cauchy stress derive in Equation 3.61 consists of two distinct components.

$$\dot{\bar{\boldsymbol{\sigma}}} = \bar{\boldsymbol{\sigma}}^{GN} + \boldsymbol{\Omega}\bar{\boldsymbol{\sigma}} + \bar{\boldsymbol{\sigma}}\boldsymbol{\Omega}^{T} \tag{7.122}$$

The (objective) Green–Naghdi rate of Cauchy stress $\bar{\boldsymbol{\sigma}}^{GN}$ (which can be related to the deformation rate $\mathbf{D}$) contains all the material constitutive information.

The material rotation effects are "removed" via the integration of

$$\dot{\bar{\boldsymbol{\sigma}}}^{R} = \boldsymbol{\Omega}\bar{\boldsymbol{\sigma}} + \bar{\boldsymbol{\sigma}}\boldsymbol{\Omega}^{T} \tag{7.123}$$

We consider this aspect first. The numerical integration of Equation 7.123 over a small but finite strain increment is not as simple as it might appear to be at first glance. The integration must be, in effect, a similarity transformation over the finite step so that the invariants of the Cauchy stress are preserved.

Hughes and Winget (1980) developed a numerical method that satisfies this incremental objectivity requirement for the Jaumann stress rate version of Equation 7.123, in which the spin tensor $\mathbf{W}$ replaces the material rotation rate $\boldsymbol{\Omega}$. Their algorithm specifically requires for its validity that midpoint numerical integration be used for the stress update, which may be a slight disadvantage because of the additional computational effort required.

Alternatively, if the formulation is expressed in terms of the (objective) material rate of unrotated Cauchy stress $\dot{\bar{\boldsymbol{\sigma}}}'$ and the unrotated deformation rate $\mathbf{d}$, the rotational effects are automatically accounted for, so that the constitutive portion of the update is carried out separately.

For the numerical integration of the rate equations, we use a fully implicit scheme that is unconditionally stable and does not require midpoint evaluations. So, for example,

$$\begin{aligned} \dot{\bar{\boldsymbol{\sigma}}}'\Delta t &\approx \bar{\boldsymbol{\sigma}}'_{n+1} - \bar{\boldsymbol{\sigma}}'_{n} \\ \mathbf{d} &\approx \mathbf{d}_{n+1} \end{aligned} \tag{7.124}$$

where $\mathbf{d}_{n+1}$ is ultimately determined from the structural displacement increment $\Delta U_{n+1} \equiv U_{n+1} - U_n$ and might therefore be more properly considered a value at the midpoint of the interval.

We note in passing that the numerical integration of the rotational effects (Equation 7.123) can be carried out quite simply using the material rotation tensor $\mathbf{R}$ as follows:

$$\Delta\bar{\boldsymbol{\sigma}}^{R}_{n+1} = \mathbf{R}_{n+1}\mathbf{R}^{T}_{n}\ \bar{\boldsymbol{\sigma}}_{n}\ \mathbf{R}_{n}\mathbf{R}^{T}_{n+1} \equiv \mathbf{Q}_{n+1}\bar{\boldsymbol{\sigma}}_{n}\mathbf{Q}^{T}_{n+1} \tag{7.125}$$

in which the matrix $\mathbf{Q}_{n+1}$ is orthonormal as required.

We next utilize the simple shear problem to demonstrate the complete stress update procedure for von Mises plasticity with isotropic hardening.

7.9.1 von Mises plasticity in simple shear problem

The simple shear problem, for which the kinematics are completely known, essentially provides a microcosm of the overall structural-level process in which the stress correction procedure can be isolated for study.

We imagine, for purposes of illustration, that the simple shear "element" is a representative material point in a structural-level incremental-iterative finite element analysis.

We first briefly outline the overall structural solution process here in order to provide a context for what follows: Loads are applied incrementally, and corresponding (nodal) displacement increments are computed at the structural level, and returned to the material points for stress updating. Iterations are performed at the structural level (at a fixed load) until convergence is deemed to have occurred. We remark as an aside that it is important that the state variables are updated only after the iterations have converged.

We assume that the structural-level solution has converged at discrete load level $\mathbf{P}_n$, and that all state variables have been updated at each material point. The next load increment $\Delta\mathbf{P}_{n+1} = \mathbf{P}_{n+1} - \mathbf{P}_n$ is then applied and the nodal displacement increment $\Delta\mathbf{U}_{n+1}^{(1)}$ is computed. Iterations are performed until global equilibrium is satisfied to within a tolerance, and then the next load increment is applied, and so on. At each iteration at the fixed load level $\mathbf{P}_{n+1}$, stresses must be updated to provide information for the next iteration.

In our example, these iterations at the fixed load level $\mathbf{P}_{n+1}$ do not occur, because the exact displacement increment $\Delta\mathbf{U}_{n+1}$ is known (specified); that is, this example is a displacement (strain)-driven process.

We recall some kinematic information that was developed in Chapter 2.

Figure 7.38 shows the geometry of the simple shear problem. The displacements at load level n are

$$
\begin{aligned}
\overline{x}_{1n} &= x_1 + \kappa_n \; x_2 \\
\overline{x}_{2n} &= x_2 \\
\overline{x}_{3n} &= x_3
\end{aligned}
\tag{7.126}
$$

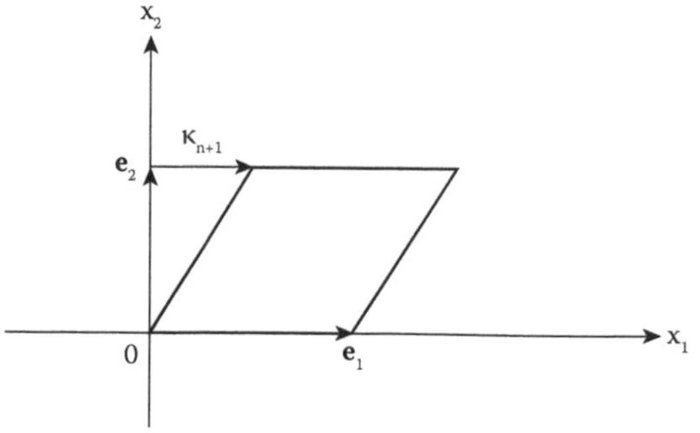

Figure 7.38 Geometry of simple shear problem.

The deformation gradient at load level n is

$$\mathbf{F}_n = \begin{bmatrix} 1 & \kappa_n & 0 \\ 0 & 1 & 0 \\ 0 & 0 & 1 \end{bmatrix} \tag{7.127}$$

and at (n + 1),

$$\mathbf{F}_{n+1} = \begin{bmatrix} 1 & \kappa_{n+1} & 0 \\ 0 & 1 & 0 \\ 0 & 0 & 1 \end{bmatrix} \tag{7.128}$$

which are assumed known. (Equation 7.128 would be determined from $\Delta \mathbf{U}_{n+1}$, which is computed at the structural level.) The remaining kinematic information follows directly.

$$\mathbf{D}_{n+1} = \frac{\Delta\kappa_{n+1}}{2}\begin{bmatrix} 0 & 1 & 0 \\ 1 & 0 & 0 \\ 0 & 0 & 0 \end{bmatrix} \tag{7.129}$$

The task is now to update the stress corresponding to the updated deformation increment. The overall process is

1. Compute the polar decomposition of the updated deformation gradient: $\mathbf{F}_{n+1} = \mathbf{R}_{n+1}\mathbf{U}_{n+1}$.
2. Compute the unrotated deformation rate: $\mathbf{d}_{n+1} = \mathbf{R}^T_{n+1}\mathbf{D}_{n+1}\mathbf{R}_{n+1}$.
3. Invoke the (von Mises) stress update algorithm to update the unrotated Cauchy stress $\bar{\boldsymbol{\sigma}}'_{n+1}$.
4. Update the Cauchy stress at (n + 1): $\bar{\boldsymbol{\sigma}}_{n+1} = \mathbf{R}_{n+1}\bar{\boldsymbol{\sigma}}'_{n+1}\mathbf{R}^T_{n+1}$.
5. Update the second Piola–Kirchhoff stress at (n + 1): $\boldsymbol{\sigma}_{n+1} = |\mathbf{F}|_{n+1}\bar{\mathbf{F}}_{n+1}\bar{\boldsymbol{\sigma}}_{n+1}\bar{\mathbf{F}}^T_{n+1}$.

In step 3, we are effectively extending the von Mises small strain plasticity model to the realm of large plastic strains by relating unrotated Cauchy stress $\bar{\boldsymbol{\sigma}}'$ to the unrotated deformation rate d, which are work conjugate, as we have seen previously.

This may appear questionable, but there is some rationale for doing so. If the material principal strain directions were to remain constant, this would amount to relating true (Cauchy) stress to logarithmic (natural) strain. Although uniaxial test results are often reported as the measured load versus elongation response of a test specimen, the test data can be (at least approximately) converted to a true stress versus natural strain relation to provide a more accurate hardening modulus h versus equivalent plastic strain relation.

7.9.2 Stress update algorithm

The general stress update algorithm is illustrated schematically in Figure 7.39. The process is compatible with the concepts and notation introduced previously, in that a trial elastic stress is first calculated, which determines the loading regime (elastic loading and unloading, plastic loading, etc.). The stresses shown in the figure are in this case unrotated Cauchy stresses $\bar{\boldsymbol{\sigma}}'$, and the (incremental) strain is the unrotated deformation rate **d**.

If plastic loading occurs, a stress correction is determined, such that the updated stress $\boldsymbol{\sigma}_{n+1}$ is "precisely" on the updated yield surface. The noteworthy feature of the particular stress update procedure illustrated in the figure is that it involves the normal $\mathbf{n}_{n+1}$ at the *updated*

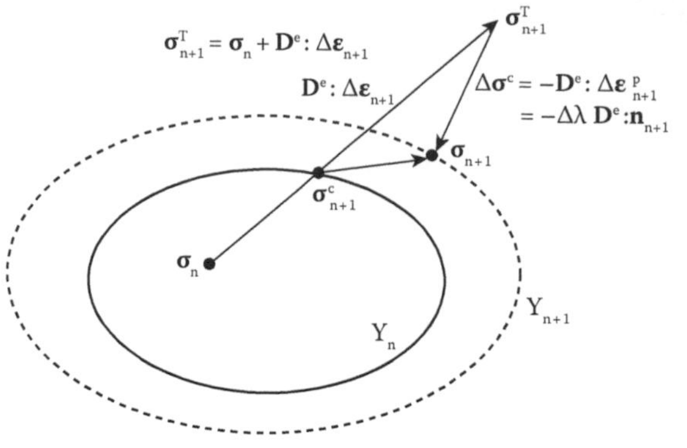

Figure 7.39 Stress update procedure (schematic).

stress point; it is therefore fully *implicit*, and in some cases may require the solution of simultaneous equations, although not for the von Mises yield function.

7.9.2.1 *Calculation of updated stress*

The updated stress $\boldsymbol{\sigma}^{n+1}$ is

$$\sigma_{ij}^{n+1} = \sigma_{ij}^{n} + D_{ijkl}^{e} \Delta\varepsilon_{kl}^{n+1} + \Delta\sigma_{ij}^{c} = \sigma_{ij}^{T} + \Delta\sigma_{ij}^{c} \tag{7.130}$$

In Equation 7.130, we note that all variables are referred to the fixed spatial coordinate system, so that components can be directly added or subtracted as needed.

7.9.2.2 *Elastic trial stress at step (n + 1)*

Computation of the elastic trial stress requires the use of a (hypoelastic) material model connecting suitable stress and strain rates. As before, we choose the isotropic fourth-order tensor given in Equation 7.66.

In the current configuration, the Green–Naghdi rate of Cauchy stress can be related to the deformation rate as follows:

$$\overline{\sigma}_{ij}^{GN} = D_{ijkl}^{e} D_{kl} = 2GD_{ij} \tag{7.131}$$

Since the hypoelastic elastic property tensor is a (fourth-order) isotropic tensor, Equation 7.131 can be "unrotated" to yield the equivalent expression,

$$\dot{\overline{\sigma}}_{ij}' = 2Gd_{ij} \tag{7.132}$$

since $\dot{\overline{\boldsymbol{\sigma}}}' = \mathbf{R}^{T}\overline{\boldsymbol{\sigma}}^{GN}\mathbf{R}$ and $\mathbf{d} = \mathbf{R}^{T}\mathbf{D}\mathbf{R}$, as previously noted.

The elastic trial stress is then

$$\overline{\boldsymbol{\sigma}}_{n+1}^{T} = \overline{\boldsymbol{\sigma}}_{n}' + 2G\mathbf{d}_{n+1} \tag{7.133}$$

The next step is the determination of the stress correction.

Before we proceed, we first evaluate the accuracy of the elastic response in simple shear. Dienes (1979) provided (lengthy) closed-form expressions for the Cauchy stresses assuming the constitutive relation shown in Equation 7.131.

We choose a step size of $\Delta\kappa = 0.002$ and carry the calculations out to $\kappa/2 = 1$.

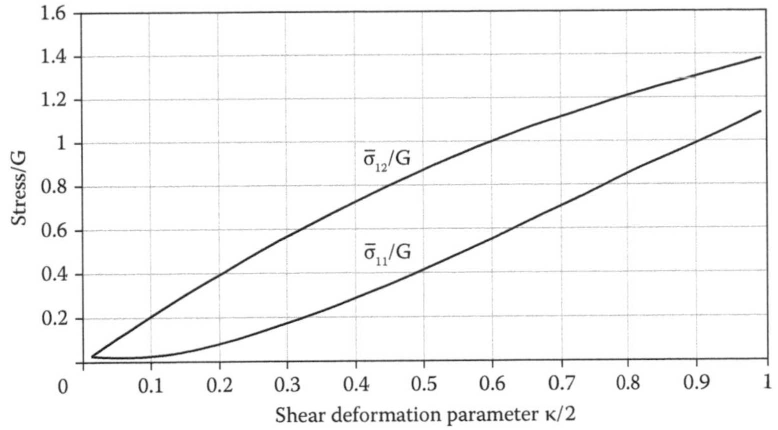

Figure 7.40 Elastic Cauchy stresses in simple shear.

The Cauchy stresses normalized by the shear modulus G are shown in Figure 7.40.

We observe a significant rotation of the Cauchy stress during the elastic deformation. In the early stages, the state of stress is essentially one of pure shear with $\bar{\sigma}_{11} \equiv -\bar{\sigma}_{22} \approx 0$. As the deformation proceeds, the principal stress axes gradually rotate clockwise until, at $\kappa/2 = 1$, $\bar{\sigma}_{11}$ exceeds 80% of $\bar{\sigma}_{12}$.

The calculated values agree closely with the analytical solution given by Dienes (1979) and cannot be distinguished in the plot. At $\kappa/2 = 1$, the calculated values of $\bar{\sigma}_{12}/G$ and $\bar{\sigma}_{11}/G$ are (1.3873, 1.1406) versus those of Dienes (1.3863, 1.1416).

7.9.2.3 Stress correction

In Equation 7.130, the finite stress correction increment is

$$\Delta\sigma^c_{ij} = -D^e_{ijkl}\Delta\varepsilon^p_{kl} \tag{7.134}$$

and the finite increment of plastic strain is

$$\Delta\varepsilon^p_{ij} = \Delta\lambda \; n^{n+1}_{ij} = \Delta\lambda \; \frac{s^{n+1}_{ij}}{Y^{n+1}} \tag{7.135}$$

Therefore,

$$\Delta\sigma^c_{ij} = -D^e_{ijkl} \; \Delta\lambda \; n^{n+1}_{kl} = -2G \; \Delta\lambda \; \frac{s^{n+1}_{ij}}{Y^{n+1}} \tag{7.136}$$

We remark that in Equation 7.136, the following intermediate result has been used:

$$D^e_{ijkl}s_{kl} = 2G\left[\frac{1}{2}(\delta_{ik}\delta_{jl} + \delta_{il}\delta_{jk}) + \frac{\nu}{1-2\nu}(\delta_{ij}\delta_{kl})\right]s_{kl} = 2G \; s_{ij} \tag{7.137}$$

It is interesting that in this case (i.e., for isotropic elasticity and von Mises plasticity), the correction is actually a *radial* return in stress space. This is not true in general.

Using Equation 7.136 in Equation 7.130, we obtain the stress update equation

$$\sigma_{ij}^{n+1} = \sigma_{ij}^{T} - 2G\ \Delta\lambda \frac{s_{ij}^{n+1}}{Y^{n+1}} \tag{7.138}$$

We now have to apply the isotropic hardening rule to determine the (finite) plastic strain increment over the stress path. We first take the trace of Equation 7.138 to obtain

$$\sigma_{kk}^{n+1} = \sigma_{kk}^{T} \tag{7.139}$$

and thereby restate Equation 7.138 in terms of deviator stresses,

$$s_{ij}^{n+1} = s_{ij}^{T} - 2G\ \Delta\lambda \frac{s_{ij}^{n+1}}{Y^{n+1}} \tag{7.140}$$

This equation can be rearranged as follows:

$$\left[1 + \frac{2G}{Y^{n+1}} \Delta\lambda\right] s_{ij}^{n+1} = s_{ij}^{T} \tag{7.141}$$

We note that the bracketed term is a scalar quantity. We form tensor inner products on both sides of Equation 7.141 and then take the square root of each side to obtain the scalar relation

$$\left[1 + \frac{2G}{Y^{n+1}} \Delta\lambda\right] \sigma_{e}^{n+1} = \sigma_{e}^{T} \tag{7.142}$$

which can be written as

$$\sigma_{e}^{n+1} + 2G\Delta\lambda = \sigma_{e}^{T} \tag{7.143}$$

The total (i.e., integrated) plastic strain increment $\Delta\lambda$ (or $\Delta\varepsilon_e^p$) takes place on the actual stress path from the contact point C to the updated stress point n + 1. On this stress path,

$$d\sigma_e = h\ d\varepsilon_e^p = h\ d\lambda \tag{7.144}$$

and if we assume that the hardening coefficient h is constant over the small, but finite, load step, then $\Delta\sigma_e = h\Delta\varepsilon_e^p$.

Therefore, we have

$$\begin{cases} \sigma_e^{n+1} = \sigma_e^{C} + h\ \Delta\varepsilon_e^p \\ Y^{n+1} = Y^{n} + h\ \Delta\lambda \end{cases} \tag{7.145}$$

since the contact point C lies on the yield surface at step n.

From the second equation above,

$$\Delta\lambda = \frac{(Y^{n+1} - Y^{n})}{h}, \quad h > 0 \tag{7.146}$$

and

$$\sigma_e^{n+1} = \frac{h\sigma_e^{T} + 2GY^{n}}{h + 2G}, \quad h \geq 0 \tag{7.147}$$

The last equation is also valid for ideal plasticity ($h=0$), as it then merely states that $\sigma_e^{n+1} = Y^{n}$.

Using Equations 7.146 and 7.147 in Equation 7.141, we obtain

$$s_{ij}^{n+1} = \alpha\ s_{ij}^{T}, \quad \text{where } \alpha \equiv \left[\frac{h\sigma_e^T + 2GY^n}{(h+2G)\sigma_e^T}\right] \tag{7.148}$$

and finally,

$$\sigma_{ij}^{n+1} = \alpha\ s_{ij}^{T} + \frac{1}{3}\sigma_{kk}^{T}\delta_{ij} \tag{7.149}$$

We note that no special treatment is needed for a strain increment during which purely elastic loading is followed by elastoplastic loading. Early numerical stress update procedures typically involved determination of the contact point on the yield surface and subdivision of the strain increment into elastic loading and elastoplastic loading segments.

The corrected Cauchy stress from Equation 7.149 is then rotated back into the current configuration at step (n+1) via

$$\overline{\boldsymbol{\sigma}}_{n+1} = \mathbf{R}_{n+1}\overline{\boldsymbol{\sigma}}'_{n+1}\mathbf{R}_{n+1}^{T} \tag{7.150}$$

7.9.2.4 Elastoplastic response

We choose the following material properties for the numerical calculations:

$$\left[\begin{array}{l} E = 30{,}000 \text{ ksi} \\ \nu = 0.3 \\ \text{Initial Uniaxial Yield Stress } \sigma_Y = 60 \text{ ksi} \\ \text{Hardening Modulus } h = 0.001\ E \end{array}\right. \tag{7.151}$$

The "load" step is $\Delta\kappa = 0.002$, so that (only) the first step is elastic. This would therefore be considered a relatively large strain increment. The calculations described above are carried out for 1000 equal load steps to a final value of $\kappa/2 = 1$.

Figure 7.41 shows the calculated normalized Cauchy stresses $\overline{\sigma}_{12}/\sigma_Y$ and $\overline{\sigma}_{11}/\sigma_Y$. First yielding theoretically occurs in a state of pure shear at $\sigma_Y/\sqrt{3} = 0.577\sigma_Y$ according to the von Mises yield criterion, as shown in the figure.

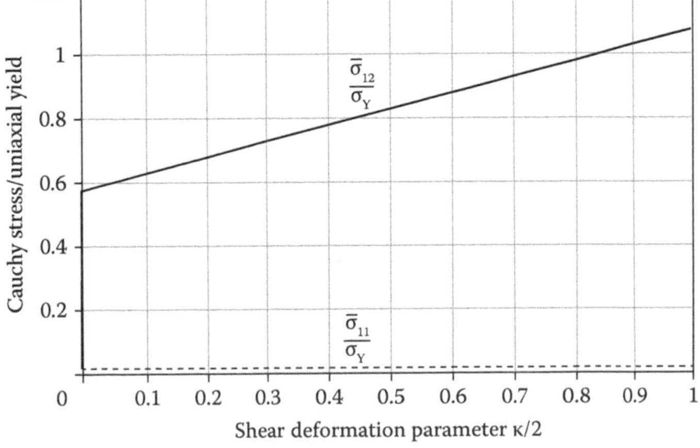

Figure 7.41 Cauchy stresses normalized by uniaxial yield stress σ_Y.

We observe that during the subsequent elastoplastic response, the principal stress axes essentially do not rotate, in contrast to the purely elastic response. In the elastoplastic case, there is a very small rotation of the Cauchy stress only during the first two load steps. Once yielding has occurred, and the (pure shear) Cauchy stress is then on the yield surface, the elastic trial stress increments and the stress corrections are both normal to the von Mises yield surface. The former is in the outward normal direction, and the latter in the opposite direction, but of slightly smaller magnitude due to the isotropic hardening, which causes the yield surface to slowly expand.

7.10 PLASTICITY MODELS FOR FRICTIONAL AND PRESSURE-SENSITIVE MATERIALS

In the previous sections, we discussed material models for ductile materials and we observed that ductile yielding is the dominant feature of the inelastic behavior of metals. As mentioned previously, it has been observed that ductile yielding is not sensitive to changes in the mean (compressive) pressure.

The inelastic behavior of frictional materials, on the other hand, is strongly influenced by the mean pressure. We consider geomaterials, soils, rocks, concrete, and granular media as frictional materials.

The behavior of frictional materials differs from that of metals in significant ways. Unlike in metals, there is not a distinct transition from elastic to elastoplastic material behavior. Nearly elastic behavior occurs in frictional materials only in the early stages of unloading and reloading. The well-known yield surfaces that have been proposed can be more accurately described as failure surfaces. In some frictional materials, such as rock and concrete, failure is a more clearly identifiable state. The failure surfaces can be considered as perfectly plastic, or limiting, surfaces within which a yield surface can also be defined.

The methods of testing frictional materials are also significantly different than the testing methods used for metals. As we have seen, a uniaxial tension test is the most common strength test for metals. Frictional materials typically have very little tensile strength; they are strong in compression and relatively weak in shear. Tests of frictional materials are primarily aimed at determining their shear strength. The preferred test methods are often triaxial tests; however, direct shear, simple shear, and unconfined compression tests are sometimes employed. The most commonly used tests for frictional materials have been discussed in Chapter 6.

The simplest conceptual friction model is a rough surface between a block and a solid, as shown in Figure 7.42. The block is pressed against the surface with a constant normal stress σ and is subjected to an increasing shear stress τ (i.e., a normalized horizontal force). If we

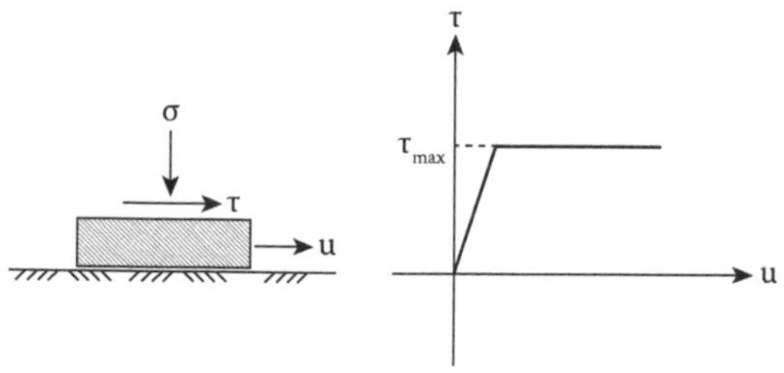

Figure 7.42 Simple conceptual frictional model.

assume that the block is relatively stiff, then the displacement u of the block will be primarily due to shear deformation in the thin layer at the contact surfaces. The shear deformation will be initially "elastic"; that is, the horizontal displacement will increase linearly with increasing shear stress, until the shear stress reaches a limiting value τ_{max}, whereupon the shear deformation (i.e., the horizontal relative displacement u) increases at a constant value of shear stress.

The shear stress–strain curve shown in Figure 7.42 is similar in appearance to the elastoplastic ductile stress–strain behavior of metals, although the underlying mechanism in a frictional model is different from the ductile behavior of metals. We remark in passing that one would have to introduce a length scale representing the (arbitrary) thickness of a deformable layer between the block and the surface in order to properly interpret the horizontal relative displacement as a shear *strain*.

It is due to this apparent similarity that the terminology and some of the modeling methodology for frictional materials are carried over from the modeling of ductile materials. For example, the onset of frictional sliding can be termed yielding and the envelope of such points in the stress space can be termed a yield surface.

In this simple conceptual model, the shear stress at which frictional sliding begins depends critically on the applied normal stress. This dependence is described by the Coulomb relation

$$\tau_{max} = c - \sigma \tan \Phi \tag{7.152}$$

In Equation 7.152, c is the apparent cohesion and Φ is the angle of friction.

An important point to keep in mind in generalizing the contact surface model to a multiaxial stress state is that frictional sliding (or yielding or failure) often occurs along a distinct failure surface or plane, whose location and orientation are determined by the state of stress in the material.

The failure criterion of Equation 7.152 is shown on the three-dimensional Mohr's circle (Figure 7.43). The maximum shear occurs in the plane of maximum and minimum principal stress (σ_1, σ_3), assuming that the principal stresses are ordered algebraically such that $\sigma_1 \geq \sigma_2 \geq \sigma_3$. Since in frictional materials the stresses are primarily compressive, the algebraically largest stress σ_1 will normally be the smallest in absolute value, as shown in Figure 7.43. We will continue to use the "tension-positive" sign convention for normal stress.

The maximum shear is $\tau_{max} = (\sigma_1 - \sigma_3)/2$, and the normal stress on the plane of maximum shear is $\sigma = (\sigma_1 + \sigma_3)/2$.

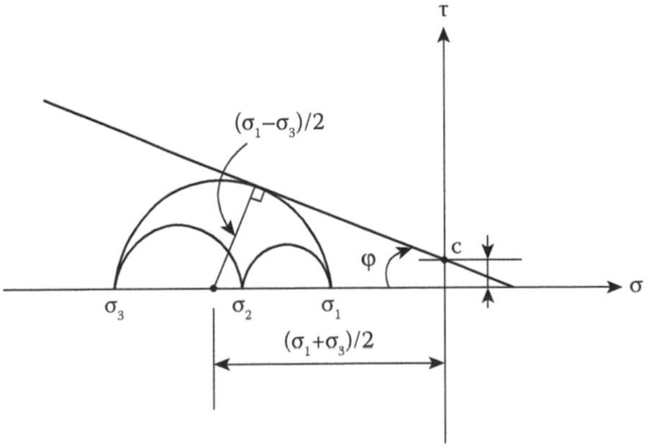

Figure 7.43 Mohr–Coulomb failure criterion illustrated in a Mohr diagram.

The failure criterion depicted in this figure for frictional materials under a multiaxial stress state is known as the Mohr–Coulomb failure criterion and has the following equation in terms of principal stresses:

$$\sigma_1 - \sigma_3 = 2c\ \cos\Phi - (\sigma_1 + \sigma_3)\sin\Phi \tag{7.153}$$

7.10.1 Mohr–Coulomb yield surface

The Mohr–Coulomb failure surface can be expressed in a more general form by rewriting Equation 7.153 in terms of stress invariants p, q, and θ. From expressions derived earlier for principal stresses, we have the following relations:

$$\begin{cases} \sigma_1 + \sigma_3 = 2p - \dfrac{2}{3}\ q\ \sin\theta \\ \sigma_1 - \sigma_3 = \dfrac{2}{\sqrt{3}} q\ \cos\theta \end{cases} \tag{7.154}$$

Substitution of the above relations in Equation 7.153 results in the following expression for the Mohr–Coulomb yield surface:

$$f(\sigma) = \frac{1}{3} q\left[\sqrt{3}\cos\theta - \sin\Phi\sin\theta\right] - [c\ \cos\Phi - p\ \sin\Phi] = 0 \tag{7.155}$$

where $-\pi/6 \leq \theta \leq \pi/6$ and θ is measured counterclockwise from the $s_2 = 0$ (pure shear) line. We note again that the above equation is derived for *tension-positive* normal stresses, so that the term $[c\ \cos\Phi - p\ \sin\Phi]$ will always be positive for compressive mean stress. The shape of the yield surface is conical with axis along the space diagonal, since the yield surface is linearly dependent on p. The cross section of the yield surface on the octahedral plane is shown in Figure 7.44. The equation for its shape may be found as follows.

The first bracketed term in Equation 7.155 can be written as

$$\left[\sqrt{3}\cos\theta - \sin\Phi\ \sin\theta\right] = \sqrt{3 + \sin^2\Phi}\cos(\theta + \alpha) \tag{7.156}$$

in which $\tan\alpha = (1/\sqrt{3})\sin\Phi$.

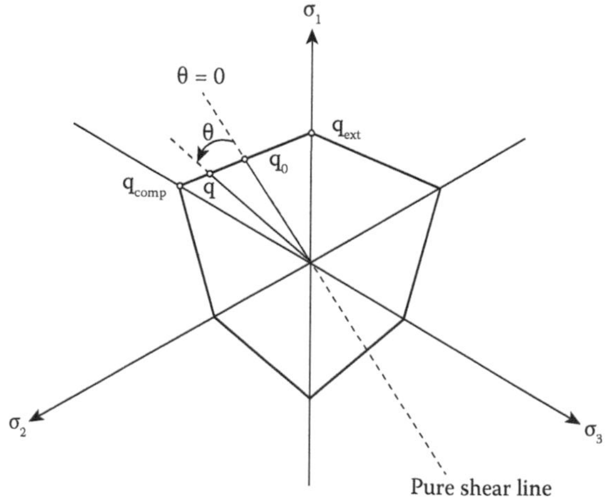

Figure 7.44 Mohr–Coulomb yield surface.

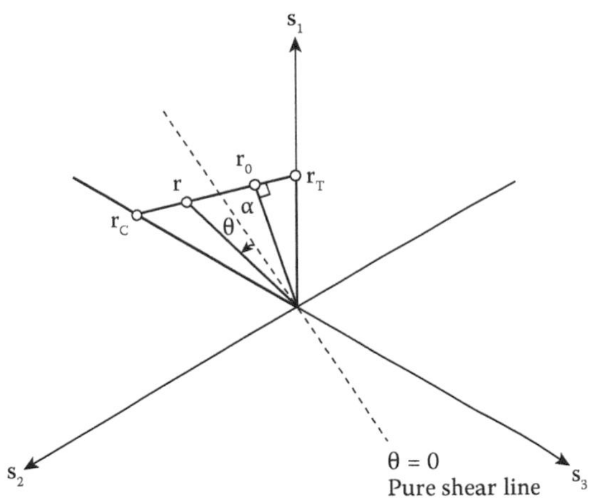

Figure 7.45 Segment of Mohr–Coulomb yield surface on octahedral plane.

Equation 7.155 can then be written more compactly as

$$r\cos(\theta + \alpha) = r_0 \tag{7.157}$$

where

$$r_0 \equiv \frac{\sqrt{6}}{\sqrt{3+\sin^2\Phi}}[c\cos\Phi - p\ \sin\Phi] \tag{7.158}$$

In Equation 7.157, $r \equiv \sqrt{2J_2}$ is the true radius in the octahedral plane and r_0, the radius at $\theta = -\alpha$, is perpendicular to the straight line that forms the segment of the yield surface in the range $-\pi/6 \le \theta \le \pi/6$ (Figure 7.45). We see that the Mohr–Coulomb yield surface resembles the Tresca yield surface except that it is not a *regular* hexagon.

On the octahedral plane, triaxial extension and triaxial compression paths start at the hydrostatic axis ($s_1 = s_2 = s_3 = 0$) and proceed along the s_1 axis ($s_1 > s_2 = s_3$). Yielding (failure) in triaxial extension corresponds to $\theta = -\pi/6$ and yielding in triaxial compression corresponds to $\theta = \pi/6$ (the latter because of the symmetries of the yield surface that are required to allow for renumbering of the principal stress directions).

On the triaxial extension and compression stress paths,

$$r = \sqrt{\frac{2}{3}}|\sigma_1 - \sigma_3| \tag{7.159}$$

so that the shear strengths in triaxial extension and triaxial compression, according to the Mohr–Coulomb yield criterion, are as follows:

$$\begin{aligned}
\left[\frac{|\sigma_1 - \sigma_3|}{2}\right]_T &= \frac{3}{3+\sin\Phi}[c\cos\Phi - p\sin\Phi] \\
\left[\frac{|\sigma_1 - \sigma_3|}{2}\right]_C &= \frac{3}{3-\sin\Phi}[c\cos\Phi - p\sin\Phi]
\end{aligned} \tag{7.160}$$

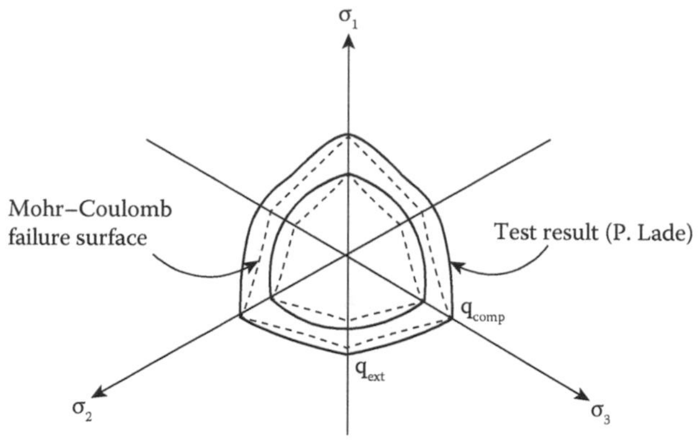

Figure 7.46 Failure surface (compression positive) for sand from true triaxial tests.

The shear strength in triaxial extension is therefore predicted to be a factor of $(3 - \sin\Phi)/(3 + \sin\Phi)$ less than that in triaxial compression, or about 0.7 for an angle of friction Φ equal to 30°.

Because of this particular feature, the Mohr–Coulomb yield surface is a far more realistic model for frictional materials than the Drucker–Prager yield surface, which will be discussed next. Most frictional materials, such as concrete and geomaterials, have high compressive strength and low tensile strength, which is one of their main features. Tests conducted on sands (Figure 7.46) have produced convex yield (failure) surfaces that resemble the Mohr–Coulomb yield surface, but with smooth corners (Lade 1977, 2014).

We note, finally, that if in the Mohr–Coulomb yield criterion the angle of friction $\Phi = 0$, then $\alpha = 0$. The yield stress in triaxial extension and compression would then be predicted to be the same, and we would have the Tresca yield surface.

7.10.2 Drucker–Prager yield surface

The Drucker–Prager yield surface was originally developed primarily for soil mechanics applications. It and enhanced versions thereof have been widely used for pressure-sensitive materials, including polymers, powders, foams, and adhesives, as well as for soils and concrete.

It is basically a generalization of the von Mises yield surface that accounts in the simplest way possible for the dependence of yielding on mean pressure. This is accomplished by including a term in the yield function that is linearly dependent on the first invariant of stress.

Therefore, the Drucker–Prager yield surface is a cone with a circular cross section and with its axis along the space diagonal in principal stress space, as shown in Figure 7.47. Because of its circular shape in the octahedral plane, we realize at the outset that the Drucker–Prager yield surface will not recognize the difference in shear strengths in triaxial extension and triaxial compression stress paths.

The primary reason is, of course, that the Drucker–Prager yield surface is a simple extension of a yield surface for ductile metals (von Mises) that does not account for the underlying frictional nature of material behavior.

However, this capability may not be essential in particular applications where the material's pressure sensitivity does not necessarily arise from a frictional mechanism.

The Drucker–Prager yield criterion may be expressed in various equivalent forms, for example,

$$f(\boldsymbol{\sigma}) = r + \beta\ I_1 - Y = 0 \tag{7.161}$$

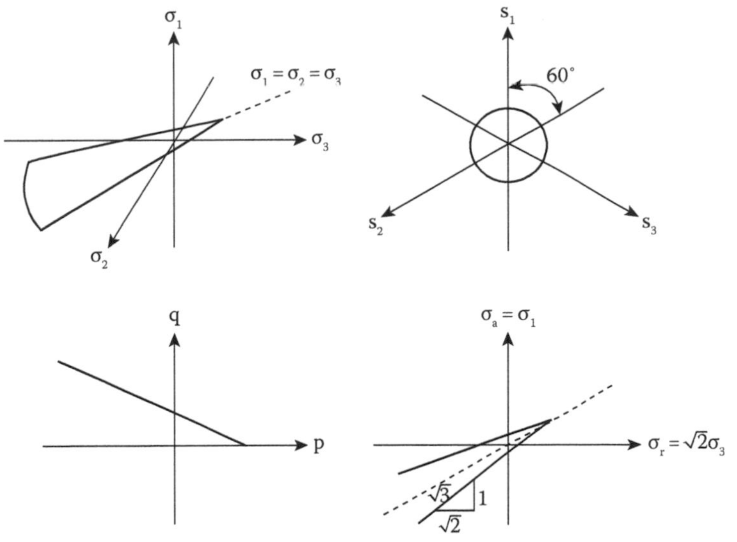

Figure 7.47 Drucker–Prager yield surface for pressure-sensitive materials.

in which $r = \sqrt{2J_2}$ and $I_1 = \sigma_{kk}$. In this equation, I_1 is the first invariant of the stress tensor and J_2 is the second invariant of the deviatoric stress tensor, as usual; β and Y are material parameters to be determined from tests.

The apex of the Drucker–Prager cone is located where $r = 0$, that is, at a distance along the hydrostatic axis equal to $\sqrt{3}p_0 = Y/\sqrt{3}\beta$, $(\beta > 0)$. For $\beta = 0$, the Drucker–Prager circular cone reduces to the von Mises circular cylinder.

The development of the basic relationships is quite similar to that for the von Mises yield surface, so we merely summarize them here.

Equivalent stress σ_e

$$\sigma_e = r + \beta\ I_1 \tag{7.162}$$

Increment of equivalent stress $d\sigma_e$

$$d\sigma_e = n_{ij} d\sigma_{ij} \tag{7.163}$$

Normal n_{ij}

$$n_{ij} = \frac{\partial f}{\partial \sigma_{ij}} = \frac{s_{ij}}{r} + \beta\ \delta_{ij}$$
$$\sqrt{n_{ij} n_{ij}} = 1 + 3\beta^2 \tag{7.164}$$

Plastic strain increment $d\varepsilon_{ij}^p$

$$d\varepsilon_{ij}^p = d\lambda\ n_{ij} = d\varepsilon_e^p n_{ij} \tag{7.165}$$

We note that the plastic volume change is nonzero, as expected,

$$d\varepsilon_{kk}^p = 3\beta\ d\varepsilon_e^p \tag{7.166}$$

due to the inclusion of the first invariant I_1 in the Drucker–Prager yield criterion. This is, of course, also the case for the Mohr–Coulomb criterion.

The Drucker–Prager model parameters Y and β can be calibrated from the results of two independent tests, for example, uniaxial compression and pure shear.

In pure shear, $r=\sqrt{2}\tau$ and $I_1=0$, from which $Y=\sqrt{2}\tau_y$, where τ_y is the yield stress in pure shear. In uniaxial compression, $r=\sqrt{2/3}|\sigma_1|$ and $I_1=-|\sigma_1|$. The Drucker–Prager parameters calibrated from these two tests are then

$$Y=\sqrt{2}\tau_y$$
$$\beta=\sqrt{\frac{2}{3}}\left[1-\frac{\sqrt{3}\tau_y}{\sigma_C}\right] \tag{7.167}$$

where σ_C is the magnitude of the uniaxial yield stress in compression.

With these parameters, the Drucker–Prager yield criterion then predicts the ratio of yield in uniaxial tension to that in uniaxial compression as

$$\frac{\sigma_T}{\sigma_C}=\left[\frac{\sqrt{3}\tau_y/\sigma_C}{2-\sqrt{3}\tau_y/\sigma_C}\right] \tag{7.168}$$

As a consistency check, if we assume $\tau_y=\sigma_C/\sqrt{3}$, which is implicit in the von Mises criterion, then $\beta=0$ and $\sigma_T=\sigma_C=\sigma_y$ and we recover the von Mises yield criterion.

7.10.3 Model refinements

Additional features have been added to these simple material models, to reflect more realistically certain aspects of observed material behavior.

7.10.3.1 Cap models

Both Mohr–Coulomb and Drucker–Prager models predict an elastic response for large principal stress combinations that lie close to the compressive hydrostatic axis, which is not reasonable for most materials.

As a result, various so-called cap models have been proposed to close off the yield surface at high compressive mean stress (Sture et al. 1989). A simple version of this concept is shown in Figure 7.48.

In the case of the particular Drucker–Prager cap model shown in the figure, the limiting surface in hydrostatic compression is a spherical cap, and the tension cutoff is a flat circular

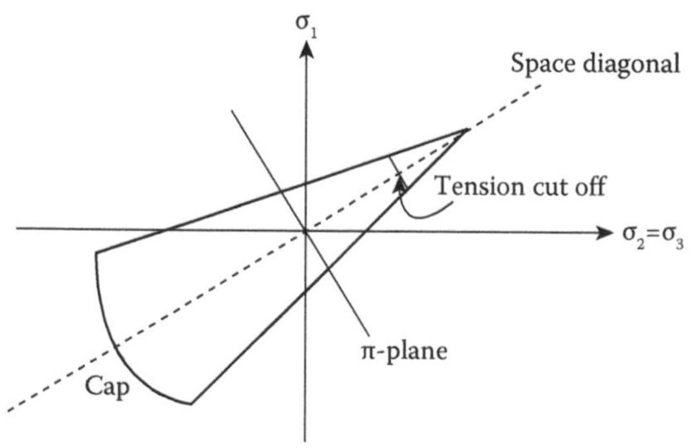

Figure 7.48 Drucker–Prager cap model with optional tension cutoff.

surface (plate). The original Drucker–Prager yield surface of course has a built-in limit in hydrostatic tension, but the tension cutoff can provide additional modeling flexibility.

For hardening plasticity models, the composite cone–cap–tension cutoff yield surface would usually be treated as an integral unit.

7.10.3.2 Refined (q,θ) shape in octahedral plane

The Mohr–Coulomb yield surface is an implicit function of the Lode angle on the octahedral plane. Its geometry is completely determined by two parameters, for example, triaxial extension and triaxial compression yield stresses. Other explicit three-invariant (p, q, θ) plasticity models that provide an adjustable smooth yield surface shape on the octahedral plane have been proposed, for applications to cohesionless soils (i.e., sands) (Lade 1977) and to concrete (Willam and Warnke 1975). These models can be used in conjunction with caps and tension cutoffs as well. Typical shapes are shown in Figure 7.46.

Chapter 8

Nonlinear solid finite elements

8.1 INTRODUCTION

In this chapter, we first briefly review the finite element discretization of one-dimensional (1D), two-dimensional (2D), and three-dimensional (3D) models of solids, which use only displacements as nodal degrees of freedom. In Chapter 9, we will consider "structural" elements, such as beams, plates, and shells, which in addition employ rotations as nodal degrees of freedom.

8.2 FINITE ELEMENT DISCRETIZATION

The element degrees of freedom are taken as the displacements at element nodes, referred to a common global Cartesian coordinate system. Nodes may be located at corners, on exterior sides or surfaces, or in the element interior, depending on the element type. Consider the simple 1D, 2D, and 3D elements shown in Figure 8.1, with two, four, and eight nodes, respectively.

8.2.1 Shape functions

For the 3D cube or "brick" element (which has eight nodes and 24 degrees of freedom), for example, the element interior displacements are related to the nodal displacements via shape functions (i.e., interpolation polynomials) as follows:

$$\begin{cases} u_x = \mathbf{N}\mathbf{U}_x \\ u_y = \mathbf{N}\mathbf{U}_y \\ u_z = \mathbf{N}\mathbf{U}_z \end{cases} \tag{8.1}$$

where u_x is the x direction displacement at a generic point (x, y, z) in the element. The vector $\mathbf{U}_x$ collects all the x direction nodal displacements of the element and is (8 × 1) in this case, and $\mathbf{N}$(x, y, z) is a (1 × 8) matrix of shape functions.

For the 2D quadrilateral "Q4" element (Figure 8.2), with four nodes and 8 degrees of freedom, the bilinear shape functions are

$$\begin{cases} N_1 = (1-\xi)(1-\eta)/4 \\ N_2 = (1+\xi)(1-\eta)/4 \\ N_3 = (1+\xi)(1+\eta)/4 \\ N_4 = (1-\xi)(1+\eta)/4 \end{cases} \tag{8.2}$$

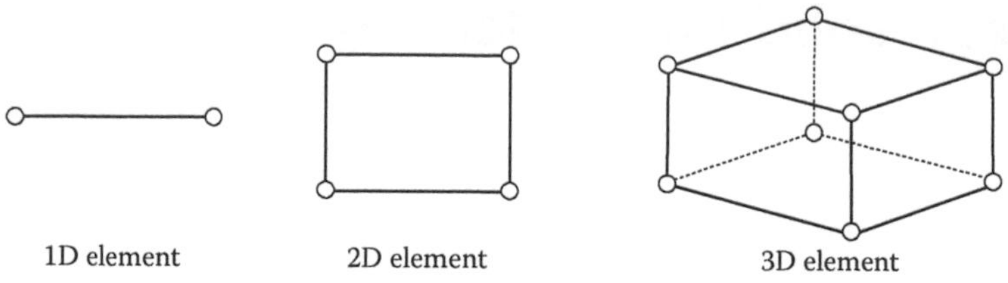

Figure 8.1 Simple 1D, 2D, and 3D solid finite elements.

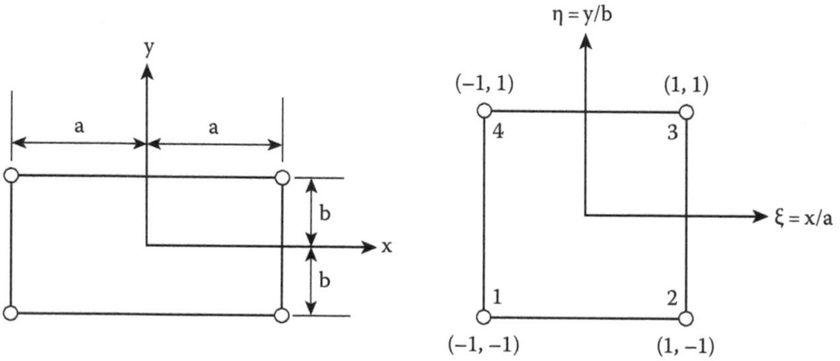

Figure 8.2 2D quadrilateral element Q4.

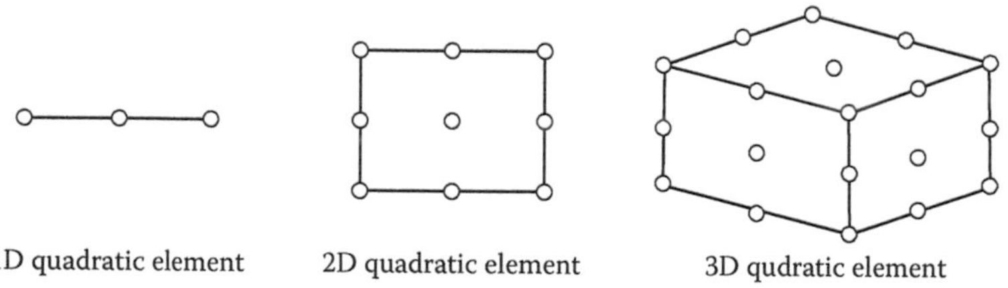

Figure 8.3 Quadratic Lagrangian elements.

where $\xi \equiv x/a$ and $\eta \equiv y/b$ are nondimensional Cartesian coordinates. These equations can be expressed in the compact form

$$N_i = \frac{1}{4}(1 + \xi_i \xi)(1 + \eta_i \eta), \quad i = 1, L, 4 \tag{8.3}$$

where (ξ_i, η_i) are the nondimensional coordinates $(\pm 1, \pm 1)$ of node i.

For the 3D brick element, with nondimensional coordinates (ξ, η, ζ) the shape functions are

$$N_i = \frac{1}{8}(1 + \xi_i \xi)(1 + \eta_i \eta)(1 + \zeta_i \zeta), \quad i = 1, L, 8 \tag{8.4}$$

Biquadratic displacement approximations can be introduced into these basic shapes as shown in Figure 8.3.

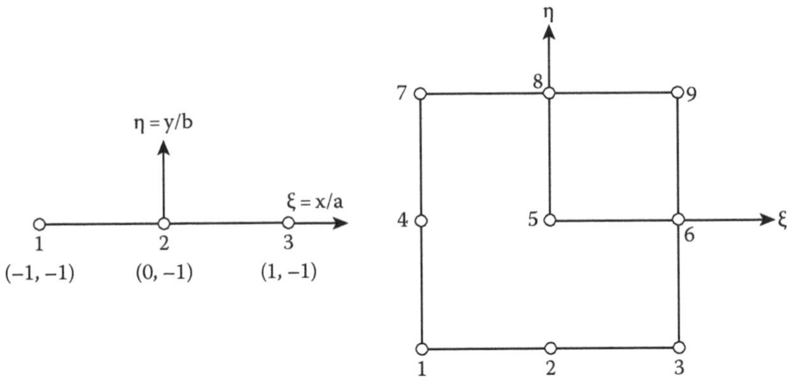

Figure 8.4 Lagrangian shape functions for 2D element Q9.

The 1D, 2D, and 3D Lagrangian elements shown in Figure 8.3 have 3, 9 (3 × 3), and 27 (3 × 3 × 3) nodes, respectively.

The term *Lagrangian* in this context refers to Lagrangian interpolation polynomials that use values of the variable at discrete points, in contrast to Hermitian polynomials, which use in addition derivatives of the variable for interpolation (e.g., as in flexural elements).

The shape functions for the 2D and 3D Lagrangian elements are simply products of the shape functions of the 1D element.

For example, consider Figure 8.4. The Lagrangian shape functions for the 1D element are parabolic functions given by

$$\begin{cases} N_1 = f_1(\xi) \equiv -\dfrac{1}{2}\xi(1-\xi) \\ N_2 = f_2(\xi) \equiv 1-\xi^2 \\ N_3 = f_3(\xi) \equiv \dfrac{1}{2}\xi(1+\xi) \end{cases} \tag{8.5}$$

The corresponding shape functions for the "Q9" 2D element are given by products of pairs of the parabolic functions f_i:

$$\begin{cases} N_1 = f_1(\xi)f_1(\eta) & N_4 = f_1(\xi)f_2(\eta) & N_7 = f_1(\xi)f_3(\eta) \\ N_2 = f_2(\xi)f_1(\eta) & N_5 = f_2(\xi)f_2(\eta) & N_8 = f_2(\xi)f_3(\eta) \\ N_3 = f_3(\xi)f_1(\eta) & N_6 = f_3(\xi)f_2(\eta) & N_9 = f_3(\xi)f_3(\eta) \end{cases} \tag{8.6}$$

The shape functions for the 27-node 3D Lagrangian element are obtained in the same way.

For example, for node 1 at $(\xi, \eta, \zeta) = (-1, -1, -1)$,

$$N_1 = f_1(\xi)f_1(\eta)f_1(\zeta) \tag{8.7}$$

8.2.1.1 Serendipity elements

The most widely used 2D and 3D elements are probably the so-called "serendipity" elements shown in Figure 8.5. They have only corner and edge nodes (8 and 20, respectively).

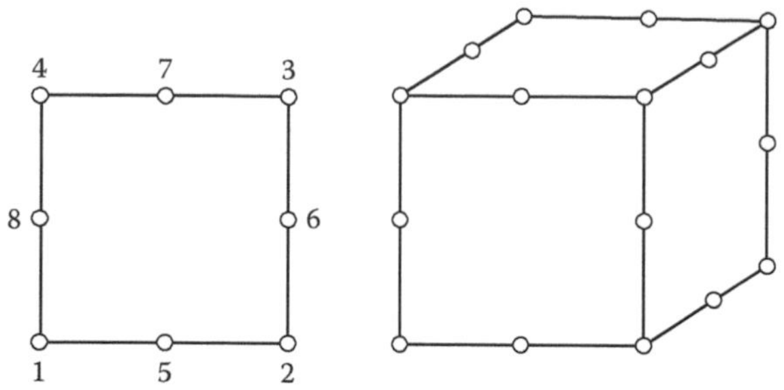

Figure 8.5 2D and 3D serendipity elements.

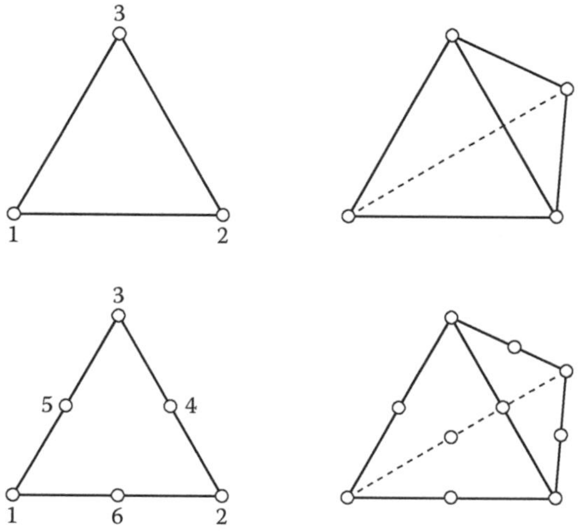

Figure 8.6 2D (triangle) and 3D (tetrahedron) simplex elements.

For the node numbering shown in the figure, the shape functions for the Q8 element are

$$\begin{cases} N_1 = (1-\xi)(1-\eta)/4 - N_5/2 - N_8/2 \\ N_2 = (1+\xi)(1-\eta)/4 - N_5/2 - N_6/2 \\ N_3 = (1+\xi)(1+\eta)/4 - N_6/2 - N_7/2 \\ N_4 = (1-\xi)(1+\eta)/4 - N_7/2 - N_8/2 \\ N_5 = (1-\xi^2)(1-\eta)/2 \\ N_6 = (1+\xi)(1-\eta^2)/2 \\ N_7 = (1-\xi^2)(1+\eta)/2 \\ N_8 = (1-\xi)(1-\eta^2)/2 \end{cases} \quad (8.8)$$

8.2.1.2 Simplex elements

The linear and quadratic simplex elements (triangles in 2D and tetrahedrons in 3D) shown in Figure 8.6, together with their hierarchy of nodal patterns (similar to that described above for the Lagrangian elements), may on occasion be useful for modeling complex geometric shapes, although they are generally of lower accuracy.

The nondimensional Cartesian coordinates (ξ, η, ζ) are an example of *natural coordinates*. They are termed "natural" for rectangles and cubes because element boundaries can be simply described by, for example, ξ = constant or η = constant, so that area and volume integrals and element shape functions can be expressed easily.

However, rectangular Cartesian coordinates are not well suited for triangular (simplex) elements. In Chapter 5, we briefly introduced triangular (area) coordinates (L_1, L_2, L_3), which are natural coordinates for that shape. For 3D simplex elements (tetrahedrons), the corresponding natural coordinates are tetrahedral (or volume) coordinates (L_1, L_2, L_3, L_4), which are natural for that shape. The tetrahedral coordinates are defined in terms of volume ratios, that is, $L_i = V_i/V$, i = 1, … , 4, which can be related to the local Cartesian coordinates (x, y, z) of a generic point, in a completely analogous fashion.

In terms of the triangular coordinates defined in Chapter 5, the shape functions for the three-node "constant strain triangle" are

$$\begin{cases} N_1 = L_1 \\ N_2 = L_2 \\ N_3 = L_3 \end{cases} \tag{8.9}$$

and the shape functions for the six 6-node "linear strain triangle" are

$$\begin{cases} N_1 = L_1(2L_1 - 1) \\ N_2 = L_2(2L_2 - 1) \\ N_3 = L_3(2L_3 - 1) \\ N_4 = 4L_2L_3 \\ N_5 = 4L_1L_3 \\ N_6 = 4L_1L_2 \end{cases} \tag{8.10}$$

These shape functions are "isotropic" in the sense that the deformed geometry they describe is independent of the orientation of the triangle. This is a result of the fact that the shape functions of Equation 8.9 contain a complete linear polynomial in (x, y), and those of Equation 8.10 contain a complete quadratic polynomial in (x, y).

We mention in passing that admissible shape functions must satisfy certain well-known basic conditions (Zienkiewicz 1977) in order to guarantee convergence with mesh refinement; that is, they must be capable of exactly representing rigid body motions and spatially constant strain states.

8.2.2 Isoparametric mapping

One of the most important early milestones in the development of finite element technology was the introduction of *isoparametric* elements, pioneered by Taig (1961) and Irons (1966), a powerful numerical mapping technique that provided the capability to accurately discretize (i.e., mesh) complex irregular geometric shapes for analysis.

The defining characteristic of isoparametric finite elements is that the same shape functions are used to interpolate both geometry and displacements, which is a simple effective idea. In other words, the element reference coordinates (x, y, z) and deformed coordinates $(\bar{x}, \bar{y}, \bar{z})$ are interpolated from the initial and deformed nodal coordinates, respectively, using the same shape functions. As a consequence, the displacements at a generic point (x, y, z) in the element interior are also interpolated from the nodal displacements using the same shape functions.

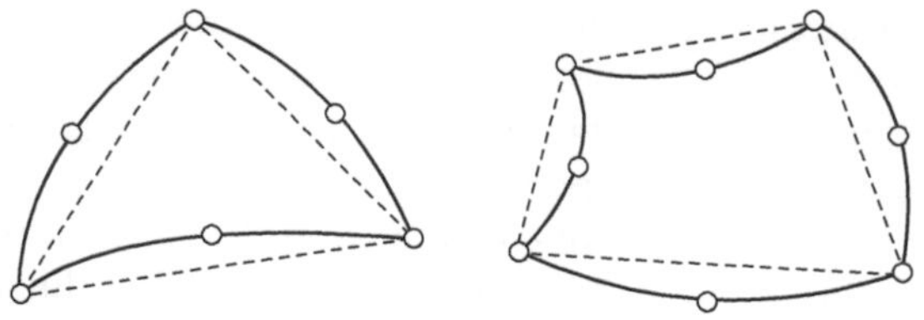

Figure 8.7 Quadratic isoparametric elements.

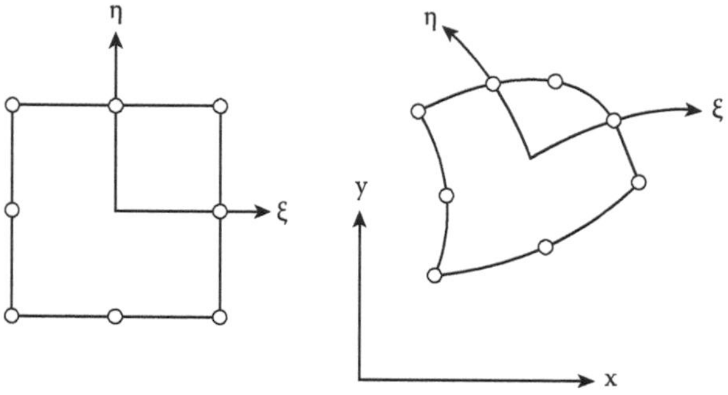

Figure 8.8 Parent and mapped Q8 isoparametric element.

Figure 8.7 shows the isoparametric versions of the linear strain triangle and Q8 serendipity elements described previously. Because these elements have quadratic shape functions, they can be mapped into shapes like those shown in which the element boundaries can take up general parabolic shapes defined by the locations of the nodes.

Therefore, the element displacements and coordinates are interpolated as follows:

$$\begin{cases} \{\mathbf{u}\} = [\mathbf{N}]\{\mathbf{U}\} \\ \{\mathbf{x}\} = [\mathbf{N}]\{\mathbf{X}\} \end{cases} \tag{8.11}$$

where {X} are the nodal coordinates and {U} the nodal displacements. Figure 8.8 shows the *parent* Q8 element on the left defined in the (ξ, η) coordinate system, and the *mapped* Q8 element on the right in the finite element mesh. We note that the shape functions $N_i(\xi, \eta)$ are defined in the parent element coordinate system.

In order to calculate the properties of the mapped geometry, such as area and volume, and eventually the kinematics of the subsequent deformation, we need to determine the derivatives of the shape functions in the (x, y) coordinate system. Area or volume integrals, needed to determine tangent stiffness matrices and internal resisting force vectors, are carried out numerically, usually using very accurate Gaussian numerical integration, in which the integrand is sampled at optimal locations ("Gauss points" or "material points") in the element. We first consider the determination of mapped element area (or volume) as an introduction to numerical integration.

8.2.2.1 Numerical (Gaussian) quadrature

A coordinate mapping from (ξ, η) to (x(ξ, η), y(ξ, η)) in two dimensions is described by

$$\begin{Bmatrix} dx \\ dy \end{Bmatrix} = \begin{bmatrix} \frac{\partial x}{\partial \xi} & \frac{\partial x}{\partial \eta} \\ \frac{\partial y}{\partial \xi} & \frac{\partial y}{\partial \eta} \end{bmatrix} \begin{Bmatrix} d\xi \\ d\eta \end{Bmatrix} \tag{8.12}$$

in which the (2 × 2) matrix is the transpose of a matrix known as the *Jacobian*. Occasionally, this term is used instead to denote the determinant of the Jacobian matrix. The extension to three dimensions is straightforward.

For a 2D solid element with n nodes, the Jacobian is computed numerically at each integration point from

$$\begin{bmatrix} \frac{\partial x}{\partial \xi} & \frac{\partial y}{\partial \xi} \\ \frac{\partial x}{\partial \eta} & \frac{\partial y}{\partial \eta} \end{bmatrix} = \begin{bmatrix} \frac{\partial N_1}{\partial \xi} & \cdots & \frac{\partial N_n}{\partial \xi} \\ \frac{\partial N_1}{\partial \eta} & \cdots & \frac{\partial N_n}{\partial \eta} \end{bmatrix}_{(2\times n)} \begin{bmatrix} x_1 & y_1 \\ \vdots & \vdots \\ x_n & y_n \end{bmatrix}_{(n\times 2)} \tag{8.13}$$

where the (2 × n) matrix of shape function derivatives is a function of (ξ, η). A differential element of area dξ dη in the parent element is transformed into the differential area dA of the mapped element in the (x, y) coordinate system by

$$dA = \det(\mathbf{J})\, d\xi\, d\eta \tag{8.14}$$

and if so desired, the area of the mapped element could then be computed from

$$A = \iint \det(\mathbf{J})\, d\xi\, d\eta \tag{8.15}$$

where the integration takes place in the parent element geometry. For an isoparametric mapping, integrations are carried out numerically, using in this case a product rule:

$$\int_{-1}^{1}\int_{-1}^{1} f(\xi,\eta)\,d\xi\,d\eta = \sum_k \sum_l W_k W_l\, f(\xi_k, \eta_l) \tag{8.16}$$

where f(ξ, η) is the function that is to be integrated, (ξ_k, η_l) are sampling locations, and (W_k, W_l) are weights. The Gaussian quadrature weights and sampling points are optimal for the integration of polynomials. For example, in one dimension, Gaussian quadrature of order N (i.e., there are N sampling points) integrates a polynomial of degree $p = 2N - 1$ exactly. Figure 8.9 shows weights and sampling locations for (1 × 1), (2 × 2), and (3 × 3) Gaussian quadrature in two dimensions.

It is not necessary that product rules be (N × N); in some situations (e.g., plate and shell elements), it may be desirable to use a product rule that preferentially treats one direction over another. Furthermore, it may be advantageous in certain situations to use suboptimal product rules (e.g., Lobatto integration) in order to sample material behavior at critical locations, for example, at extreme fibers in flexurally dominated problems.

We also note that other symmetric quadrature rules have been developed for finite element applications, in particular for triangular domains (Cowper 1973).

For isoparametric triangles, the Jacobian is computed from

$$\begin{bmatrix} \frac{\partial x}{\partial \xi} & \frac{\partial y}{\partial \xi} \\ \frac{\partial x}{\partial \eta} & \frac{\partial y}{\partial \eta} \end{bmatrix} = \frac{1}{2A}\begin{bmatrix} y_{23} & y_{31} & y_{12} \\ x_{32} & x_{13} & x_{21} \end{bmatrix} \begin{bmatrix} \frac{\partial N_1}{\partial L_1} & \cdots & \frac{\partial N_n}{\partial L_1} \\ \frac{\partial N_1}{\partial L_2} & \cdots & \frac{\partial N_n}{\partial L_2} \\ \frac{\partial N_1}{\partial L_3} & \cdots & \frac{\partial N_n}{\partial L_3} \end{bmatrix}_{(3\times n)} \begin{bmatrix} x_1 & y_1 \\ \vdots & \vdots \\ x_n & y_n \end{bmatrix}_{(n\times 2)} \tag{8.17}$$

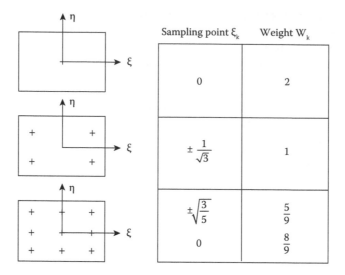

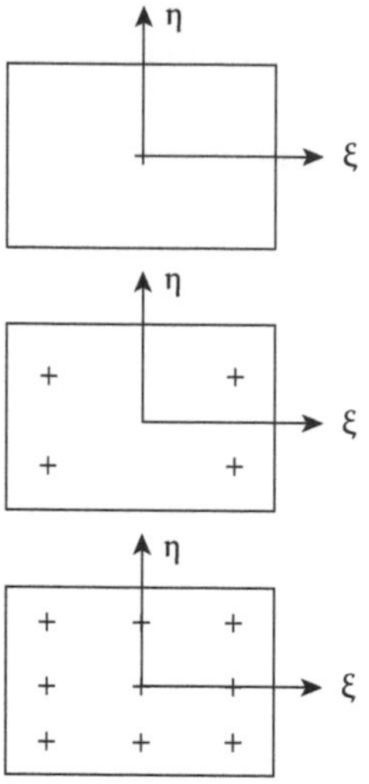

Sampling point ξ_k	Weight W_k
0	2
$\pm\frac{1}{\sqrt{3}}$	1
$\pm\sqrt{\frac{3}{5}}$ 0	$\frac{5}{9}$ $\frac{8}{9}$

Figure 8.9 Sampling points and weights for (1 × 1), (2 × 2), and (3 × 3) Gauss quadrature.

where A is the area of the parent triangle (see Chapter 5). We point out that (ξ, η) in this equation are *dimensional* Cartesian coordinates of the parent element (the natural coordinates are L_1, L_2, L_3) and that the Cartesian coordinates of the vertex nodes 1, 2, and 3 are the same in the parent and mapped element.

8.2.2.2 *Shape function derivatives*

The computation of the derivatives of each element shape function in the (x, y) coordinate system also employs the Jacobian matrix. They are computed at each integration point from

$$\begin{bmatrix} \frac{\partial x}{\partial \xi} & \frac{\partial y}{\partial \xi} \\ \frac{\partial x}{\partial \eta} & \frac{\partial y}{\partial \eta} \end{bmatrix} \begin{Bmatrix} \frac{\partial N_i}{\partial x} \\ \frac{\partial N_i}{\partial y} \end{Bmatrix} = \begin{Bmatrix} \frac{\partial N_i}{\partial \xi} \\ \frac{\partial N_i}{\partial \eta} \end{Bmatrix} \tag{8.18}$$

In this form, Equation 8.18 would be solved n times, once for each shape function N_i. However, the set of equations for (i = 1, … , n) can be stacked together to form

$$\begin{bmatrix} \frac{\partial N_1}{\partial x} & \cdots & \frac{\partial N_n}{\partial x} \\ \frac{\partial N_1}{\partial y} & \cdots & \frac{\partial N_n}{\partial y} \end{bmatrix}_{(2\times n)} = \begin{bmatrix} \frac{\partial x}{\partial \xi} & \frac{\partial y}{\partial \xi} \\ \frac{\partial x}{\partial \eta} & \frac{\partial y}{\partial \eta} \end{bmatrix}^{-1} \begin{bmatrix} \frac{\partial N_1}{\partial \xi} & \cdots & \frac{\partial N_n}{\partial \xi} \\ \frac{\partial N_1}{\partial \eta} & \cdots & \frac{\partial N_n}{\partial \eta} \end{bmatrix}_{(2\times n)} \tag{8.19}$$

8.3 TOTAL LAGRANGIAN FORMULATION

We focus first on the formulation for geometric nonlinearity in a total Lagrangian (TL) analysis (Zienkiewicz and Nayak 1971). We subsequently consider combined geometric and material nonlinearity.

8.3.1 Geometric nonlinearity

In order to develop the Lagrangian strain–displacement matrix for finite element analysis, we start with the vector form of Green strain, in which we now replace axes (1, 2, 3) with (x, y, z). The (x, y, z) Cartesian reference system should be understood here as an *element*-specific reference system that may be different for each element in the finite element mesh.

The vector forms of stress and strain are shown below.

$$\{\varepsilon\} = \begin{Bmatrix} \varepsilon_{xx} \\ \varepsilon_{yy} \\ \varepsilon_{zz} \\ 2\varepsilon_{xy} \\ 2\varepsilon_{xz} \\ 2\varepsilon_{yz} \end{Bmatrix}; \quad \{\sigma\} = \begin{Bmatrix} \sigma_{xx} \\ \sigma_{yy} \\ \sigma_{zz} \\ \sigma_{xy} \\ \sigma_{xz} \\ \sigma_{yz} \end{Bmatrix} \tag{8.20}$$

As mentioned previously, these vectorial representations of stress and strain are defined so as to preserve the inner product relationship

$$\boldsymbol{\varepsilon} : \boldsymbol{\sigma} \equiv \{\varepsilon\}^{\mathrm{T}}\{\sigma\} \tag{8.21}$$

We mention here in passing that the fourth-order constitutive tensors D_{ijkl}, whether elastic or inelastic, that relate stress to strain or stress rate to strain rate, are now represented as (6 × 6) matrices [D] via the mapping shown in Figure 8.10.

	ε_{11}	ε_{22}	ε_{33}	$2\varepsilon_{12}$	$2\varepsilon_{13}$	$2\varepsilon_{23}$
σ_{11}	(1111)	(1122)	(1133)	(1112+ 1121)/2	(1113+ 1131)/2	(1123+ 1132)/2
σ_{22}	(2211)	(2222)	(2233)	(2212+ 2221)/2	(2213+ 2231)/2	(2223+ 2232)/2
σ_{33}	(3311)	(3322)	(3333)	(3312+ 3321)/2	(3313+ 3331)/2	(3323+ 3332)/2
σ_{12}	(1211)	(1222)	(1233)	(1212+ 1221)/2	(1213+ 1231)/2	(1223+ 1232)/2
σ_{13}	(1311)	(1322)	(1333)	(1312+ 1321)/2	(1313+ 1331)/2	(1323+ 1332)/2
σ_{23}	(2311)	(2322)	(2333)	(2312+ 2321)/2	(2313+ 2331)/2	(2323+ 2332)/2

Figure 8.10 Mapping from (3 × 3 × 3 × 3) **D** tensor to (6 × 6) **[D]** matrix.

The 6 × 6 table contains the entries for [D] mapped from the fourth-order tensor D_{ijkl} that are indicated as (i, j, k, l). For example, the expression for σ_{12} can be expanded in terms of D_{ijkl} as

$$D_{1211}\varepsilon_{11} + D_{1222}\varepsilon_{22} + D_{1233}\varepsilon_{33} + D_{1212}\varepsilon_{12} + D_{1221}\varepsilon_{21} + D_{1213}\varepsilon_{13} + D_{1231}\varepsilon_{31} + D_{1223}\varepsilon_{23} + D_{1232}\varepsilon_{32}$$

Using the symmetry of the strain tensor this can be rewritten as

$$D_{1211}\varepsilon_{11} + D_{1222}\varepsilon_{22} + D_{1233}\varepsilon_{33} + \frac{1}{2}(D_{1212} + D_{1221})2\varepsilon_{12} + \frac{1}{2}(D_{1213} + D_{1231})2\varepsilon_{13} + \frac{1}{2}(D_{1223} + D_{1232})2\varepsilon_{23}$$

Therefore, the corresponding entries of the fourth row of the [D] matrix are as follows:

$$\left[D_{1211} \quad D_{1222} \quad D_{1233} \quad \frac{1}{2}(D_{1212} + D_{1221}) \quad \frac{1}{2}(D_{1213} + D_{1231}) \quad \frac{1}{2}(D_{1223} + D_{1232}) \right]$$

As an example, we consider the isotropic elastic material model. The fourth-order tensor

$$D^e_{ijkl} = \lambda\delta_{ij}\ \delta_{kl} + \mu(\delta_{ik}\ \delta_{jl} + \delta_{il}\ \delta_{jk}) \tag{8.22}$$

maps to the (6 × 6) matrix as follows:

$$[\mathbf{D}]^e = \begin{bmatrix} \lambda+2\mu & \lambda & \lambda & 0 & 0 & 0 \\ \lambda & \lambda+2\mu & \lambda & 0 & 0 & 0 \\ \lambda & \lambda & \lambda+2\mu & 0 & 0 & 0 \\ 0 & 0 & 0 & 2\mu & 0 & 0 \\ 0 & 0 & 0 & 0 & 2\mu & 0 \\ 0 & 0 & 0 & 0 & 0 & 2\mu \end{bmatrix} \tag{8.23}$$

8.3.1.1 Green strain

The Cartesian tensor form of the Green strain–displacement relations that was developed earlier is repeated below.

$$\varepsilon_{ij} = \frac{1}{2}[u_{i,j} + u_{j,i} + u_{k,i}\ u_{k,j}] \tag{8.24}$$

Expanding this relation in the 3D Cartesian coordinate system (x, y, z), with displacements (u_1, u_2, u_3) now denoted as (u, v, w), leads to the following explicit expressions for Green strains:

$$\begin{cases} \varepsilon_{xx} = u_{,x} + \frac{1}{2}[u_{,x}^2 + v_{,x}^2 + w_{,x}^2] \\ \varepsilon_{yy} = v_{,y} + \frac{1}{2}[u_{,y}^2 + v_{,y}^2 + w_{,y}^2] \\ \varepsilon_{zz} = w_{,z} + \frac{1}{2}[u_{,z}^2 + v_{,z}^2 + w_{,z}^2] \\ 2\varepsilon_{xy} = u_{,y} + v_{,x} + [u_{,x}u_{,y} + v_{,x}v_{,y} + w_{,x}w_{,y}] \\ 2\varepsilon_{xz} = u_{,z} + w_{,x} + [u_{,z}u_{,x} + v_{,z}v_{,x} + w_{,z}w_{,x}] \\ 2\varepsilon_{yz} = v_{,z} + w_{,y} + [u_{,y}u_{,z} + v_{,y}v_{,z} + w_{,y}w_{,z}] \end{cases} \tag{8.25}$$

The Green strains can be divided into a linear portion ε^l and a nonlinear portion ε^{nl} containing the (bracketed) quadratic terms as

$$\{\varepsilon\} = \{\varepsilon\}^l + \{\varepsilon\}^{nl} \tag{8.26}$$

The linear portion of Green strain is related to the element nodal displacement vector $\{\mathbf{u}\}$ by a (6 × 3n) matrix $\mathbf{B}^l$ (where n is the number of element nodes), which contains the appropriate material derivatives of the shape functions.

$$\{\varepsilon\}^l = \mathbf{B}^l\mathbf{u} \tag{8.27}$$

The element nodal displacement vector $\mathbf{u} = \{u_1, v_1, w_1, \ldots, u_n, v_n, w_n\}^T$.

The linear strain–displacement matrix $[\mathbf{B}]^l$ can be partitioned into n (6 × 3) matrices, one for each element node as follows:

$$\mathbf{B}^l = \left[\mathbf{B}_1^l \quad \ldots \quad \mathbf{B}_n^l\right]_{(6\times 3n)} \tag{8.28}$$

where for the i-th node,

$$\mathbf{B}_i^l = \begin{bmatrix} N_{i,x} & 0 & 0 \\ 0 & N_{i,y} & 0 \\ 0 & 0 & N_{i,z} \\ N_{i,y} & N_{i,x} & 0 \\ N_{i,z} & 0 & N_{i,x} \\ 0 & N_{i,z} & N_{i,y} \end{bmatrix} \tag{8.29}$$

The nonlinear portion of Green strain can be written in terms of displacement gradients $\boldsymbol{\theta}_x$, $\boldsymbol{\theta}_y$, and $\boldsymbol{\theta}_z$ which are defined by

$$\boldsymbol{\theta}_x = \begin{Bmatrix} u_{,x} \\ v_{,x} \\ w_{,x} \end{Bmatrix}; \quad \boldsymbol{\theta}_y = \begin{Bmatrix} u_{,y} \\ v_{,y} \\ w_{,y} \end{Bmatrix}; \quad \boldsymbol{\theta}_z = \begin{Bmatrix} u_{,z} \\ v_{,z} \\ w_{,z} \end{Bmatrix} \tag{8.30}$$

In terms of these displacement gradients, the nonlinear portion of the Green strain is then

$$\{\varepsilon\}^{nl} = \frac{1}{2}\begin{Bmatrix} \boldsymbol{\theta}_x^T\boldsymbol{\theta}_x \\ \boldsymbol{\theta}_y^T\boldsymbol{\theta}_y \\ \boldsymbol{\theta}_z^T\boldsymbol{\theta}_z \\ \boldsymbol{\theta}_x^T\boldsymbol{\theta}_y + \boldsymbol{\theta}_y^T\boldsymbol{\theta}_x \\ \boldsymbol{\theta}_x^T\boldsymbol{\theta}_z + \boldsymbol{\theta}_z^T\boldsymbol{\theta}_x \\ \boldsymbol{\theta}_z^T\boldsymbol{\theta}_y + \boldsymbol{\theta}_y^T\boldsymbol{\theta}_z \end{Bmatrix} \tag{8.31}$$

which can be written as

$$\{\varepsilon\}^{nl} = \frac{1}{2}\mathbf{A}\boldsymbol{\theta} \tag{8.32}$$

where $\mathbf{A}$ is a function of the displacement gradients, given by

$$\mathbf{A}=\begin{bmatrix} \boldsymbol{\theta}_x^T & \mathbf{0}_3^T & \mathbf{0}_3^T \\ \mathbf{0}_3^T & \boldsymbol{\theta}_y^T & \mathbf{0}_3^T \\ \mathbf{0}_3^T & \mathbf{0}_3^T & \boldsymbol{\theta}_z^T \\ \boldsymbol{\theta}_y^T & \boldsymbol{\theta}_x^T & \mathbf{0}_3^T \\ \boldsymbol{\theta}_z^T & \mathbf{0}_3^T & \boldsymbol{\theta}_x^T \\ \mathbf{0}_3^T & \boldsymbol{\theta}_z^T & \boldsymbol{\theta}_y^T \end{bmatrix} \tag{8.33}$$

in which $\mathbf{0}_3$ is a (3×1) null vector.

The displacement gradients in turn are related to nodal displacements through matrices that contain the material derivatives of shape functions.

$$\begin{cases} \boldsymbol{\theta}_x = \mathbf{G}_x\mathbf{u} \\ \boldsymbol{\theta}_y = \mathbf{G}_y\mathbf{u} \\ \boldsymbol{\theta}_z = \mathbf{G}_z\mathbf{u} \end{cases} \tag{8.34}$$

where

$$\begin{cases} \mathbf{G}_x = [N_{1,x}\mathbf{I}_3 \quad \cdots \quad N_{n,x}\mathbf{I}_3]_{(3\times 3n)} \\ \mathbf{G}_y = [N_{1,y}\mathbf{I}_3 \quad \cdots \quad N_{n,y}\mathbf{I}_3]_{(3\times 3n)} \\ \mathbf{G}_z = [N_{1,z}\mathbf{I}_3 \quad \cdots \quad N_{n,z}\mathbf{I}_3]_{(3\times 3n)} \end{cases} \tag{8.35}$$

and $\mathbf{I}_3$ is the (3×3) identity matrix.

The displacement gradients are therefore expressed as follows:

$$\boldsymbol{\theta} = \mathbf{G}\mathbf{u} \tag{8.36}$$

$$\boldsymbol{\theta} = \begin{Bmatrix} \boldsymbol{\theta}_x \\ \boldsymbol{\theta}_y \\ \boldsymbol{\theta}_z \end{Bmatrix}; \quad \mathbf{G} = \begin{bmatrix} \mathbf{G}_x \\ \mathbf{G}_y \\ \mathbf{G}_z \end{bmatrix} \tag{8.37}$$

and the nonlinear portion of the Green strain in terms of element nodal displacements is

$$\{\varepsilon\}^{nl} = \frac{1}{2}\mathbf{A}\mathbf{G}\mathbf{u} \equiv \frac{1}{2}\mathbf{B}^{nl}\mathbf{u} \tag{8.38}$$

where

$$\mathbf{B}^{nl} \equiv \mathbf{A}\mathbf{G} \tag{8.39}$$

We note that $\mathbf{B}^{nl}$, the nonlinear portion of the strain–displacement matrix, is linearly dependent on nodal displacements, since the matrix $\mathbf{A}$ is a linear function of nodal displacements, while the matrix $\mathbf{G}$ is independent of nodal displacements and contains only the material derivatives of the shape functions.

The total Green strain is written as

$$\{\varepsilon\} = \mathbf{B}_T\mathbf{u} \tag{8.40}$$

in which

$$\mathbf{B}_T \equiv \mathbf{B}^l + \frac{1}{2}\mathbf{B}^{nl} \tag{8.41}$$

In summary, the equation above describes a total Green strain–displacement relation in which the matrix $\mathbf{B}_T$ is a linear function of the element nodal displacements.

An attractive feature of the TL formulation is obvious. The matrix **G** contains material derivatives of shape functions defined in the reference configuration. Therefore, they are not affected by the deformation or motion of the solid system. This matrix **G** needs to be computed only at the beginning of the analysis and remains unchanged during the motion.

8.3.1.2 Green strain rate

We can also develop a matrix relation between the material rate of Green strain and the material rate of nodal displacements. We start with the total Green strain–displacement relation and take its material derivative.

$$\{\varepsilon\} = \mathbf{B}^l\mathbf{u} + \frac{1}{2}\mathbf{A}\boldsymbol{\theta} \tag{8.42}$$

$$\{\dot{\varepsilon}\} = \mathbf{B}^l\dot{\mathbf{u}} + \frac{1}{2}\mathbf{A}\dot{\boldsymbol{\theta}} + \frac{1}{2}\dot{\mathbf{A}}\boldsymbol{\theta} \tag{8.43}$$

The following identity is easily verified:

$$\dot{\mathbf{A}}\boldsymbol{\theta} \equiv \mathbf{A}\dot{\boldsymbol{\theta}} \tag{8.44}$$

Therefore, Equation 8.43 becomes

$$\{\dot{\varepsilon}\} = \mathbf{B}^l\dot{\mathbf{u}} + \mathbf{A}\dot{\boldsymbol{\theta}} \tag{8.45}$$

Because the matrix **G** contains only material derivatives of the shape functions, its material (time) rate is zero. Therefore, the Green strain–displacement relation in rate form is

$$\{\dot{\varepsilon}\} = (\mathbf{B}^l + \mathbf{A}\mathbf{G})\dot{\mathbf{u}} = (\mathbf{B}^l + \mathbf{B}^{nl})\dot{\mathbf{u}} = \mathbf{B}\dot{\mathbf{u}} \tag{8.46}$$

$$\mathbf{B} \equiv \mathbf{B}^l + \mathbf{B}^{nl} \tag{8.47}$$

The incremental form of the Green strain–displacement relation follows directly by multiplying both sides of the rate form by a small time increment Δt to yield

$$\{\Delta\varepsilon\} = \mathbf{B}\,\Delta\mathbf{u} \tag{8.48}$$

This expression can also be considered a variational form relating virtual Green strains to virtual displacements.

$$\{\delta\varepsilon\} = \mathbf{B}\,\delta\mathbf{u} \tag{8.49}$$

We start with the Lagrangian internal virtual work expression for a single element,

$$\delta W_{int,e} = -\int_{Ve} \{\delta\varepsilon\}^T\{\boldsymbol{\sigma}\}dV_e \tag{8.50}$$

where $\{\boldsymbol{\sigma}\}$ is the second Piola–Kirchhoff (2PK) stress in the current deformed configuration and the virtual Green strains are due to (infinitesimal) virtual nodal displacements in the deformed configuration. The integration is over the initial undeformed (i.e., reference) configuration of the element. The sign convention for internal virtual work that is used in Equation 8.50 is explained in Chapter 4.

Using Equation 8.49, we can write Equation 8.50 as

$$\delta W_{int,e} = -\delta\mathbf{u}_e^T\mathbf{I}_e \tag{8.51}$$

where

$$\mathbf{I}_e = \int_{Ve} \mathbf{B}^T\{\boldsymbol{\sigma}\}dV_e \tag{8.52}$$

is the (element) internal resisting force vector introduced in Chapter 4. It is work equivalent to the element stresses.

For the sake of simplicity, we temporarily assume that the only external loads acting on the structure are concentrated loads $\mathbf{P}$ at the structural degrees of freedom $\mathbf{U}$. The virtual work of the external nodal loads is therefore

$$\delta W_{ext} = \delta\mathbf{U}^T\mathbf{P} \tag{8.53}$$

In order to write the total virtual work for the entire finite element model, we must transform (i.e., rotate) each element internal resisting force vector into the global Cartesian reference system, and then assemble them to form the internal resisting force vector for the structure. These operations can be *formally* represented by

$$\mathbf{I} = \sum_e \mathbf{L}_e^T\mathbf{T}_e^T\mathbf{I}_e \tag{8.54}$$

We have used here a subscript e to designate element-specific quantities where confusion might otherwise arise. The orthogonal matrix $\mathbf{T}_e$ rotates the element nodal displacements from the global Cartesian coordinates to the element Cartesian reference system (x, y, z). The (very sparse) Boolean localizing matrix $\mathbf{L}_e$ maps the structure nodal displacements to the nodal displacements of element e. These operations are displayed explicitly below.

$$\begin{cases} u_e = T_e U_e \\ U_e = L_e U \end{cases} \tag{8.55}$$

The virtual work equation for the complete finite element model is then

$$\delta U^T \left[P - \sum_e L_e^T T_e^T I_e \right] = 0 \tag{8.56}$$

Operations involving localizing matrices are of course carried out computationally via direct stiffness algorithms rather than matrix multiplications. Since the independent virtual nodal displacements are arbitrary, the equilibrium equations (weak form) in the current configuration can be written as

$$P - I = 0 \tag{8.57}$$

or

$$\left[\sum_e \int_{V_e} B_e^T \{\sigma\}_e \, dV_e \right] U = P \tag{8.58}$$

where here and henceforth we interpret the symbol $\sum_e$ to imply both coordinate rotation and direct stiffness assembly operations for matrices (where appropriate), as well as vectors.

For situations in which a total (2PK) stress versus total (Green) strain relation can be specified, Equation 8.58 could be used to develop a set of (nonlinear) equilibrium equations relating total nodal displacements to total nodal loads, that is, to develop a secant stiffness matrix (Zienkiewicz and Nayak 1971). However, even in those rare cases where such a total formulation might be possible, this approach is not of great utility since the set of equations is still nonlinear and would require some form of iterative solution.

8.3.1.3 Tangent stiffness matrix

To obtain the incremental form of the equilibrium equations, we take the material (time) derivative of the virtual work expression given in Equation 8.58,

$$\sum_e \int_V \left[\dot{B}^T \{\sigma\} + B^T \{\dot{\sigma}\} \right] dV = \dot{P} \tag{8.59}$$

where we have now dropped the element subscript e to simplify the notation.

We consider the first term in this equation by recalling Equation 8.47 and observing that the matrix **A** is a function of "time" through its dependence on nodal displacements, whereas B^l and G are independent of the nodal displacements. Therefore,

$$B = B^l + AG \tag{8.60}$$

$$\dot{B} = \dot{A}G \tag{8.61}$$

Using this expression for the material rate of the Green strain–displacement matrix, we can develop the following expression for the first term of Equation 8.59:

$$\sum_e \int_V \dot{B}^T \{\sigma\} dV = \sum_e \int_V G^T \dot{A}^T \{\sigma\} dV = \left[\sum_e \int_V G^T M_\sigma G dV \right] \dot{U} \tag{8.62}$$

To arrive at the above expression, we observe the following:

$$\dot{A}^T\{\sigma\} = \begin{bmatrix} \dot{\theta}_x & 0_3 & 0_3 & \dot{\theta}_y & 0_3 & \dot{\theta}_z \\ 0_3 & \dot{\theta}_y & 0_3 & \dot{\theta}_x & \dot{\theta}_z & 0_3 \\ 0_3 & 0_3 & \dot{\theta}_z & 0_3 & \dot{\theta}_y & \dot{\theta}_x \end{bmatrix} \begin{Bmatrix} \sigma_{xx} \\ \sigma_{yy} \\ \sigma_{zz} \\ \sigma_{xy} \\ \sigma_{yz} \\ \sigma_{xz} \end{Bmatrix} = \begin{Bmatrix} \sigma_{xx}\dot{\theta}_x + \sigma_{xy}\dot{\theta}_y + \sigma_{xz}\dot{\theta}_z \\ \sigma_{yy}\dot{\theta}_y + \sigma_{xy}\dot{\theta}_x + \sigma_{yz}\dot{\theta}_z \\ \sigma_{zz}\dot{\theta}_z + \sigma_{yz}\dot{\theta}_y + \sigma_{xz}\dot{\theta}_x \end{Bmatrix} \quad (8.63)$$

The last term in Equation 8.63 can be written as the product shown below, in which $\mathbf{I}_3$ is the 3 × 3 identity matrix.

$$\dot{A}^T\{\sigma\} = \begin{bmatrix} \sigma_{xx}\mathbf{I}_3 & \sigma_{xy}\mathbf{I}_3 & \sigma_{xz}\mathbf{I}_3 \\ \sigma_{xy}\mathbf{I}_3 & \sigma_{yy}\mathbf{I}_3 & \sigma_{yz}\mathbf{I}_3 \\ \sigma_{xz}\mathbf{I}_3 & \sigma_{yz}\mathbf{I}_3 & \sigma_{zz}\mathbf{I}_3 \end{bmatrix} \begin{Bmatrix} \dot{\theta}_x \\ \dot{\theta}_y \\ \dot{\theta}_z \end{Bmatrix} = \mathbf{M}_\sigma \dot{\theta} \quad (8.64)$$

where

$$\mathbf{M}_\sigma = \begin{bmatrix} \sigma_{xx}\mathbf{I}_3 & \sigma_{xy}\mathbf{I}_3 & \sigma_{xz}\mathbf{I}_3 \\ \sigma_{xy}\mathbf{I}_3 & \sigma_{yy}\mathbf{I}_3 & \sigma_{yz}\mathbf{I}_3 \\ \sigma_{xz}\mathbf{I}_3 & \sigma_{yz}\mathbf{I}_3 & \sigma_{zz}\mathbf{I}_3 \end{bmatrix} \quad (8.65)$$

The matrix $\mathbf{M}_\sigma$ is symmetric; therefore, $\mathbf{G}^T\mathbf{M}_\sigma\mathbf{G}$ is symmetric. The quantity in the parentheses in Equation 8.62 is called the geometric stiffness matrix.

$$\mathbf{K}_G = \sum_e \int_V \mathbf{G}^T\mathbf{M}_\sigma\mathbf{G}dV \quad (8.66)$$

We note that the geometric stiffness matrix is a function of the 2PK stress. In a prestressed structure, the geometric stiffness matrix is therefore initially nonzero. The geometric stiffness matrix is also used in linearized buckling analysis.

To further develop the second term of Equation 8.59, the specific rate form of the material constitutive law is needed. Here, we simply represent the material model as the relation between the symmetric Piola–Kirchhoff stress rate and the Green strain rate in matrix–vector format.

$$\{\dot{\sigma}\} = [D]\{\dot{\varepsilon}\} \quad (8.67)$$

Substituting this general representation for the (tangent) constitutive matrix, we obtain the following expression for the second term of Equation 8.59:

$$\sum_e \int_V \mathbf{B}^T\{\dot{\sigma}\}dV = \left[\sum_e \int_V \mathbf{B}^T[D]\mathbf{B}dV\right]\dot{\mathbf{U}} \quad (8.68)$$

Therefore, the tangent stiffness matrix is

$$K_t = \sum_e \int_V \left[B^T[D]B + G^T M_\sigma G\right] dV \tag{8.69}$$

This is the form that is normally used in computation; however, to examine its structure more closely, we can decompose $B = B^l + B^{nl}$ in Equation 8.69 and expand as follows:

$$\left[\sum_e \int_V B^T[D]BdV\right]\dot{U} = \left[K_0 + K_1 + K_1^T + K_2\right]\dot{U} \tag{8.70}$$

where

$$\begin{cases} K_0 = \sum \int_V B^{l^T}[D]B^l dV \\ K_1 = \sum \int_V B^{l^T}[D]B^{nl} dV \\ K_2 = \sum \int_V B^{nl^T}[D]B^{nl} dV \end{cases} \tag{8.71}$$

If D is assumed constant, then K_0 is constant, K_1 is linear, and K_2 is quadratic in the displacement gradients. In this particular case, the complete tangent stiffness is

$$K_t = K_0 + K_1 + K_1^T + K_2 + K_G \tag{8.72}$$

and the secant stiffness is

$$K_S = K_0 + \frac{1}{2}K_1 + K_1^T + \frac{1}{2}K_2 \tag{8.73}$$

The secant stiffness displayed in Equation 8.73 is not symmetric because K_1 is not symmetric, as is evident from Equation 8.71. However, the secant stiffness is not unique and alternate forms, including a symmetric form, can be developed if required. We do not pursue that further here.

The tangent stiffness matrix, on the other hand, is symmetric if the material constitutive matrix [D] is symmetric. Nonassociated plasticity (plastic flow direction not normal to the yield surface) is an example of a material model leading to an unsymmetric [D].

The equilibrium equations in rate form are

$$K_t \dot{U} = \dot{P} \tag{8.74}$$

For small increments of displacements and nodal loads, an initial tangential approximation is given by

$$K_t \Delta U = \Delta P \tag{8.75}$$

However, even for very small load increments, a straightforward application of this equation for moderately nonlinear systems produces results of unacceptably low accuracy. Iterations within a load step are essential (Chapter 11).

Example 8.1 Hyperelastic bar element

In the preceding sections, the development of element matrices was presented for a 3D solid continuum element. The reduction of these matrices to 2D (plane strain and plane stress) elements is obvious and is not further elaborated here.

Instead, we consider the simpler example of a 1D finite element model of a slender hyperelastic bar. The cross-sectional dimensions of the bar are as usual assumed to be small compared with its length. The element model is 1D in the sense that the deformation is a function of only a single material coordinate x measured along the bar axis in the undeformed configuration. Such a bar can be connected to other elements of the same type to form an articulated 2D or 3D space lattice structure that can undergo large displacements (and deformations).

The element shape functions are defined in terms of the natural material coordinate $\xi = x/l$ as

$$\begin{cases} u = N_1(\xi)u_1 + N_2(\xi)u_2 \\ v = N_1(\xi)v_1 + N_2(\xi)v_2 \\ w = N_1(\xi)w_1 + N_2(\xi)w_2 \end{cases} \tag{8.76}$$

where (u, v, w) are the Cartesian components of the axial displacement of the bar, (u_1, v_1, w_1) are the nodal displacements of node 1 ($\xi = 0$), (u_2, v_2, w_2) are the displacements of node 2 ($\xi = 1$), and l is the undeformed length. The linear shape functions defined in the material coordinate $x = \xi/l$ are

$$\begin{cases} N_1(\xi) = 1 - \xi \\ N_2(\xi) = \xi \end{cases} \tag{8.77}$$

The only nonzero displacement gradient vector in this case is $\boldsymbol{\theta}_x$, which we denote here simply as $\theta \equiv \{\theta_1, \theta_2, \theta_3\}^T$, where $\theta_1 = u_{,x}$ $\theta_2 = v_{,x}$ and $\theta_3 = w_{,x}$. With these definitions, the axial Green strain is

$$\varepsilon_{xx} = u_{,x} + \frac{1}{2}[u_{,x}^2 + v_{,x}^2 + w_{,x}^2] = u_{,x} + \frac{1}{2}\boldsymbol{\theta}^T\boldsymbol{\theta} \tag{8.78}$$

and

$$\mathbf{B}^l = \frac{1}{l}[-1 \quad 0 \quad 0 \quad 1 \quad 0 \quad 0] \tag{8.79}$$

$$\mathbf{A} = [\theta_1 \quad \theta_2 \quad \theta_3] \tag{8.80}$$

$$\mathbf{G} = \frac{1}{l}[-\mathbf{I}_3 \quad \mathbf{I}_3] \tag{8.81}$$

$$\mathbf{B}^{nl} = \frac{1}{l}[-\theta_1 \quad -\theta_2 \quad -\theta_3 \quad \theta_1 \quad \theta_2 \quad \theta_3] \tag{8.82}$$

We make further simplifying assumptions: The bar is prismatic with cross-sectional area A_0 that remains constant during the deformation, and the 2PK stress σ_{xx} is linearly related to the Green strain ε_{xx}, that is, $\sigma_{xx} = E\,\varepsilon_{xx}$. The latter assumption implies $\dot{\sigma}_{xx} = E\,\dot{\varepsilon}_{xx}$.

The integrations required to determine the various finite element matrices reduce here to multiplication by $A_0 l$.

$$\mathbf{K}_0 = \frac{A_0 E}{l}\begin{bmatrix} 1 & 0 & 0 & -1 & 0 & 0 \\ 0 & 0 & 0 & 0 & 0 & 0 \\ 0 & 0 & 0 & 0 & 0 & 0 \\ -1 & 0 & 0 & 1 & 0 & 0 \\ 0 & 0 & 0 & 0 & 0 & 0 \\ 0 & 0 & 0 & 0 & 0 & 0 \end{bmatrix} \tag{8.83}$$

We note that the zero terms in the linear stiffness matrix $\mathbf{K}_0$ are associated with the lateral displacements, which do not generate axial stiffness in linear analysis.

The higher-order (nonlinear) stiffness terms that account for coupling due to element rotation are

$$\mathbf{K}_1 = \frac{A_0 E}{l}\begin{bmatrix} \theta_1 & \theta_2 & \theta_3 & -\theta_1 & -\theta_2 & -\theta_3 \\ 0 & 0 & 0 & 0 & 0 & 0 \\ 0 & 0 & 0 & 0 & 0 & 0 \\ -\theta_1 & -\theta_2 & -\theta_3 & \theta_1 & \theta_2 & \theta_3 \\ 0 & 0 & 0 & 0 & 0 & 0 \\ 0 & 0 & 0 & 0 & 0 & 0 \end{bmatrix} \tag{8.84}$$

$$\mathbf{K}_2 = \frac{A_0 E}{l}\begin{bmatrix} \theta_1^2 & \theta_1\theta_2 & \theta_1\theta_3 & -\theta_1^2 & -\theta_1\theta_2 & -\theta_1\theta_3 \\ \theta_1\theta_2 & \theta_2^2 & \theta_2\theta_3 & -\theta_1\theta_2 & -\theta_2^2 & -\theta_2\theta_3 \\ \theta_1\theta_3 & \theta_2\theta_3 & \theta_3^2 & -\theta_1\theta_3 & -\theta_2\theta_3 & -\theta_3^2 \\ -\theta_1^2 & -\theta_1\theta_2 & -\theta_1\theta_3 & \theta_1^2 & \theta_1\theta_2 & \theta_1\theta_3 \\ -\theta_1\theta_2 & -\theta_2^2 & -\theta_2\theta_3 & \theta_1\theta_2 & \theta_2^2 & \theta_2\theta_3 \\ -\theta_1\theta_3 & -\theta_2\theta_3 & -\theta_3^2 & \theta_1\theta_3 & \theta_2\theta_3 & \theta_3^2 \end{bmatrix} \tag{8.85}$$

$$\mathbf{K}_G = \frac{A_0 \sigma_{xx}}{l}\begin{bmatrix} 1 & 0 & 0 & -1 & 0 & 0 \\ 0 & 1 & 0 & 0 & -1 & 0 \\ 0 & 0 & 1 & 0 & 0 & -1 \\ -1 & 0 & 0 & 1 & 0 & 0 \\ 0 & -1 & 0 & 0 & 1 & 0 \\ 0 & 0 & -1 & 0 & 0 & 1 \end{bmatrix} \tag{8.86}$$

The term $A_0\sigma_{xx}$ appearing in the geometric stiffness $\mathbf{K}_G$ is not the true axial force in the bar. We recall from Chapter 3 that the force vector in the deformed configuration on the plane originally normal to the x_1 axis is

$$d\bar{f} = [\sigma_{11}G_1 + \sigma_{21}G_2 + \sigma_{31}G_3]dA \tag{8.87}$$

Here, the convected base vector $G_1 = (1 + \theta_x)e_1 + \theta_y e_2 + \theta_z e_3$. Therefore, the true axial force $\bar{F}$ is

$$\bar{F} = A_0\sigma_{xx}\lambda_x \tag{8.88}$$

where the axial stretch λ_x is given by

$$\lambda_x = \sqrt{(1+\theta_x)^2 + \theta_y^2 + \theta_z^2} \tag{8.89}$$

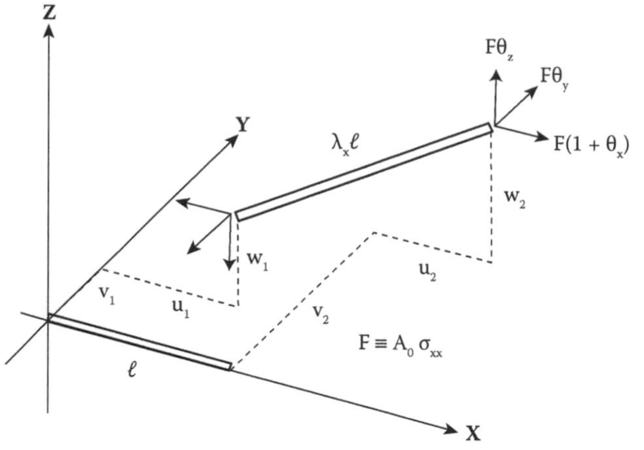

Figure 8.11 Finite deformation of hyperelastic bar element.

The geometric stiffness can also be put in the form

$$K_G = \frac{\bar{F}}{\bar{l}} \begin{bmatrix} 1 & 0 & 0 & -1 & 0 & 0 \\ 0 & 1 & 0 & 0 & -1 & 0 \\ 0 & 0 & 1 & 0 & 0 & -1 \\ -1 & 0 & 0 & 1 & 0 & 0 \\ 0 & -1 & 0 & 0 & 1 & 0 \\ 0 & 0 & -1 & 0 & 0 & 1 \end{bmatrix} \tag{8.90}$$

where $\bar{l} \equiv l\lambda_x$ is the deformed length of the bar. Figure 8.11 shows the deformed bar in the element Cartesian reference coordinate system.

If the axial deformation is small, that is, $\theta_x \ll 1$, the axial stretch can be approximated as

$$\lambda_x \approx 1 + \theta_x + \frac{1}{2}(\theta_x^2 + \theta_y^2 + \theta_z^2) \approx 1 + \theta_x + \frac{1}{2}(\theta_y^2 + \theta_z^2) \tag{8.91}$$

Then the axial (engineering) strain $e_{xx} \approx \theta_x$ and θ_y, θ_z are first approximations to the rotational effects. This type of approximation is often made for structural elements (beams, plates, and shells) in which bending effects are important.

8.3.2 Nonlinear material behavior

The element tangent stiffness for combined geometric–material nonlinearity is

$$K_t = \int_V B^T [D] B dV + K_G \tag{8.92}$$

In this equation, **B** is the "tangent" strain–displacement matrix defined in Equation 8.60 that depends on element displacement gradients. The matrix **[D]** relates the 2PK stress rate to the Green strain rate as determined from the nonlinear material model. The geometric stiffness $\mathbf{K_G}$ is a function of the current total 2PK stress. It accounts for the change in the element internal force vector due to the current total stresses undergoing an incremental geometry change $\dot{\mathbf{U}}$.

Therefore, the only additional feature compared with the previous section on geometric nonlinearity is that the **[D]** matrix now represents inelastic rather than elastic material behavior. We focus here on elastoplastic material behavior combined with geometric nonlinearity.

8.3.2.1 Elastoplastic material behavior

We first recall the material ("continuum") tangent stiffness developed in Chapter 7 in Cartesian tensor form, and show how to convert it to the matrix–vector form used for computation.

$$D^{ep}_{ijkl} = \left[D^{e}_{ijkl} - \frac{D^{e}_{ijrs} n_{rs} n_{mn} D^{e}_{mnkl}}{h + k^{e}_{nn}} \right] \tag{8.93}$$

The normal $n_{ij} = \partial f / \sigma_{ij}$ is written as a (6×1) vector $\mathbf{n}$ in three dimensions and must preserve the relation

$$df = \frac{\partial f}{\partial \sigma_{ij}} d\sigma_{ij} \equiv \mathbf{n}^T \{d\sigma\} \tag{8.94}$$

Therefore, $\mathbf{n}$ must be defined as

$$\mathbf{n} = \left\{ \frac{\partial f}{\partial \sigma_{11}} \quad \frac{\partial f}{\partial \sigma_{22}} \quad \frac{\partial f}{\partial \sigma_{33}} \quad 2\frac{\partial f}{\partial \sigma_{12}} \quad 2\frac{\partial f}{\partial \sigma_{13}} \quad 2\frac{\partial f}{\partial \sigma_{23}} \right\}^T \tag{8.95}$$

in the same manner as the strain vector $\{\varepsilon\}$ is defined in terms of its tensor components ε_{ij}. We note here that as a result, $n_{ij}n_{ij} \neq \mathbf{n}^T\mathbf{n}$, so that "normalizing" the tensor, that is, $n_{ij}n_{ij} = 1$, does not produce a unit vector $\mathbf{n}$. The tensor form of the elastoplastic continuum tangent D^{ep}_{ijkl} of course maps to the matrix form $\mathbf{D}^{ep}$, as shown in Figure 8.10.

The continuum tangent stiffness $\mathbf{D}^{ep}$ of Chapter 7 relates the unrotated Cauchy stress rate $\dot{\bar{\sigma}}'$ to the unrotated deformation rate $\mathbf{d}$. What we ultimately need for the TL formulation is a relation between the 2PK stress rate and Green strain rate.

Before addressing that topic, we remark that in an incremental-iterative Newton–Raphson solution technique, it is not necessary to update the (element or structure) tangent stiffness at every iteration. In fact, the frequency of tangent stiffness updates is a significant (and problem-dependent) parameter option in a cost-effective solution strategy. Fewer tangent stiffness updates (i.e., using a previous, and therefore less accurate, tangent stiffness) usually results in slower convergence, requiring more, but less costly, iterations.

8.3.2.2 Consistent tangent stiffness

It is well known that Newton methods for the solution of nonlinear equations exhibit asymptotic quadratic convergence under certain favorable conditions. Simo and Taylor (1985) and Nagtegaal (1982) observed that in order to retain full quadratic convergence, the tangent stiffness used for resolving residual forces during iterations at the structural level must be *consistent* to first order with the local (i.e., material point) stress update algorithm. The consistent tangent is also called the *algorithmic* tangent.

We recall here the backward Euler stress update scheme for a material point that was described in Chapter 7:

$$\boldsymbol{\sigma}_{n+1} = \boldsymbol{\sigma}^T_{n+1} - \Delta\lambda \mathbf{D}^e \mathbf{n}_{n+1} = \boldsymbol{\sigma}_n + \mathbf{D}^e(\Delta\boldsymbol{\varepsilon}_{n+1} - \Delta\lambda \mathbf{n}_{n+1}) \tag{8.96}$$

where $\mathbf{n}_{n+1}$ is the updated normal (i.e., at $\boldsymbol{\sigma}_{n+1}$) and $\boldsymbol{\sigma}^T_{n+1}$ is the elastic trial stress. To obtain the consistent tangent, we take the derivative of the stress update equation to obtain

$$\dot{\sigma}_{ij} = D^e_{ijkl}(\dot{\varepsilon}_{kl} - \dot{\lambda} n_{kl}) - \Delta\lambda D^e_{ijkl} \frac{\partial n_{kl}}{\partial \sigma_{mn}} \dot{\sigma}_{mn} \tag{8.97}$$

where $\Delta\lambda$ has previously been determined as part of the material point stress update. This equation can be rearranged as

$$\left[I_{ijmn} + \Delta\lambda D^{e}_{ijkl}\frac{\partial n_{kl}}{\partial \sigma_{mn}}\right]\dot{\sigma}_{mn} = D^{e}_{ijkl}(\dot{\varepsilon}_{kl} - \Delta\lambda n_{kl}) \tag{8.98}$$

in which the fourth-order symmetric unit tensor I_{ijmn} is given by

$$I_{ijmn} = \frac{1}{2}(\delta_{im}\delta_{jn} + \delta_{in}\delta_{jm}) \tag{8.99}$$

Equation 8.98 now becomes

$$\dot{\sigma}_{ij} = \hat{D}^{e}_{ijkl}(\dot{\varepsilon}_{kl} - \dot{\lambda}n_{kl}) \tag{8.100}$$

in which the modified elastic constitutive tensor $\hat{D}^{e}_{ijkl}$ is defined by

$$\hat{D}^{e}_{ijkl} = \left[I_{ijmn} + \Delta\lambda D^{e}_{ijrs}\frac{\partial n_{rs}}{\partial \sigma_{mn}}\right]^{-1} D^{e}_{mnkl} \tag{8.101}$$

In this equation, we have assumed as an approximation that the change in direction of the normal vector depends only on the stress change and have omitted any (presumably small) effect of material hardening.

It is evident that the derivative of the normal vector is needed for the consistent tangent. This will of course depend on the particular yield criterion that is used, and in general, a matrix inversion will be necessary (de Borst and Groen 1994), as indicated in Equation 8.101.

8.3.2.3 von Mises yield criterion

For the von Mises criterion, the yield function is $f = \sigma_e - Y$, where

$$\begin{aligned} \sigma_e &\equiv \sqrt{s_{ij}s_{ij}} \\ n_{ij} &= \frac{s_{ij}}{\sigma_e} \end{aligned} \tag{8.102}$$

The derivative of the normal vector is found to be

$$\frac{\partial n_{rs}}{\partial \sigma_{mn}} = \frac{1}{\sigma_e}\left[\frac{1}{2}(\delta_{rm}\delta_{sn} + \delta_{rn}\delta_{sm}) - \frac{1}{3}\delta_{rs}\delta_{mn} - n_{rs}n_{mn}\right] \tag{8.103}$$

and the modified elastic constitutive tensor for the von Mises yield criterion is then

$$\hat{D}^{e}_{ijkl} = \left[\left(1 + \frac{\Delta\lambda\ G}{\sigma_e}\right)I_{ijmn} - \frac{\Delta\lambda\ G}{\sigma_e}\left(\frac{1}{3}I_{ij}I_{mn} + n_{ij}n_{mn}\right)\right]^{-1} D^{e}_{mnkl} \tag{8.104}$$

Equation 8.104 can be expressed in direct tensor notation as

$$\hat{\mathbf{D}}^{e} = \left[\left(1 + \frac{\Delta\lambda\ G}{\sigma_e}\right)\mathbf{I}_4 - \frac{\Delta\lambda\ G}{\sigma_e}\left(\frac{1}{3}\mathbf{I}_2 \otimes \mathbf{I}_2 + \mathbf{n} \otimes \mathbf{n}\right)\right]^{-1} \mathbf{D}^{e} \tag{8.105}$$

The development of the elastoplastic consistent tangent from Equation 8.100 then follows a path similar to the one that led to the continuum tangent in Chapter 7, with the modified elastic constitutive tensor $\hat{\mathbf{D}}^e$ simply substituted for $\mathbf{D}^e$. Thus, we have for the consistent tangent

$$d\sigma_{ij} = \left(\hat{D}^e_{ijkl} - \frac{\hat{D}^e_{ijrs} n_{rs} n_{mn} \hat{D}^e_{mnkl}}{h + k^e_{nn}} \right) d\varepsilon_{kl} \tag{8.106}$$
$$k^e_{nn} \equiv n_{\alpha\beta} \hat{D}^e_{\alpha\beta\gamma\delta n_{\gamma\delta}} n_{\gamma\delta}$$

In direct tensor notation,

$$d\sigma = \left(\hat{\mathbf{D}}^e - \frac{\hat{\mathbf{D}}^e : \mathbf{n} \otimes \mathbf{n} : \hat{\mathbf{D}}^e}{h + k^e_{nn}} \right) : d\varepsilon \tag{8.107}$$
$$k^e_{nn} \equiv \mathbf{n} : \hat{\mathbf{D}}^e : \mathbf{n}$$

8.3.2.4 Discussion

We emphasize first that ultimate convergence of the incremental-iterative Newton solution is not adversely affected by use of the continuum tangent. However, the consistent tangent matrix in many cases substantially accelerates convergence. While this is an undoubted advantage, the matrix inversion (at each material point) that is often required for the consistent tangent entails a counterbalancing computational cost.

As we have alluded to previously, the continuum tangent matrix may be unsymmetric, most notably for nonassociated plasticity. The consistent tangent matrix also may be unsymmetric, even in those instances where the continuum tangent is symmetric. Unless one wishes to use an unsymmetric equation solver, symmetric versions of the element tangent stiffness matrices have to be used at the structural level for resolution of applied and residual loads.

These are some of the competing factors that must be considered in devising an effective overall solution strategy.

8.3.2.5 2PK stress rate versus Green strain rate for TL formulation

In the TL formulation that has been presented thus far, the elastoplastic material tangent stiffness (whether continuum or consistent) relates the unrotated Cauchy stress rate $\dot{\bar{\boldsymbol{\sigma}}}'$ to the unrotated deformation rate $\mathbf{d}$. For the TL formulation, we need to transform this relation into a relation between the 2PK stress rate $\dot{\sigma}$ and Green strain rate $\dot{\varepsilon}$. We first recall from Chapter 3 the relation between 2PK stress and Cauchy stress,

$$\boldsymbol{\sigma} = |\mathbf{F}| \bar{\mathbf{F}} \bar{\boldsymbol{\sigma}} \bar{\mathbf{F}}^T \tag{8.108}$$

and also that taking the material rate of both sides of this equation provides the definition of the Truesdell rate of Cauchy stress. Here, we first employ the polar decomposition to relate the 2PK stress to unrotated Cauchy stress as follows:

$$\boldsymbol{\sigma} = |\mathbf{U}| \bar{\mathbf{U}} (\mathbf{R}^T \bar{\boldsymbol{\sigma}} \mathbf{R}) \bar{\mathbf{U}} = |\mathbf{U}| \bar{\mathbf{U}} \bar{\boldsymbol{\sigma}}' \bar{\mathbf{U}} \tag{8.109}$$

where $\bar{\mathbf{F}} \equiv \mathbf{F}^{-1}$ and $\bar{\mathbf{U}} \equiv \mathbf{U}^{-1}$, as before. We take the material rate of both sides to obtain

$$\dot{\boldsymbol{\sigma}} = |\mathbf{U}| \bar{\mathbf{U}} [\dot{\bar{\boldsymbol{\sigma}}}' - \dot{\mathbf{U}} \bar{\mathbf{U}} \bar{\boldsymbol{\sigma}}' - \bar{\boldsymbol{\sigma}}' \bar{\mathbf{U}} \dot{\mathbf{U}} + \bar{\boldsymbol{\sigma}}' \ \mathrm{tr}(\mathbf{d})] \bar{\mathbf{U}} \tag{8.110}$$

From the material rate of the polar decomposition $\mathbf{F} = \mathbf{RU}$, we find the kinematic relation

$$\mathbf{D} + \mathbf{W} = \boldsymbol{\Omega} + \mathbf{R} \dot{\mathbf{U}} \bar{\mathbf{U}} \mathbf{R}^T \tag{8.111}$$

Although $\dot{\mathbf{U}}$ and $\overline{\mathbf{U}}$ are both symmetric, the term $\dot{\mathbf{U}}\overline{\mathbf{U}}$ is not necessarily so. However, it can be written as the sum of its symmetric and antisymmetric components as

$$\dot{\mathbf{U}}\overline{\mathbf{U}} = (\dot{\mathbf{U}}\overline{\mathbf{U}})_s + (\dot{\mathbf{U}}\overline{\mathbf{U}})_a \tag{8.112}$$

where

$$\begin{cases} (\dot{\mathbf{U}}\overline{\mathbf{U}})_s \equiv \dfrac{1}{2}[(\dot{\mathbf{U}}\overline{\mathbf{U}}) + (\dot{\mathbf{U}}\overline{\mathbf{U}})^T] \\ (\dot{\mathbf{U}}\overline{\mathbf{U}})_a \equiv \dfrac{1}{2}[(\dot{\mathbf{U}}\overline{\mathbf{U}}) - (\dot{\mathbf{U}}\overline{\mathbf{U}})^T] \end{cases} \tag{8.113}$$

Equation 8.111 then yields

$$\begin{cases} (\dot{\mathbf{U}}\overline{\mathbf{U}})_s = \mathbf{R}^T\mathbf{D}\mathbf{R} = \mathbf{d} \\ (\dot{\mathbf{U}}\overline{\mathbf{U}})_a = \mathbf{R}^T(\mathbf{W} - \mathbf{\Omega})\mathbf{R} \end{cases} \tag{8.114}$$

Due to the symmetry of the stress tensor, we have the following relation:

$$\dot{\mathbf{U}}\overline{\mathbf{U}}\bar{\boldsymbol{\sigma}}' + \bar{\boldsymbol{\sigma}}'\overline{\mathbf{U}}\dot{\mathbf{U}} = \mathbf{d}\bar{\boldsymbol{\sigma}}' + \bar{\boldsymbol{\sigma}}'\mathbf{d} \tag{8.115}$$

Equation 8.110 therefore simplifies to

$$\dot{\boldsymbol{\sigma}} = |\mathbf{U}|\overline{\mathbf{U}}[\dot{\bar{\boldsymbol{\sigma}}}' - \mathbf{d}\,\bar{\boldsymbol{\sigma}}' - \bar{\boldsymbol{\sigma}}'\mathbf{d} + \bar{\boldsymbol{\sigma}}' \ \mathrm{tr}(\mathbf{d})]\overline{\mathbf{U}} \tag{8.116}$$

where it is now more evident that the bracketed term is symmetric.

Incidentally, this equation also provides the relation between the Truesdell rate of Cauchy stress and the material rate of unrotated Cauchy stress, namely,

$$\bar{\boldsymbol{\sigma}}^{Tr} = \mathbf{R}[\dot{\bar{\boldsymbol{\sigma}}}' - \mathbf{d}\bar{\boldsymbol{\sigma}}' - \bar{\boldsymbol{\sigma}}'\mathbf{d} + \bar{\boldsymbol{\sigma}}' \ \mathrm{tr}(\mathbf{d})]\mathbf{R}^T \tag{8.117}$$

from which a relation between the Green–Naghdi and Truesdell rates of Cauchy stress could easily be obtained.

The Green strain rate is related to the unrotated deformation rate by

$$\dot{\boldsymbol{\varepsilon}} = \mathbf{F}^T\mathbf{D}\mathbf{F} = \mathbf{U}(\mathbf{R}^T\mathbf{D}\mathbf{R})\mathbf{U} = \mathbf{U}^T\mathbf{d}\mathbf{U} \tag{8.118}$$

We seek to transform Equation 8.116 into the form

$$\dot{\boldsymbol{\sigma}} = \mathbf{C} : \dot{\boldsymbol{\varepsilon}} \tag{8.119}$$

or

$$\dot{\sigma}_{ij} = C_{ijkl} \ \dot{\varepsilon}_{kl} \tag{8.120}$$

The tensor C consists of four terms, which we consider separately. The first term, which contains the constitutive information, is written in Cartesian tensor notation as

$$|\mathbf{U}|\overline{\mathbf{U}} \ \dot{\bar{\boldsymbol{\sigma}}}'\overline{\mathbf{U}} = |\mathbf{U}|\overline{U}_{i\alpha}\dot{\bar{\sigma}}'_{\alpha\beta}\overline{U}_{\beta j} \tag{8.121}$$

The unrotated Cauchy stress rate is given by

$$\dot{\bar{\sigma}}'_{\alpha\beta} = D^{ep}_{\alpha\beta\gamma\delta}d_{\gamma\delta} = D^{ep}_{\alpha\beta\gamma\delta}\overline{U}_{\gamma k}\overline{U}_{\delta l} \ \dot{\varepsilon}_{kl} \tag{8.122}$$

which yields for the first term

$$|\mathbf{U}|\overline{\mathbf{U}}\dot{\bar{\boldsymbol{\sigma}}}'\overline{\mathbf{U}} = |\mathbf{U}|[\overline{U}_{i\alpha}\overline{U}_{j\beta}D^{ep}_{\alpha\beta\gamma\delta}\overline{U}_{\gamma k}\overline{U}_{\delta l}]\dot{\varepsilon}_{kl} \tag{8.123}$$

Considering the second and third terms together, we first use Equation 8.118 to write them as

$$|U|\overline{U}(d\overline{\sigma}' + \overline{\sigma}'d)\overline{U} = |U|\overline{U}(\overline{U}\ \dot{\varepsilon}\ \overline{U}\ \overline{\sigma}' + \overline{\sigma}'\overline{U}\ \dot{\varepsilon}\ \overline{U})\overline{U} \tag{8.124}$$

and then use Equation 8.109 to express them in terms of 2PK stress and Green strain rate as

$$|U|\overline{U}(d\overline{\sigma}' + \overline{\sigma}'d)\overline{U} = |U|\overline{U}(\overline{U}\ \dot{\varepsilon}\ \overline{U}\ \overline{\sigma}' + \overline{\sigma}'\ \overline{U}\ \dot{\varepsilon}\ \overline{U})\overline{U} = \overline{U}^2\dot{\varepsilon}\ \sigma + \sigma\ \dot{\varepsilon}\ \overline{U}^2 \tag{8.125}$$

In Cartesian tensor notation, the second and third terms are written as

$$|U|\overline{U}\left(d\overline{\sigma}' + \overline{\sigma}'d\right)\overline{U} = \overline{U}^2\dot{\varepsilon}\ \sigma + \sigma\ \dot{\varepsilon}\ \overline{U}^2 = (\overline{U}_{ir}\overline{U}_{rk}\sigma_{jl} + \sigma_{ik}\overline{U}_{jr}\overline{U}_{rl})\dot{\varepsilon}_{kl} \tag{8.126}$$

The fourth term is

$$|U|\overline{U}\ \overline{\sigma}'\ tr(d)\overline{U} = \sigma\ tr(d) \tag{8.127}$$

$$tr(d) = tr\left(\overline{U}\ \dot{\varepsilon}\ \overline{U}\right) = \delta_{\alpha\beta}(\overline{U}_{\alpha k}\ \dot{\varepsilon}_{kl}\ \overline{U}_{l\beta}) = \overline{U}_{k\beta}\overline{U}_{\beta l}\dot{\varepsilon}_{kl} \tag{8.128}$$

Therefore,

$$|U|\overline{U}\ \overline{\sigma}'\ tr(d)\overline{U} = \sigma_{ij}\overline{U}_{k\beta}\overline{U}_{\beta l}\dot{\varepsilon}_{kl} \tag{8.129}$$

Finally, we have

$$\begin{cases} \dot{\sigma}_{ij} = C_{ijkl}\dot{\varepsilon}_{kl} \\ C_{ijkl} = \{|U|[\overline{U}_{i\alpha}\overline{U}_{j\beta}D^{ep}_{\alpha\beta\gamma\delta}\overline{U}_{\gamma k}\overline{U}_{\delta l}] - (\overline{U}_{ir}\overline{U}_{rk}\sigma_{jl} + \sigma_{ik}\overline{U}_{jr}\overline{U}_{rl}) + \sigma_{ij}\overline{U}_{k\beta}\overline{U}_{\beta l}\} \end{cases} \tag{8.130}$$

The matrix form [C] is constructed from the tensor form given above as described earlier (Figure 8.11).

8.3.3 Plane stress

When appropriate, 2D idealizations of a full 3D continuum, that is, plane strain and plane stress, are often utilized in practice. The objective of both idealizations is to reduce unneeded computational effort and also, in the case of plane stress, to eliminate the numerical ill conditioning that would arise from modeling a thin-membrane-like body as a 3D continuum.

The plane strain idealization in which it is assumed that the normal strain $\varepsilon_{zz} = 0$ (and therefore $\sigma_{zz} \neq 0$) can be obtained via straightforward reduction of the 3D formulation. Plane stress in which it is assumed that the normal stress $\sigma_{zz} = 0$ (and therefore $\varepsilon_{zz} \neq 0$) demands additional consideration. We emphasize that the 2D models (finite element meshes) for plane stress and plane strain are identical. In both cases, the models contain only in-plane degrees of freedom u = u(x,y), v = v(x,y), and applied loadings are likewise restricted to be in plane (flexural effects in thin bodies, i.e., plates and shells, are therefore not included here).

The additional complication in plane stress problems arises from the fact that there is no kinematic information contained in the model that can be used to calculate the nonzero normal strain ε_{zz} (analogous to the situation for incompressible material behavior). Several procedures have been proposed to address this issue. Here, we describe a simple version of a procedure outlined by de Borst (1991) for nonlinear plane stress that is similar in concept to the linear elastic approach.

In linear elastic plane stress problems, the normal strain is determined from constitutive information, that is, it is a "Poisson effect," and it is eliminated or "condensed out" *a priori*, leading to well-known effective elastic moduli that transform plane stress to plane strain or vice versa.

In the nonlinear case, the material moduli change throughout the loading process, so that a simple *a priori* transformation between plane stress and plane strain cannot be made. We first review some relevant basic relationships in a linear elastic context.

8.3.3.1 Partial inversion and condensation of material stiffness

For some plane stress applications, it is convenient to reorder the stress and strain vectors as follows:

$$\{\sigma\} = \begin{Bmatrix} \sigma_{xx} \\ \sigma_{yy} \\ \sigma_{xy} \\ \sigma_{zz} \\ \sigma_{xz} \\ \sigma_{yz} \end{Bmatrix}; \quad \{\varepsilon\} = \begin{Bmatrix} \varepsilon_{xx} \\ \varepsilon_{yy} \\ 2\varepsilon_{xy} \\ \varepsilon_{zz} \\ 2\varepsilon_{xz} \\ 2\varepsilon_{yz} \end{Bmatrix} \tag{8.131}$$

With this new ordering, an original (6 × 6) [D] matrix is then rearranged by simply interchanging the third and fourth rows and then interchanging the third and fourth columns. The (6 × 1) stress and strain vectors can then be partitioned into (3 × 1) vectors as follows:

$$\{\sigma\} = \begin{Bmatrix} \sigma_i \\ \sigma_o \end{Bmatrix}; \quad \{\varepsilon\} = \begin{Bmatrix} \varepsilon_i \\ \varepsilon_o \end{Bmatrix} \tag{8.132}$$

in which

$$\left\{ \begin{array}{ll} \sigma_i \equiv \begin{Bmatrix} \sigma_{xx} \\ \sigma_{yy} \\ \sigma_{xy} \end{Bmatrix}; & \varepsilon_i \equiv \begin{Bmatrix} \varepsilon_{xx} \\ \varepsilon_{yy} \\ 2\varepsilon_{xy} \end{Bmatrix} \\ \sigma_o \equiv \begin{Bmatrix} \sigma_{zz} \\ \sigma_{xz} \\ \sigma_{yz} \end{Bmatrix}; & \varepsilon_o \equiv \begin{Bmatrix} \varepsilon_{zz} \\ 2\varepsilon_{xz} \\ 2\varepsilon_{yz} \end{Bmatrix} \end{array} \right. \tag{8.133}$$

The subscripts i and o indicate in plane and out of plane, respectively. The out-of-plane stresses would correspond to tractions on the top and bottom surfaces of a thin membrane lying in the (x, y) plane (Figure 8.12). If the membrane were to be stacked together with, and bonded to, other thin-membrane layers (having differing elastic properties) to form a laminate, equilibrium would require that the out-of-plane stresses (i.e., tractions) be continuous across the interfaces, and interface displacement continuity would require that the in-plane strains be continuous across the interfaces.

Consider a general linear elastic stress–strain relation of the membrane, partitioned as follows:

$$\begin{Bmatrix} \sigma_i \\ \sigma_o \end{Bmatrix} = \begin{bmatrix} A & B \\ B^T & C \end{bmatrix} \begin{Bmatrix} \varepsilon_i \\ \varepsilon_o \end{Bmatrix} + \begin{Bmatrix} p_i \\ p_o \end{Bmatrix} \tag{8.134}$$

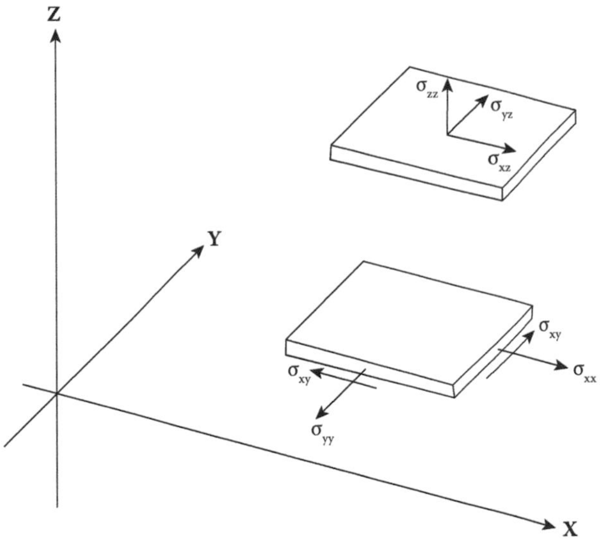

Figure 8.12 In-plane and out-of-plane stresses on a thin membrane in the (x, y) plane.

where **A**, **B**, and **C** are (3 × 3) matrices containing elastic moduli, and **A** and **C** are symmetric. The (3 × 1) vectors $\mathbf{p}_i$ and $\mathbf{p}_o$ represent initial stresses. Solving the lower partition for the out-of-plane strains, we find

$$\varepsilon_o = C^{-1}(\sigma_o - p_o - B^T\varepsilon_i) \tag{8.135}$$

Substituting this relation into the upper partition, we obtain

$$\sigma_i = (A - BC^{-1}B^T)\varepsilon_i + BC^{-1}\sigma_o + p_i - C^{-1}B\ p_o \tag{8.136}$$

As a result of this *part inversion* of Equation 8.134, we may express it in the form

$$\begin{Bmatrix} \sigma_i \\ \varepsilon_o \end{Bmatrix} = \begin{bmatrix} A^* & BC^{-1} \\ -C^{-1}B^T & C^{-1} \end{bmatrix} \begin{Bmatrix} \varepsilon_i \\ \sigma_o \end{Bmatrix} + \begin{Bmatrix} p_i - BC^{-1}p_o \\ -C^{-1}p_o \end{Bmatrix} \tag{8.137}$$

The (3 × 3) (symmetric) *condensed stiffness* **A*** is defined as

$$A^* \equiv (A - BC^{-1}B^T) \tag{8.138}$$

8.3.3.2 Nonlinear material properties

The motivation for de Borst's approach is the desire to directly utilize 3D (elastoplasticity) algorithms for plane stress (and shells as well). This is a very significant consideration for general-purpose finite element codes, but perhaps less so for special-purpose codes. An obvious alternative approach is to employ a plane stress version of the plasticity yield criterion by eliminating the third principal stress σ_3, which can here be identified with σ_{zz}. For the von Mises yield criterion, for example, the 2D yield function in principal stress space is the projection of the von Mises yield cylinder on the σ_3 = 0 plane, which is, as we have seen previously in Chapter 7, an ellipse.

Recognizing that for plane stress, ε_{xz} and ε_{yz} are zero, as are σ_{xz} and σ_{yz}, we need to only deal with (4 × 1) stress and strain vectors as follows:

$$\{\sigma\} = \begin{Bmatrix} \sigma_i \\ \sigma_{zz} \end{Bmatrix}; \quad \{\varepsilon\} = \begin{Bmatrix} \varepsilon_i \\ \varepsilon_{zz} \end{Bmatrix} \tag{8.139}$$

At the conclusion of the stress update at the previous iteration (n), the (consistent) elasto-plastic tangent stiffness relations can be expressed as

$$\begin{Bmatrix} d\sigma_i \\ d\sigma_{zz} \end{Bmatrix} = \begin{bmatrix} \mathbf{D}_{ii} & \mathbf{D}_{io} \\ \mathbf{D}_{oi} & D_{oo} \end{bmatrix} \begin{Bmatrix} d\varepsilon_i \\ d\varepsilon_{zz} \end{Bmatrix} \tag{8.140}$$

In incremental form,

$$\begin{Bmatrix} \sigma_i^{n+1} \\ \sigma_{zz}^{n+1} \end{Bmatrix} = \begin{bmatrix} \mathbf{D}_{ii} & \mathbf{D}_{io} \\ \mathbf{D}_{oi} & D_{oo} \end{bmatrix} \begin{Bmatrix} \Delta\varepsilon_i \\ \Delta\varepsilon_{zz} \end{Bmatrix} + \begin{Bmatrix} \sigma_i^n \\ \sigma_{zz}^n \end{Bmatrix} \tag{8.141}$$

In this equation, $\mathbf{D}_{ii}$ is a (3 × 3) matrix, $\mathbf{D}_{io}$ is a (3 × 1) matrix, $\mathbf{D}_{oi} = \mathbf{D}_{io}^T$ is (1 × 3), and D_{oo} is a scalar. We omit the superscript n on the material moduli, as no confusion should arise thereby.

We wish to enforce the constraint σ_{zz}^{n+1}, noting that σ_{zz}^n will in general be nonzero as (global) iterations proceed. With this constraint, the second partition of Equation 8.141 gives the scalar relation

$$\Delta\varepsilon_{zz} = D_{oo}^{-1}(-\mathbf{D}_{oi}\Delta\varepsilon_i + \sigma_{zz}^n) \tag{8.142}$$

The strain increments appearing in this equation are

$$\begin{cases} \Delta\varepsilon_i \equiv \varepsilon_i^{n+1} - \varepsilon_i^n \\ \Delta\varepsilon_{zz} \equiv \varepsilon_{zz}^{n+1} - \varepsilon_{zz}^n \end{cases} \tag{8.143}$$

Substitution of $\Delta\varepsilon_{zz}$ into the upper partition of Equation 8.141 would yield

$$\Delta\sigma_i^{n+1} = \mathbf{D}_{ii}^*\Delta\varepsilon_i - \mathbf{D}_{oi}D_{oo}^{-1}\sigma_{zz}^n \tag{8.144}$$

where

$$\mathbf{D}_{ii}^* \equiv \mathbf{D}_{ii} - \mathbf{D}_{io}D_{oo}^{-1}\mathbf{D}_{oi} \tag{8.145}$$

is the condensed (3 × 3) tangent stiffness at iteration (n) that has already been computed and stored at the end of iteration (n), as have the quantities D_{oo} and $\mathbf{D}_{io}$.

The simple procedure we describe here is a "once-through" stress update at the material point for each global iteration. There is no local "iteration within an iteration" to update the material moduli and thereby improve the approximation of Equation 8.142, although that would certainly be feasible.

8.3.3.3 Expansion of strain increment and stress update

With $\Delta\varepsilon_{zz}$ now determined from Equation 8.142, the local (4 × 1) strain vector increment is expanded as

$$\{\Delta\varepsilon\}^{n+1} = \begin{Bmatrix} \Delta\varepsilon_i^{n+1} \\ \Delta\varepsilon_{zz}^{n+1} \end{Bmatrix} \tag{8.146}$$

and it is processed through the regular 3D plasticity algorithms. The constituents of the updated (4 × 1) stresses are denoted by $\boldsymbol{\sigma}_i, \sigma_{zz}$ and the updated (4 × 4) material consistent tangent matrix by **D**.

We point out that at the end of this stress update, the (4 × 1) stress vector will lie "precisely" on the 3D yield surface but not necessarily on the 2D projection of the 3D yield surface. In the case of the von Mises yield criterion, the 2D projection of the (4 × 1) stress will lie on a shifted ellipse in the (σ_1, σ_2) plane. As global iterations continue, the shift of the ellipse (at each material point) approaches zero. We naturally expect that a nonzero value of the local $\boldsymbol{\sigma}_{zz}$ stress will be reflected in the material point contribution to the element internal resisting force vector, namely, Equation 8.137.

8.3.3.4 *Condensed tangent stiffness and internal resisting force vector*

The local (4 × 4) consistent tangent matrix is condensed to a (3 × 3) in-plane material tangent stiffness matrix as described previously. The result is

$$\mathbf{D}_{ii}^{*} \equiv \mathbf{D}_{ii} - \mathbf{D}_{io}\mathbf{D}_{oo}^{-1}\mathbf{D}_{oi} \tag{8.147}$$

which is the updated in-plane tangent stiffness that is returned to the element-level processors for the next global iteration.

The material point contribution **I** to the element internal resisting force vector for the next global iteration is Equation 8.137:

$$\mathbf{I} = \sigma_i - \mathbf{D}_{io}\mathbf{D}_{oo}^{-1}\sigma_{zz} \tag{8.148}$$

We note that de Borst introduced an additional equilibrium correction to this basic procedure that appears to significantly improve the convergence rate of Newton iterations.

8.3.4 Pressure loading

We briefly consider here one important type of "follower load," that is, a constant magnitude pressure that acts on, and remains normal to, a surface that undergoes large rotations and deformations. Follower loads are deformation dependent and lead to stiffness-type terms, sometimes termed "load stiffness," in the incremental equilibrium equations. The load stiffness matrices add to the element (and structure) tangent stiffness matrices.

The load stiffness is symmetric if the follower loads are conservative, that is, if they can be derived from a potential function. Nonconservative loads lead in general to unsymmetric tangent stiffness matrices.

The calculation of nodal equivalent loads for tractions applied on an element surface naturally involves integration over that surface. We recall from Chapter 3 that the force vector $d\bar{\mathbf{f}}$ on a differential surface element $d\bar{\mathbf{A}}$ in the deformed configuration is defined in terms of the traction vector by

$$d\bar{\mathbf{f}} = \bar{\mathbf{t}}\ d\bar{A} \tag{8.149}$$

If the applied traction vector $\bar{\mathbf{t}}$ acts in the direction of the normal to the deformed surface, that is, $\bar{\mathbf{t}} = t^0\bar{\mathbf{n}}$, then

$$d\bar{\mathbf{f}} = t^0\bar{\mathbf{n}}d\bar{A} = t^0 d\bar{\mathbf{A}} \tag{8.150}$$

For the case of a follower pressure load, t^0 ($\equiv -p^0$) is constant. The external virtual work due to the applied surface traction can be calculated for the TL formulation as

$$\delta W_{ext} = t^0 \int_{\bar{A}} \delta u^T \bar{n}\, d\bar{A} = t^0 \int_{A} \delta u^T \bar{n} \left(\frac{d\bar{A}}{dA}\right) dA \tag{8.151}$$

The finite element displacement interpolation on the surface must be introduced into this equation, as well as the description of the element surface geometry. The equivalent nodal loads are then evaluated via numerical integration over the surface.

The element surface description is convenient for isoparametric elements. For example, in three dimensions, on an isoparametric element surface described by (ξ, η, ± 1), the parameters (ξ, η) constitute a natural surface coordinate system from which surface properties are easily computed (the numerical integration is performed in the parent element geometry).

We now consider as an example a follower pressure load on the simplest isoparametric element (other than a 1D bar).

8.3.4.1 *Load stiffness for 2D simplex element*

Consider the simple three-node triangular element introduced previously. Figure 8.13 shows a generic 2D simplex element. Nodes are numbered (1, 2, 3) in a counterclockwise direction. Element edges are numbered (1, 2, 3) opposite to the corresponding node.

Edge ("surface") lengths are designated (l_1, l_2, l_3), and corresponding outward unit normal vectors are labeled ($\mathbf{n}_1$, $\mathbf{n}_2$, $\mathbf{n}_3$), again proceeding around the element boundary in a counterclockwise direction. Element nodal coordinates and displacements are (x_i, y_i) and (u_i, v_i), respectively. The convenient notation $x_j - x_k \equiv x_{jk}$, $u_j - u_k \equiv u_{jk}$ for nodal variables is also used, where (i, j, k) are taken in cyclic (1, 2, 3) order. We omit the $\{\cdot\}$ notation for vectors where it is not needed for clarity.

The figure shows a uniform normal traction t_1 (tension positive) acting on edge 1 in the reference configuration. In the deformed configuration, the magnitude of the traction is unchanged, but it now acts over the length $\bar{l}_1$ and in the direction of the normal vector $\bar{\mathbf{n}}_1$.

Figure 8.14 shows the direction of a vector $\mathbf{n}_i l_i$ normal to element edge i, where $\mathbf{n}_i$ is the *unit* normal.

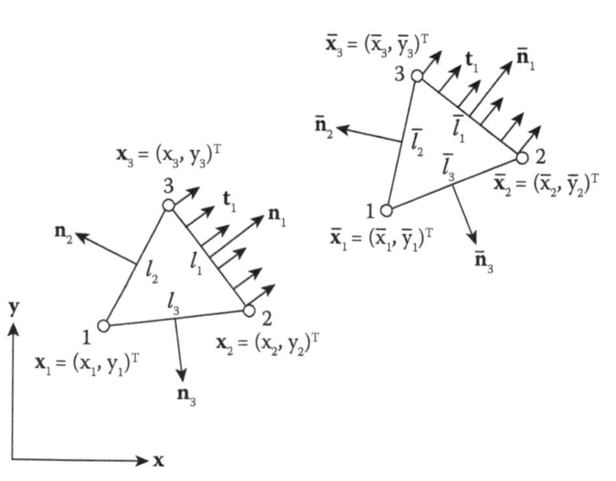

Figure 8.13 2D simplex element with follower traction on one edge.

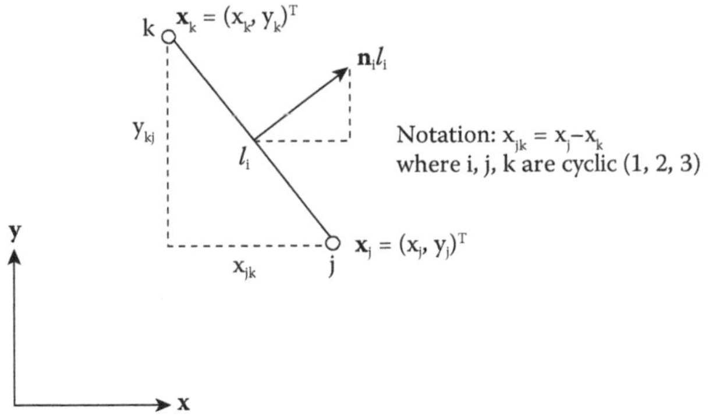

Figure 8.14 Vector normal to element edge i.

$$\mathbf{n}_i l_i = \begin{Bmatrix} y_{kj} \\ x_{jk} \end{Bmatrix} \tag{8.152}$$

We note here that $\mathbf{n}_i l_i$ is the 2D version of the 3D area vector dA. The total force vector on edge i is $t_i \mathbf{n}_i l_i$ (no sum).

It is easily verified that

$$\sum_{i=1,2,3} \mathbf{n}_i l_i \equiv 0_2 \tag{8.153}$$

which demonstrates, as a check, that if a uniform pressure field, say $t_i = -p_0$, acts on all three edges of the element, the vector sum of the external forces is zero. In this equation, 0_2 is a (2×1) zero vector.

Incidentally, Equation 8.153 is the 2D counterpart of

$$\oint_A \mathbf{n}\ dA = 0_3 \tag{8.154}$$

for a closed surface A, which follows from the divergence theorem (Chapter 4).

In the deformed configuration, the element nodal coordinates are

$$\begin{aligned} \bar{x}_i &= x_i + u_i \\ \bar{y}_i &= y_i + v_i \end{aligned} \tag{8.155}$$

and

$$\begin{aligned} \bar{x}_{jk} &= x_{jk} + u_j - u_k \\ \bar{y}_{jk} &= y_{jk} + v_j - v_k \end{aligned} \tag{8.156}$$

The corresponding normal vectors are

$$\bar{\mathbf{n}}_i \bar{l}_i = \begin{Bmatrix} \bar{y}_{kj} \\ \bar{x}_{jk} \end{Bmatrix} = \begin{Bmatrix} y_{kj} \\ x_{jk} \end{Bmatrix} + \begin{Bmatrix} v_k - v_j \\ u_j - u_k \end{Bmatrix} = \begin{Bmatrix} y_{kj} \\ x_{jk} \end{Bmatrix} + \begin{bmatrix} 0 & -1 & 0 & 1 \\ 1 & 0 & -1 & 0 \end{bmatrix} \begin{Bmatrix} u_j \\ v_j \\ u_k \\ v_k \end{Bmatrix} \tag{8.157}$$

To calculate the external virtual work, we need expressions for the displacement interpolation on the element edge. Throughout the element, the displacements are interpolated as

$$\begin{Bmatrix} u \\ v \end{Bmatrix} = \begin{bmatrix} L_1 & 0 & L_2 & 0 & L_3 & 0 \\ 0 & L_1 & 0 & L_2 & 0 & L_3 \end{bmatrix} \begin{Bmatrix} u_1 \\ v_1 \\ u_2 \\ v_2 \\ u_3 \\ v_3 \end{Bmatrix} \tag{8.158}$$

In this last equation, the (6 × 1) element nodal displacement vector in the element local coordinate system is denoted as

$$\mathbf{u}_e \equiv \begin{bmatrix} u_1 & v_1 & u_2 & v_2 & u_3 & v_3 \end{bmatrix}^T \tag{8.159}$$

On edge 1, $L_1 = 0$ and $L_2 + L_3 = 1$. Therefore, on edge 1, $L_3 = 1 - L_2$ and L_2 serves as a nondimensional edge coordinate running from 0 at node 2 to 1 at node 3. The distance along edge 1, measured from node 2, is therefore $l_1 L_2$ and the differential element of length is $l_1 dL_2$. The displacements along edge 1 are therefore interpolated as

$$\begin{Bmatrix} u \\ v \end{Bmatrix} = \begin{bmatrix} 0 & 0 & L_2 & 0 & 1-L_2 & 0 \\ 0 & 0 & 0 & L_2 & 0 & 1-L_2 \end{bmatrix} \mathbf{u}_e \tag{8.160}$$

The external virtual work done during a nodal virtual displacement $\delta\mathbf{u}_e$ is

$$\delta W_{ext} = \int \delta\mathbf{u}^T \mathbf{n}_1 t_1 l_1 dL_2 = \delta\mathbf{u}_e^T \left[t_1 \int_0^1 \begin{bmatrix} 0 & 0 \\ 0 & 0 \\ L_2 & 0 \\ 0 & L_2 \\ 1-L_2 & 0 \\ 0 & 1-L_2 \end{bmatrix} \begin{Bmatrix} y_{32} \\ x_{23} \end{Bmatrix} dL_2 \right] \tag{8.161}$$

In the reference configuration, the external virtual work is therefore

$$\delta W_{ext} = \delta\mathbf{u}_e^T \mathbf{p}_{ext} \tag{8.162}$$

where

$$\mathbf{p}_{ext} = \frac{1}{2} t_1 \begin{Bmatrix} 0 \\ 0 \\ y_{32} \\ x_{23} \\ y_{32} \\ x_{23} \end{Bmatrix} \tag{8.163}$$

which is the expected result.

8.3.4.2 *Secant load stiffness*

In the deformed configuration, the normal vectors are given by Equation 8.157. For the case of a uniform traction on edge 1, the external virtual work is

$$\delta W_{ext} = \int \delta \mathbf{u}^T \bar{\mathbf{n}}_1 t_1 \bar{l}_1 dL_2 = \delta \mathbf{u}_e^T \left[t_1 \int_0^1 \begin{bmatrix} 0 & 0 \\ 0 & 0 \\ L_2 & 0 \\ 0 & L_2 \\ 1-L_2 & 0 \\ 0 & 1-L_2 \end{bmatrix} \begin{Bmatrix} \bar{y}_{32} \\ \bar{x}_{23} \end{Bmatrix} dL_2 \right] \tag{8.164}$$

where

$$\begin{Bmatrix} \bar{y}_{32} \\ \bar{x}_{23} \end{Bmatrix} = \begin{Bmatrix} y_{32} \\ x_{23} \end{Bmatrix} + \begin{bmatrix} 0 & 0 & 0 & -1 & 0 & 1 \\ 0 & 0 & 1 & 0 & -1 & 0 \end{bmatrix} \mathbf{u}_e \tag{8.165}$$

Thus, we see that the external virtual work in the deformed configuration consists of two terms. The first term results in the load vector of Equation 8.163, and the second (displacement-dependent) term gives

$$\delta W_{ext}^L = \delta \mathbf{u}_e^T \mathbf{k}_L \mathbf{u}_e \tag{8.166}$$

where

$$\mathbf{k}_L = t_1 \int_0^1 \begin{bmatrix} 0 & 0 \\ 0 & 0 \\ L_2 & 0 \\ 0 & L_2 \\ 1-L_2 & 0 \\ 0 & 1-L_2 \end{bmatrix} \begin{bmatrix} 0 & 0 & 0 & -1 & 0 & 1 \\ 0 & 0 & 1 & 0 & -1 & 0 \end{bmatrix} dL_2 \tag{8.167}$$

$$\mathbf{k}_L = \frac{t_1}{2} \begin{bmatrix} 0 & 0 & 0 & 0 & 0 & 0 \\ 0 & 0 & 0 & 0 & 0 & 0 \\ 0 & 0 & 0 & -1 & 0 & 1 \\ 0 & 0 & 1 & 0 & -1 & 0 \\ 0 & 0 & 0 & -1 & 0 & 1 \\ 0 & 0 & 1 & 0 & -1 & 0 \end{bmatrix} \tag{8.168}$$

for a pressure load $t_1 = -p_1$. If a secant stiffness matrix can be computed, independent of follower load contributions (e.g., for hyperelastic material properties), the load stiffness matrix is simply taken to the other side of the equilibrium equation and adds to the total secant stiffness.

8.3.4.3 *Tangent load stiffness*

The load stiffness contribution to the element tangent stiffness is determined by taking the time rate of $\mathbf{k}_L \mathbf{u}_e$. In this case, with a constant pressure, the element tangent pressure stiffness contribution is $\mathbf{k}_L$. Therefore, the same load stiffness matrix given in Equation 8.168, which we observe is unsymmetric, adds to the tangent stiffness matrix, which is therefore also unsymmetric for a follower pressure load.

8.4 UPDATED LAGRANGIAN METHODS

An updated Lagrangian (UDL) method is one in which the reference configuration is updated, either at selected intervals or at every converged equilibrium configuration, throughout the analysis.

Figure 8.15 illustrates three successive configurations (C_0, C_1, C_2) occupied by a deforming body along its equilibrium (i.e., solution) path: the initial configuration C_0, the current deformed configuration C_2, and an intermediate reference configuration C_1.

The reference configuration is intermittently updated to the current configuration and used for the next several increments (e.g., ∂C_2) until it is again updated. At any given instant, we therefore have all three distinct configurations (C_0, C_1, C_2), as shown in the figure. The update frequency can be adjusted based on diagnostic parameters monitored during the solution process to allow more or less frequent updates in much the same way as load increments can be adjusted in response to experienced convergence rates.

We remark that in the TL method, C_1 is identical to the initial configuration C_0, and therefore no distinction between them need be made.

8.4.1 Basic UDL formulation

The basic UDL formulation, illustrated in Figure 8.16, is one in which the reference configuration is updated at every incremental step along the equilibrium path, so that in this case C_1 always coincides with C_2.

This basic UDL formulation is similar to that previously described for the TL method, with several important distinctions. We concentrate on those aspects of UDL here.

First, the matrices $\mathbf{K}_1$ and $\mathbf{K}_2$, which appear as part of the TL tangent stiffness, contain nonlinear terms that depend on the total displacements from the reference (initial) configuration

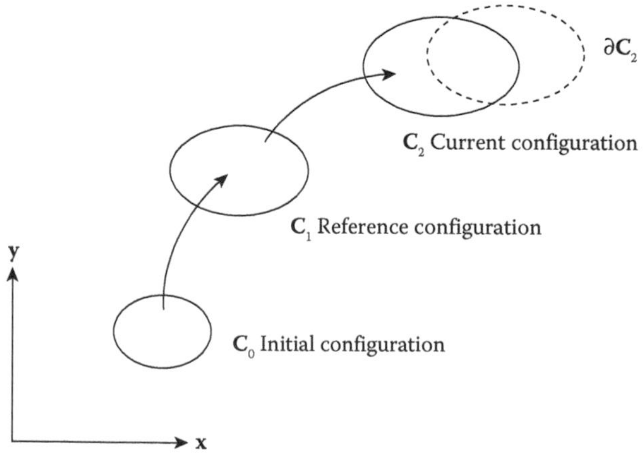

Figure 8.15 Initial, reference, and current configurations in the general UDL method.

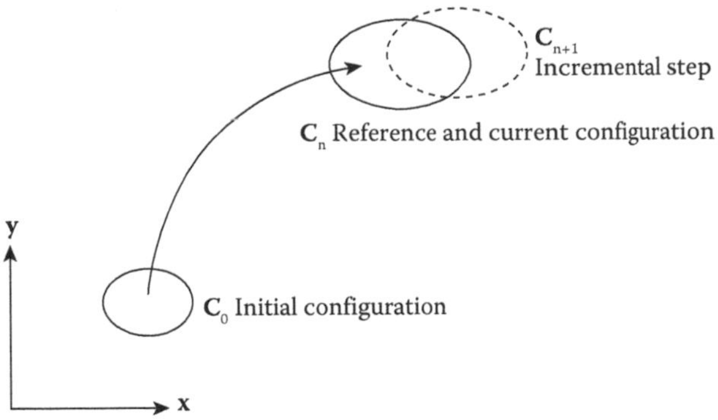

Figure 8.16 Initial and reference or current configurations in basic UDL method.

to the current configuration. In the basic UDL formulation, these two configurations coincide. As a result, the UDL tangent stiffness has the simpler form

$$\mathbf{K}_t = \mathbf{K}_0 + \mathbf{K}_G + \mathbf{K}_L \tag{8.169}$$

This $\mathbf{K}_0$ matrix contains the current constitutive moduli. It differs from the TL $\mathbf{K}_0$ matrix in another way: it is a function of the geometry of the deformed body, whereas the TL $\mathbf{K}_0$ matrix is a function of the undeformed geometry. We first consider the equilibrium equations.

8.4.1.1 *Virtual work in deformed configuration*

The virtual work expression for a single element is written in general as

$$\delta W_{int,e} = -\int_{Ve} \{\delta\varepsilon\}^T\{\sigma\}dV_e = -\delta\mathbf{u}_e^T\mathbf{I}_e \tag{8.170}$$

where $\mathbf{I}_e$ is the element internal resisting force vector.

Virtual work in the deformed configuration (in terms of 2PK stress and virtual Green strain), but with variables referred to the reference configuration, provides

$$\mathbf{I}_e + \Delta\mathbf{I}_e = \int_{Ve} [\mathbf{B}^l + \mathbf{B}^{nl}]^T(\{\sigma\}_n + \{\Delta\sigma\}_{n+1})dV_e \tag{8.171}$$

Virtual work in the *reference* configuration (with variables referred to that same configuration) provides

$$\mathbf{I}_e = \int_{Ve} \mathbf{B}^{l^T}\{\sigma\}_n dV_e \tag{8.172}$$

Subtracting this last equation from the previous one gives (Nagtegaal 1982)

$$\Delta\mathbf{I}_e = \int_{V_e} [\mathbf{B}^l + \mathbf{B}^{nl}]^T\{\Delta\sigma\}_{n+1}dV_e + \int_{V_e} \mathbf{B}^{nl^T}\{\sigma\}_n dV_e \tag{8.173}$$

The second term on the right in this equation is linear in the incremental nodal displacements through the $\mathbf{B}^{nl}$ matrix and gives rise to the geometric tangent stiffness matrix.

Either the 2PK stress or the Cauchy stress in the reference configuration may be used in the geometric stiffness because they are identical in that updated reference configuration. The first term on the right requires a relation between the increment of 2PK stress and the increment of Green strain analogous to that developed for the TL formulation.

8.4.1.2 Coordinate systems

Element nodal displacements are referred to the initial element Cartesian coordinate system, which does not change during the UDL analysis. This coordinate system may or may not differ from the common global Cartesian system for the finite element assembly (refer to Equation 8.55). The element reference geometry is updated at every converged equilibrium configuration using the nodal displacement increments and used to compute the deformation gradient (and related quantities) at material sampling point locations within the element, again referred to the element Cartesian coordinate axes. This must be done at the element level since no spatial gradient information is available at the material point level. At each material point, a Cartesian coordinate system aligned with the local principal stretch directions rotates during the UDL analysis as dictated by the incremental deformation. As in TL, the nonlinear (e.g., elastoplastic) material behavior is described in this material coordinate system, thereby removing local rigid body effects from the constitutive relations.

Figure 8.17 illustrates the updating process at the material point level. C_n is the current (converged) reference configuration, and C_{n+1} represents an incremental step from C_n. As we have seen from the TL formulation, the rotation of the Cartesian material coordinate system $(\overline{x}', \overline{y}')$ in C_n relative to the element Cartesian coordinates is described by

$$\overline{\mathbf{x}}'_n = \mathbf{R}_n^T \overline{\mathbf{x}}_n \tag{8.174}$$

The unrotated Cauchy stress at C_n is given by

$$\overline{\boldsymbol{\sigma}}'_n = \mathbf{R}_n^T \; \overline{\boldsymbol{\sigma}}_n \; \mathbf{R}_n \tag{8.175}$$

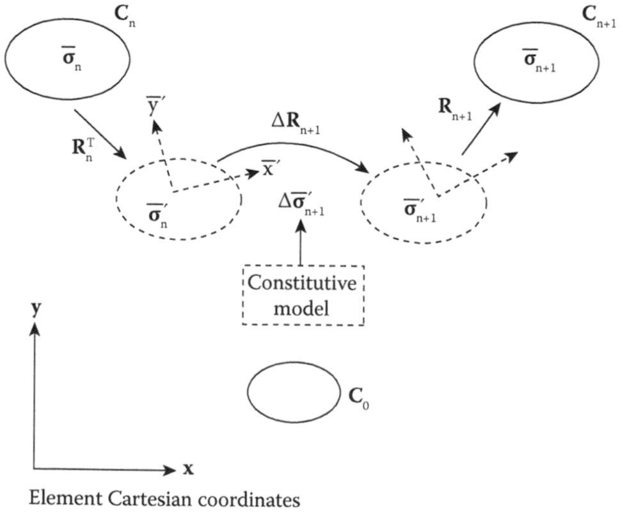

Figure 8.17 UDL update at the material point level.

and the stress update is

$$\bar{\sigma}'_{n+1} = \bar{\sigma}'_n + \Delta\bar{\sigma}'_{n+1} \tag{8.176}$$

The stress increment $\Delta\bar{\sigma}'_{n+1}$ is provided by the constitutive model. The updated Cauchy stress in C_{n+1} is then determined from

$$\bar{\sigma}_{n+1} = \mathbf{R}_{n+1}\ \bar{\sigma}'_{n+1}\ \mathbf{R}^T_{n+1} \tag{8.177}$$

which consists of two components. It may be written as

$$\bar{\sigma}_{n+1} = \mathbf{R}_{n+1}\ \Delta\bar{\sigma}'_{n+1}\ \mathbf{R}^T_{n+1} + \Delta\mathbf{R}_{n+1}\ \bar{\sigma}'_{n+1}\ \Delta\mathbf{R}^T_{n+1} \tag{8.178}$$

where the incremental rotation of the material reference frame during the step is

$$\Delta\mathbf{R}_{n+1} \equiv \mathbf{R}_{n+1}\mathbf{R}^T_n \tag{8.179}$$

Thus far, the stress update procedure is identical to that described earlier for TL. The polar decompositions of the deformation gradients at C_n and C_{n+1} provide the "exact" incremental rotation via Equation 8.179.

However, to obviate the need for polar decompositions, an approximate incremental rotation (Hughes and Winget 1980) is generally used in UDL.

8.4.1.3 Hughes–Winget incremental update

The rigid body rotation of the (material) coordinate frame is described by the matrix differential equation

$$\frac{d}{dt}\mathbf{R}(t) = \mathbf{W}(t)\mathbf{R}(t) \tag{8.180}$$

where $\mathbf{W}(t)$ is the spin. $\mathbf{R}(0) = \mathbf{I}$ (i.e., $\dot{\mathbf{R}}(0) = \mathbf{W}(0)$) because the initial rotation takes place with respect to a coordinate frame that is already rotated from the fixed Cartesian spatial axes. The solution of Equation 8.180 can be written in the form of a matrix exponential, which we will discuss in Chapter 9.

We label the Hughes–Winget incremental rotation matrix $\Delta\mathbf{Q}_{n+1}$ to distinguish it from the exact incremental rotation $\Delta\mathbf{R}_{n+1}$ determined from polar decompositions. The fundamental requirement for any such approximation is that $\Delta\mathbf{Q}_{n+1}$ be orthonormal, that is, $\Delta\mathbf{Q}^T_{n+1}\Delta\mathbf{Q}_{n+1} = \mathbf{I}$ and that tensor updates, for example,

$$\bar{\sigma}_{n+1} = \Delta\mathbf{Q}^T_{n+1}\ \bar{\sigma}_n\ \Delta\mathbf{Q}_{n+1} \tag{8.181}$$

are incrementally objective, that is, tensor invariants are preserved. The incremental rotation updates are simply accumulated over the load steps to determine the (approximate) material reference frame orientation at each configuration, that is,

$$\begin{cases} \mathbf{R}_1 = \mathbf{I} \\ \mathbf{R}_{n+1} \approx \Delta\mathbf{Q}_{n+1}\mathbf{R}_n \end{cases} \tag{8.182}$$

Hughes and Winget obtained the incremental rotation $\Delta\mathbf{Q}_{n+1}$ by numerical integration of Equation 8.180 over the pseudo-time increment Δt using the generalized midpoint rule (Ortiz and Popov 1985), that is,

$$\mathbf{Q}_{n+1} - \mathbf{Q}_n = \mathbf{W}[\alpha\ \mathbf{Q}_{n+1} + (1-\alpha)\mathbf{Q}_n] \tag{8.183}$$

which leads to

$$
\begin{aligned}
&[\mathbf{I}-\alpha\ \mathbf{W}]\mathbf{Q}_{n+1} = [\mathbf{I}+(1-\alpha)\mathbf{W}]\mathbf{Q}_n \\
&\mathbf{Q}_{n+1} = [\mathbf{I}-\alpha\ \mathbf{W}]^{-1}[\mathbf{I}+(1-\alpha)\mathbf{W}]\mathbf{Q}_n
\end{aligned}
\tag{8.184}
$$

For $\alpha = 1/2$, that is, the sampling point $t = t_n + \Delta t/2$ (the actual midpoint of the interval),

$$
\Delta\mathbf{Q}_{n+1} = \left[\mathbf{I}-\frac{1}{2}\mathbf{W}\right]^{-1}\left[\mathbf{I}+\frac{1}{2}\mathbf{W}\right] \tag{8.185}
$$

It can be shown that the two matrices on the right of Equation 8.185 commute and that $\Delta\mathbf{Q}_{n+1}$ is orthonormal and incrementally objective. These essential properties hold only for midinterval ($\alpha = 1/2$) sampling. This equation is actually an instance of the Cayley transform (1846) mentioned later in Chapter 9.

The spin $\mathbf{W}$ is calculated as the antisymmetric part of the incremental deformation gradient

$$
G_{ij} = \frac{\partial \Delta u_i}{\partial \overline{x}_{jj}} \tag{8.186}
$$

$$
\overline{\mathbf{x}}_{n+1} = \overline{\mathbf{x}}_n + \mathbf{u}_{n+1} \tag{8.187}
$$

$\mathbf{W}$ must be expressed in the material point Cartesian coordinate system $(\overline{x}', \overline{y}')$ in C_n for use in Equation 8.185. The 2D velocity gradient $\mathbf{V}_{n+1}$ in the element Cartesian system is

$$
\mathbf{V}_{n+1} = \begin{bmatrix} \dfrac{\partial \dot{\overline{x}}_1}{\partial \overline{x}_1} & \dfrac{\partial \dot{\overline{x}}_1}{\partial \overline{x}_2} \\ \dfrac{\partial \dot{\overline{x}}_2}{\partial \overline{x}_1} & \dfrac{\partial \dot{\overline{x}}_2}{\partial \overline{x}_2} \end{bmatrix} = \begin{bmatrix} \dfrac{\partial \dot{\overline{x}}_1}{\partial \overline{x}'_1} & \dfrac{\partial \dot{\overline{x}}_1}{\partial \overline{x}'_2} \\ \dfrac{\partial \dot{\overline{x}}_2}{\partial \overline{x}'_1} & \dfrac{\partial \dot{\overline{x}}_2}{\partial \overline{x}'_2} \end{bmatrix} \begin{bmatrix} \dfrac{\partial \overline{x}'_1}{\partial \overline{x}_1} & \dfrac{\partial \overline{x}'_1}{\partial \overline{x}_2} \\ \dfrac{\partial \overline{x}'_2}{\partial \overline{x}_1} & \dfrac{\partial \overline{x}'_2}{\partial \overline{x}_2} \end{bmatrix} \tag{8.188}
$$

Generalization of this relationship to 3D is obvious. Therefore, the coordinate transformation for the velocity gradient is

$$
\begin{cases} \mathbf{V}_{n+1} = \mathbf{V}'_{n+1}\mathbf{R}_n^T \\ \mathbf{V}'_{n+1} = \mathbf{V}_{n+1}\mathbf{R}_n \end{cases} \tag{8.189}
$$

The spin in the material point Cartesian coordinate system then follows as

$$
\mathbf{W}'_{n+1} = \frac{1}{2}[\mathbf{V}'_{n+1} - \mathbf{V}'^{T}_{n+1}] \tag{8.190}
$$

Example 8.2 Rotation of initial Cauchy stress

We naturally expect some accumulation of error due to this approximation for large material axis rotations. In order to examine the characteristics of the approximation, we consider a simple numerical example in which simple shear kinematics are used to drive the rotation.

From previous examples, we recall that at a shear parameter $\kappa/2 = 1$, the material axes reach a final rotation of 45°, at which point the in-plane principal stretches are (2.4142, 0.4142). Figure 8.18 shows the calculated (2D) rotation angle using a very large deformation increment, $\Delta\kappa/2 = 0.05$ (corresponding to perhaps 20 times the uniaxial yield strain for a mild steel).

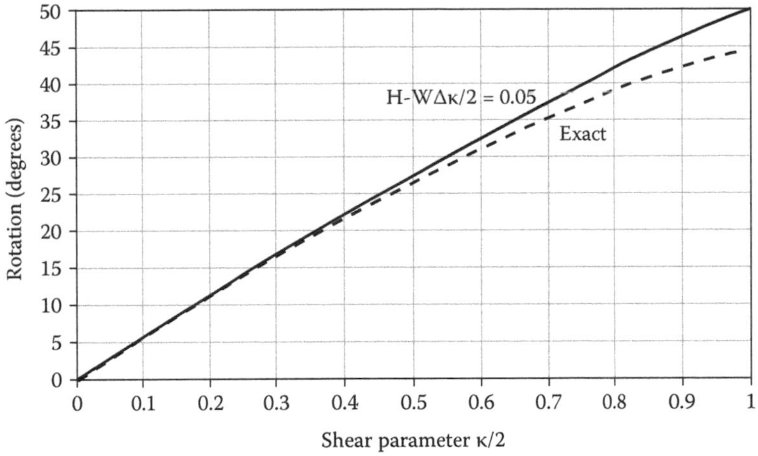

Figure 8.18 Material axis rotation using Hughes–Winget approximation.

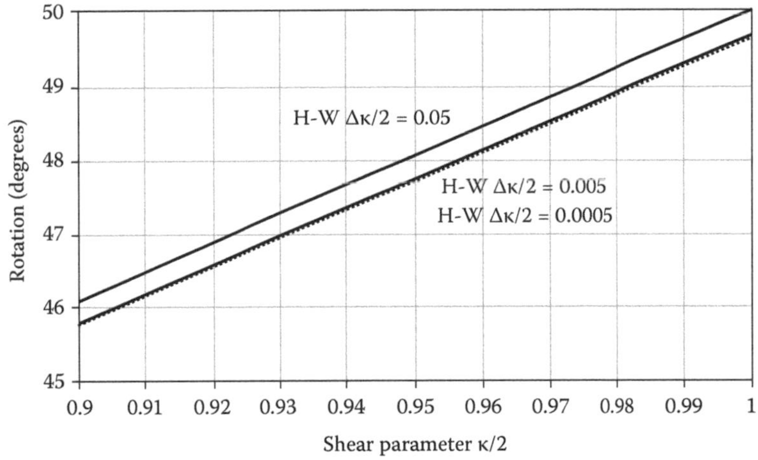

Figure 8.19 Effect of deformation increment $\Delta\kappa/2$ on calculated material axis rotation.

We observe that up to about $\kappa/2 = 0.25$ (five deformation increments), the approximation appears to be quite good. The final rotation at $\kappa/2 = 1$ is $\theta = 50.003°$ compared with the exact value of 45°.

Figure 8.19 shows the results of similar calculations for two smaller deformation increments, 1/10 and 1/100, respectively, of the deformation increment of Figure 8.18. Note the smaller range of Figure 8.19 ($0.90 \leq \kappa/2 \leq 1.0$, $45° \leq \theta \leq 50°$); the plot of the exact rotation angle falls just outside the range of Figure 8.19. The final rotation angles for $\kappa/2 = 1$ are shown in Table 8.1. Drastically increasing the number of increments (N) has very little effect.

We now explore numerically the (incremental) objectivity characteristics of the Hughes–Winget approximation. Assume an initial pure shear Cauchy stress (arbitrary units) in the (x, y) plane,

Table 8.1 Computed rotation angle θ at κ/2 = 1 vs. number of integration increments N

N	θ (deg)
20	50.034
200	49.645
2000	49.609

$$\bar{\boldsymbol{\sigma}} = \begin{bmatrix} 0 & 100 & 0 \\ 100 & 0 & 0 \\ 0 & 0 & 0 \end{bmatrix} \tag{8.191}$$

rotated in plane with simple shear kinematics. Of course, for a 45° rotation the final Cauchy stress is exactly

$$\text{Rotated } \bar{\boldsymbol{\sigma}} = \begin{bmatrix} 100 & 0 & 0 \\ 0 & -100 & 0 \\ 0 & 0 & 0 \end{bmatrix} \tag{8.192}$$

that is, a state of equibiaxial tension–compression in the fixed Cartesian coordinate system. We realize that since the approximate update does not rotate the Cauchy stress tensor precisely through 45°, we cannot expect the result shown above. For the largest increment considered above ($\Delta\kappa/2 = 0.05$, N = 20), the rotated Cauchy stress is

$$\text{Approx } \bar{\boldsymbol{\sigma}} = \begin{bmatrix} 98.4787 & -17.3767 & 0 \\ -17.3767 & -98.4787 & 0 \\ 0 & 0 & 0 \end{bmatrix} \tag{8.193}$$

Despite the observed error accumulation in the computed material axis rotation tensor (and in individual components of the rotated Cauchy stress), the J_2 norm of the rotated Cauchy stress is preserved to very high accuracy, greater than 10 significant figures for the numerical examples considered here. This result is actually expected because (as discussed in Chapter 9) a transformation of a skew-symmetric matrix like that shown in Equation 8.185 always produces a proper orthogonal matrix. In this case, it's just not the precisely correct orthogonal matrix.

We remark here that in a UDL implementation using the Hughes–Winget update, it would be a simple matter to monitor material axis rotation and update (i.e., correct) it periodically as and if required, using polar decomposition.

8.4.1.4 Element geometry updating

Figure 8.20 illustrates the geometry updating process using the 2D simplex element as an example. In this case, there is only one material point in the element that may be considered as located at the element centroid. In other respects, the process is very much the same for other 2D isoparametric elements.

Consider an element incremental nodal displacement vector $\Delta\mathbf{u}_e$ for load step (n + 1) resulting from a structural-level equilibrium iteration. We assume that at this stage, all variables have been previously updated at configuration (n). Discrete approximations for the kinematic variables are as follows, in which the pseudo–time increment Δt is suppressed for rate quantities. The incremental deformation gradient $\Delta\mathbf{F}$ is determined

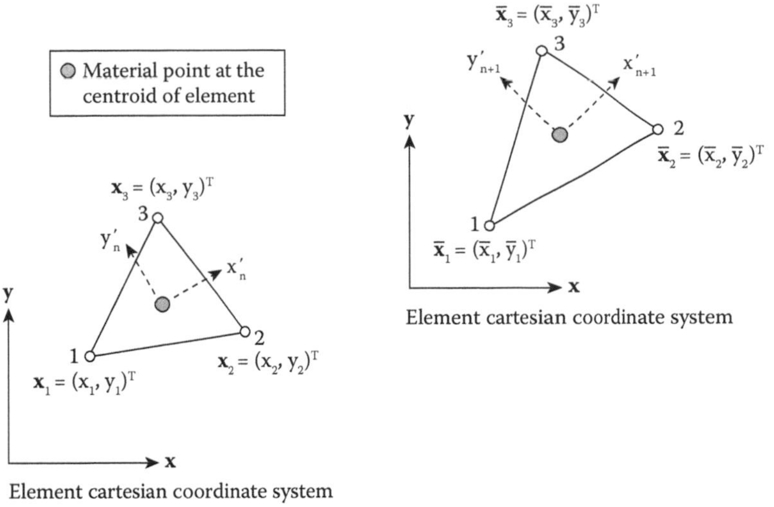

Figure 8.20 Updating of element stresses for nodal displacement increment.

from the updated nodal coordinates $[\bar{x}_i, \bar{y}_i]$ at (n + 1). The transpose of $\Delta \mathbf{F}$ is computed from

$$\Delta \mathbf{F}^T = \begin{bmatrix} \dfrac{\partial \bar{x}}{\partial x} & \dfrac{\partial \bar{y}}{\partial x} \\ \dfrac{\partial \bar{x}}{\partial y} & \dfrac{\partial \bar{y}}{\partial y} \end{bmatrix} = \begin{bmatrix} \dfrac{\partial L_1}{\partial x} & \dfrac{\partial L_2}{\partial x} & \dfrac{\partial L_3}{\partial x} \\ \dfrac{\partial L_1}{\partial y} & \dfrac{\partial L_2}{\partial y} & \dfrac{\partial L_3}{\partial y} \end{bmatrix} \begin{bmatrix} \bar{x}_1 & \bar{y}_1 \\ \bar{x}_2 & \bar{y}_2 \\ \bar{x}_3 & \bar{y}_3 \end{bmatrix} \tag{8.194}$$

For the 2D simplex element, because the shape function derivatives are constants, the Jacobian is constant throughout the element. For higher-order isoparametrics (e.g., the eight-node serendipity element), the shape function derivatives, and therefore the Jacobian, vary with location. Therefore, the Jacobian and all other variables have to be evaluated at each material point. For these higher-order isoparametrics, one additional step to determine the shape function derivatives with respect to the element Cartesian coordinates is required, as shown previously in Equations 8.18 and 8.19. For the 2D simplex element, the (constant) shape function derivatives are given by

$$\frac{\partial L_i}{\partial x} = \frac{1}{2A} y_{jk}, \quad \frac{\partial L_i}{\partial y} = \frac{1}{2A} x_{kj} \tag{8.195}$$

The notation $x_{ij} \equiv x_i - x_j$, $y_{ij} \equiv y_i - y_j$ for the nodal coordinates is again used with i = 1, 2, 3 and i, j, k cyclic. With the above relations for the shape function derivatives, we find

$$\Delta \mathbf{F}^T = \mathbf{T} \begin{bmatrix} \bar{x}_1 & \bar{y}_1 \\ \bar{x}_2 & \bar{y}_2 \\ \bar{x}_3 & \bar{y}_3 \end{bmatrix} \tag{8.196}$$

In the above equation, T is the gradient operator with respect to configuration (n), namely,

$$\mathbf{T} = \frac{1}{2A_n}\begin{bmatrix} y_{23} & y_{31} & y_{12} \\ x_{32} & x_{13} & x_{21} \end{bmatrix} \tag{8.197}$$

T is evaluated at the element centroid ($L_1 = L_2 = L_3 = 1/3$) for the 2D simplex element. In higher-order elements, this series of computations is carried out at each material sampling point in turn.

It is easily verified that the determinant of the Jacobian of the incremental deformation is

$$|\Delta\mathbf{F}| = \frac{A_{n+1}}{A_n} \tag{8.198}$$

where A_n and A_{n+1} are element areas in C_n and C_{n+1}, respectively. The displacement increments within the element are interpolated as usual using the nodal displacement increments as

$$\begin{Bmatrix} \Delta u \\ \Delta v \end{Bmatrix} = \begin{bmatrix} L_1 & 0 & L_2 & 0 & L_3 & 0 \\ 0 & L_1 & 0 & L_2 & 0 & L_3 \end{bmatrix} \begin{Bmatrix} \Delta u_1 \\ \Delta v_1 \\ \Delta u_2 \\ \Delta v_2 \\ \Delta u_3 \\ \Delta v_3 \end{Bmatrix} \tag{8.199}$$

The gradient of the incremental displacement is

$$\Delta\mathbf{G} = \begin{bmatrix} \dfrac{\partial(\Delta u)}{\partial x} & \dfrac{\partial(\Delta u)}{\partial y} \\ \dfrac{\partial(\Delta v)}{\partial x} & \dfrac{\partial(\Delta v)}{\partial y} \end{bmatrix} \tag{8.200}$$

Its transpose is computed from

$$\Delta\mathbf{G}^T = \mathbf{T}\begin{bmatrix} \Delta u_1 & \Delta v_1 \\ \Delta u_2 & \Delta v_2 \\ \Delta u_3 & \Delta v_3 \end{bmatrix} \tag{8.201}$$

The velocity gradient in configuration (n + 1) is

$$\mathbf{V}_{n+1} = \begin{bmatrix} \dfrac{\partial(\Delta u)}{\partial \overline{x}} & \dfrac{\partial(\Delta u)}{\partial \overline{y}} \\ \dfrac{\partial(\Delta v)}{\partial \overline{x}} & \dfrac{\partial(\Delta v)}{\partial \overline{y}} \end{bmatrix} \tag{8.202}$$

Therefore,

$$\mathbf{V}_{n+1} = \Delta\mathbf{G}\Delta\mathbf{F}^{-1} \tag{8.203}$$

We note that for small increments, the deformation gradient $\Delta\mathbf{F} \approx \mathbf{I}$ and $\mathbf{V}_{n+1} \approx \Delta\mathbf{G}$. The deformation rate in configuration (n + 1) is

$$\mathbf{D}_{n+1} = \frac{1}{2}(\mathbf{V}_{n+1} + \mathbf{V}_{n+1}^T) \tag{8.204}$$

All these computations are of course carried out in the (unchanging) element Cartesian coordinate reference system. The material point coordinate system update (i.e., incremental rotation) is carried out as described in the previous section, using the Hughes–Winget update after the velocity gradient is transformed into the rotated material coordinate system via

$$\mathbf{V}'_{n+1} = \mathbf{V}_{n+1}\mathbf{R}_n^T \tag{8.205}$$

and used to compute the spin $\mathbf{W}'_{n+1}$ in the rotated coordinate system (Equation 8.190). The updated rotation $\mathbf{R}_{n+1}^T$ is then used to calculate the unrotated deformation rate via

$$\mathbf{d}_{n+1} = \mathbf{R}_{n+1}^T\mathbf{D}_{n+1}\mathbf{R}_{n+1} \tag{8.206}$$

for use in the constitutive model. As in the TL formulation, the constitutive relations are expressed in the material coordinate system. The relation between the unrotated Cauchy stress rate $\dot{\bar{\boldsymbol{\sigma}}}'$ and the unrotated deformation rate $\mathbf{d}$ is expressed as

$$\dot{\bar{\boldsymbol{\sigma}}}' = \mathbf{D} : \mathbf{d} \tag{8.207}$$

which is "integrated" over the time step to provide the discretized relations

$$\Delta\bar{\boldsymbol{\sigma}}'_{n+1} \equiv \bar{\boldsymbol{\sigma}}'_{n+1} - \bar{\boldsymbol{\sigma}}'_n = \mathbf{D}^{ep} : \mathbf{d}_{n+1} \tag{8.208}$$

in a fashion similar to that of TL. We have again suppressed a Δt factor here; that is, $\mathbf{d}_{n+1}\Delta t$ is used as an approximation to an increment of logarithmic strain. The consistent tangent may be used, if appropriate, as the material tangent tensor $\mathbf{D}^{ep}$ for the increment.

Chapter 9

Kinematics of large rotations

9.1 INTRODUCTION

Subsequently (in Chapter 10), we will discuss nonlinear finite element formulations for structural elements, that is, beams, plates, and shells, which necessarily have rotations, as well as displacements as nodal degrees of freedom. It is well known that large rotations in three dimensions are not vector quantities, and they cannot therefore be combined vectorially as can nodal displacement degrees of freedom. In the following sections, we present a summary of representations of large rotations in three dimensions, with several applications in mind: (1) updating (combining) large *nodal* rotational degrees of freedom, (2) initial orientation of element coordinate systems in the global coordinate system via rotation matrices, and (3) separating large *element* rigid body motions from small deformations via a *corotational (CR) finite element formulation.*

9.2 FINITE ROTATIONS IN THREE DIMENSIONS

In the previous chapters, we have already encountered the treatment of rigid body rotation at the continuum (material point) level via polar decomposition ($\mathbf{F} = \mathbf{RU}$) of the deformation gradient. In that context, the rotation matrix $\mathbf{R}$ produced a rigid body rotation ($d\overline{\mathbf{x}} = \mathbf{R}d\overline{\mathbf{x}'}$) of the orthogonal material coordinate system (i.e., principal stretch axes).

9.2.1 Rotation matrix R

Rigid body rotation of a vector, say $\mathbf{r}$, to a new position $\mathbf{r}'$ can be described in general by a rotation matrix $\mathbf{R}$,

$$\mathbf{r}' = \mathbf{R}\mathbf{r} \tag{9.1}$$

where vectors $\mathbf{r}$ and $\mathbf{r}'$ are both referred to the same fixed Cartesian coordinate frame. The (3×3) or (2×2) rotation matrix $\mathbf{R}$ is proper orthogonal, that is,

$$\begin{cases} \mathbf{R}\mathbf{R}^{\mathrm{T}} = \mathbf{I} \\ \det(\mathbf{R}) = 1 \end{cases} \tag{9.2}$$

Two sequential rigid body rotations can be expressed as

$$\begin{cases} \mathbf{r}' = \mathbf{R}_1\ \mathbf{r} \\ \mathbf{r}'' = \mathbf{R}_2\ \mathbf{r}' \end{cases} \tag{9.3}$$

or

$$\mathbf{r}'' = \mathbf{R}_2\mathbf{R}_1\mathbf{r} \tag{9.4}$$

Since matrix multiplication is not commutative, we see that the order in which the rotations physically take place is important; that is, (finite) rotations in three dimensions are not vectors.

To investigate the properties of **R**, consider the identity

$$(\mathbf{I} - \mathbf{R})\mathbf{R}^T = \mathbf{R}^T - \mathbf{I} \tag{9.5}$$

Therefore,

$$\det(\mathbf{I} - \mathbf{R})\det(\mathbf{R}^T) = \det(\mathbf{R}^T - \mathbf{I}) \tag{9.6}$$

Since $\det(\mathbf{R}^T) = \det(\mathbf{R}) = 1$ and $\det(\mathbf{R}^T - \mathbf{I}) = \det(\mathbf{R} - \mathbf{I})$, we have

$$\begin{aligned} \det(\mathbf{I} - \mathbf{R}) &= \det(\mathbf{R} - \mathbf{I}) \\ &= (-1)^n \det(\mathbf{I} - \mathbf{R}) \end{aligned} \tag{9.7}$$

where n is the dimension (= 2 or 3). For two dimensions (n = 2), the equation above is an identity. For n = 3, we have

$$\det(\mathbf{I} - \mathbf{R}) = -\det(\mathbf{I} - \mathbf{R}) \tag{9.8}$$

that is, $\det(\mathbf{I} - \mathbf{R}) = 0$. Therefore, **R** has one eigenvalue (say λ_3) = 1. Since $\det(\mathbf{R}) = \lambda_1\lambda_2\lambda_3$, the remaining two eigenvalues, which must satisfy $\lambda_1\lambda_2 = 1$, are either both real, in which case **R** does not represent a rotation at all, or λ_1 and λ_2 are complex conjugates that can be represented as

$$\begin{cases} \lambda_1 = \cos\theta + i\ \sin\theta \\ \lambda_2 = \cos\theta - i\ \sin\theta \end{cases} \tag{9.9}$$

In two dimensions, straightforward expansion of $\det(\mathbf{R} - \lambda\mathbf{I}) = 0$ leads to a quadratic equation whose solutions for the eigenvalues λ_1 and λ_2 are given above.

9.2.1.1 *Euler's theorem*

Goldstein (1980) phrases Euler's theorem on rigid body rotation in the following way: "The general displacement of a rigid body with one point fixed is a rotation about some axis." That is, any three-dimensional (3D) rigid body rotation can be effected by a plane rotation about some axis.

We note in passing that Chasles (1830) later proved that a general rigid body motion is equivalent to a rigid body translation plus a (plane) rotation about some axis, and further that it is possible to choose a body axis coordinate system so that the translation is in the direction of the rotation axis.

The early results on the representation of rigid body rotation were proved using spherical trigonometry, before matrices or vectors had been conceived. Nevertheless, Euler's theorem is essentially proved by the fact that the rotation matrix **R** has one real eigenvalue $\lambda_3 = 1$ in three dimensions, that is,

$$(\mathbf{R} - \mathbf{I})\mathbf{n} = 0 \tag{9.10}$$

where **n** is the eigenvector corresponding to the eigenvalue $\lambda_3 = 1$.

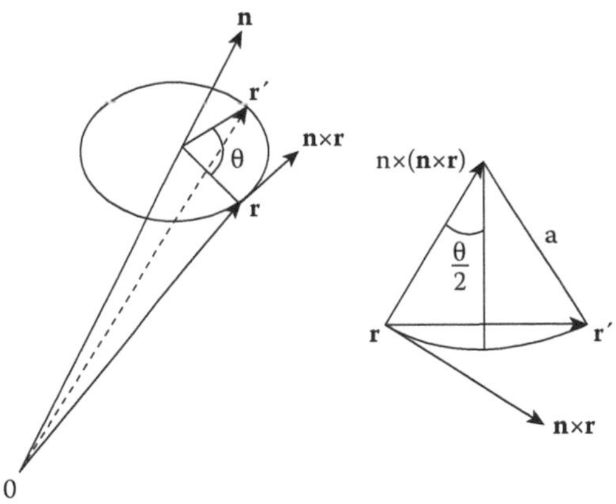

Figure 9.1 Rigid body rotation about a fixed point.

9.2.1.2 Rodrigues's formula

Figure 9.1 illustrates the derivation of an explicit expression for **R** in terms of the axial unit vector (sometimes called the "pseudovector") **n** and the rotation angle θ that may be found in various sources (e.g., Goldstein 1980; Argyris 1982).

Given **R**, the corresponding axial unit vector **n** is easily determined. In the figure, the vector r with origin at the fixed point O is embedded in the rigid body and is rotated through angle θ to a new position r′. The length of r is, by definition, unchanged by the rigid body rotation.

The projection of **r** on the axis of rotation is (**n**·**r**) **n**, and therefore **r**−(**n**·**r**) **n** lies in the plane of rotation. A vector tangent to the circular orbit described by the tip of r as the rotation angle θ increases is given by **n** × **r**. A third vector orthogonal to both **n** and **n** × **r**, together forming a right-handed system, is **n** × (**n** × **r**).

While **n** is (chosen to be) a unit vector, **n** × **r** and **n** × (**n** × **r**) both have length a, the length of r projected on the plane of rotation, that is, the radius of the circular orbit. A unit vector in the direction of **r**′ −**r** is

$$\frac{1}{a}\left[\cos\frac{\theta}{2}(\mathbf{n}\times\mathbf{r})+\sin\frac{\theta}{2}[\mathbf{n}\times(\mathbf{n}\times\mathbf{r})]\right]$$

The length of **r**′ −**r** is

$$|\mathbf{r}'-\mathbf{r}|=2a\sin\frac{\theta}{2}$$

Yielding

$$\mathbf{r}'-\mathbf{r}=2\sin\frac{\theta}{2}\left[\cos\frac{\theta}{2}(\mathbf{n}\times\mathbf{r})+\sin\frac{\theta}{2}[\mathbf{n}\times(\mathbf{n}\times\mathbf{r})]\right] \tag{9.11}$$

which may also be written as

$$\mathbf{r}'-\mathbf{r}=\sin\theta\,(\mathbf{n}\times\mathbf{r})+(1-\cos\theta)[\mathbf{n}\times(\mathbf{n}\times\mathbf{r})] \tag{9.12}$$

In matrix form, $\mathbf{r}' = \mathbf{Rr}$,

$$\mathbf{R} = \mathbf{I} + \sin\theta\ \mathbf{N} + (1-\cos\theta)\mathbf{N}^2 \tag{9.13}$$

where $\mathbf{n} \times \mathbf{r} \Rightarrow \mathbf{Nr}$ and $\mathbf{n} \times (\mathbf{n} \times \mathbf{r}) \Rightarrow \mathbf{N}^2\mathbf{r}$.

The skew-symmetric spin matrix $\mathbf{N}$ is defined by

$$\mathbf{N} \equiv \begin{bmatrix} 0 & -n_3 & n_2 \\ n_3 & 0 & -n_1 \\ -n_2 & n_1 & 0 \end{bmatrix} \tag{9.14}$$

Since the axial vector $\mathbf{n}$ is of unit length, $\mathbf{N}^2$ can be written as

$$\mathbf{N}^2 = \mathbf{nn}^{\mathrm{T}} - \mathbf{I}, (\text{where } \mathbf{n}^{\mathrm{T}}\mathbf{n} = 1) \tag{9.15}$$

and Equation 9.13 can then be put in the form

$$\mathbf{R} = \mathbf{I}\ \cos\theta + (1-\cos\theta)\mathbf{nn}^{\mathrm{T}} + \sin\theta\ \mathbf{N} \tag{9.16}$$

This latter equation is often called Rodrigues's (Rodrigues 1840) formula.

For a plane rotation about the z axis,

$$\mathbf{N} \equiv \begin{bmatrix} 0 & -1 & 0 \\ 1 & 0 & 0 \\ 0 & 0 & 0 \end{bmatrix} \text{ and } \mathbf{nn}^{\mathrm{T}} = \begin{bmatrix} 0 & 0 & 0 \\ 0 & 0 & 0 \\ 0 & 0 & 1 \end{bmatrix}$$

yielding the familiar

$$\mathbf{R} = \begin{bmatrix} \cos\theta & -\sin\theta & 0 \\ \sin\theta & \cos\theta & 0 \\ 0 & 0 & 1 \end{bmatrix} \tag{9.17}$$

9.2.1.3 Extraction of the axial vector from R

We observe from this last equation the relation

$$\mathrm{tr}(\mathbf{R}) = 1 + 2\cos\theta \tag{9.18}$$

which is true in general. This equation alone is not sufficient for determining the angle θ from the rotation matrix, even for small angles, since $\cos\theta = \cos(-\theta)$, and the equation does not therefore determine the sign of θ. However, we observe that the Rodrigues formula nicely separates the rotation matrix into symmetric and antisymmetric components, yielding an expression for $\sin\theta$:

$$\sin\theta\ \mathbf{N} = \frac{1}{2}(\mathbf{R} - \mathbf{R}^{\mathrm{T}}) \tag{9.19}$$

From Equation 9.19, we find an expression for the axial vector $\mathbf{n}$ in terms of the components of $\mathbf{R}$ as

$$\sin\theta \begin{Bmatrix} n_1 \\ n_2 \\ n_3 \end{Bmatrix} = \frac{1}{2} \begin{Bmatrix} R_{32} - R_{23} \\ R_{13} - R_{31} \\ R_{21} - R_{12} \end{Bmatrix} \tag{9.20}$$

We note that this expression yields $\sin\theta = \frac{1}{2}[R(2,1) - R(1,2)]$ for a plane rotation about the z axis. Introducing $1 + \text{tr}(\mathbf{R}) = 2(1 + \cos\theta)$ from Equation 9.18 into Equation 9.19, we find

$$\tan\frac{\theta}{2}\mathbf{N} = \frac{1}{1 + \text{tr}(\mathbf{R})}(\mathbf{R} - \mathbf{R}^T) \tag{9.21}$$

which provides a direct means of extracting an axial vector from a given rotation matrix. Additionally, if a scaled axial pseudovector **e** defined by

$$\mathbf{e} = \tan\frac{\theta}{2}\ \mathbf{n} \tag{9.22}$$

is introduced (Cayley 1843), Equation 9.13 can be written as

$$\mathbf{R} = \mathbf{I} + 2\cos^2\frac{\theta}{2}(\mathbf{N}_e + \mathbf{N}_e{}^2) \tag{9.23}$$

Where

$$\mathbf{N}_e \equiv \begin{bmatrix} 0 & -e_3 & e_2 \\ e_3 & 0 & -e_1 \\ -e_2 & e_1 & 0 \end{bmatrix} \tag{9.24}$$

Several different ways of scaling the axial vector have been proposed, but the scaling (Equation 9.22) introduced by Cayley appears to be the most useful. Alternate scalings are usually expressed in terms of the half-angle $\theta/2$, as above. The half-angle arises naturally in a spherical trigonometry derivation and was originally presented by Rodrigues in that form.

We note that this particular scaling of the pseudovector becomes singular at $\theta = \pi$. A quaternion representation of rotations (Hamilton 1844), involving four parameters rather than three, is free of singularities.

9.2.1.4 Eigenstructure of the spin matrix

The eigenvalues of the skew-symmetric spin matrix $\mathbf{N}_e$ satisfy

$$(\mathbf{N}_e - \lambda_i\mathbf{I})\mathbf{v}_i = 0 \tag{9.25}$$

which leads to the characteristic equation

$$\lambda_i^3 + e^2\lambda_i = 0 \tag{9.26}$$

in which $e^2 \equiv \mathbf{e}^T\mathbf{e} = e_1^2 + e_2^2 + e_3^2$. The three eigenvalues are the complex conjugate pair $\lambda_{1,2} = \pm ie$ and $\lambda_3 = 0$. The product $\mathbf{N}_e{}^2 = \mathbf{e}\mathbf{e}^T - e^2\mathbf{I}$, as may be directly verified.

Further, from the Cayley–Hamilton theorem we find $\mathbf{N}_e^3 = -e^2\mathbf{N}_e$, which enables us to express all higher powers of the spin matrix in terms of $\mathbf{N}_e$ and $\mathbf{N}_e^2$.

9.2.1.5 Matrix exponential

The matrix exponential function, denoted exp **A**, is defined for all square matrices **A** by the exponential power series

$$\exp\mathbf{A} \equiv \mathbf{I} + \mathbf{A} + \frac{1}{2!}\mathbf{A}^2 + \frac{1}{3!}\mathbf{A}^3 + \frac{1}{4!}\mathbf{A}^4 + \ldots \tag{9.27}$$

which always converges. Furthermore, exp **A** is always invertible.

9.2.1.6 *Exponential map*

The rotation matrix **R** can also be expressed as a matrix exponential known as the exponential map. As we know, for any orthogonal matrix Q, the product $\dot{\mathbf{Q}}\mathbf{Q}^T$ is skew symmetric. In the particular case of the rotation matrix, $\dot{\mathbf{R}}\mathbf{R}^T = \theta\mathbf{N}$, where **N** is the skew-symmetric matrix representation of the unit axial pseudovector **n**. The (pseudotime) rate of **R** is then $\dot{\mathbf{R}} = \theta\mathbf{N}\mathbf{R}$. This represents a (infinitesimal) tangential change at **R**(0), that is, the matrix representation of the vector θ (**n** × **r**).

For **R**(0) = **I**, the solution of this matrix differential equation is

$$\mathbf{R} = \exp(\theta\mathbf{N}) \tag{9.28}$$

Repeated use of the recurrence relation $\mathbf{N}^3 = -\mathbf{N}$ for the spin matrix **N** in the power series yields

$$\begin{aligned} \exp\theta\mathbf{N} &\equiv \mathbf{I} + \theta\mathbf{N} + \frac{\theta^2}{2!}\mathbf{N}^2 + \frac{\theta^3}{3!}\mathbf{N}^3 + \frac{\theta^4}{4!}\mathbf{N}^4 + \frac{\theta^5}{5!}\mathbf{N}^5 + \frac{\theta^6}{6!}\mathbf{N}^6 + \dots \\ &= \mathbf{I} + \theta\mathbf{N} + \frac{\theta^2}{2!}\mathbf{N}^2 - \frac{\theta^3}{3!}\mathbf{N} - \frac{\theta^4}{4!}\mathbf{N}^2 + \frac{\theta^5}{5!}\mathbf{N} + \frac{\theta^6}{6!}\mathbf{N}^2 + \dots \\ &= \mathbf{I} + \left(\theta - \frac{\theta^3}{3!} + \frac{\theta^5}{5!} - \dots\right)\mathbf{N} + \left(\frac{\theta^2}{2!} - \frac{\theta^4}{4!} + \frac{\theta^6}{6!} - \dots\right)\mathbf{N}^2 \\ &= \mathbf{I} + \sin\theta\mathbf{N} + (1 - \cos\theta)\mathbf{N}^2 \end{aligned} \tag{9.29}$$

Therefore, Equation 9.13 is essentially the exponential map of the rotation matrix **R** and is exact for finite rotations. One interesting identity involving the exponential map is det [exp A] = exp[tr(A)], which, applied to the orthogonal rotation matrix, yields det(**R**) = 1 since tr(**N**) = 0.

9.2.2 Cayley transform

Cayley's theorem (Cayley 1846, 1848a), as it is also sometimes called, is a mapping between orthogonal matrices Q and skew-symmetric matrices **W**, as follows:

$$\begin{cases} \mathbf{Q} = (\mathbf{I} + \mathbf{W})(\mathbf{I} - \mathbf{W})^{-1} \\ \mathbf{W} = (\mathbf{Q} + \mathbf{I})^{-1}(\mathbf{Q} - \mathbf{I}) \end{cases} \tag{9.30}$$

Given any skew-symmetric matrix **W**, the first equation above produces an orthogonal matrix Q, and similarly, given any orthogonal matrix Q, the second equation produces a skew-symmetric matrix **W**. We are of course reminded of the Hughes–Winget rotation update formula from Chapter 8.

Our specific objective here is to develop a mapping between the rotation matrix **R** defining a large rotation and the spin matrix $\mathbf{N}_e$ that will be of use in updating rotations. We note that in order for Q + **I** to be invertible, Q must not have an eigenvalue of –1. This condition is satisfied for the proper orthogonal rotation matrix Q = **R**. The identities above are easily verified using the definitions $\mathbf{Q}\mathbf{Q}^T = \mathbf{I}$ for an orthogonal matrix and $\mathbf{W} = -\mathbf{W}^T$ for a skew-symmetric matrix.

We note also that the matrix products in Equation 9.30 commute. For example, taking the second of these equations and using the definitions $\mathbf{QQ}^T = \mathbf{I}$ and $\mathbf{W} = -\mathbf{W}^T$, we obtain

$$\begin{aligned} &\mathbf{W} = (\mathbf{Q}+\mathbf{I})^{-1}(\mathbf{Q}-\mathbf{I}) \\ &(\mathbf{Q}+\mathbf{I})\mathbf{W} = \mathbf{Q}-\mathbf{I} \\ &\mathbf{W}^T(\mathbf{Q}^T+\mathbf{I}) = \mathbf{Q}^T-\mathbf{I} \\ &-\mathbf{W}(\mathbf{Q}^T+\mathbf{I}) = \mathbf{Q}^T-\mathbf{I} \\ &-\mathbf{W}(\mathbf{I}+\mathbf{Q}) = \mathbf{I}-\mathbf{Q} \\ &\mathbf{W} = (\mathbf{Q}-\mathbf{I})(\mathbf{Q}+\mathbf{I})^{-1} \end{aligned}$$

In order to link the two equations (9.30), we choose $\mathbf{N}_e$ (Equation 9.24) as the skew-symmetric matrix $\mathbf{W}$ in the Cayley transform and evaluate the orthogonal matrix $\mathbf{Q}$ (i.e., $\mathbf{R}$) therefrom as follows. First,

$$\mathbf{I}-\mathbf{N}_e \equiv \begin{bmatrix} 1 & e_3 & -e_2 \\ -e_3 & 1 & e_1 \\ e_2 & -e_1 & 1 \end{bmatrix} \tag{9.31}$$

and we then find

$$(\mathbf{I}-\mathbf{N}_e)^{-1} \equiv \frac{1}{1+e^2}\begin{bmatrix} 1+e_1^2 & e_1e_2-e_3 & e_1e_3+e_2 \\ e_1e_2+e_3 & 1+e_2^2 & e_2e_3-e_1 \\ e_1e_3-e_2 & e_2e_3+e_1 & 1+e_3^3 \end{bmatrix} \tag{9.32}$$

We recognize three distinct components in this equation:

$$(\mathbf{I}-\mathbf{N}_e)^{-1} = \frac{1}{1+e^2}(\mathbf{I}+\mathbf{ee}^T+\mathbf{N}_e) = \mathbf{I}+\frac{1}{1+e^2}(\mathbf{N}_e+\mathbf{N}_e{}^2) \tag{9.33}$$

in which we have used the identity $\mathbf{ee}^T \equiv e^2\mathbf{I}+\mathbf{N}_e{}^2$.

The first of Equations 9.30,

$$\mathbf{R} = (\mathbf{I}+\mathbf{N}_e)(\mathbf{I}-\mathbf{N}_e)^{-1} \tag{9.34}$$

then yields the general result

$$\mathbf{R} = \mathbf{I}+\frac{2}{1+e^2}(\mathbf{N}_e+\mathbf{N}_e^2) \tag{9.35}$$

Only with $e^2 = \tan^2\frac{\theta}{2}$ in Equation 9.35 do we arrive at the representation for $\mathbf{R}$ given in Equation 9.23. Thus, the rationale for the particular scaling of the pseudovector in Equation 9.22 is apparent.

Equation 9.35 may be expressed explicitly as follows:

$$\mathbf{R} = \frac{1}{1+e^2}\begin{bmatrix} 1-e^2+2e_1^2 & 2(e_1e_2-e_3) & 2(e_1e_3+e_2) \\ 2(e_1e_2+e_3) & 1-e^2+2e_2^2 & 2(e_2e_3-e_1) \\ 2(e_1e_3-e_2) & 2(e_2e_3+e_1) & 1-e^2+2e_3^2 \end{bmatrix} \tag{9.36}$$

As a check, consider again the special case of a plane rotation about the z axis, for which $e_1 = e_2 = 0$, $e_3 = \tan\frac{\theta}{2}$.

$$N_e = \tan\frac{\theta}{2}\begin{bmatrix} 0 & -1 & 0 \\ 1 & 0 & 0 \\ 0 & 0 & 0 \end{bmatrix} \tag{9.37}$$

Equation 9.23 then yields the familiar result

$$\mathbf{R} = \begin{bmatrix} \cos\theta & -\sin\theta & 0 \\ \sin\theta & \cos\theta & 0 \\ 0 & 0 & 1 \end{bmatrix} \tag{9.38}$$

9.2.3 Composition of finite rotations

In the CR formulation, the common, that is, "structural," nodes are each assumed to have 6 degrees of freedom, three translational displacements that are accumulated (combined) in a nodal displacement vector and three (possibly large) rotations that are stored and processed as a nodal rotation matrix referred to the common global Cartesian coordinate system.

At each incremental load step, a set of incremental nodal displacements and rotations are determined. The updating of the nodal rotation matrix involves superimposing these small, but finite, increments of rotation on the accumulated large rotations of the node.

9.2.3.1 *Infinitesimal rotations*

For an *infinitesimal* 3D rotation, say $\mathbf{dr} = \mathbf{r}' - \mathbf{r}$, produced by a plane rotation $d\theta$ about an axis $\mathbf{n}$, we find from Equation 9.12, to first order in $d\theta$,

$$\mathbf{dr} = d\theta(\mathbf{n} \times \mathbf{r}) = \mathbf{d\boldsymbol{\theta}} \times \mathbf{r} \tag{9.39}$$

where the infinitesimal rotation *vector* $\mathbf{d\boldsymbol{\theta}} \equiv d\theta\,\mathbf{n}$, and the cross product may be expressed in matrix form by a spin tensor, as we have seen. Thus, infinitesimal rotations are combined as vectors in which their sequence is immaterial.

9.2.3.2 *Two successive finite rotations*

While small (but finite) incremental rotations are sometimes treated as vectors as an approximation, it is important that the updated (nodal) rotation matrix remains orthogonal as a series of incremental rotations is accumulated. Therefore, we consider the effect of two successive finite rotations, which we envision as the updating of an accumulated finite nodal rotation. Many sources treat the composition of two finite rotations. See Cheng and Gupta (1989) for a concise summary of the original developments in this area. Rodrigues (1840), using spherical trigonometry, is acknowledged to be the first to give a complete solution for the composition of finite rotations.

We consider now the successive orientations of a vector, say $\mathbf{r}$, as it undergoes rigid body rotations, first from $\mathbf{r}$ to $\mathbf{r}'$ and then from $\mathbf{r}'$ to $\mathbf{r}''$. We have already determined the rotation matrix $\mathbf{R}_1$ for the first rotation, in terms of the angle of rotation θ and the (unit) axial pseudovector $\mathbf{n}$, or scaled axial pseudovector $\mathbf{e} = \tan\theta/2\ \mathbf{n}$ in Equation 9.23.

These successive rotations are described by

$$\begin{cases} \mathbf{r}' = \mathbf{R}_1\ \mathbf{r} \\ \mathbf{r}'' = \mathbf{R}_2\ \mathbf{r}' \end{cases} \tag{9.40}$$

and the composite rotation by

$$\mathbf{r}'' = \mathbf{R}_2\ \mathbf{r}' = (\mathbf{R}_2\mathbf{R}_1)\mathbf{r} \equiv \mathbf{R}_3\ \mathbf{r} \tag{9.41}$$

The simplest approach (Argyris 1982) is to update the skew-symmetric spin matrix $\mathbf{N}_e$, which appears in Equation 9.23, from which the updated rotation matrix is easily constructed. Consider the first rotation from $\mathbf{r}$ to $\mathbf{r}'$, which can be rewritten as

$$\mathbf{r}' - \mathbf{r} = (\mathbf{R}_1 - \mathbf{I})\mathbf{r} \tag{9.42}$$

The displacement vector $\mathbf{r}' - \mathbf{r}$ lies in the plane of the first rotation (Figure 9.1).
The first rotation can also be rewritten as

$$\mathbf{r}' + \mathbf{r} = (\mathbf{R}_1 + \mathbf{I})\mathbf{r} \tag{9.43}$$

from which

$$\mathbf{r} = (\mathbf{R}_1 + \mathbf{I})^{-1}(\mathbf{r}' + \mathbf{r}) \tag{9.44}$$

and Equation 9.42 is then recast as

$$\mathbf{r}' - \mathbf{r} = (\mathbf{R}_1 - \mathbf{I})(\mathbf{R}_1 + \mathbf{I})^{-1}(\mathbf{r}' + \mathbf{r}) \tag{9.45}$$

We note that the matrix product (which commutes) in this equation is the spin matrix $\mathbf{W}$ ($\equiv \mathbf{N}_e$) corresponding to $\mathbf{Q}$ ($\equiv \mathbf{R}$) in the Cayley transform when the axial pseudovector $\mathbf{e}$ in $\mathbf{N}_e$ is scaled according to $\mathbf{e} = \tan \theta/2\ \mathbf{n}$.

Therefore, the two successive finite rotations can be written in a more convenient vector form as

$$\begin{cases} \mathbf{r}' - \mathbf{r} = \mathbf{W}_1(\mathbf{r}' + \mathbf{r}) \\ \mathbf{r}'' - \mathbf{r}' = \mathbf{W}_2(\mathbf{r}'' + \mathbf{r}') \end{cases} \tag{9.46}$$

The vector $\mathbf{r}''-\mathbf{r}'$ of course lies in the plane of the second rotation. Adding these two equations, we obtain an expression for the vector $\mathbf{r}'' - \mathbf{r}$, resulting from the combined rotation as follows:

$$\mathbf{r}'' - \mathbf{r} = \mathbf{W}_1\ \mathbf{r} + (\mathbf{W}_1 + \mathbf{W}_2)\mathbf{r}' + \mathbf{W}_2\ \mathbf{r}'' \tag{9.47}$$

In order to extract the spin matrix $\mathbf{W}_3$ corresponding to the compound rotation $\mathbf{R}_3 = \mathbf{R}_2\mathbf{R}_1$, we have to transform Equation 9.47 into the form

$$\mathbf{r}'' - \mathbf{r} = \mathbf{W}_3(\mathbf{r}'' + \mathbf{r}) \tag{9.48}$$

by eliminating the vector $\mathbf{r}'$. Before doing so, we first establish some necessary identities.

Identities

We denote the associated axial vectors of $\mathbf{W}_1$ and $\mathbf{W}_2$ as $\boldsymbol{\omega}_1$ and $\boldsymbol{\omega}_2$, respectively, and consider first the action of the product $\mathbf{W}_1\mathbf{W}_2$ on a generic position vector $\mathbf{v}$.

$$\mathbf{W}_1\mathbf{W}_2\mathbf{v} = \boldsymbol{\omega}_1 \times (\boldsymbol{\omega}_2 \times \mathbf{v}) = (\boldsymbol{\omega}_1 \cdot \mathbf{v})\boldsymbol{\omega}_2 - (\boldsymbol{\omega}_1 \cdot \boldsymbol{\omega}_2)\mathbf{v} = \left[\boldsymbol{\omega}_2\boldsymbol{\omega}_1^T - (\boldsymbol{\omega}_1 \cdot \boldsymbol{\omega}_2)\mathbf{I}\right]\mathbf{v} \tag{9.49}$$

We note that $\mathbf{W}_1\mathbf{W}_2$ is not skew symmetric. Further premultiplication by $\mathbf{W}_2$ results in

$$\mathbf{W}_2\mathbf{W}_1\mathbf{W}_2\mathbf{v} = \boldsymbol{\omega}_2 \times \left[(\boldsymbol{\omega}_1 \cdot \mathbf{v})\boldsymbol{\omega}_2 - (\boldsymbol{\omega}_1 \cdot \boldsymbol{\omega}_2)\mathbf{v}\right] = -(\boldsymbol{\omega}_1 \cdot \boldsymbol{\omega}_2)(\boldsymbol{\omega}_2 \times \mathbf{v}) = -(\boldsymbol{\omega}_1 \cdot \boldsymbol{\omega}_2)\mathbf{W}_2\mathbf{v} \tag{9.50}$$

Similarly, $\mathbf{W}_1\mathbf{W}_2\mathbf{W}_1\mathbf{v} = -(\boldsymbol{\omega}_1 \cdot \boldsymbol{\omega}_2)\mathbf{W}_1\mathbf{v}$.

The action of the operator $\mathbf{W}_1\mathbf{W}_2 - \mathbf{W}_2\mathbf{W}_1$ (which will appear subsequently) on a generic position vector is

$$\left[\mathbf{W}_1\mathbf{W}_2 - \mathbf{W}_2\mathbf{W}_1\right]\mathbf{v} = \left[\boldsymbol{\omega}_2\boldsymbol{\omega}_1^T - \boldsymbol{\omega}_1\boldsymbol{\omega}_2^T\right]\mathbf{v} \equiv (\boldsymbol{\omega}_1 \times \boldsymbol{\omega}_2) \times \mathbf{v} \tag{9.51}$$

We now proceed to evaluate the right-hand side of Equation 9.47 by premultiplying the first of Equations 9.46 by $\mathbf{W}_2$ and the second by $\mathbf{W}_1$ to obtain

$$\begin{aligned} \mathbf{W}_2\ \mathbf{r}' &= \mathbf{W}_2\ \mathbf{r} + \mathbf{W}_2\mathbf{W}_1(\mathbf{r}' + \mathbf{r}) \\ \mathbf{W}_1\ \mathbf{r}' &= \mathbf{W}_1\ \mathbf{r}'' - \mathbf{W}_1\mathbf{W}_2(\mathbf{r}'' + \mathbf{r}') \end{aligned} \tag{9.52}$$

Adding these two equations and inserting the result in Equation 9.47 leads, after some rearranging, to

$$\mathbf{r}'' - \mathbf{r} = \left[\mathbf{W}_1 + \mathbf{W}_2 - (\mathbf{W}_1\mathbf{W}_2 - \mathbf{W}_2\mathbf{W}_1)\right](\mathbf{r}'' + \mathbf{r}) - \mathbf{W}_1\mathbf{W}_2(\mathbf{r}' - \mathbf{r}) - \mathbf{W}_2\mathbf{W}_1(\mathbf{r}'' - \mathbf{r}') \tag{9.53}$$

In the last two terms, we substitute for $\mathbf{r}' - \mathbf{r}$ and $\mathbf{r}'' - \mathbf{r}'$ their equivalent expressions from Equations 9.46 to obtain

$$-\mathbf{W}_1\mathbf{W}_2(\mathbf{r}' - \mathbf{r}) - \mathbf{W}_2\mathbf{W}_1(\mathbf{r}'' - \mathbf{r}') = -\mathbf{W}_1\mathbf{W}_2\mathbf{W}_1(\mathbf{r}' + \mathbf{r}) - \mathbf{W}_2\mathbf{W}_1\mathbf{W}_2(\mathbf{r}'' + \mathbf{r}') \tag{9.54}$$

which results in the following expression for the last two terms in Equation 9.53:

$$-\mathbf{W}_1\mathbf{W}_2\mathbf{W}_1(\mathbf{r}' + \mathbf{r}) - \mathbf{W}_2\mathbf{W}_1\mathbf{W}_2(\mathbf{r}'' + \mathbf{r}') = (\boldsymbol{\omega}_1 \cdot \boldsymbol{\omega}_2)(\mathbf{r}'' - \mathbf{r}) \tag{9.55}$$

Therefore,

$$\mathbf{r}'' - \mathbf{r} = \frac{1}{1 - (\boldsymbol{\omega}_1 \cdot \boldsymbol{\omega}_2)}\left[\mathbf{W}_1 + \mathbf{W}_2 - (\mathbf{W}_1\mathbf{W}_2 - \mathbf{W}_2\mathbf{W}_1)\right](\mathbf{r}'' + \mathbf{r}) \tag{9.56}$$

That is, the spin matrix corresponding to the composite rotation (Argyris 1982) is

$$\mathbf{W}_3 = \frac{1}{1 - (\boldsymbol{\omega}_1 \cdot \boldsymbol{\omega}_2)}\left[\mathbf{W}_1 + \mathbf{W}_2 - (\mathbf{W}_1\mathbf{W}_2 - \mathbf{W}_2\mathbf{W}_1)\right] \tag{9.57}$$

and its axial vector is

$$\boldsymbol{\omega}_3 = \frac{1}{1 - (\boldsymbol{\omega}_1 \cdot \boldsymbol{\omega}_2)}\left[\boldsymbol{\omega}_1 + \boldsymbol{\omega}_2 - (\boldsymbol{\omega}_1 \times \boldsymbol{\omega}_2)\right] \tag{9.58}$$

9.2.3.3 *Update of finite nodal rotations*

Consider a structural node with 6 degrees of freedom in a global 3D finite element mesh. We assume that the node has translated and rotated from its original position and orientation under the prescribed loading. Each load increment produces an incremental nodal displacement and ("incremental") nodal spin, say $\Delta \mathbf{U}_{n+1}$ and $\Delta \boldsymbol{\omega}_{n+1}$. The (3 × 1) nodal displacement vector is simply updated according to $\mathbf{U}_{n+1} = \mathbf{U}_n + \Delta \mathbf{U}_{n+1}$. The (3 × 1) nodal spin $\Delta \boldsymbol{\omega}_{n+1}$ is used to generate an orthogonal incremental rotation matrix $\Delta \mathbf{R}_{n+1}$, which is then used to update the current nodal rotation matrix via $\mathbf{R}_{n+1} = \Delta \mathbf{R}_{n+1} \mathbf{R}_n$. This procedure could be carried out using the formulas just developed for compound finite rotations. Here, we briefly describe a simple alternative procedure. Given the calculated spin vector $\Delta \boldsymbol{\omega}_{n+1}$ and the previous total rotation matrix $\mathbf{R}_n$, first normalize the spin vector to express it as

$$\Delta \boldsymbol{\omega} = \Delta \omega \ \mathbf{n}, \text{where } \Delta \omega \equiv (\Delta \boldsymbol{\omega} \cdot \Delta \boldsymbol{\omega})^{1/2} \tag{9.59}$$

where we have dropped the (n + 1) load increment subscript. Then (Cayley) scale the unit normal by

$$\mathbf{e} = \tan \frac{\Delta \omega}{2} \mathbf{n} \tag{9.60}$$

Construct the scaled spin matrix $\mathbf{N}_e$ from $\mathbf{e}$ as usual (Equation 9.24). Finally, use the Cayley transform to compute the *incremental* rotation matrix from

$$\Delta \mathbf{R} = (\mathbf{I} + \mathbf{N}_e)(\mathbf{I} - \mathbf{N}_e)^{-1} \tag{9.61}$$

and update

$$\mathbf{R} \leftarrow \Delta \mathbf{R} \ \mathbf{R} \tag{9.62}$$

We emphasize that the scaling shown in Equation 9.60 is essential. In terms of computational cost, there are trade-offs to be made, as usual. The first approach entails extracting an axial vector from the current rotation matrix, which may involve sign ambiguities that have to be addressed (or storage of the current axial vector from the previous update), while the alternate procedure just described appears to involve inversion of a (3 × 3) matrix. The inverse is available in closed form, however (Equation 9.33).

9.3 ELEMENT LOCAL COORDINATE SYSTEMS

In this section, we describe two methods for orienting element local coordinate systems within the common global 3D Cartesian coordinate system. The first method, described in the context of a 3D space frame element (i.e., a line element), uses three successive plane rotations (Euler angles) to construct the orthogonal transformation matrix relating global to local coordinates. The second method, described in the context of a triangular flat or curved shell element, works directly with the global Cartesian coordinates of the element nodes in its rotated orientation.

9.3.1 Euler angles

A well-known method of parameterizing 3D rotations is by means of Euler angles, which orient the rotated rigid "body" (i.e., the rotated Cartesian body frame), by specifying a plane rotation about each embedded body axis in turn rather than about the fixed global Cartesian axes.

These rotations must be carried out in a specific predefined (but otherwise arbitrary) order. In aeronautics and other attitude orientation applications, these three plane rotations are often dubbed "yaw, pitch, roll." Here, we adopt the order "pitch, yaw, roll," as explained subsequently. Thus, the rotation matrix is of the form

$$\mathbf{R} = \mathbf{R}_3\mathbf{R}_2\mathbf{R}_1 \tag{9.63}$$

The rotation matrix constructed in this form often contains rather complex-appearing entries consisting of combinations of sines and cosines. Apart from computational cost, in some applications deterioration of accuracy, multivaluedness, and singularities (e.g., for angles near 0°, 90°, or 180°) may be problematic. As a result, other parameterizations of rotations, particularly via quaternions (Hamilton 1844; Cayley 1845, 1848b), are often used, for example, in computer graphics and gaming, as well as in aerospace applications.

These particular issues do not concern us here. We use Euler angles to determine the (element) rotation matrix that corresponds to its orientation in terms of the global Cartesian coordinates of its nodes. That is, the element initial or updated nodal coordinates are known, and the task is to determine the corresponding rotation matrix so that element quantities (e.g., tangent stiffness matrices) can be transformed to the global reference system. As it turns out, this can be done easily without evaluations of trigonometric functions.

The element with two nodes (e.g., space frame element) shown in Figure 9.2 has nodal coordinates (X_1, Y_1, Z_1) and (X_2, Y_2, Z_2) in the global (X, Y, Z) Cartesian coordinate system. The element (x, y, z) coordinate system is visualized as being initially aligned with the global system. In the element coordinate system, the x axis is the longitudinal axis with origin at node 1. The y and z axes are aligned with the principal geometric axes of the element cross section. For plate and shell elements, the z axis is normal to the element midsurface, and the (x, y) plane may be determined by an additional dummy node.

Three successive plane rotations carry the element into its initial orientation in the global element assembly. As already mentioned, the order of these rotations is arbitrary but fixed.

Figure 9.3 shows the three plane rotations. Prior to the first plane rotation, we label the orientation of the element coordinate axes as (x′, y′, z′). That is, (x′, y′, z′) = (X, Y, Z).

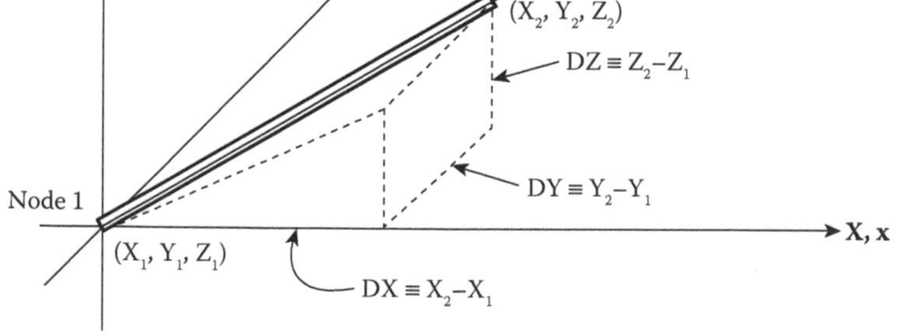

Figure 9.2 Line element oriented in global Cartesian coordinate system.

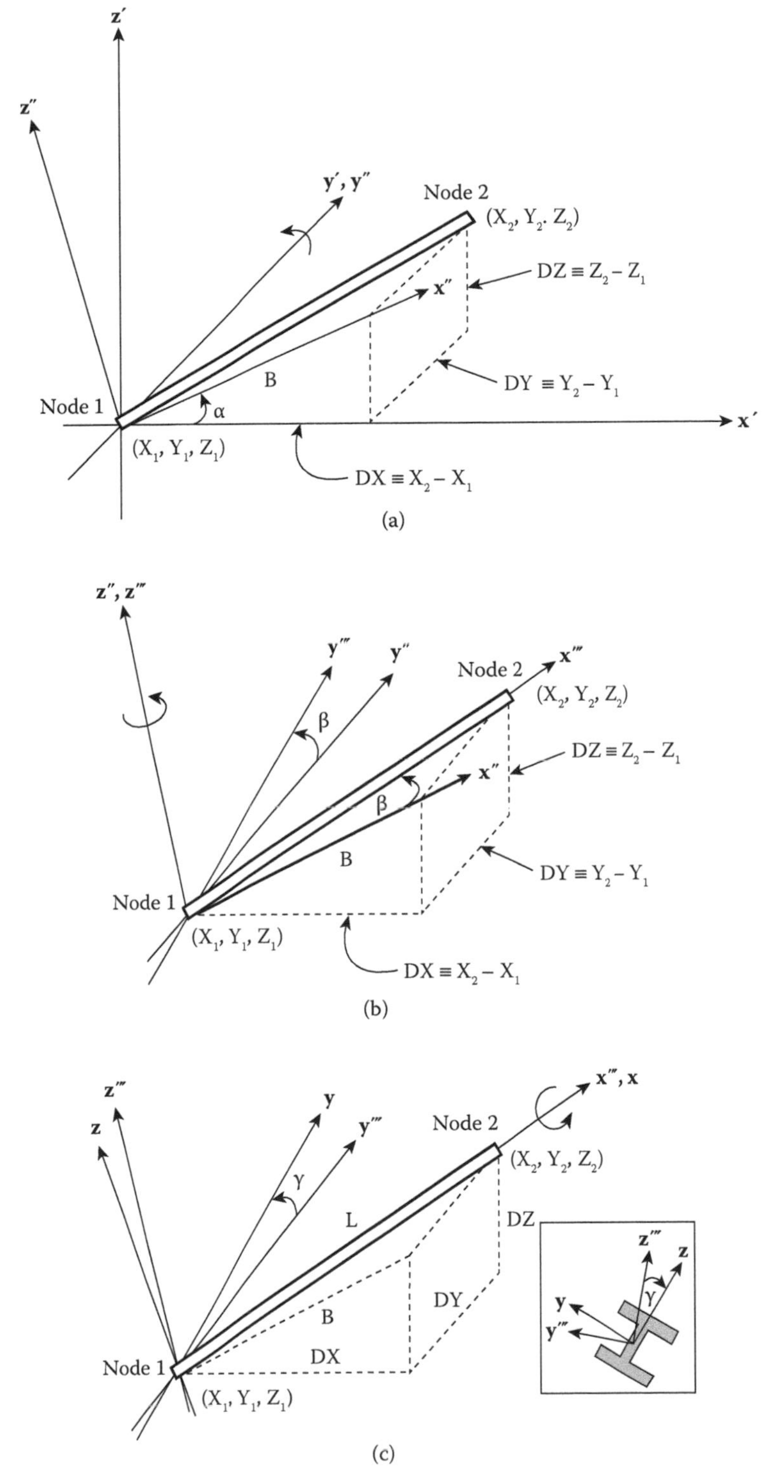

Figure 9.3 (a–c) Rotation of space frame element into the global Cartesian coordinate system. (a) Rotation around (−y′)-axis: (x′, y′, z′) to (x″, y″, z″); (b) Rotation around (z″)-axis: (x″, y″, z″) to (x‴, y‴, z‴); (c) Rotation around (x‴)-axis: (x‴, y‴, z‴) to (x, y, z).

The first rotation (pitch) is in the "vertical" (x′, z′) plane about the (−y′) axis,

$$\begin{Bmatrix} x'' \\ y'' \\ z'' \end{Bmatrix} = \begin{bmatrix} \cos\alpha & 0 & \sin\alpha \\ 0 & 1 & 0 \\ -\sin\alpha & 0 & \cos\alpha \end{bmatrix} \begin{Bmatrix} x' \\ y' \\ z' \end{Bmatrix} \tag{9.64}$$

Note that there are no trigonometric function evaluations required since

$$\begin{cases} \sin\alpha \equiv DZ/DB \\ \cos\alpha \equiv DX/DB \\ DB \equiv (DX^2 + DZ^2)^{1/2} \end{cases}$$

The second rotation (Figure 9.3) is a positive rotation (yaw) about the (z″) body axis,

$$\begin{Bmatrix} x''' \\ y''' \\ z''' \end{Bmatrix} = \begin{bmatrix} \cos\beta & \sin\beta & 0 \\ -\sin\beta & \cos\beta & 0 \\ 0 & 0 & 1 \end{bmatrix} \begin{Bmatrix} x'' \\ y'' \\ z'' \end{Bmatrix} \tag{9.65}$$

where

$$\begin{cases} \sin\beta \equiv DY/DL \\ \cos\beta \equiv DB/DL \\ DL \equiv (DY^2 + DB^2)^{1/2} \end{cases} \tag{9.66}$$

The third rotation is a rotation (roll) about the longitudinal axis, shown in the positive direction in Figure 9.3.

$$\begin{Bmatrix} x \\ y \\ z \end{Bmatrix} = \begin{bmatrix} 1 & 0 & 0 \\ 0 & \cos\gamma & \sin\gamma \\ 0 & -\sin\gamma & \cos\gamma \end{bmatrix} \begin{Bmatrix} x''' \\ y''' \\ z''' \end{Bmatrix} \tag{9.67}$$

This last rotation (γ) is not determined from the nodal coordinates for line elements. It is an independent piece of information that orients the principal geometric axes of the element cross section.

Therefore, the global-to-local transformation is the product of the three rotation matrices given above:

$$\mathbf{R} = \mathbf{R}(\gamma)\mathbf{R}(\beta)\mathbf{R}(\alpha) \tag{9.68}$$

A special case occurs if DB = 0 above (i.e., the element axis lies in the global Y direction). In that case,

$$\mathbf{R}(\alpha) = \mathbf{I}_3, \text{ and } \mathbf{R}(\beta) = \begin{bmatrix} 0 & 0 & 1 \\ 0 & 1 & 0 \\ -1 & 0 & 0 \end{bmatrix} \tag{9.69}$$

We restate Equation 9.68 as

$$\begin{Bmatrix} x \\ y \\ z \end{Bmatrix} = \mathrm{T} \begin{Bmatrix} X \\ Y \\ Z \end{Bmatrix} \tag{9.70}$$

and note that the rows of the (3 × 3) transformation matrix $\mathbf{T}$ are unit vectors directed along the element reference axes (x, y, z) in the rotated configuration.

This process can be used to initially orient an element within the global element assembly. As the analysis proceeds and nodal coordinates are updated, this procedure could also be used to update the element CR Cartesian frame. For a line element, the initial roll (angle γ) of the longitudinal axis orients the principal geometric axes of the cross section, and updating of the roll angle requires information on the nodal rotations.

9.3.2 Orienting plate and shell elements in three dimensions

We briefly describe here a general method for orienting plate and shell element coordinate systems in a global 3D mesh. The method can also be used for line (space frame) elements by introducing an extra dummy node to provide the same information as the roll angle described in the previous section. Figure 9.4 shows a three-node flat triangle oriented in a global (X, Y, Z) Cartesian coordinate system. The flat triangle shape shown in the figure is the reference plane formed by the three nodes, although the element itself need not necessarily be flat.

The orthogonal unit vectors ($\mathbf{e}_1$, $\mathbf{e}_2$, $\mathbf{e}_3$) define the element local Cartesian coordinate axes.

We assume that the origin of the element coordinate system is at node 1, which may be chosen arbitrarily, but nodes 2 and 3 then follow in a cyclic order that serves to define the direction of the normal to the plane according to the right-hand rule.

The vectors that lie along the sides of the element are

$$\begin{cases} \mathbf{X}_{21} \equiv \mathbf{X}_2 - \mathbf{X}_1 \\ \mathbf{X}_{32} \equiv \mathbf{X}_3 - \mathbf{X}_2 \\ \mathbf{X}_{13} \equiv \mathbf{X}_1 - \mathbf{X}_3 \end{cases} \tag{9.71}$$

The notation $\mathbf{X}_i \equiv (X_i, Y_i, Z_i)^T$, i = 1, 2, 3 is used for nodal position vectors, where (X_i, Y_i, Z_i) are the corresponding (scalar) global coordinates. The compact notation $\mathbf{X}_{ij} \equiv \mathbf{X}_i\text{-}\mathbf{X}_j$ is also employed here.

The unit vector $\mathbf{e}_1$ is chosen to lie along the line from the element origin at node 1 to node 2 and is therefore defined by

$$\mathbf{e}_1 = \frac{\mathbf{X}_{21}}{|\mathbf{X}_{21}|} \tag{9.72}$$

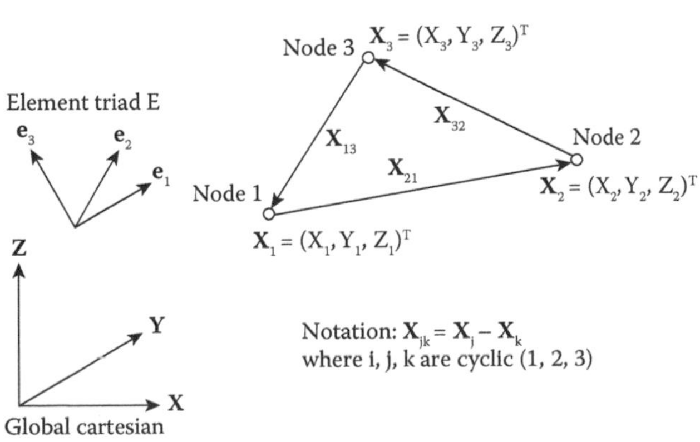

Figure 9.4 Element local coordinate systems for plate and shell elements in three dimensions.

The unit vector $\mathbf{e}_3$ normal to the plane is then defined as the cross product,

$$\mathbf{e}_3 = \frac{\mathbf{X}_{21} \times \mathbf{X}_{31}}{\left|\mathbf{X}_{21} \times \mathbf{X}_{31}\right|} \tag{9.73}$$

which may be expressed as

$$\mathbf{e}_3 = \frac{1}{2A}\left[\mathbf{X}_{21} \times \mathbf{X}_{31}\right] \tag{9.74}$$

where A is the area of the flat triangular reference plane.

The Cartesian components of $\mathbf{e}_3$ are

$$\mathbf{e}_3 = \frac{1}{A}\begin{Bmatrix} A_X \\ A_Y \\ A_Z \end{Bmatrix} \tag{9.75}$$

where (A_X, A_Y, A_Z) are the signed (i.e., positive or negative) areas of the projections of the flat triangular reference plane on the Y–Z, X–Z, and X–Y coordinate planes, respectively. In terms of global Cartesian coordinates of the nodes, these projected areas are

$$\begin{cases} 2A_X = \sum\limits_{i=1,2,3} Y_i Z_{jk} \\ 2A_Y = \sum\limits_{i=1,2,3} Z_i X_{jk} \quad (i, j, k \text{ cyclic}) \\ 2A_Z = \sum\limits_{i=1,2,3} X_i Y_{jk} \end{cases} \tag{9.76}$$

The right-handed orthogonal coordinate system is then completed by defining $\mathbf{e}_2$ as

$$\mathbf{e}_2 = \mathbf{e}_3 \times \mathbf{e}_1 \tag{9.77}$$

9.4 COROTATIONAL FINITE ELEMENT FORMULATION

The concept of a CR (Lagrangian) formulation is intuitively appealing, since its basic premise is that the small deformation (strain) of a structural element embedded within a flexible structure undergoing large displacements and rotations can be most effectively treated by first separating out the rigid body component of its overall motion (Wempner 1969; Belytschko and Hsieh 1973).

In terms of the polar decomposition of the deformation gradient for a continuum, the closest comparison is the left polar decomposition ($\mathbf{F} = \mathbf{VR}$) in which the principal material axes at a material point are first rotated ($d\mathbf{x}' = \mathbf{R}\, d\mathbf{x}$), followed by the left stretch ($d\bar{\mathbf{x}} = \mathbf{V}d\mathbf{x}'$). Of course, in a continuum the required rotation $\mathbf{R}$ varies in general from point to point. In a CR formulation, an orthogonal element reference frame ("element triad") is rotated rigidly as dictated by the displacements of the structural nodes to which the element is connected. Only small deformations and rotations then occur (in theory) relative to the moving element triad.

Rankin and Brogan (1986) introduced the important concept of an element-independent "CR" computational procedure in which the effects of large nodal rotations are handled essentially independently of, and prior to, element-specific computations. This of course has enormous significance in the context of large general-purpose finite element systems.

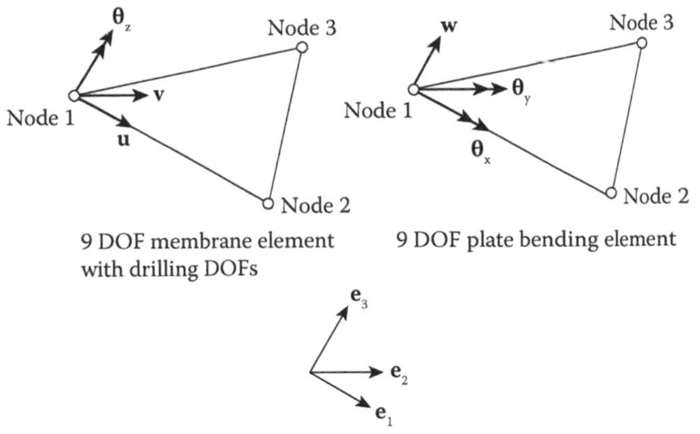

Figure 9.5 Eighteen degrees of freedom flat shell element constructed by overlaying membrane and plate bending triangles.

Although termed a CR procedure there, in hindsight it might more properly be considered an updated Lagrangian (UDL) procedure.

Rankin and Nour-Omid (1988) and Nour-Omid and Rankin (1991) later introduced true CR (moving frame) features into their formulation. They also significantly improved computational performance with the systematic use of projectors to filter out effects of possible deficiencies in element formulations (i.e., lack of invariance to small rigid body motion). Haugen (1994) and Felippa and Haugen (2005) introduced further enhancements and presented a comprehensive treatment of CR foundations.

Crisfield (1990) appears to have been the first to point out the importance of consistent development of the CR tangent stiffness in his work on space frame elements with large rotations.

We concentrate here on the kinematics of a CR formulation and assume that all finite elements used in such a formulation have 6 degrees of freedom (three displacements and three rotations) at all nodal connectors. A simple flat triangular (18 degrees of freedom) shell element constructed by overlaying a membrane triangle with corner "drilling" degrees of freedom (e.g., Allman 1984; Bergan and Felippa 1985) with a triangular (9 degrees of freedom) plate bending element (e.g., Dhatt et al. 1986) is a good exemplar to keep in mind (Figure 9.5).

9.4.1 Corotational coordinate systems

As mentioned above, a fundamental component of a CR formulation is an *element* triad with orthogonal unit vectors ($\mathbf{e}_1$, $\mathbf{e}_2$, $\mathbf{e}_3$) that rotate with the element—thus the term corotational (CR). This is distinct from a convected coordinate system whose base vectors do not remain orthonormal after deformation, and also distinct from a UDL formulation in which the current reference, although updated, remains fixed for the current incremental step.

The second fundamental component of a CR formulation is a *nodal* triad with orthogonal unit vectors ($\mathbf{r}_1$, $\mathbf{r}_2$, $\mathbf{r}_3$) that rotate with the node.

9.4.1.1 Notation

Element nodal degrees of freedom (displacements and finite rotations) are referred to global Cartesian coordinates. The 6 degrees of freedom at each node consist of a (3 × 1) displacement vector $\mathbf{U}_i$ (the subscript i is a node label) and three rotational freedoms encoded in an orthogonal rotation matrix $\mathbf{R}_i$. Incremental (or virtual) nodal freedoms consist of an incremental displacement vector $\delta\mathbf{U}_i$ and an instantaneous spin vector $\delta\boldsymbol{\omega}_i^R$ of the nodal rotation matrix. We denote the hybrid nodal freedom quantities that collect the displacement and rotational freedoms as $\mathbf{V}_i$ and $\delta\mathbf{V}_i$, respectively (Felippa and Haugen 2005). That is,

$$\mathbf{V}_i \equiv \begin{bmatrix} \mathbf{U}_i \\ \mathbf{R}_i \end{bmatrix} \text{ and } \delta\mathbf{V}_i \equiv \begin{Bmatrix} \delta\mathbf{U}_i \\ \delta\boldsymbol{\omega}_i^R \end{Bmatrix} \tag{9.78}$$

The square brackets in Equation 9.78 are used to emphasize that $\mathbf{V}_i$ is not a vector.

In the local (corotated) element coordinate system, these nodal freedoms are denoted as

$$\mathbf{v}_i \equiv \begin{Bmatrix} \mathbf{u}_i \\ \boldsymbol{\theta}_i \end{Bmatrix} \text{ and } \delta\mathbf{v}_i \equiv \begin{Bmatrix} \delta\mathbf{u}_i \\ \delta\mathbf{w}_i^R \end{Bmatrix} \tag{9.79}$$

where $\boldsymbol{\theta}_i$ is the axial vector of $\mathbf{R}_i$ transformed to the element coordinate system.

9.4.1.2 Element triads

In a CR formulation, we must transform global nodal freedoms, both displacements and rotations, to the current element reference system, in order to isolate the (small) deformational effects that are to be seen by the element. In the following discussion, we follow closely the original proposal of Rankin and Brogan (1986), and the subsequent significant developments (Rankin and Nour-Omid 1988; Nour-Omid and Rankin 1991) cited above.

The displacement vector of a generic reference point $\mathbf{x}$ in the element is denoted by $\mathbf{u}$. By hypothesis, the motion can be decomposed into

$$\mathbf{u} = \bar{\mathbf{u}} + \mathbf{u}^r \tag{9.80}$$

where $\mathbf{u}^r$ is due to an overall rigid body motion of the element and $\bar{\mathbf{u}} = \mathbf{u} - \mathbf{u}^r$ produces only small deformations in the element. The overbar on a quantity such as $\bar{\mathbf{u}}$ marks it as a variable that produces deformation in the corotated element. Generally, uppercase indicates a quantity expressed in the global Cartesian coordinate system and lowercase indicates a quantity expressed in a local element coordinate system.

We first address the transformation of position and displacement vectors from global Cartesian to the corotating element coordinate system.

Figure 9.6 is a (schematic) representation of the initial and deformed positions of an element. The (3×3) matrix $\mathbf{E}$ is the transformation from (element) local to global coordinates, that is,

$$\begin{Bmatrix} X \\ Y \\ Z \end{Bmatrix} = \mathbf{E} \begin{Bmatrix} x \\ y \\ z \end{Bmatrix} \tag{9.81}$$

We note that $\mathbf{E}^T \equiv \mathbf{T}$ (the transformation matrix developed in Section 9.3.1 via Euler angles, Equation 9.70). Therefore, the columns of $\mathbf{E}$ contain the unit vectors of the element triad $(\mathbf{e}_1, \mathbf{e}_2, \mathbf{e}_3)$ in a given configuration. As the deformation proceeds, the succession of element triads can be described by a series of orthogonal transformation matrices $\mathbf{E}_k$, $k = 0, 1, 2, \ldots$, where $k = 0$ corresponds to the initial undeformed configuration. The orthogonal element transformation matrices can be represented explicitly in terms of their respective unit base vectors as

$$\begin{cases} \mathbf{E}_0 = [\mathbf{e}_1 \quad \mathbf{e}_2 \quad \mathbf{e}_3] \\ \mathbf{E}_k = [\mathbf{e}_1' \quad \mathbf{e}_2' \quad \mathbf{e}_3'] \end{cases} \tag{9.82}$$

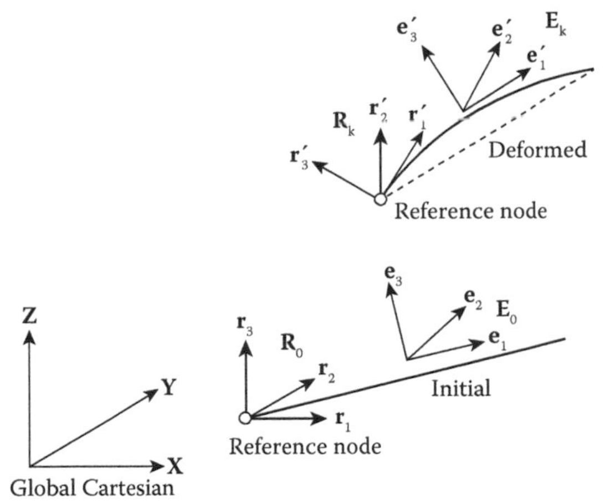

Figure 9.6 Element and nodal triads (schematic).

A position vector $\mathbf{x}$ in the undeformed reference configuration C_0 can be represented as

$$\mathbf{X} = x_1\mathbf{e}_1 + x_2\mathbf{e}_2 + x_3\mathbf{e}_3 \tag{9.83}$$

or in matrix notation as

$$\mathbf{X} = \mathbf{E}_0\mathbf{x} \tag{9.84}$$

where the vector $\mathbf{X}$ contains the global Cartesian components of $\mathbf{x}$, and the vector $\mathbf{x} = [x_1, x_2, x_3]^T$ contains its components in the local element coordinate system.

A rigid body rotation from C_o to C_k takes $\mathbf{X}$ to $\mathbf{X}'$, where

$$\mathbf{X}' = x_1\mathbf{e}_1' + x_2\mathbf{e}_2' + x_3\mathbf{e}_3' \tag{9.85}$$

The scalar components (x_1, x_2, x_3) are unchanged since the position vector and the base vectors rotate together. Equation 9.85 can also be written as

$$\mathbf{X}' = \begin{bmatrix} \mathbf{e}_1' & \mathbf{e}_2' & \mathbf{e}_3' \end{bmatrix} \begin{Bmatrix} x_1 \\ x_2 \\ x_3 \end{Bmatrix} \tag{9.86}$$

or

$$\mathbf{X}' = \mathbf{E}_k\mathbf{x} \tag{9.87}$$

Using Equation 9.84, we then have

$$\mathbf{X}' = \mathbf{E}_k\mathbf{E}_0^T\mathbf{X} \tag{9.88}$$

This equation represents a rigid rotation of the position vector $\mathbf{X}$ in the reference configuration to the updated configuration C_k, in which its global Cartesian coordinates are $\mathbf{X}'$ (i.e., the rotation $\mathbf{x} \rightarrow \mathbf{x}'$ illustrated in Figure 9.7). Therefore, the global components of the rigid body portion $\mathbf{U}^r$ of the total displacement from C_0 to C_k are

$$\mathbf{U}^r = \mathbf{X}' - \mathbf{X} = (\mathbf{E}_k\mathbf{E}_0^T - \mathbf{I})\mathbf{X}. \tag{9.89}$$

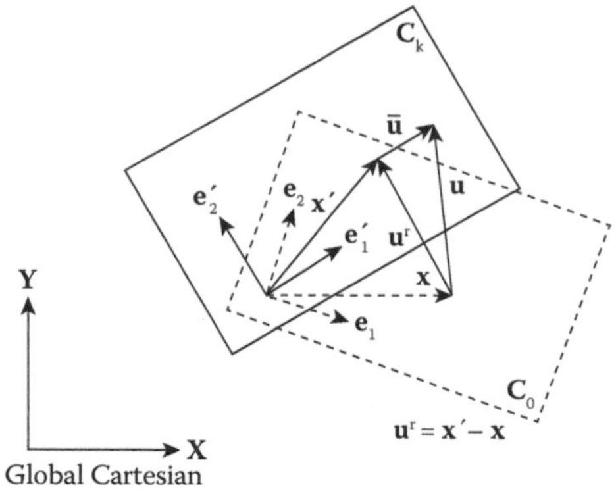

Figure 9.7 CR displacement vector transformation.

(i.e., $\mathbf{u}' = \mathbf{x}' - \mathbf{x}$ in Figure 9.7), and the deformation-producing displacement is then simply

$$\bar{\mathbf{U}} = \mathbf{U} - \mathbf{U}^r \tag{9.90}$$

In the local element coordinate system, in C_k

$$\bar{\mathbf{u}} = \mathbf{E}_k^T \bar{\mathbf{U}} \tag{9.91}$$

which can be put in the form (Rankin and Brogan 1986)

$$\bar{\mathbf{u}} = \mathbf{E}_k^T (\mathbf{U} + \mathbf{X}) - \mathbf{E}_0^T \mathbf{X} \tag{9.92}$$

We note that $\mathbf{E}_0^T \mathbf{X} = \mathbf{x}$, the local coordinates of the position vector in the reference configuration C_0, so that Equation 9.92 may be written as

$$\bar{\mathbf{u}} + \mathbf{x} = \mathbf{E}_k^T (\mathbf{U} + \mathbf{X}) \tag{9.93}$$

9.4.1.3 Nodal triads

Figure 9.6 also shows the nodal triad **R** that rotates rigidly with a global node. Similar to the element triads, a series of orthogonal transformation matrices $\mathbf{R}_k$, k = 0, 1, 2, … describes the orientation of the global node as the deformation proceeds. In the undeformed configuration C_0, the nodal triad is aligned with the global Cartesian directions, that is, $\mathbf{R}_0 = \mathbf{I}_3$. The columns of $\mathbf{R}_k$ contain the unit base vectors of the nodal triad, that is, $\mathbf{R}_k = [\mathbf{r}'_1, \mathbf{r}'_2, \mathbf{r}'_3]$.

In some cases, an additional "surface" triad **S** is required, as illustrated in Figure 9.8. The surface triad also rotates rigidly with the node. Its initial orientation, described by the orthogonal matrix $\mathbf{S}_0$ [whose columns consist of the unit vectors $(\mathbf{s}_1, \mathbf{s}_2, \mathbf{s}_3)$], can be used to define the local tangent plane to a surface. In the case of a faceted finite element model of a shell structure (i.e., flat shell elements that approximate the curved surface), $\mathbf{S}_0$ represents an average tangent plane for the contiguous elements meeting at the node (Figure 9.8). Since the surface triad is uniquely defined at each global node (and rotates with the node), it could be used in lieu of the nodal triad. However, in what follows we use the nodal triad **R**, which is initially aligned with the global Cartesian coordinate system.

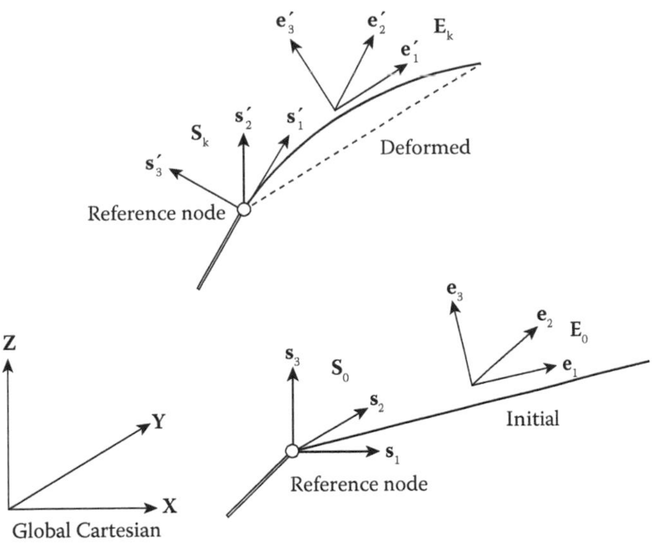

Figure 9.8 Surface triad for faceted models of curved surfaces.

We now proceed to address the CR transformation of nodal rotational freedoms to the small deformational rotational freedoms seen by the element. We assume here that the (large) nodal rotations are stored as (3 × 3) matrices **R** referred to the common global Cartesian coordinate system, although other formats, such as axial pseudovectors or quaternions, could be used. The (3 × 3) orthogonal matrix **R** transforms vectors from local to global coordinates, that is,

$$\begin{Bmatrix} X \\ Y \\ Z \end{Bmatrix} = \mathbf{R} \begin{Bmatrix} x \\ y \\ z \end{Bmatrix} \tag{9.94}$$

The nodal triad must be expressed in the element coordinate system in order to determine the rotation of the node relative to the rotation of the element frame. In the undeformed configuration, characterized by $\mathbf{R}_0$ and $\mathbf{E}_0$, respectively, there is an initial angular "mismatch" between these two triads that has to be accounted for.

We first note that the transformation of a tensor from global coordinates to an element reference system follows directly from Equation 9.81 by considering the transformation of a generic dyad (tensor) $\mathbf{XY}^T$to $\mathbf{xy}^T$ in the element reference system, that is, $\mathbf{xy}^T = \mathbf{E}^T\,\mathbf{XY}^T\,\mathbf{E}$.

In the undeformed configuration C_0, the angular mismatch between nodal and element triads referred to the global coordinate system is $\mathbf{R}_0\mathbf{E}_0^T$. Referred to the element coordinate system in C_0, the initial mismatch $\mathbf{D}_0$ is therefore $\mathbf{E}_0^T(\mathbf{R}_0\mathbf{E}_0^T)\mathbf{E}_0$ or

$$\mathbf{D}_0 = \mathbf{E}_0^T\,\mathbf{R}_0 \tag{9.95}$$

In the configuration C_k, the relative rotation referred to the updated element coordinate system is

$$\mathbf{D}_k = \mathbf{E}_k^T\,\mathbf{R}_k \tag{9.96}$$

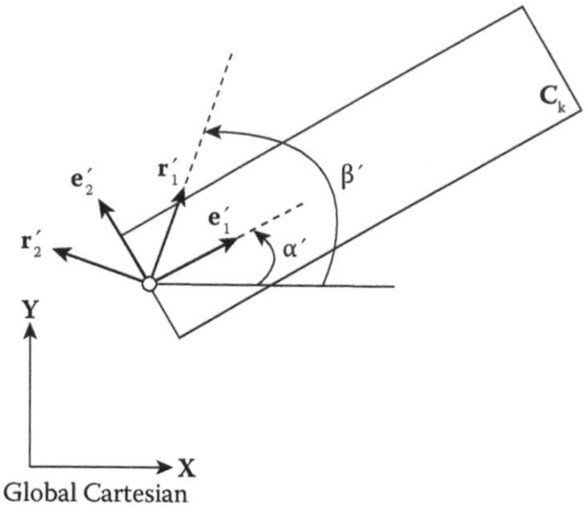

Figure 9.9 CR transformation of nodal rotations.

The explicit representation of $\mathbf{D}_k$ in terms of the basis vectors of the element and nodal triads is

$$\mathbf{D}_k = \begin{bmatrix} \mathbf{e}_1' \cdot \mathbf{r}_1' & \mathbf{e}_1' \cdot \mathbf{r}_2' & \mathbf{e}_1' \cdot \mathbf{r}_3' \\ \mathbf{e}_2' \cdot \mathbf{r}_1' & \mathbf{e}_2' \cdot \mathbf{r}_2' & \mathbf{e}_2' \cdot \mathbf{r}_3' \\ \mathbf{e}_3' \cdot \mathbf{r}_1' & \mathbf{e}_3' \cdot \mathbf{r}_2' & \mathbf{e}_3' \cdot \mathbf{r}_3' \end{bmatrix} \tag{9.97}$$

For the rotations about the z axis illustrated in Figure 9.9, Equation 9.96 gives

$$\begin{aligned} \mathbf{E}_k^T\mathbf{R}_k &= \begin{bmatrix} \cos\alpha' & \sin\alpha' & 0 \\ -\sin\alpha' & \cos\alpha' & 0 \\ 0 & 0 & 1 \end{bmatrix} \begin{bmatrix} \cos\beta' & -\sin\beta' & 0 \\ \sin\beta' & \cos\beta' & 0 \\ 0 & 0 & 1 \end{bmatrix} \\ &= \begin{bmatrix} \cos(\beta'-\alpha') & -\sin(\beta'-\alpha') & 0 \\ \sin(\beta'-\alpha') & \cos(\beta'-\alpha') & 0 \\ 0 & 0 & 1 \end{bmatrix} \end{aligned} \tag{9.98}$$

The relative element nodal rotation that contributes to element deformation is (in the element coordinate system) therefore

$$\overline{\mathbf{R}}_k = \mathbf{D}_k\mathbf{D}_0^T = \mathbf{E}_k^T\mathbf{R}_k\mathbf{R}_0^T\mathbf{E}_0 = \mathbf{E}_k^T\mathbf{R}_k\mathbf{E}_0 \tag{9.99}$$

9.4.1.4 *Nodal rotational freedoms*

In what follows, subscripts (or in some cases, superscripts) i, j will indicate element nodes. At an element node i, given the deformational rotation $\overline{\mathbf{R}}_k^i$, its axial pseudovector $\overline{\boldsymbol{\theta}}_k^i$ (which constitutes the element nodal rotational freedoms) can be found as follows.

Notation

For a vector $\mathbf{z} = [z_1, z_2, z_3]^T$ corresponding to a skew-symmetric matrix $\mathbf{Z}$, where

$$\mathbf{Z} \equiv \begin{bmatrix} 0 & -z_3 & z_2 \\ z_3 & 0 & -z_1 \\ -z_2 & z_1 & 0 \end{bmatrix} \tag{9.100}$$

the standard notations $\mathbf{z}$ = axial $\mathbf{Z}$ and $\mathbf{Z}$ = Spin $\mathbf{z}$ are used herein.

First calculate (Equation 9.21) the Cayley axial vector (which we denote here by $\mathbf{c}$ to avoid confusion with the base vectors of the element triad) from

$$\mathbf{c}_k^i = \frac{1}{1 + \mathrm{tr}(\overline{\mathbf{R}}_k^i)} \mathrm{axial}\left[\overline{\mathbf{R}}_k^i - \overline{\mathbf{R}}_k^{i\,T}\right] \tag{9.101}$$

which is related to the unit axial vector $\mathbf{n}$ of $\overline{\mathbf{R}}_k^i$ by

$$\mathbf{c}_k^i = \tan\frac{\overline{\theta}_k^i}{2}\mathbf{n} \tag{9.102}$$

The magnitude of the axial pseudovector $\overline{\boldsymbol{\theta}}_k^i$ can therefore be found from

$$\tan\frac{\overline{\theta}_k^i}{2} = (\mathbf{c}_k^i \cdot \mathbf{c}_k^i)^{1/2} \tag{9.103}$$

and the pseudovector $\overline{\boldsymbol{\theta}}_k^i$ is then given by

$$\overline{\boldsymbol{\theta}}_k^i \equiv \overline{\theta}_k^i \mathbf{n} = \overline{\theta}_k^i \cot\frac{\overline{\theta}_k^i}{2}\mathbf{c}_k^i \tag{9.104}$$

9.4.2 Incremental nodal degrees of freedom

In order to formulate the incremental equilibrium equations in the moving CR system via virtual work, we must establish expressions for incremental (virtual) nodal displacements and rotations. These relationships must account for the motion of the reference frames, as well as that of the elements.

9.4.2.1 Incremental nodal displacements

We recall from Equation 9.93 the expression for the CR transformation of a general position vector $\mathbf{X}$ and its displacement $\mathbf{U}$ (in global coordinates):

$$\overline{\mathbf{u}} + \mathbf{x} = \mathbf{E}_k^T(\mathbf{U} + \mathbf{X}) \tag{9.105}$$

where $\mathbf{x}$ is the corotated position vector of a generic point in local coordinates and $\overline{\mathbf{u}}$ is its local corotated displacement vector as seen by the element.

The corotated *incremental* nodal displacement $\delta\overline{\mathbf{u}}_i$ at node i with local coordinates $\mathbf{x}_i = [x_i, y_i, z_i]^T$ is determined by taking the variation of the expression in Equation 9.105. We then have

$$\delta\overline{\mathbf{u}}_i = \delta\mathbf{E}_k^T(\mathbf{U}_i + \mathbf{X}_i) + \mathbf{E}_k^T\delta\mathbf{U}_i \tag{9.106}$$

Since the variation (i.e., derivative) $\delta \mathbf{Q}$ of any orthogonal matrix $\mathbf{Q}$ can be expressed as $\delta \mathbf{Q}\mathbf{Q}^T = \delta \mathbf{W} = \text{Spin}\ \delta \mathbf{w}$, the variation of the element frame can be written as $\delta \mathbf{E}_k = \text{Spin}\ (\delta \boldsymbol{\omega}^E)\mathbf{E}_k$ in global coordinates, and therefore $\delta \mathbf{E}_k^T = -\mathbf{E}_k^T \text{Spin}(\delta \boldsymbol{\omega}^E)$, which we use in Equation 9.106 to obtain

$$\delta \bar{\mathbf{u}}_i = -\mathbf{E}_k^T\ \text{Spin}(\delta \boldsymbol{\omega}^E)(\mathbf{U}_i + \mathbf{X}_i) + \mathbf{E}_k^T \delta \mathbf{U}_i \tag{9.107}$$

We use a superscript on $\delta \boldsymbol{\omega}^E$ to mark it explicitly as the spin vector of the element frame and to distinguish it from the spin vector of the nodal triad.

Transformation to element local coordinates yields

$$\begin{aligned}
\delta \bar{\mathbf{u}}_i &= -\mathbf{E}_k^T\ \text{Spin}(\delta \boldsymbol{\omega}^E)(\mathbf{E}_k \mathbf{E}_k^T)(\mathbf{U}_i + \mathbf{X}_i) + \mathbf{E}_k^T \delta \mathbf{U}_i \\
&= -\mathbf{E}_k^T\ \text{Spin}(\delta \boldsymbol{\omega}^E)\mathbf{E}_k \mathbf{E}_k^T(\mathbf{U}_i + \mathbf{X}_i) + \mathbf{E}_k^T \delta \mathbf{U}_i \\
&= -\text{Spin}(\delta \mathbf{w}^E)\hat{\mathbf{x}}_i + \delta \mathbf{u}_i \\
&= \text{Spin}(\hat{\mathbf{x}}_i)\delta \mathbf{w}^E + \delta \mathbf{u}_i
\end{aligned} \tag{9.108}$$

In this last equation, $\hat{\mathbf{x}}_i \equiv \mathbf{x}_i + \mathbf{u}_i$ is the current (i.e., updated) position vector of node i in local coordinates. We have yet to establish an expression for the incremental spin vector $\delta \boldsymbol{\omega}^E$ of the element triad that appears in Equation 9.108. It of course depends on the element incremental nodal displacements $\delta \mathbf{u}_i$.

9.4.2.2 *Incremental nodal rotations*

We also recall the expression for the relative rotation at an element node (in local element coordinates), given in Equation 9.99 and repeated below.

$$\overline{\mathbf{R}}_k^i = \mathbf{E}_k^T \mathbf{R}_k^i \mathbf{E}_0 \tag{9.109}$$

Taking the variation, we find

$$\delta \overline{\mathbf{R}}_k^i = (\delta \mathbf{E}_k^T \mathbf{R}_k^i + \mathbf{E}_k^T \delta \mathbf{R}_k^i)\mathbf{E}_0 \tag{9.110}$$

We write the variation of the nodal triad as $\delta \mathbf{R}_k^i = \text{Spin}(\delta \boldsymbol{\omega}_i^R)\mathbf{R}_k^i$ in global coordinates (which is also yet to be determined). Equation 9.110 then becomes

$$\begin{aligned}
\delta \overline{\mathbf{R}}_k^i &= \mathbf{E}_k^T \left[\text{Spin}(\delta \boldsymbol{\omega}_i^R) - \text{Spin}(\delta \boldsymbol{\omega}^E)\right]\mathbf{R}_k^i \mathbf{E}_0 \\
&= \mathbf{E}_k^T \left[\text{Spin}(\delta \boldsymbol{\omega}_i^R) - \text{Spin}(\delta \boldsymbol{\omega}^E)\right]\mathbf{E}_k \overline{\mathbf{R}}_k^i
\end{aligned} \tag{9.111}$$

in which we have used Equation 9.109. Therefore,

$$\delta \overline{\mathbf{R}}_k^i \overline{\mathbf{R}}_k^{i,T} \equiv \text{Spin}(\delta \overline{\mathbf{w}}_i^R) \tag{9.112}$$

where

$$\delta \overline{\mathbf{w}}_i^R = \delta \mathbf{w}_i^R - \delta \mathbf{w}^E \tag{9.113}$$

in element coordinates. We next develop an expression for the spin vector $\delta \boldsymbol{\omega}^E$.

9.4.3 Incremental variation of element frame

The orientation of an element frame generally depends only on the global coordinates of the nodes to which the element is connected, and is independent of nodal rotations. An exception occurs in the case of a space frame element where the local e_3 axis may be determined by the axial rotations of its end nodes. For the sake of simplicity, we omit consideration of this special case in our general discussion of CR kinematics.

It therefore follows that the *incremental* rotation of the element frame depends only on the *incremental* nodal displacements and is independent of incremental nodal rotations.

We remark here that, as a result, there is no mechanism in the CR procedure that operates to make the element local relative *rotations* small as there is for local relative displacements. That issue must be addressed via mesh refinement in specific applications.

Completely arbitrary incremental nodal displacements of course produce an incremental rigid body rotation, as well as incremental strains. In order to determine the incremental rotation $\delta\mathbf{E}_k$ of the orthogonal element triad $\mathbf{E}_k$, given a set of arbitrary nodal virtual displacements $\delta\mathbf{U}_e$, we seek to determine a "best-fit" rigid body motion for the element frame that minimizes in some sense the *deformations* produced by the nodal virtual displacements. $\delta\mathbf{U}_e$ denotes the set of nodal virtual *displacement* freedoms (in global coordinates) applied to the element. We drop the element subscript e in what follows to simplify the notation.

In the previous section, we in essence defined the incremental rotation (i.e., spin) of the element frame in global coordinates via

$$\delta\mathbf{E}_k\mathbf{E}_k^T = \delta\boldsymbol{\Omega}^E \tag{9.114}$$

where $\delta\boldsymbol{\Omega}^E \equiv \text{Spin}\,(\delta\boldsymbol{\omega}^E)$ and

$$\delta\boldsymbol{\Omega}^E = \begin{bmatrix} 0 & -\delta\omega_3^E & \delta\omega_2^E \\ \delta\omega_3^E & 0 & -\delta\omega_1^E \\ -\delta\omega_2^E & \delta\omega_1^E & 0 \end{bmatrix} \tag{9.115}$$

We again suppress the load-level subscript k on $\delta\boldsymbol{\Omega}$. The incremental spin vector $\delta\boldsymbol{\omega}^E \equiv$ axial $\delta\boldsymbol{\Omega}^E$ is

$$\delta\boldsymbol{\omega}^E \equiv \begin{Bmatrix} \delta\omega_1^E \\ \delta\omega_2^E \\ \delta\omega_3^E \end{Bmatrix} \tag{9.116}$$

In the element coordinate system, the incremental spin tensor is denoted $\delta\mathbf{W}^E$, where

$$\delta\mathbf{W}^E = \begin{bmatrix} 0 & -\delta w_3^E & \delta w_2^E \\ \delta w_3^E & 0 & -\delta w_1^E \\ -\delta w_2^E & \delta w_1^E & 0 \end{bmatrix} \tag{9.117}$$

The corresponding incremental spin vector $\delta\mathbf{w}^E \equiv$ axial $\delta\mathbf{W}^E$ in local element coordinates is

$$\delta\mathbf{w}^E \equiv \begin{Bmatrix} \delta w_1^E \\ \delta w_2^E \\ \delta w_3^E \end{Bmatrix} \tag{9.118}$$

The generic vector and tensor transformations from global to element coordinate systems are restated below.

$$\begin{cases} \delta \mathbf{w} = \mathbf{E}_k^T \delta \boldsymbol{\omega} \\ \delta \mathbf{W} = \mathbf{E}_k^T \ \delta \boldsymbol{\Omega} \ \mathbf{E}_k \end{cases} \tag{9.119}$$

Similarly, virtual displacements $\delta \mathbf{u}$ in the element local coordinate system are related to virtual displacements $\delta \mathbf{U}$ in the global coordinate system by

$$\delta \mathbf{u} = \mathbf{E}_k^T \delta \mathbf{U} \tag{9.120}$$

Consider now the triangular element shown in Figure 9.4 and the incremental rotation of its element triad $\mathbf{E}_k$.

For simplicity, we assume as a convention that the origin of the element coordinate system is located at node 1, although other locations (e.g., the centroid) may in fact be preferable (Felippa and Haugen 2005).

In this section, we again employ subscripts (or superscripts as needed) i or j as nodal identifiers. Therefore, $\delta \mathbf{u}_i$ (i = 1, 2, or 3) denote the (3 × 1) element nodal displacement vectors. The (9 × 1) vector of (incremental) nodal *displacement* freedoms may be written

$$\delta \mathbf{u} = \begin{Bmatrix} \delta \mathbf{u}_1 \\ \delta \mathbf{u}_2 \\ \delta \mathbf{u}_3 \end{Bmatrix} \tag{9.121}$$

These incremental nodal displacement vectors are decomposed as follows:

$$\begin{Bmatrix} \delta \mathbf{u}_1 \\ \delta \mathbf{u}_2 \\ \delta \mathbf{u}_3 \end{Bmatrix} = \begin{Bmatrix} \delta \bar{\mathbf{u}}_1 \\ \delta \bar{\mathbf{u}}_2 \\ \delta \bar{\mathbf{u}}_3 \end{Bmatrix} + \begin{Bmatrix} \delta \mathbf{u}_1^r \\ \delta \mathbf{u}_2^r \\ \delta \mathbf{u}_3^r \end{Bmatrix} \tag{9.122}$$

where $\delta \mathbf{u}_i^r$ represents the effect of the rigid body frame rotation and $\delta \bar{\mathbf{u}}_i \equiv \delta \mathbf{u}_i - \delta \mathbf{u}_i^r$ produces (virtual) deformations.

A rigid body rotation with respect to the element origin (node 1) produces the nodal virtual displacements

$$\begin{cases} \delta \mathbf{u}_1^r = 0 \\ \delta \mathbf{u}_2^r = \delta \mathbf{W}^E \mathbf{x}_{21} \\ \delta \mathbf{u}_3^r = \delta \mathbf{W}^E \mathbf{x}_{31} \end{cases} \tag{9.123}$$

where $\mathbf{x}_{21}$ and $\mathbf{x}_{31}$ are the local position vectors of nodes 2 and 3, respectively, relative to the element origin at node 1. Specifically, in this case,

$$\mathbf{x}_{21} = \begin{Bmatrix} x_{21} \\ 0 \\ 0 \end{Bmatrix} \text{ and } \mathbf{x}_{31} = \begin{Bmatrix} x_{31} \\ y_{31} \\ 0 \end{Bmatrix} \tag{9.124}$$

Equation 9.123 can be put in the form

$$\begin{cases} \delta \mathbf{u}_1^r = 0 \\ \delta \mathbf{u}_2^r = \text{Spin}(\delta \mathbf{w}^E)\mathbf{x}_{21} = -\text{Spin}(\mathbf{x}_{21})\delta \mathbf{w}^E \\ \delta \mathbf{u}_3^r = \text{Spin}(\delta \mathbf{w}^E)\mathbf{x}_{31} = -\text{Spin}(\mathbf{x}_{31})\delta \mathbf{w}^E \end{cases} \tag{9.125}$$

which follows from the property of the vector cross product, that is, $\mathbf{a} \times \mathbf{b} = -(\mathbf{b} \times \mathbf{a})$.

As a result, Equation 9.122 may be written compactly as

$$\delta \mathbf{u} = \delta \bar{\mathbf{u}} + \mathbf{S}\delta \mathbf{w}^E \tag{9.126}$$

where $\delta \mathbf{u}^r = \mathbf{S}\ \delta \mathbf{w}^E$. The (9 × 3) matrix **S** is defined as

$$\mathbf{S} \equiv \begin{bmatrix} \mathbf{O}_3 \\ -\mathrm{Spin}(\mathbf{x}_{21}) \\ -\mathrm{Spin}(\mathbf{x}_{31}) \end{bmatrix} \tag{9.127}$$

The explicit form of **S** in this case is

$$\mathbf{S} = -\begin{bmatrix} 0 & 0 & 0 \\ 0 & 0 & 0 \\ 0 & 0 & 0 \\ 0 & 0 & 0 \\ 0 & 0 & -x_{21} \\ 0 & x_{21} & 0 \\ 0 & 0 & y_{31} \\ 0 & 0 & -x_{31} \\ -y_{31} & x_{31} & 0 \end{bmatrix} \tag{9.128}$$

In order to "minimize" the virtual deformations resulting from arbitrary nodal virtual displacements, we make the rigid body nodal virtual displacement $\delta \mathbf{u}^r$ orthogonal to the nodal virtual displacement $\delta \bar{\mathbf{u}} = \delta \mathbf{u} - \delta \mathbf{u}^r$ seen by the element, that is,

$$\delta \mathbf{u}^{r,T}(\delta \mathbf{u} - \delta \mathbf{u}^r) = 0 \tag{9.129}$$

which leads to

$$\delta \mathbf{w}^E = (\mathbf{S}^T\mathbf{S})^{-1}\mathbf{S}^T\delta \mathbf{u} \tag{9.130}$$

The explicit expression of the (3 × 3) matrix $\mathbf{SS}^T$ in this case is

$$\mathbf{S}^T\mathbf{S} = \begin{bmatrix} y_{31}^2 & -x_{31}y_{31} & 0 \\ -x_{31}y_{31} & x_{21}^2 + x_{31}^2 & 0 \\ 0 & 0 & x_{21}^2 + x_{31}^2 + y_{31}^2 \end{bmatrix} \tag{9.131}$$

and its inverse is

$$(\mathbf{S}^T\mathbf{S})^{-1} = \frac{1}{x_{21}^2 y_{31}^2}\begin{bmatrix} x_{21}^2 + x_{31}^2 & x_{31}y_{31} & 0 \\ x_{31}y_{31} & y_{31}^2 & 0 \\ 0 & 0 & x_{21}^2 y_{31}^2/d^2 \end{bmatrix}, \tag{9.132}$$

where $d^2 = x_{21}^2 + x_{31}^2 + y_{31}^2$

Equation 9.130 may be rewritten as

$$\delta \mathbf{w}^E = \mathbf{G}\delta \mathbf{u} \tag{9.133}$$

or equivalently as

$$\delta w^E = \sum_i G_i \delta u_i \tag{9.134}$$

where the (3 × 3) nodal blocks G_i of the (3 × 9) matrix **G** are

$$G_i \equiv (S^T S)^{-1} S_i^T \tag{9.135}$$

The explicit forms of G_i in this case are

$$G_1 = O_3, \quad G_2 = \begin{bmatrix} 0 & 0 & -\dfrac{x_{31}}{x_{21}y_{31}} \\ 0 & 0 & -\dfrac{1}{x_{21}} \\ 0 & \dfrac{x_{21}}{d^2} & 0 \end{bmatrix}, \quad G_3 = \begin{bmatrix} 0 & 0 & \dfrac{1}{y_{31}} \\ 0 & 0 & 0 \\ -\dfrac{y_{31}}{d^2} & \dfrac{x_{31}}{d^2} & 0 \end{bmatrix} \tag{9.136}$$

We note that $GS = I_3$, as may be verified directly. The deformational incremental displacement vector $\delta\bar{u}_i$ at node i is then

$$\delta\bar{u}_i = \sum_j \left[\delta_{ij} I_3 - S_i G_j\right] \delta u_j \tag{9.137}$$

Collecting the incremental nodal displacements together in a (9 × 1) vector, we have

$$\delta\bar{\mathbf{u}} = \mathbf{P}\,\delta\mathbf{u} \tag{9.138}$$

in which the (9 × 9) projection matrix $\mathbf{P} \equiv \mathbf{I} - \mathbf{P}_R = \mathbf{I} - \mathbf{SG}$.

In global coordinates, the spin of the element triad $\delta\omega^E = E_k \delta w^E$ and $\delta\Omega^E = \text{Spin}\,(\delta\omega^E)$.

9.4.4 Incremental variation of the nodal triad

We now address the incremental variation of a nodal triad **R**. Whereas the incremental rotation of the element frame was chosen on a best-fit criterion in order to minimize the incremental (virtual) strains, there are no similar constraints to be placed on the spins of the nodal triads.

The incremental variation $\delta\mathbf{R}$ of any orthogonal matrix **R** is described by

$$\delta\mathbf{R}\mathbf{R}^T = \delta\boldsymbol{\Omega} \tag{9.139}$$

where $\delta\boldsymbol{\Omega}$ is a skew-symmetric matrix.

We also know that the representation of the orthogonal matrix **R** in terms of the spin matrix **N**,

$$\mathbf{R} = \mathbf{I} + \sin\theta\;\mathbf{N} + (1 - \cos\theta)\mathbf{N}^2 \tag{9.140}$$

which is often represented as $\mathbf{R} = \exp(\theta\mathbf{N})$, is exact for a finite rotation. However, $\delta\mathbf{R} \neq \exp(\delta\theta\,\mathbf{N})$ in general because that would preclude an incremental change in the *axis* of rotation; that is, it would assume that **N** is constant. However, the incremental spin vector $\delta\boldsymbol{\omega}^R \equiv \text{axial}\;\delta\boldsymbol{\Omega}^R$ is an arbitrary nodal freedom that can result in general in incremental changes in both the rotation axis and the rotation angle.

We briefly outline here the derivation of $\delta\boldsymbol{\omega}^R$ given by Nour-Omid and Rankin (1991), which had first appeared in Simo and Vu-Quoc (1986) and Szwabowicz (1986).

We temporarily suppress the superscript R and the nodal identifiers i (e.g., on $\delta\boldsymbol{\omega}_i^R$) in this section to simplify the notation, and will reintroduce them subsequently when needed.

Starting from Equation 9.140 for **R**, we first take independent variations $\delta\theta$ and $\delta\mathbf{N}$ and then postmultiply by $\mathbf{R}^T$ to arrive at

$$\delta\mathbf{R}\mathbf{R}^T \equiv \delta\boldsymbol{\Omega} = \delta\theta\ \mathbf{N} + \sin\theta\ \delta\mathbf{N} + (1-\cos\theta)(\mathbf{N}\ \delta\mathbf{N} - \delta\mathbf{N}\ \mathbf{N}) \tag{9.141}$$

in which we have used the identities $\mathbf{N}^3 \equiv -\mathbf{N}$ and $\mathbf{N}\ \delta\mathbf{N}\ \mathbf{N} \equiv 0$.

In order to facilitate extraction of the incremental axial vector $\delta\boldsymbol{\omega}$, we now make the substitution $\mathbf{N}_\theta \equiv \theta\mathbf{N}$, that is,

$$\mathbf{N}_\theta \equiv \begin{bmatrix} 0 & -\theta_3 & \theta_2 \\ \theta_3 & 0 & -\theta_1 \\ -\theta_2 & \theta_1 & 0 \end{bmatrix} \tag{9.142}$$

The variation $\delta\mathbf{N}$ in Equation 9.141 is found in terms of $\delta\mathbf{N}_\theta$ by differentiating $\mathbf{N}_\theta \equiv \theta\mathbf{N}$:

$$\delta\mathbf{N} = \frac{1}{\theta}\left[\delta\mathbf{N}_\theta - \frac{\delta\theta}{\theta}\mathbf{N}_\theta\right] \tag{9.143}$$

Equation 9.141 then becomes

$$\delta\boldsymbol{\Omega} = \frac{\delta\theta}{\theta}\left(1-\frac{\sin\theta}{\theta}\right)\mathbf{N}_\theta + \frac{\sin\theta}{\theta}\ \delta\mathbf{N}_\theta + \frac{(1-\cos\theta)}{\theta^2}(\mathbf{N}_\theta\ \delta\mathbf{N}_\theta - \delta\mathbf{N}_\theta\ \mathbf{N}_\theta) \tag{9.144}$$

We use the identity $\boldsymbol{\theta}^T\delta\boldsymbol{\theta} \equiv \theta\ \delta\theta$ to replace the first term above by

$$\frac{\delta\theta}{\theta}\left(1-\frac{\sin\theta}{\theta}\right)\mathbf{N}_\theta = \frac{1}{\theta^3}(\theta-\sin\theta)\mathbf{N}_\theta(\boldsymbol{\theta}^T\delta\boldsymbol{\theta}) \tag{9.145}$$

and, taking note that axial $(\mathbf{N}_\theta\delta\mathbf{N}_\theta - \delta\mathbf{N}_\theta\mathbf{N}_\theta) \equiv \mathbf{N}_\theta\delta\boldsymbol{\theta}$, then find $\delta\boldsymbol{\omega}$ = axial $(\delta\boldsymbol{\Omega})$, resulting in the relation between $\delta\boldsymbol{\omega}$ and $\delta\boldsymbol{\theta}$ below.

$$\begin{aligned}\delta\boldsymbol{\omega} &= \left[\frac{1}{\theta^3}(\theta-\sin\theta)\boldsymbol{\theta}\boldsymbol{\theta}^T + \frac{\sin\theta}{\theta}\ \mathbf{I} + \frac{(1-\cos\theta)}{\theta^2}\mathbf{N}_\theta\right]\delta\boldsymbol{\theta} \\ &= \left[\mathbf{I} + \frac{(1-\cos\theta)}{\theta^2}\mathbf{N}_\theta + \frac{(\theta-\sin\theta)}{\theta^3}\mathbf{N}_\theta^2\right]\delta\boldsymbol{\theta}\end{aligned} \tag{9.146}$$

This last expression results from substitution of the identity $\boldsymbol{\theta}\boldsymbol{\theta}^T \equiv \mathbf{N}_\theta^2 + \theta^2\mathbf{I}$ in the first expression.

Equation 9.146 resolves the arbitrary spin vector $\delta\boldsymbol{\omega}$ into a component $\delta\boldsymbol{\theta}$ in the direction of the current axial vector $\boldsymbol{\theta}$ and a vector normal to $\boldsymbol{\theta}$.

We require the inverse of Equation 9.146 for the transformation of the internal resisting force vector and for the tangent stiffness matrix. We expect that the inverse will also involve the same basic independent building blocks, that is, $\mathbf{I}, \mathbf{N}_\theta$, and $\mathbf{N}_\theta^2$. Accordingly, we premultiply Equations 9.146 first by $\mathbf{N}_\theta$ and then by $\mathbf{N}_\theta^2$ to obtain the two independent vector equations,

$$\begin{cases} \mathbf{N}_\theta \delta\boldsymbol{\omega} = \dfrac{\sin\theta}{\theta}\mathbf{N}_\theta\delta\boldsymbol{\theta} + \dfrac{(1-\cos\theta)}{\theta^2}\mathbf{N}_\theta^2\delta\boldsymbol{\theta} \\ \mathbf{N}_\theta^2\delta\boldsymbol{\omega} = -(1-\cos\theta)\mathbf{N}_\theta\delta\boldsymbol{\theta} + \dfrac{\sin\theta}{\theta}\mathbf{N}_\theta^2\delta\boldsymbol{\theta} \end{cases} \tag{9.147}$$

in which we have again used the identity $\mathbf{N}_\theta^3 \equiv -\theta^2\mathbf{N}_\theta$. These two equations are easily solved to yield expressions for the vectors $\mathbf{N}_\theta\delta\boldsymbol{\theta}$ and $\mathbf{N}_\theta^2\delta\boldsymbol{\theta}$ in terms of $\mathbf{N}_\theta\delta\boldsymbol{\omega}_\theta$ and $\mathbf{N}_\theta^2\delta\boldsymbol{\omega}$, and these results are then substituted back into Equation 9.146 to finally yield

$$\delta\boldsymbol{\theta} = \left[\mathbf{I} - \frac{1}{2}\mathbf{N}_\theta + \eta(\theta)\mathbf{N}_\theta^2\right]\delta\boldsymbol{\omega} \equiv \boldsymbol{\Lambda}\,\delta\boldsymbol{\omega} \tag{9.148}$$

Now reintroducing the superscripts and subscripts, we have

$$\delta\boldsymbol{\theta}_i^R = \boldsymbol{\Lambda}_i(\theta_i^R)\delta\boldsymbol{\omega}_i^R \tag{9.149}$$

The coefficient $\eta(\theta)$ in the expression for $\boldsymbol{\Lambda}_i$ above is

$$\eta(\theta) = \frac{1}{\theta^2}\left(1 - \frac{\theta}{2}\cot\frac{\theta}{2}\right) \tag{9.150}$$

Care must be taken in numerically evaluating many of the foregoing coefficients for very small values of θ, as they approach 0/0 as $\theta \to 0$. Felippa and Haugen (2005) provide a four-term power series expression for θ, the first term of which gives $\eta \to 1/12$ as $\theta \to 0$.

9.4.4.1 *Corotated incremental nodal rotations*

We now relate the incremental nodal pseudovector $\delta\bar{\boldsymbol{\theta}}_i$ to the spin of the nodal triad. From Equation 9.113, that is,

$$\delta\bar{\mathbf{w}}_i^R = \delta\mathbf{w}_i^R - \delta\mathbf{w}^E$$

we first find

$$\delta\bar{\mathbf{w}}_i^R = \delta\mathbf{w}_i^R - \sum_j \mathbf{G}_j\delta\mathbf{u}_j \tag{9.151}$$

using Equation 9.134. Combining this last equation with

$$\delta\bar{\boldsymbol{\theta}}_i = \boldsymbol{\Lambda}_i(\bar{\theta}_i)\delta\bar{\mathbf{w}}_i^R \tag{9.152}$$

we then have

$$\delta\bar{\boldsymbol{\theta}}_i = \boldsymbol{\Lambda}_i(\bar{\theta}_i)[\delta\mathbf{w}_i^R - \sum_j \mathbf{G}_j\mathbf{u}_j] \tag{9.153}$$

9.4.5 Corotated incremental element freedoms combined

We have thus far considered separately the effects of incremental nodal displacements and nodal spins on corotated element displacements and rotations. We now combine these two types of nodal freedoms into a (6×1) element nodal freedom vector $\mathbf{v}_i$ in the element coordinate system or $\mathbf{V}_i$ in the global coordinate system. We again suppress the element index e on the vector $\mathbf{V}_i$.

We first recapitulate the notations for total and incremental element nodal freedoms as follows.

The total element nodal freedoms, in global and local coordinates, respectively, are

$$\mathbf{V}_i \equiv \begin{bmatrix} \mathbf{U}_i \\ \mathbf{R}_i \end{bmatrix}, \quad \mathbf{v}_i \equiv \begin{Bmatrix} \mathbf{u}_i \\ \boldsymbol{\theta}_i \end{Bmatrix} \tag{9.154}$$

where $\boldsymbol{\theta}_i$ is the axial pseudovector of $\mathbf{R}_i$ transformed to element coordinates, and the square brackets indicate that the hybrid nodal quantity $\mathbf{V}_i$ is not a vector.

The incremental element nodal freedoms, in global and local coordinates, respectively, are

$$\delta\mathbf{V}_i \equiv \begin{Bmatrix} \delta\mathbf{U}_i \\ \delta\boldsymbol{\omega}_i^R \end{Bmatrix}, \quad \delta\mathbf{v}_i \equiv \begin{Bmatrix} \delta\mathbf{u}_i \\ \delta\mathbf{w}_i^R \end{Bmatrix} \tag{9.155}$$

where $\delta\boldsymbol{\omega}_i^R$ is the spin of the nodal triad $\mathbf{R}_i$, and $\delta\mathbf{w}_i^R$ is that same quantity in local coordinates. The total and incremental corotated element nodal freedoms, which are expressed only in local coordinates, are

$$\bar{\mathbf{v}}_i \equiv \begin{Bmatrix} \bar{\mathbf{u}}_i \\ \bar{\boldsymbol{\theta}}_i \end{Bmatrix}, \quad \delta\bar{\mathbf{v}}_i \equiv \begin{Bmatrix} \delta\bar{\mathbf{u}}_i \\ \delta\bar{\mathbf{w}}_i^R \end{Bmatrix} \tag{9.156}$$

where $\bar{\boldsymbol{\theta}}_i$ is the axial pseudovector of the relative nodal rotation $\bar{\mathbf{R}}_i$ (Equation 9.113).

The increments of corotated (deformational) element nodal freedoms are then expressed in terms of (6×6) nodal block matrices as

$$\begin{Bmatrix} \delta\bar{\mathbf{u}}_i \\ \delta\bar{\mathbf{w}}^R{}_i \end{Bmatrix} = \sum_j \left[\delta_{ij} \begin{bmatrix} \mathbf{I}_3 & \mathbf{O}_3 \\ \mathbf{O}_3 & \mathbf{I}_3 \end{bmatrix} - \begin{bmatrix} \mathbf{S}_i\mathbf{G}_j & \mathbf{O}_3 \\ \mathbf{G}_j & \mathbf{O}_3 \end{bmatrix} \right] \begin{Bmatrix} \delta\mathbf{u}_j \\ \delta\mathbf{w}_j^R \end{Bmatrix} \tag{9.157}$$

and with

$$\begin{Bmatrix} \delta\bar{\mathbf{u}}_i \\ \delta\bar{\boldsymbol{\theta}}_i \end{Bmatrix} = \mathbf{H}_i(\bar{\boldsymbol{\theta}}_i) \begin{Bmatrix} \delta\bar{\mathbf{u}}_i \\ \delta\bar{\mathbf{w}}_i^R \end{Bmatrix} \tag{9.158}$$

where

$$\mathbf{H}_i(\bar{\boldsymbol{\theta}}_i) \equiv \begin{bmatrix} \mathbf{I}_3 & \mathbf{O}_3 \\ \mathbf{O}_3 & \boldsymbol{\Lambda}_i(\bar{\boldsymbol{\theta}}_i) \end{bmatrix} \tag{9.159}$$

we have finally

$$\begin{aligned} \begin{Bmatrix} \delta\bar{\mathbf{u}}_i \\ \delta\bar{\boldsymbol{\theta}}_i \end{Bmatrix} &= \mathbf{H}_i(\bar{\boldsymbol{\theta}}_i) \sum_j \left[\delta_{ij} \begin{bmatrix} \mathbf{I}_3 & \mathbf{O}_3 \\ \mathbf{O}_3 & \mathbf{I}_3 \end{bmatrix} - \begin{bmatrix} \mathbf{S}_i\mathbf{G}_j & \mathbf{O}_3 \\ \mathbf{G}_j & \mathbf{O}_3 \end{bmatrix} \right] \begin{Bmatrix} \delta\mathbf{u}_j \\ \delta\mathbf{w}_j^R \end{Bmatrix} \\ \text{or} \quad \delta\bar{\mathbf{v}}_i &= \mathbf{H}_i(\bar{\boldsymbol{\theta}}_i) \sum_j \left[\delta_{ij} \begin{bmatrix} \mathbf{I}_3 & \mathbf{O}_3 \\ \mathbf{O}_3 & \mathbf{I}_3 \end{bmatrix} - \begin{bmatrix} \mathbf{S}_i\mathbf{G}_j & \mathbf{O}_3 \\ \mathbf{G}_j & \mathbf{O}_3 \end{bmatrix} \right] \delta\mathbf{v}_j \end{aligned} \tag{9.160}$$

The transformation to the global nodal incremental freedoms is given by

$$\begin{Bmatrix} \delta u_i \\ \delta w_i^R \end{Bmatrix} = \begin{bmatrix} E_3^T & O_3 \\ O_3 & E_3^T \end{bmatrix} \begin{Bmatrix} \delta U_i \\ \delta \omega_i^R \end{Bmatrix} \tag{9.161}$$

We introduce the notation $\mathbf{E}_3$ here for the (3 × 3) element triad, to distinguish it from the (18 × 18) block diagonal matrix **E** consisting of matrices $\mathbf{E}_3$ on its diagonal.

We note that the more common notation for the transformation from local to global coordinates is $\mathbf{u} = \mathbf{TU}$ (where $\mathbf{E}^T \equiv \mathbf{T}$).

In expanded form, these relationships can be represented as

$$\delta\bar{\mathbf{v}} = \mathbf{HPE}^T\delta\mathbf{V} \tag{9.162}$$

where the (18 × 18) matrices **H**, **P**, and $\mathbf{E}^T$ are built block by block from the nodal matrices given above. The (18×18) block diagonal matrix $\mathbf{H} \equiv \text{diag}[\mathbf{H}_1\ \mathbf{H}_2\ \mathbf{H}_3]$, in which the (6×6) nodal blocks $\mathbf{H}_i$ are each themselves block diagonal, according to Equation 9.159.

The element internal nodal forces that are work conjugate to the displacements are denoted by

$$\mathbf{f}_i = \begin{Bmatrix} \mathbf{n}_i \\ \mathbf{m}_i \end{Bmatrix} \text{ and } \bar{\mathbf{f}}_i = \begin{Bmatrix} \bar{\mathbf{n}}_i \\ \bar{\mathbf{m}}_i \end{Bmatrix} \tag{9.163}$$

in the element coordinate system. We denote the element internal resisting force vector in global coordinates as **I**, and here use **f** (rather than **i**) for the same vector in the element coordinate system. We remark in passing that the nodal moment $\mathbf{m}_i$ is a bivector (i.e., an axial vector).

We recall that in both the total Lagrangian (TL) and UDL formulations discussed in Chapter 8, the incremental equilibrium equations involved a geometric stiffness term that arose from the effect of an incremental geometry change (i.e., a virtual rotation) on the current internal resisting force vector. In summary, in the matrix–vector format used in the TL finite element formulation for solids in Chapter 8, the element internal resisting force vector was expressed as

$$\mathbf{I}_e = \int_{Ve} \mathbf{B}^T\boldsymbol{\sigma}dV_e$$

in which the tangent strain–displacement matrix $\mathbf{B}$(i.e., $\dot{\boldsymbol{\varepsilon}} = \mathbf{B}\dot{\mathbf{u}}_e$) is a function of element nodal displacements. Here we dispense with the notation $\{\boldsymbol{\sigma}\}$ to indicate the vector form of the tensor $\boldsymbol{\sigma}$. The incremental (rate) form then gave

$$\dot{\mathbf{I}}_e = \int_{Ve} \mathbf{B}^T\dot{\boldsymbol{\sigma}}dV_e + \int_{Ve} \dot{\mathbf{B}}^T\boldsymbol{\sigma}dV_e$$

The first term above involves a "material" tangent stiffness relation $\dot{\boldsymbol{\sigma}} = \mathbf{D}\dot{\boldsymbol{\varepsilon}}$ arising from the strain rate $\dot{\boldsymbol{\varepsilon}}$, and the second term leads to a geometric stiffness matrix that is a function of the current state of total stress $\boldsymbol{\sigma}$.

In a CR formulation, only a material stiffness appears explicitly. The effect of the incremental geometry change (rotation) on the internal resisting force rate is implicitly accounted for by CR moving frame kinematics.

We assume that the nodal force variables $\bar{\mathbf{f}}_i$ (= $[\bar{\mathbf{n}}_i^T, \bar{\mathbf{m}}_i^T]^T$) in the local element coordinate system are related to the deformational nodal variables by an elastic (material) stiffness

$$\bar{\mathbf{f}}_i = \sum_j \bar{\mathbf{k}}_{ij} \bar{\mathbf{v}}_j \tag{9.164}$$

where $\bar{\mathbf{k}}_{ij}$ is a (6 × 6) nodal stiffness. This linear elastic relation may be written in full as

$$\bar{\mathbf{f}} = \bar{\mathbf{k}} \bar{\mathbf{v}} \tag{9.165}$$

where $\bar{\mathbf{f}}$ and $\bar{\mathbf{v}}$ are (18 × 1) and $\bar{\mathbf{k}}$ is (18 × 18). We note that the material stiffness here is not a fundamental linear elastic relation between stress and strain in the sense referred to above, but is instead a discretized matrix relationship that implicitly incorporates, in addition to constitutive (elastic) properties, equilibrium relationships involving the nodal force variables.

In general, if the incremental nodal displacements are related by $\delta\bar{\mathbf{v}} = \hat{\mathbf{T}}\delta\mathbf{V}$, where $\hat{\mathbf{T}}$ is some transformation matrix, and in addition the element stiffness matrix is insensitive to infinitesimal rigid body motions (i.e., it passes the patch test), then it follows from the virtual work relation $\delta\bar{\mathbf{v}}^T\bar{\mathbf{f}} = \delta\mathbf{V}^T\mathbf{I}$ that

$$\mathbf{I} = \hat{\mathbf{T}}^T \bar{\mathbf{f}} \tag{9.166}$$

which transforms the internal force vector from the one that is work conjugate to the deformational freedoms (i.e., $\bar{\mathbf{f}}$) to one that is work conjugate to the nodal freedoms (i.e., **I**).

Therefore, in the CR formulation the element internal force vector **I** in global coordinates is given by

$$\mathbf{I} = \mathbf{E}\mathbf{P}^T\mathbf{H}^T\bar{\mathbf{f}} \tag{9.167}$$

and the effect of the current internal force vector $\bar{\mathbf{f}}$ undergoing an incremental geometry change (the geometric stiffness) is found by taking the variation (e.g., $\delta\mathbf{I}_G$) in the equation above, while holding $\bar{\mathbf{f}}$ constant. Therefore, the geometric stiffness terms arise from

$$\delta\mathbf{I}_G = \delta\mathbf{E}\mathbf{P}^T\mathbf{H}^T\bar{\mathbf{f}} + \mathbf{E}\delta\mathbf{P}^T\mathbf{H}^T\bar{\mathbf{f}} + \mathbf{E}\mathbf{P}^T\delta\mathbf{H}^T\bar{\mathbf{f}} \tag{9.168}$$

whereas the material stiffness arises from deformation (e.g., $\delta\mathbf{I}_M$) in which only $\bar{\mathbf{f}}$ is varied, that is,

$$\begin{aligned}\delta\mathbf{I}_M &= \mathbf{E}\mathbf{P}^T\mathbf{H}^T\delta\bar{\mathbf{f}} \\ &= \mathbf{E}\mathbf{P}^T\mathbf{H}^T\bar{\mathbf{k}}\delta\bar{\mathbf{v}} \\ &= \left[(\mathbf{H}\mathbf{P}\mathbf{E}^T)^T\bar{\mathbf{k}}(\mathbf{H}\mathbf{P}\mathbf{E}^T)\right]\delta\mathbf{V} \\ &\equiv \mathbf{K}_M\delta\mathbf{V}\end{aligned} \tag{9.169}$$

Therefore, the CR material stiffness matrix $\mathbf{K}_M$ is symmetric (if $\bar{\mathbf{k}}$ is symmetric).

9.4.5.1 Geometric stiffness

The geometric stiffness matrix (containing three terms) arising out of Equation 9.168 is unsymmetric. However, symmetry is recovered as an equilibrium configuration is approached. Nour-Omid and Rankin (1991) proved that a symmetrized geometric stiffness matrix can be employed while still retaining the characteristic quadratic rate of convergence in an incremental-iterative Newton–Raphson procedure.

Haugen (1994) briefly discusses this asymmetry and relates it to the fact that incremental nodal rotation variables are not integrable in a moving reference frame.

The three geometric stiffness terms in Equation 9.168 have been developed and studied in detail by Nour-Omid and Rankin (1991) and Haugen (1994).

The first contribution above to the geometric stiffness, that is,

$$\delta I_{GR} = \delta \mathbf{E}\mathbf{P}^T\mathbf{H}^T\bar{\mathbf{f}} \tag{9.170}$$

termed by Haugen the "rotational geometric stiffness," is a familiar quantity. It reflects the reorientation (in the global coordinate system) of the current element nodal forces $\mathbf{n}_i$ and nodal bivectors $\mathbf{m}_i$ due to the incremental rigid body rotation (via $\delta\mathbf{E}$) of the element triad and is the most important contribution to the geometric stiffness. It may first be written compactly as

$$\delta I_{GR} = \delta \mathbf{E}\mathbf{f} \tag{9.171}$$

where $\mathbf{f} \equiv \mathbf{P}^T\mathbf{H}^T\bar{\mathbf{f}}$. Then the contribution of the nodal forces and moments at node i is

$$\delta I_{GRi} = \begin{bmatrix} \delta \mathbf{E}_3 & \mathbf{O}_3 \\ \mathbf{O}_3 & \delta \mathbf{E}_3 \end{bmatrix} \begin{Bmatrix} \mathbf{n}_i \\ \mathbf{m}_i \end{Bmatrix} \tag{9.172}$$

where $\mathbf{f}_i^T \equiv [\mathbf{n}_i^T \mathbf{m}_i^T]$. The variation of the element triad is $\delta\mathbf{E}_3 = \delta\mathbf{\Omega}^E\mathbf{E}_3$ in global coordinates. In local coordinates, $\delta\mathbf{E}_3 = \mathbf{E}_3\delta\mathbf{W}^E$. Therefore,

$$\delta I_{GRi} = \begin{Bmatrix} \delta \mathbf{E}_3\mathbf{n}_i \\ \delta \mathbf{E}_3\mathbf{m}_i \end{Bmatrix} = \begin{Bmatrix} \mathbf{E}_3\mathrm{Spin}(\delta\mathbf{w}^E)\mathbf{n}_i \\ \mathbf{E}_3\mathrm{Spin}(\delta\mathbf{w}^E)\mathbf{m}_i \end{Bmatrix} = -\begin{Bmatrix} \mathbf{E}_3\mathrm{Spin}(\mathbf{n}_i)\delta\mathbf{w}^E \\ \mathbf{E}_3\mathrm{Spin}(\mathbf{m}_i)\delta\mathbf{w}^E \end{Bmatrix} \tag{9.173}$$

In combined form,

$$\delta I_{GR} = -\mathbf{E}\mathbf{F}_{GR}\mathbf{G}\ \delta\mathbf{v} = -\mathbf{E}\mathbf{F}_{GR}\mathbf{G}\mathbf{E}^T\delta\mathbf{V} \equiv \mathbf{K}_{GR}\delta\mathbf{V} \tag{9.174}$$

$\mathbf{E}$ is (18 × 18) block diagonal, and $\mathbf{F}_{GR}$ is (18 × 3) defined as follows:

$$\mathbf{F}_{GR} \equiv \begin{bmatrix} \mathrm{Spin}(\mathbf{n}_1) \\ \mathrm{Spin}(\mathbf{m}_1) \\ \mathrm{Spin}(\mathbf{n}_2) \\ \mathrm{Spin}(\mathbf{m}_2) \\ \mathrm{Spin}(\mathbf{n}_3) \\ \mathrm{Spin}(\mathbf{m}_3) \end{bmatrix} \tag{9.175}$$

G is (3 × 18) defined as

$$\mathbf{G} \equiv [\mathbf{G}_1\ \mathbf{O}_3\ \mathbf{G}_2\ \mathbf{O}_3\ \mathbf{G}_3\ \mathbf{O}_3] \tag{9.176}$$

The second contribution to the geometric stiffness in Equation 9.168, namely,

$$\delta I_{GP} = \mathbf{E}\delta\mathbf{P}^T\mathbf{H}^T\bar{\mathbf{f}} \tag{9.177}$$

is termed the "equilibrium projection geometric stiffness" by Haugen. As it arises from variations of the projector $\mathbf{P} = \mathbf{I} - \mathbf{P}_R$ that depend on element strains, it would seem to be of minor importance in our summary. Accordingly, we omit detailed discussion of it.

The third contribution to the geometric stiffness in Equation 9.168,

$$\delta \mathbf{I}_{GM} = \mathbf{E}\mathbf{P}^T \delta \mathbf{H}^T \bar{\mathbf{f}} \tag{9.178}$$

is termed the "moment correction geometric stiffness" by Haugen. Consider the (18 × 1) vector appearing in this last equation:

$$\delta \mathbf{H}^T \bar{\mathbf{f}} \equiv \begin{bmatrix} \delta \mathbf{H}_1^T & \mathbf{O}_6 & \mathbf{O}_6 \\ \mathbf{O}_6 & \delta \mathbf{H}_2^T & \mathbf{O}_6 \\ \mathbf{O}_6 & \mathbf{O}_6 & \delta \mathbf{H}_3^T \end{bmatrix} \begin{Bmatrix} \bar{\mathbf{f}}_1 \\ \bar{\mathbf{f}}_2 \\ \bar{\mathbf{f}}_3 \end{Bmatrix} = \begin{Bmatrix} \delta \mathbf{H}_1^T \bar{\mathbf{f}}_1 \\ \delta \mathbf{H}_2^T \bar{\mathbf{f}}_2 \\ \delta \mathbf{H}_3^T \bar{\mathbf{f}}_3 \end{Bmatrix} \tag{9.179}$$

The (6 × 1) vector acting at node i is

$$\delta \mathbf{H}_i^T \bar{\mathbf{f}}_i = \begin{bmatrix} \mathbf{O}_3 & \mathbf{O}_3 \\ \mathbf{O}_3 & \delta \mathbf{\Lambda}_i^T \end{bmatrix} \begin{Bmatrix} \bar{\mathbf{n}}_i \\ \bar{\mathbf{m}}_i \end{Bmatrix} = \begin{Bmatrix} \mathbf{0}_{3x1} \\ \delta \mathbf{\Lambda}_i^T \bar{\mathbf{m}}_i \end{Bmatrix} \tag{9.180}$$

9.4.5.2 *Variation of Λ_i^T contracted with a nodal moment vector $\bar{\mathbf{m}}_i$*

From Equation 9.148, we recall the expression for $\mathbf{\Lambda}(\theta)$, the transformation from a nodal spin vector $\delta\boldsymbol{\omega}$ to the corresponding incremental nodal axial vector $\delta\boldsymbol{\theta}$, that is, $\delta\boldsymbol{\theta} = \mathbf{\Lambda}(\theta)\, \delta\boldsymbol{\omega}$. In this section, we temporarily omit nodal subscripts and the overbars on the nodal moment axial vector $\bar{\mathbf{m}}_i$ and the conjugate deformational axial rotation $\bar{\boldsymbol{\theta}}_i$.

The expression for $\mathbf{\Lambda}(\theta)$ given earlier is

$$\mathbf{\Lambda} = \left[\mathbf{I} - \frac{1}{2}\mathbf{N}_\theta + \eta(\theta)\mathbf{N}_\theta^2\right]$$

Therefore,

$$\delta \mathbf{\Lambda}^T \mathbf{m} = \frac{1}{2}\delta \mathbf{N}_\theta\ \mathbf{m} + \eta'\delta\theta\ \mathbf{N}_\theta^2\ \mathbf{m} + \eta\left[\mathbf{N}_\theta \delta \mathbf{N}_\theta\ \mathbf{m} + \delta \mathbf{N}_\theta \mathbf{N}_\theta\ \mathbf{m}\right] \tag{9.181}$$

The first term above can be rewritten as

$$\frac{1}{2}\delta \mathbf{N}_\theta \mathbf{m} = \frac{1}{2}\mathrm{Spin}(\delta\boldsymbol{\theta})\mathbf{m} = \left[-\frac{1}{2}\mathrm{Spin}(\mathbf{m})\right]\delta\boldsymbol{\theta}$$

The second term can be rewritten as

$$\eta'\delta\theta\ \mathbf{N}_\theta^2 \mathbf{m} = \nu\left[\mathrm{Spin}^2(\boldsymbol{\theta})\mathbf{m}\boldsymbol{\theta}^T\right]\delta\boldsymbol{\theta}, \quad \text{where } \nu \equiv \eta'/\theta$$

and the third term as

$$\eta\left[\mathbf{N}_\theta \delta \mathbf{N}_\theta \mathbf{m} + \delta \mathbf{N}_\theta \mathbf{N}_\theta \mathbf{m}\right] = \eta\left\{\left[(\mathbf{m}^T\boldsymbol{\theta})\mathbf{I} - \mathbf{m}\boldsymbol{\theta}^T\right] + \left[\boldsymbol{\theta}\mathbf{m}^T - \mathbf{m}\boldsymbol{\theta}^T\right]\right\}\delta\boldsymbol{\theta} = \left[(\mathbf{m}^T\boldsymbol{\theta})\mathbf{I} - 2\mathbf{m}\boldsymbol{\theta}^T + \boldsymbol{\theta}\mathbf{m}^T\right]\delta\boldsymbol{\theta}$$

Combining these three terms, we have

$$\delta\mathbf{\Lambda}^T\mathbf{m} = \left[-\frac{1}{2}\text{Spin}(\mathbf{m}) + \nu\text{Spin}^2(\boldsymbol{\theta})\mathbf{m}\boldsymbol{\theta}^T + \eta\left[(\mathbf{m}^T\boldsymbol{\theta})\mathbf{I} - 2\mathbf{m}\boldsymbol{\theta}^T + \boldsymbol{\theta}\mathbf{m}^T\right]\right]\delta\boldsymbol{\theta} \tag{9.182}$$

Now reintroducing the nodal labels i and the overbars to denote deformational element nodal variables, we may write this expression as

$$\delta\mathbf{\Lambda}_i^T\overline{\mathbf{m}}_i = \overline{\mathbf{M}}_i\delta\overline{\mathbf{w}}_i^R \tag{9.183}$$

in which we also use the relation $\delta\overline{\boldsymbol{\theta}}_i = \mathbf{\Lambda}_i(\overline{\theta}_i)\delta\overline{\mathbf{w}}^R{}_i$. The (3×3) nodal matrix $\overline{\mathbf{M}}_i$ (given by Nour-Omid and Rankin [1991]) is then

$$\overline{\mathbf{M}}_i \equiv \left[-\frac{1}{2}\text{Spin}(\overline{\mathbf{m}}_i) + \nu\text{Spin}^2(\overline{\boldsymbol{\theta}}_i)\overline{\mathbf{m}}_i\overline{\boldsymbol{\theta}}_i^T + \eta\left[(\overline{\mathbf{m}}_i^T\overline{\boldsymbol{\theta}}_i)\mathbf{I} - 2\overline{\mathbf{m}}_i\overline{\boldsymbol{\theta}}_i^T + \overline{\boldsymbol{\theta}}_i\overline{\mathbf{m}}_i^T\right]\right]\mathbf{\Lambda}_i(\overline{\theta}_i) \tag{9.184}$$

The coefficient η is given explicitly in Equation 9.150. The coefficient $\nu \equiv \eta'/\theta$ is

$$\nu(\theta) = \frac{\theta(\theta + \sin\theta) - 8\sin^2\frac{\theta}{2}}{4\theta^4\sin^2\frac{\theta}{2}} \tag{9.185}$$

For $\theta \to 0$, this expression approaches 0/0, as was the case with η. Haugen (1994) gives a three-term power series for ν from which it is found that $\nu \to 1/360$ as $\theta \to 0$.

The "moment correction geometric stiffness" can finally be expressed as

$$\delta\mathbf{I}_{GM} = \mathbf{E}\mathbf{P}^T\mathbf{M}\mathbf{P}\mathbf{E}^T\delta\mathbf{V} = \mathbf{K}_{GM}\delta\mathbf{V} \tag{9.186}$$

where the (18×18) block diagonal matrix $\mathbf{M} = \text{diag}\left[\mathbf{O}_3\overline{\mathbf{M}}_1\mathbf{O}_3\overline{\mathbf{M}}_2\mathbf{O}_3\overline{\mathbf{M}}_3\right]$.

9.4.5.3 Tangent stiffness summary

The CR element tangent stiffness $\mathbf{K}_t$ in global coordinates can be summarized as follows:

$$\mathbf{K}_t = \mathbf{K}_M + \mathbf{K}_G \tag{9.187}$$

where the symmetric material tangent stiffness $\mathbf{K}_M$ is

$$\mathbf{K}_M = \mathbf{E}\mathbf{P}^T\mathbf{H}^T[\overline{\mathbf{k}}]\mathbf{H}\mathbf{P}\mathbf{E}^T \tag{9.188}$$

and the (unsymmetric) geometric tangent stiffness $\mathbf{K}_G$ is composed of two contributions:

$$\mathbf{K}_G = \mathbf{K}_{GR} + \mathbf{K}_{GM} \tag{9.189}$$

The first contribution in this equation is the "rotational geometric stiffness" (in Haugen's terminology)

$$\mathbf{K}_{GR} = -\mathbf{E}[\mathbf{F}_{GR}\mathbf{G}]\mathbf{E}^T \tag{9.190}$$

It is a familiar effect, appearing in other formulations (e.g., TL and UDL). It accounts for the rigid body rotation in the global frame of the current nodal force vectors (due to the element frame rotation).

The second contribution is the "moment correction geometric stiffness," $\mathbf{K}_{GM}$:

$$\mathbf{K}_{GM} = \mathbf{E}[\mathbf{P}^T\mathbf{M}\mathbf{P}]\mathbf{E}^T \tag{9.191}$$

It involves only the rotations of the current nodal moment bivectors $\overline{\mathbf{m}}_i$.

As mentioned above, Haugen (1994) derived and developed a third contribution to the geometric stiffness, the "equilibrium projection geometric stiffness," which accounts for the effect of element deformations on the projection matrix **P**. We have opted to omit this last geometric stiffness term from our discussion as it is likely to be a quite small effect, given our assumption of infinitesimal strains.

9.4.6 Discussion of a CR versus UDL formulation

A convenient assumption that has been used in the rigorous treatment of CR moving frame kinematics (by the authors cited herein) is that element material behavior is elastic, or at least that a strain energy potential function exists. Strictly speaking, this precludes modeling of inelastic nonlinear material behavior in a CR formulation. As a result, published applications have mainly dealt with highly geometrically nonlinear problems, for example, cable hockling (Nour-Omid and Rankin 1991), often involving instability (e.g., snap-through and bifurcation buckling) and branching of equilibrium paths.

Despite this (perhaps only theoretical) limitation to elastic behavior, Skallerud and Haugen (1999) have incorporated a stress-resultant plasticity model for thin shells directly into a CR formulation.

A UDL formulation with the reference configuration updated at every increment to reflect an overall element rigid body motion would seem to produce the same advantages as a CR formulation, without some of the complexities introduced by a corotating element reference frame. In a UDL approach, the large 3D rigid body rotation of the element coordinate system is recognized and updated at every increment just as in a CR approach. The incremental displacements and rotations are expected to be small and to produce only small incremental deformations, again as in a CR formulation.

The significant difference is that in a UDL approach, the reference system remains constant for the current load step, and is only updated when a new equilibrium configuration is accepted.

The methodology that has been described in the foregoing sections for handling large 3D rotations would be directly applicable to a UDL approach.

Finally, it seems in retrospect that many of the early "CR" implementations were fundamentally UDL in concept.

Chapter 10

Structural elements

10.1 INTRODUCTION

In Chapter 8, we discussed nonlinear (total Lagrangian [TL] and updated Lagrangian [UDL]) finite element formulations for two-dimensional (2D) and three-dimensional (3D) solids. In this chapter, we turn our attention to structural elements, that is, beams, plates, and shells, which may be components in 2D or 3D structural systems. Curved roof shells, folded plates, long-span latticed roof structures, thin-walled open sections, box girders, heavy cables, piping systems, marine mooring lines, and risers are just a few of the many important application areas.

We focus our discussion here on an important subclass of nonlinear problems: ones involving large displacements, large rotations, and material nonlinearity, but small strains.

The geometrical form of the structural elements that we consider determines the dominant mechanisms by which they carry applied loads. Stand-alone beams and flat plates carry lateral distributed loads primarily by flexure. On the other hand, due to their surface curvature, shells are often able to carry lateral distributed loads primarily by in-plane membrane forces, except near supports or structural discontinuities (e.g., interconnections). Interconnection to other structural elements in a structural system in general activates combinations of load-carrying mechanisms in an element, for example, flexure–axial force–torsion in beams and columns, and bending and twisting moments and membrane forces in plates and shells. Therefore, the modeling of flexural response is essential for all these structural elements.

Because of their geometry (i.e., relatively small cross section and thickness dimensions compared with length and surface dimensions), reasonable idealizations can be introduced to relate through-thickness deformations to centerline or midsurface degrees of freedom (DOF). Analytically, this process amounts to imposing constraints on a 2D or 3D continuum, to reduce dimensionality and thereby simplify the problem to a reasonable level.

10.2 MODELING OF SHELL STRUCTURES

The classic thin-shell theory is formulated using coordinates, say (ξ, η, ζ), to define the position of a generic point in the shell. A point on the shell midsurface is defined by the Cartesian position vector $\mathbf{r}(\xi, \eta)$ in terms of the curvilinear surface coordinates (ξ, η). An associated generic point off the midsurface is defined by $\mathbf{r}(\xi, \eta) + \zeta\mathbf{n}$, where $\mathbf{n}$ is the unit normal to the midsurface. The midsurface tangent vectors $\mathbf{r}_{,\xi}$ and $\mathbf{r}_{,\eta}$ (which are not necessarily orthogonal), together with the normal vector $\mathbf{n}$, are the basis to which displacements and rotations are referred. Differential geometry is used to systematically characterize the curvatures of the surface. Even for simple shapes such as cylinders and spheres, the resulting partial differential equations of equilibrium are very often intractable, despite many simplifying assumptions. Moment equilibrium about the normal to the shell is identically satisfied in many of these theories due to these approximations, and does not provide an independent equilibrium equation. Therefore, in a finite element shell model there is no fundamental reason why a rotational DOF about the normal is needed. However, in a general-purpose finite element system it is most convenient (e.g., for coordinate transformations) to have the full 6 DOF (three displacements and three rotations) at each node.

We can identify two basic points of view, which are also applicable to finite element modeling in general: (1) use relatively low-order, simple, and therefore computationally inexpensive elements with as few DOF per node as possible, and achieve the desired accuracy via a fine mesh, or (2) use higher-order elements each with more nodal freedoms, together with a coarser mesh, to achieve the desired accuracy. Of course, accuracy of geometric modeling is an important factor in the trade-off. In the context of modeling of 2D solids, an extreme example of the first viewpoint is the simple three-node constant strain triangle (CST) element, in which strains (and stresses) are constant within an element. Examples of the second viewpoint are some of the higher-order serendipity or Lagrangian isoparametric elements mentioned in Chapter 8 or the hierarchical TUBA family of triangular plate elements (Argyris et al. 1968). In the case of 2D or 3D solids, the consensus seems to have settled somewhere in the middle, with eight-node serendipity or nine-node Lagrangian elements being the usual choice.

In the shell modeling context, these two basic schools of thought are again apparent:

1. Use a faceted model of the shell surface and a flat plate or membrane finite element (in which membrane and flexural behavior is uncoupled at the element level). The simplest example would be a triangular plate bending element with 9 DOF overlaid with a triangular membrane element with 6 DOF (thus no normal rotation DOF in the combined element). In some early applications along these lines, a "throwaway" normal rotation DOF was added for processing reasons, and then suppressed via a rotational spring element attached to the normal DOF, which generally proved to be unsatisfactory. Subsequently, triangular membrane elements with "drilling" DOF (rotations about the normal to the flat surface) have been proposed, in which the drilling DOF is intended not only to produce a flat shell element with the desired 6 DOF per node but also to improve the rather poor performance of the CST membrane element. Bending and membrane elements have often been developed using alternate variational principles (Horrigmoe 1978), rather than potential energy.
2. Use a curved finite element shell model, derived from a 3D isoparametric solid (Ahmad et al. 1970), to enable accurate modeling of an essentially arbitrarily shaped shell surface, based on nodal coordinate input data. This general process is usually termed "degeneration." The development, in a sense, follows the concepts of the classic shell theory described above, introducing (numerically) an orthogonal coordinate triad consisting of midsurface tangent vectors and a thickness direction "normal" vector. The resulting element has 5 DOF per node, three components of displacement and two in-plane rotations.

10.3 3D STRUCTURAL ELEMENT FORMULATIONS

We now briefly consider the formulation of 3D structural elements, that is, beams, plates, and shells. Following the format of our discussion in Chapter 8 of solid finite element formulations, we first focus on (linear) elastic material behavior with (moderate) geometric nonlinearity, and later discuss modeling of material nonlinearity. We implicitly assume here that large displacements and rotations (but small strains) are handled at the structural level via a UDL formulation. Of course, for small-strain, small-displacement applications, a TL formulation at the structural level is appropriate.

We note that an important distinguishing feature of UDL versus CR formulations is that the element tangent stiffness must now explicitly include a geometric stiffness matrix term. In addition, the geometric stiffness matrix finds extensive use in standard linearized buckling analyses.

10.3.1 Kirchhoff beam, plate, and shell finite elements

Flexural behavior of slender beam, plate, and shell elements is traditionally modeled using the Kirchhoff theory "normals remain normal" approximation. This allows deformations throughout the element to be fully described by midsurface displacements and rotations, greatly simplifying the analysis. As a result, stress resultants (bending moments and in-plane forces) rather than pointwise stresses appear in the formulation.

10.3.1.1 Flat plate and shell elements

We first consider a flat plate or shell finite element. Figure 10.1 shows a 2D differential element (dx × dy × h) and the local stresses acting on a thin lamina of thickness dz located at a distance z from the midsurface.

The Green strains in the x-y plane within the generic thin lamina (dx × dy × dz) are

$$\begin{cases} \bar{\varepsilon}_x = \bar{u}_{,x} + \dfrac{1}{2}\left(\bar{u}_{,x}^2 + \bar{v}_{,x}^2 + \bar{w}_{,x}^2\right) \\ \bar{\varepsilon}_y = \bar{v}_{,y} + \dfrac{1}{2}\left(\bar{u}_{,y}^2 + \bar{v}_{,y}^2 + \bar{w}_{,y}^2\right) \\ \bar{\gamma}_{xy} = \bar{u}_{,y} + \bar{v}_{,x} + \left(\bar{u}_{,x}\bar{u}_{,y} + \bar{v}_{,x}\bar{v}_{,y} + \bar{w}_{,x}\bar{w}_{,y}\right) \end{cases} \tag{10.1}$$

where we denote $\bar{\gamma}_{xy} = 2\bar{\varepsilon}_{xy}$.

For small strains and moderate rotations ($\bar{u}_{,x}^2 \ll |\bar{u}_{,x}|$, etc.), these expressions reduce to

$$\begin{cases} \bar{\varepsilon}_x = \bar{u}_{,x} + \dfrac{1}{2}\bar{w}_{,x}^2 \\ \bar{\varepsilon}_y = \bar{v}_{,y} + \dfrac{1}{2}\bar{w}_{,y}^2 \\ \bar{\gamma}_{xy} = \bar{u}_{,y} + \bar{v}_{,x} + \bar{w}_{,x}\bar{w}_{,y} \end{cases} \tag{10.2}$$

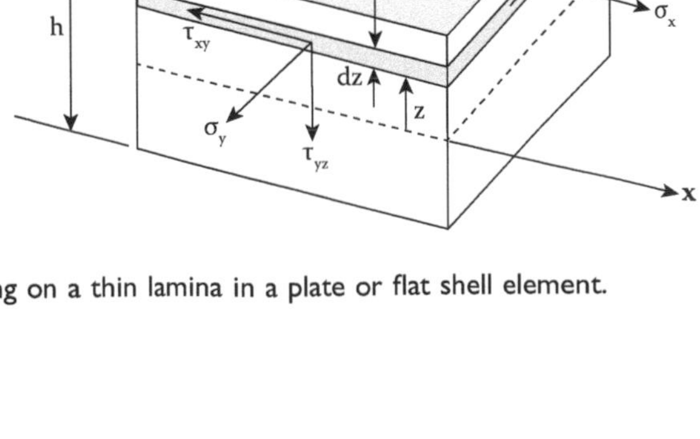

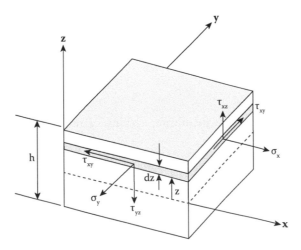

Figure 10.1 Stresses acting on a thin lamina in a plate or flat shell element.

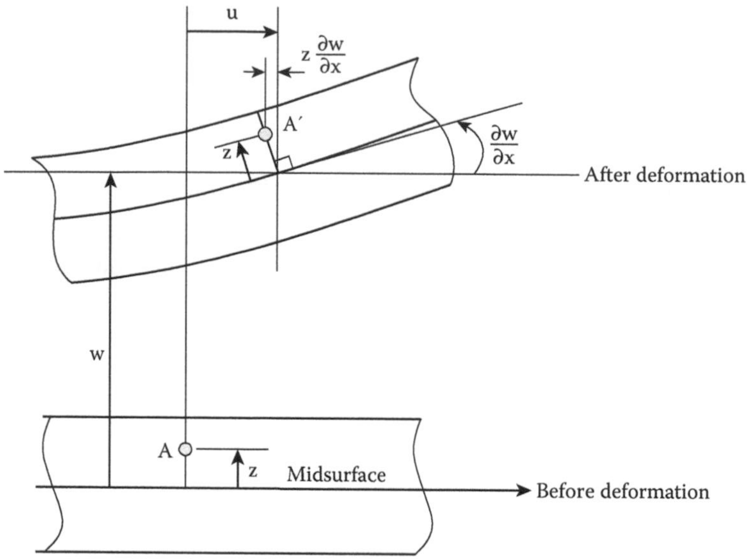

Figure 10.2 Kirchhoff hypothesis—normal rotation equal to midsurface rotation.

The in-plane displacements of the thin lamina are related to the midsurface variables using the Kirchhoff kinematic constraints illustrated in Figure 10.2, that is,

$$\begin{cases} \bar{u} = u - z\ w_{,x} \\ \bar{v} = v - z\ w_{,y} \\ \bar{w} = w \end{cases} \tag{10.3}$$

where u, v, w are displacements of the midsurface. These constraints guarantee that the normal remains straight, unextended, and perpendicular to the rotated midsurface tangent. (The Mindlin plate theory, discussed in Section 10.3.2, allows a simplified representation of transverse shear deformation, so that the normal is no longer assumed to be perpendicular to the rotated midsurface tangent.)

We first note that with $\bar{w} = w$, Equation 10.2 becomes

$$\begin{cases} \bar{\varepsilon}_x = \bar{u}_{,x} + \dfrac{1}{2} w_{,x}^2 \\ \bar{\varepsilon}_y = \bar{v}_{,y} + \dfrac{1}{2} w_{,y}^2 \\ \bar{\gamma}_{xy} = \bar{u}_{,y} + \bar{v}_{,x} + w_{,x} w_{,y} \end{cases} \tag{10.4}$$

that is, the nonlinear terms in the membrane strains of the generic lamina are *midsurface* rotations.

Now applying the first two constraints of Equations 10.3, we find the strains in the generic lamina in terms of midsurface variables as

$$\begin{cases} \bar{\varepsilon}_x = \varepsilon_x - z w_{,xx} \\ \bar{\varepsilon}_y = \varepsilon_y - z w_{,yy} \\ \bar{\gamma}_{xy} = \gamma_{xy} - 2z w_{,xy} \end{cases} \tag{10.5}$$

where the midsurface strains are nonlinear functions of the displacement gradients:

$$\begin{cases} \varepsilon_x = u_{,x} + \dfrac{1}{2} w_{,x}^2 \\ \varepsilon_y = v_{,y} + \dfrac{1}{2} w_{,y}^2 \\ \gamma_{xy} = u_{,y} + v_{,x} + w_x w_y \end{cases} \tag{10.6}$$

It is interesting to note in passing that the 2D compatibility equation for the midsurface strains becomes

$$\frac{\partial^2 \varepsilon_x}{\partial y^2} - \frac{\partial^2 \gamma_{xy}}{\partial x \partial y} + \frac{\partial^2 \varepsilon_y}{\partial x^2} = w_{,xy}^2 - w_{,xx} w_{,yy} \tag{10.7}$$

for the slightly curved midsurface. Of course, for a flat 2D surface the right-hand side of this equation is zero.

If the generic lamina is assumed to be in a state of plane stress with isotropic elastic properties, then

$$\begin{Bmatrix} \bar{\sigma}_x \\ \bar{\sigma}_y \\ \bar{\tau}_{xy} \end{Bmatrix} = \frac{E}{1-\nu^2} \begin{bmatrix} 1 & \nu & 0 \\ \nu & 1 & 0 \\ 0 & 0 & (1-\nu)/2 \end{bmatrix} \begin{Bmatrix} \bar{\varepsilon}_x \\ \bar{\varepsilon}_y \\ \bar{\gamma}_{xy} \end{Bmatrix} \tag{10.8}$$

Using Equations 10.5 in this last equation, we find

$$\begin{Bmatrix} \bar{\sigma}_x \\ \bar{\sigma}_y \\ \bar{\tau}_{xy} \end{Bmatrix} = \mathbf{E} \begin{Bmatrix} \varepsilon_x \\ \varepsilon_y \\ \gamma_{xy} \end{Bmatrix} - z\mathbf{E} \begin{Bmatrix} w_{,xx} \\ w_{,yy} \\ 2w_{,xy} \end{Bmatrix} \tag{10.9}$$

where

$$\mathbf{E} \equiv \frac{E}{1-\nu^2} \begin{bmatrix} 1 & \nu & 0 \\ \nu & 1 & 0 \\ 0 & 0 & (1-\nu)/2 \end{bmatrix} \tag{10.10}$$

The membrane stress resultants (per unit length of midsurface) are defined as

$$\mathbf{N} \equiv \begin{Bmatrix} N_x \\ N_y \\ N_{xy} \end{Bmatrix} = \int_{-h/2}^{h/2} \begin{Bmatrix} \bar{\sigma}_x \\ \bar{\sigma}_y \\ \bar{\tau}_{xy} \end{Bmatrix} dz = \mathbf{E}h \left(\begin{Bmatrix} \varepsilon_x \\ \varepsilon_y \\ \gamma_{xy} \end{Bmatrix}^{l} + \begin{Bmatrix} \varepsilon_x \\ \varepsilon_y \\ \gamma_{xy} \end{Bmatrix}^{nl} \right) \tag{10.11}$$

where

$$\begin{Bmatrix} \varepsilon_x \\ \varepsilon_y \\ \gamma_{xy} \end{Bmatrix}^{l} \equiv \begin{Bmatrix} u_{,x} \\ v_{,y} \\ u_{,y} + v_{,x} \end{Bmatrix} \quad \text{and} \quad \begin{Bmatrix} \varepsilon_x \\ \varepsilon_y \\ \gamma_{xy} \end{Bmatrix}^{nl} \equiv \begin{Bmatrix} \dfrac{1}{2} w_{,x}^2 \\ \dfrac{1}{2} w_{,y}^2 \\ w_{,x} w_{,y} \end{Bmatrix} \tag{10.12}$$

The bending moments (per unit length of midsurface) are defined as

$$\mathbf{M} \equiv \begin{Bmatrix} M_x \\ M_y \\ M_{xy} \end{Bmatrix} = \int_{-h/2}^{h/2} -z \begin{Bmatrix} \bar{\sigma}_x \\ \bar{\sigma}_y \\ \bar{\tau}_{xy} \end{Bmatrix} dz = \mathbf{E}\frac{h^3}{12} \begin{Bmatrix} w_{,xx} \\ w_{,yy} \\ 2w_{,xy} \end{Bmatrix} \tag{10.13}$$

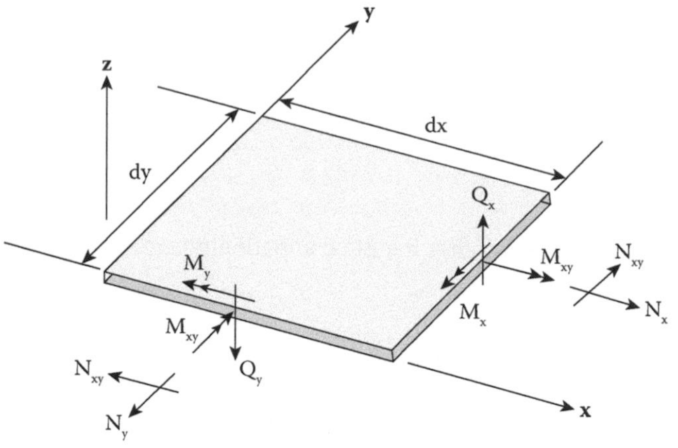

Figure 10.3 Sign conventions for bending moments and in-plane forces.

which can be written as

$$\mathbf{M} = \mathbf{D}\boldsymbol{\kappa} \tag{10.14}$$

where the curvature (tensor), written as a vector, is

$$\boldsymbol{\kappa} \equiv \begin{Bmatrix} w_{,xx} \\ w_{,yy} \\ 2w_{,xy} \end{Bmatrix} \tag{10.15}$$

and

$$\mathbf{D} = D \begin{bmatrix} 1 & \nu & 0 \\ \nu & 1 & 0 \\ 0 & 0 & (1-\nu)/2 \end{bmatrix} \tag{10.16}$$

In Equation 10.16, the scalar $D = Eh^3/12(1-\nu^2)$ is the plate bending stiffness, analogous to the flexural stiffness EI for a beam. We note that whereas the in-plane membrane strains are nonlinear, the curvatures are linear.

The sign convention for the stress resultants is shown in Figure 10.3.

Without specifying element details, we can use virtual work to investigate the form of the geometric stiffness. Considering a differential element ($dx \times dy \times h$), the internal virtual work of the membrane forces can be written as

$$-\delta W_{int}^N = N_x \delta\varepsilon_x + N_{xy} \delta\,\gamma_{xy} + N_y \delta\varepsilon_y \tag{10.17}$$

The linear portion of the in-plane virtual strain leads to the standard 2D equilibrium equations for the membrane stress resultants. Keeping only the *nonlinear membrane strain*, we obtain

$$-\delta W_{int}^N = N_x w_{,x} \delta w_{,x} + N_{xy}(w_{,x}\delta w_{,y} + w_{,y}\delta w_{,x}) + N_y w_{,y} \delta w_{,y} \tag{10.18}$$

We now integrate this relation twice by parts, discarding the "boundary" values at each stage to find

$$-\delta W_{int}^N = (N_x w_{,xx} + 2N_{xy} w_{,xy} + N_y w_{,yy})\delta w \tag{10.19}$$

The internal virtual work of the bending moments can be written as

$$-\delta W_{int}^{M} = M_x\delta w_{,xx} + 2M_{xy}\delta w_{,xy} + M_y\delta w_{,yy} \tag{10.20}$$

Again, integrating twice by parts, and discarding boundary values, we find

$$-\delta W_{int}^{M} = (M_{x,xx} + 2M_{xy,xy} + M_{y,yy})\delta w \tag{10.21}$$

which provides the coupled equilibrium equation (with distributed load p from external virtual work)

$$D\nabla^4 w = p + N_x w_{,xx} + 2N_{xy} w_{,xy} + N_y w_{,yy} \tag{10.22}$$

where ∇^4 is the biharmonic operator

$$\nabla^4 w \equiv \frac{\partial^4 w}{\partial x^4} + 2\frac{\partial^4 w}{\partial x^2 \partial y^2} + \frac{\partial^4 w}{\partial y^4} \tag{10.23}$$

We of course do not seek to satisfy this equation pointwise in a finite element formulation.

10.3.1.2 General form of geometric stiffness

We now consider the geometric stiffness of a *finite* volume of material (i.e., the geometric stiffness matrix of a finite element). We adopt a notation here that is familiar from the TL formulation for solids in Chapter 8. Accordingly, let the displacement gradients be denoted by $w_{,x} \equiv \theta_x$ and $w_{,y} \equiv \theta_y$. The displacement gradients are related to the element nodal DOF by a matrix **G**. This relationship can bc represented in general as

$$\boldsymbol{\theta} = \mathbf{G}\mathbf{u} \tag{10.24}$$

where

$$\boldsymbol{\theta} \equiv \begin{Bmatrix} \theta_x \\ \theta_y \end{Bmatrix} \tag{10.25}$$

G is (2 × N) involving derivatives of the element shape functions, and the nodal displacement vector **u** is (N × 1). We explicitly specify neither the element DOF nor the element shape functions here. The nonlinear membrane strains can be written as

$$\boldsymbol{\varepsilon}^{nl} = \frac{1}{2}\begin{bmatrix} \theta_x & 0 \\ 0 & \theta_y \\ \theta_y & \theta_x \end{bmatrix}\begin{Bmatrix} \theta_x \\ \theta_y \end{Bmatrix} \equiv \frac{1}{2}\mathbf{A}\boldsymbol{\theta} = \frac{1}{2}\mathbf{A}\mathbf{G}\mathbf{u} \tag{10.26}$$

where the (3 × 2) matrix **A** is defined as

$$\mathbf{A} \equiv \begin{bmatrix} \theta_x & 0 \\ 0 & \theta_y \\ \theta_y & \theta_x \end{bmatrix} \tag{10.27}$$

The time (material) rate of the nonlinear membrane strain is

$$\dot{\boldsymbol{\varepsilon}}^{nl} = \frac{1}{2}(\dot{\mathbf{A}}\theta + \mathbf{A}\dot{\theta}) \equiv \mathbf{A}\dot{\theta} \tag{10.28}$$

which provides the equivalent virtual strain relationship

$$\delta\boldsymbol{\varepsilon}^{nl} = \mathbf{AG}\delta\mathbf{u} \tag{10.29}$$

The corresponding internal virtual work of the membrane forces is

$$\delta\boldsymbol{\varepsilon}^{nl,T}\mathbf{N} = \delta\mathbf{u}^T\mathbf{G}^T\mathbf{A}^T\mathbf{N} \tag{10.30}$$

The term $\mathbf{A}^T\mathbf{N}$ can be transformed by rearranging the membrane forces in matrix form as follows:

$$\mathbf{A}^T\mathbf{N} = \begin{bmatrix} \theta_x & 0 & \theta_y \\ 0 & \theta_y & \theta_x \end{bmatrix} \begin{Bmatrix} N_x \\ N_y \\ N_{xy} \end{Bmatrix} = \begin{bmatrix} N_x & N_{xy} \\ N_{xy} & N_y \end{bmatrix} \begin{Bmatrix} \theta_x \\ \theta_y \end{Bmatrix} = \tilde{\mathbf{N}}\boldsymbol{\theta} \tag{10.31}$$

Equation 10.30 then gives

$$\delta\boldsymbol{\varepsilon}^{nl,T}\mathbf{N} = \delta\mathbf{u}^T\mathbf{G}^T\tilde{\mathbf{N}}\mathbf{G}\mathbf{u} \tag{10.32}$$

We integrate over the element area a to find the geometric stiffness matrix $\mathbf{k}_G$,

$$\int_a \delta\boldsymbol{\varepsilon}^{nl,T}\mathbf{N}da = \delta\mathbf{u}^T\mathbf{k}_G\mathbf{u} \tag{10.33}$$

where

$$\mathbf{k}_G = \int_a \mathbf{G}^T\tilde{\mathbf{N}}\mathbf{G}da \qquad \tilde{\mathbf{N}} \equiv \begin{bmatrix} N_x & N_{xy} \\ N_{xy} & N_y \end{bmatrix} \tag{10.34}$$

10.3.1.3 Remarks

1. For displacement continuity across element boundaries, the normals of adjacent elements must match. For Kirchhoff elements, this requires interelement continuity of midsurface rotations, a well-known difficulty for plate and shell elements.
2. There is an apparent contradiction between the Kirchhoff hypothesis (via which the strain ε_z normal to the midsurface is assumed to be zero) and the assumption of plane stress (which implies that the stress σ_z normal to the midsurface is zero). It is perhaps logically more satisfying to consider plane stress as an approximation in which the effect of the relatively small normal stress σ_z [compared with the in-plane stresses (σ_x, σ_y)] is ignored.

10.3.1.4 Synthesis of space frame stiffness matrices

A standard 3D 12 DOF space frame element is shown in Figure 10.4. The (12 × 12) linear elastic stiffness matrix **k** for an element with a doubly symmetric cross section can be obtained by overlaying four simpler elements, each one representing an uncoupled mode of behavior in

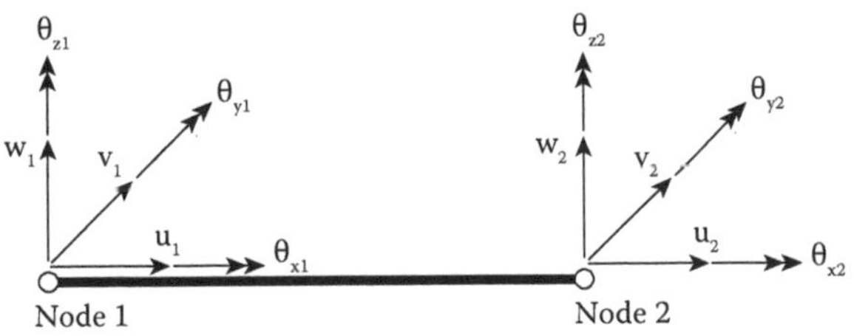

Figure 10.4 Space frame element nodal freedoms in element local reference system.

the local element x-y-z coordinate system: independent bending in the x-z and y-z planes, axial elongation, and torsional twisting about the axial x direction.

The (12 × 12) linear elastic (small displacement) stiffness matrix **k** synthesized in this manner is shown below in partitioned block matrix form. As per the usual definitions, a (6 × 6) block matrix $\mathbf{k}_{ij}$ gives the contribution to the internal force vector at node i of unit nodal freedoms at node j. Thus, $\mathbf{I}_i = \mathbf{k}_{ij}\mathbf{u}_j$, where

$$\mathbf{k} = \begin{bmatrix} \mathbf{k}_{11} & \mathbf{k}_{12} \\ \mathbf{k}_{21} & \mathbf{k}_{22} \end{bmatrix} \tag{10.35}$$

$$\mathbf{k}_{11} = \begin{bmatrix} \frac{AE}{L} & o & o & o & o & o \\ o & 12\frac{EI_z}{L^3} & o & o & o & 6\frac{EI_z}{L^2} \\ o & o & 12\frac{EI_y}{L^3} & o & -6\frac{EI_y}{L^2} & o \\ o & o & o & \frac{GJ}{L} & o & o \\ o & o & -6\frac{EI_y}{L^2} & o & 4\frac{EI_y}{L} & o \\ o & 6\frac{EI_z}{L^2} & o & o & o & 4\frac{EI_z}{L} \end{bmatrix} \tag{10.36}$$

$$\mathbf{k}_{21} = \begin{bmatrix} -\frac{AE}{L} & o & o & o & o & o \\ o & -12\frac{EI_z}{L^3} & o & o & o & -6\frac{EI_z}{L^2} \\ o & o & -12\frac{EI_y}{L^3} & o & 6\frac{EI_y}{L^2} & o \\ o & o & o & -\frac{GJ}{L} & o & o \\ o & o & -6\frac{EI_y}{L^2} & o & 2\frac{EI_y}{L} & o \\ o & 6\frac{EI_z}{L^2} & o & o & o & 2\frac{EI_z}{L} \end{bmatrix} \tag{10.37}$$

$$
\mathbf{k}_{12} = \begin{bmatrix} -\frac{AE}{L} & 0 & 0 & 0 & 0 & 0 \\ 0 & -12\frac{EI_z}{L^3} & 0 & 0 & 0 & 6\frac{EI_z}{L^2} \\ 0 & 0 & -12\frac{EI_y}{L^3} & 0 & -6\frac{EI_y}{L^2} & 0 \\ 0 & 0 & 0 & -\frac{GJ}{L} & 0 & 0 \\ 0 & 0 & 6\frac{EI_y}{L^2} & 0 & 2\frac{EI_y}{L} & 0 \\ 0 & -6\frac{EI_z}{L^2} & 0 & 0 & 0 & 2\frac{EI_z}{L} \end{bmatrix} \tag{10.38}
$$

$$
\mathbf{k}_{22} = \begin{bmatrix} \frac{AE}{L} & 0 & 0 & 0 & 0 & 0 \\ 0 & 12\frac{EI_z}{L^3} & 0 & 0 & 0 & -6\frac{EI_z}{L^2} \\ 0 & 0 & 12\frac{EI_y}{L^3} & 0 & 6\frac{EI_y}{L^2} & 0 \\ 0 & 0 & 0 & \frac{GJ}{L} & 0 & 0 \\ 0 & 0 & 6\frac{EI_y}{L^2} & 0 & 4\frac{EI_y}{L} & 0 \\ 0 & -6\frac{EI_z}{L^2} & 0 & 0 & 0 & 4\frac{EI_z}{L} \end{bmatrix} \tag{10.39}
$$

10.3.1.5 Space frame geometric stiffness

We proceed in the same way as for the plate and flat shell Kirchhoff elements, by first approximating the Green strain expression for the axial strain and then evaluating the virtual work done on the axial force by the nonlinear portion of the axial strain.

Figure 10.5 is a schematic view of a space frame cross section, shown as a solid rectangle only for illustration purposes. We assume for simplicity that the cross section is compact and is doubly symmetric with y and z axes of symmetry. We do not therefore consider, for example, warping torsion effects in thin-walled open sections.

The axial Green strain of the elemental fiber da in general is

$$
\bar{\varepsilon}_x = \bar{u}_{,x} + \frac{1}{2}(\bar{u}_{,x}^2 + \bar{v}_{,x}^2 + \bar{w}_{,x}^2) \tag{10.40}
$$

which, with $\bar{u}_{,x}^2 \ll |\bar{u}_{,x}|$, simplifies to

$$
\bar{\varepsilon}_x = \bar{u}_{,x} + \frac{1}{2}(\bar{v}_{,x}^2 + \bar{w}_{,x}^2) \tag{10.41}
$$

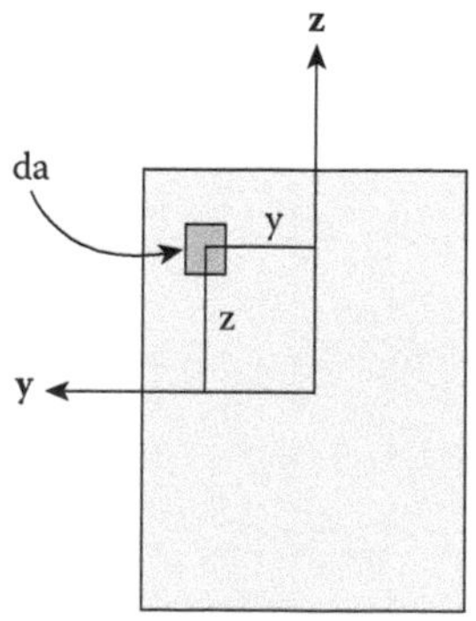

Figure 10.5 Space frame cross section (schematic).

Here, the Kirchhoff constraint becomes

$$\begin{cases} \bar{u} = u - yv_{,x} - zw_{,x} \\ \bar{v} = v \\ \bar{w} = w \end{cases} \quad (10.42)$$

Incorporating first the latter two of these conditions, we write Equation 10.41 in terms of the centerline rotations as

$$\bar{\varepsilon}_x = \bar{u}_{,x} + \frac{1}{2}(v_{,x}^2 + w_{,x}^2) \quad (10.43)$$

and then using the first condition, we have

$$\bar{\varepsilon}_x = u_{,x} - yv_{,xx} - zw_{,xx} + \frac{1}{2}(v_{,x}^2 + w_{,x}^2) = \varepsilon_x - yv_{,xx} - zw_{,xx} \quad (10.44)$$

As before, the strain of the reference surface or axis consists of a linear portion and a nonlinear portion:

$$\begin{cases} \varepsilon_x^l = u_{,x} \\ \varepsilon_x^{nl} = \dfrac{1}{2}(v_{,x}^2 + w_{,x}^2) \end{cases} \quad (10.45)$$

The (axial) fiber stress is

$$\bar{\sigma}_x = E\bar{\varepsilon}_x = E(\varepsilon_x - yv_{,xx} - zw_{,xx}) \quad (10.46)$$

The internal bending moments in the xy and xz planes are defined as

$$\begin{cases} M_y = -\displaystyle\int_a y\bar{\sigma}_x da \\ M_z = -\displaystyle\int_a z\bar{\sigma}_x da \end{cases} \quad (10.47)$$

yielding the uncoupled moment–curvature relations

$$\begin{cases} M_y = EI_y v_{,xx} \\ M_z = EI_z w_{,xx} \end{cases} \quad (10.48)$$

where

$$\begin{cases} I_y \equiv \int_a y^2 da \\ I_z \equiv \int_a z^2 da \\ I_{yz} \equiv \int_a yz da = 0 \end{cases} \quad (10.49)$$

The axial force (tension positive) is

$$P = \int_a \bar{\sigma}_x da = E \int_a \bar{\varepsilon}_x da = EA\varepsilon_x \tag{10.50}$$

The axial strain $\varepsilon_x \equiv \varepsilon_x^l + \varepsilon_x^{nl}$, where the nonlinear contribution can be written as

$$\varepsilon_x^{nl} = \frac{1}{2}\mathbf{A\theta} \tag{10.51}$$

where

$$\mathbf{A} \equiv [v_{,x} w_{,x}], \text{ and } \boldsymbol{\theta} \equiv \begin{Bmatrix} v_{,x} \\ w_{,x} \end{Bmatrix} \tag{10.52}$$

As before, we take the (time) material rate of ε_x^{nl} to find

$$\dot{\varepsilon}_x^{nl} = \frac{1}{2}(\dot{\mathbf{A}}\theta + \mathbf{A}\dot{\boldsymbol{\theta}}) = \mathbf{A}\dot{\boldsymbol{\theta}} \tag{10.53}$$

The virtual (nonlinear) strain is then

$$\delta\varepsilon_x^{nl} = \mathbf{A}\delta\boldsymbol{\theta} \tag{10.54}$$

We now need to establish the relation, stemming from the element shape functions, between the displacement gradients $\boldsymbol{\theta}$ and the element nodal displacements $\mathbf{u}$.

Figure 10.6 shows the 6 DOF subsets of the space frame element that are involved in the two bending modes. The two bending modes share only the axial DOF u_1 and u_2, which are involved in the determination of the axial force P. The displacement gradient $v_{,x}$ is completely determined from the nodal freedoms (v_1, θ_{z1}, v_2, θ_{z2}), while the displacement gradient $w_{,x}$ is completely determined from the nodal freedoms (w_1, θ_{y1}, w_2, θ_{y2}). Therefore, the displacement gradients are totally uncoupled. This means that the geometric stiffness for the space frame element can also be synthesized by simply overlaying the uncoupled geometric stiffness matrices for planar bending in the xy and xz planes.

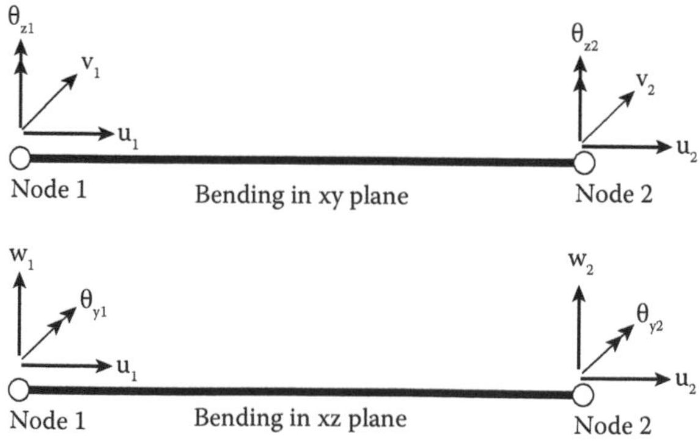

Figure 10.6 DOFs associated with bending in xy and xz planes.

10.3.1.6 *Planar bending*

We consider uncoupled bending in the xy plane (Figure 10.7).

The displacement v(x) is interpolated with cubic Hermitian shape functions that use nodal values of derivatives of the variable, as well as nodal values of the variable itself.

$$v(x) = N_1(x)v_1 + N_2(x)v_1' + N_3(x)v_2 + N_4(x)v_2' \tag{10.55}$$

where v_1 denotes $v(0)$, v_1' denotes dv/dx at $x = 0$, and so forth. The shape functions are listed and displayed graphically in Figure 10.8. The axial displacement is interpolated with linear shape functions

$$u(x) = \left(\frac{1-x}{L}\right)u_1 + \left(\frac{x}{L}\right)u_2 \tag{10.56}$$

but we will not be concerned with the interpolation of the axial displacement here.

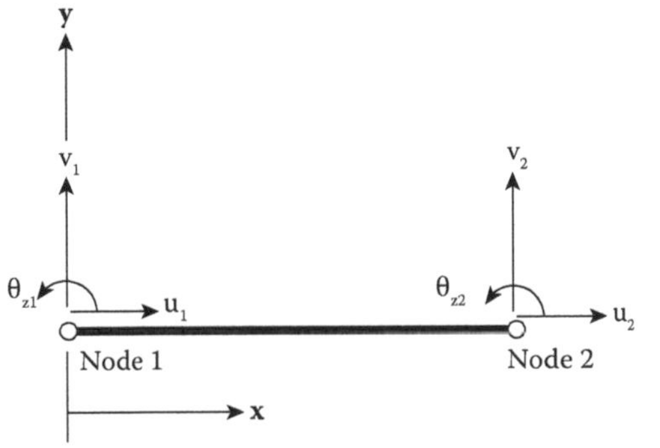

Figure 10.7 Bending in the xy plane.

$N_1 = (1 - 3\xi^2 + 2\xi^3)$

$N_2 = L(\xi - 2\xi^2 + \xi^3)$

$N_3 = (3\xi^2 - 2\xi^3)$

$N_4 = L(-\xi^2 + \xi^3)$

$\xi \equiv \dfrac{x}{L}$

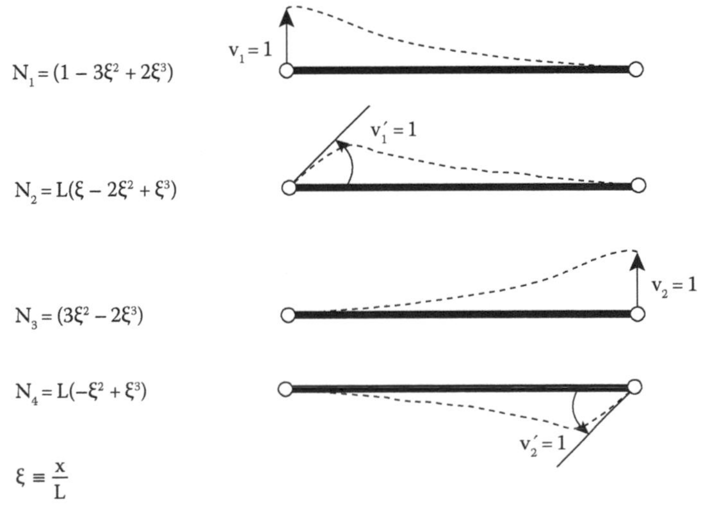

Figure 10.8 Cubic Hermitian shape functions for beam bending.

For bending in the xy plane, the nonlinear axial strain can be written simply as

$$\varepsilon_x^{nl} = \frac{1}{2}\theta^T\theta \tag{10.57}$$

where the scalar $\theta \equiv v_{,x}$. The virtual (nonlinear) axial strain is then

$$\delta\varepsilon_x^{nl} = \theta^T\delta\theta \tag{10.58}$$

Denoting the end rotations θ_{z1} and θ_{z2} as v_1' and v_2', as in Equation 10.55, we find the displacement gradient θ as

$$\theta = [N_1' \quad N_2' \quad N_3' \quad N_4'] \begin{Bmatrix} v_1 \\ v_1' \\ v_2 \\ v_2' \end{Bmatrix} = \mathbf{Gv} \tag{10.59}$$

in terms of the (4 × 1) nodal displacement vector **v**, which will subsequently be mapped onto the 12 DOF nodal displacement vector **u** of the space frame element. Note that we do not include the axial DOF u_1 and u_2 in the vector **v**.

The derivatives of the Hermitian shape functions appearing in this equation are

$$\begin{cases} N_1' = (1/L)(-6\xi + 6\xi^2) \\ N_2' = (1 - 4\xi + 3\xi^2) \\ N_3' = (1/L)(6\xi - 6\xi^2) \\ N_4' = (-2\xi + 3\xi^2) \end{cases} \tag{10.60}$$

where N_i' denotes dN_i/dx and $\xi \equiv x/L$.

The virtual work of the axial stress due to the nonlinear portion of the axial strain is then

$$\int_0^L [\delta\varepsilon_x^{nl,T} P] \, dx = \int_0^L \delta\theta^T P\theta \, dx = \delta v^T \left[\int_0^L G^T P G dx \right] v \tag{10.61}$$

where the axial force

$$P \equiv \int_a \bar{\sigma}_x da \tag{10.62}$$

is constant over the length of the element.

The (4 × 4) geometric stiffness of the flexural element lying in the xy plane is

$$k_G = P \int_0^L G^T G \, dx \tag{10.63}$$

Evaluation of the required integrals of products of shape function derivatives given in Equation 10.60 yields

$$k_G = \frac{P}{L}\begin{bmatrix} \frac{6}{5} & \frac{L}{10} & -\frac{6}{5} & \frac{L}{10} \\ \frac{L}{10} & \frac{2L^2}{15} & -\frac{L}{10} & -\frac{L^2}{30} \\ -\frac{6}{5} & -\frac{L}{10} & \frac{6}{5} & -\frac{L}{10} \\ \frac{L}{10} & -\frac{L^2}{30} & -\frac{L}{10} & \frac{2L^2}{15} \end{bmatrix} \tag{10.64}$$

We can now synthesize the space frame geometric stiffness by overlaying the geometric stiffness matrices for xy and xz flexure as described earlier. The 6 × 6 block submatrices constituting the consistent geometric stiffness for the space frame element are as follows:

$$k_{G11} = \frac{P}{L}\begin{bmatrix} 0 & 0 & 0 & 0 & 0 & 0 \\ 0 & \frac{6}{5} & 0 & 0 & 0 & \frac{L}{10} \\ 0 & 0 & \frac{6}{5} & 0 & -\frac{L}{10} & 0 \\ 0 & 0 & 0 & 0 & 0 & 0 \\ 0 & 0 & -\frac{L}{10} & 0 & \frac{2L^2}{15} & 0 \\ 0 & \frac{L}{10} & 0 & 0 & 0 & \frac{L^2}{10} \end{bmatrix} \tag{10.65}$$

$$k_{G21} = \frac{P}{L}\begin{bmatrix} 0 & 0 & 0 & 0 & 0 & 0 \\ 0 & -\frac{6}{5} & 0 & 0 & 0 & -\frac{L}{10} \\ 0 & 0 & -\frac{6}{5} & 0 & \frac{L}{10} & 0 \\ 0 & 0 & 0 & 0 & 0 & 0 \\ 0 & 0 & -\frac{L}{10} & 0 & -\frac{L^2}{30} & 0 \\ 0 & \frac{L}{10} & 0 & 0 & 0 & -\frac{L^2}{30} \end{bmatrix} \tag{10.66}$$

$$k_{G12} = \frac{P}{L}\begin{bmatrix} 0 & 0 & 0 & 0 & 0 & 0 \\ 0 & -\frac{6}{5} & 0 & 0 & 0 & \frac{L}{10} \\ 0 & 0 & -\frac{6}{5} & 0 & -\frac{L}{10} & 0 \\ 0 & 0 & 0 & 0 & 0 & 0 \\ 0 & 0 & \frac{L}{10} & 0 & -\frac{L^2}{30}\,0 & 0 \\ 0 & -\frac{L^2}{30} & 0 & 0 & 0 & -\frac{L^2}{30} \end{bmatrix} \tag{10.67}$$

$$\mathbf{k}_{G22} = \frac{P}{L}\begin{bmatrix} 0 & 0 & 0 & 0 & 0 & 0 \\ 0 & \frac{6}{5} & 0 & 0 & 0 & -\frac{L}{10} \\ 0 & 0 & \frac{6}{5} & 0 & \frac{L}{10} & 0 \\ 0 & 0 & 0 & 0 & 0 & 0 \\ 0 & 0 & \frac{L}{10} & 0 & \frac{2L^2}{15} & 0 \\ 0 & -\frac{L}{10} & 0 & 0 & 0 & \frac{2L^2}{15} \end{bmatrix} \tag{10.68}$$

10.3.1.7 Plane frame stiffness matrices

For later reference, we explicitly display here the plane frame linear and geometric stiffness matrices, which are easily obtained by extracting the appropriate terms from the (6 × 6) nodal block matrices given above for the space frame element. The (6 × 1) nodal displacement vector for a plane frame element lying in the xy plane is

$$\mathbf{u} \equiv \begin{bmatrix} u_1 & v_1 & \theta_{z1} & u_2 & v_2 & \theta_{z2} \end{bmatrix}^T \tag{10.69}$$

The (6 × 6) plane frame linear and geometric stiffness matrices obtained in this way are

$$\mathbf{k}_E = \begin{bmatrix} \frac{AE}{L} & 0 & 0 & -\frac{AE}{L} & 0 & 0 \\ 0 & \frac{12EI_y}{L^3} & \frac{6EI_y}{L^2} & 0 & -\frac{12EI_y}{L^3} & \frac{6EI_y}{L^2} \\ 0 & \frac{6EI_y}{L^2} & \frac{4EI_y}{L} & 0 & -\frac{6EI_y}{L^2} & \frac{2EI_y}{L} \\ -\frac{AE}{L} & 0 & 0 & \frac{AE}{L} & 0 & 0 \\ 0 & -\frac{12EI_y}{L^3} & -\frac{6EI_y}{L^2} & 0 & \frac{12EI_y}{L^3} & -\frac{6EI_y}{L^2} \\ 0 & \frac{6EI_y}{L^2} & \frac{2EI_y}{L} & 0 & -\frac{6EI_y}{L^2} & \frac{4EI_y}{L} \end{bmatrix} \tag{10.70}$$

$$\mathbf{k}_G = \frac{P}{L}\begin{bmatrix} 0 & 0 & 0 & 0 & 0 & 0 \\ 0 & \frac{6}{5} & \frac{L}{10} & 0 & -\frac{6}{5} & \frac{L}{10} \\ 0 & \frac{L}{10} & \frac{2L^2}{15} & 0 & -\frac{L}{10} & -\frac{L^2}{30} \\ 0 & 0 & 0 & 0 & 0 & 0 \\ 0 & -\frac{6}{5} & -\frac{L}{10} & 0 & \frac{6}{5} & -\frac{L}{10} \\ 0 & \frac{L}{10} & -\frac{L^2}{30} & 0 & -\frac{L}{10} & \frac{2L^2}{15} \end{bmatrix} \tag{10.71}$$

10.3.1.8 Remarks

We offer some general comments here, using the 2D plane frame stiffness matrices displayed above as a focus for discussion. However, some of our remarks are intended to apply in a more general structural analysis context.

Linear and geometric stiffness matrices have often been used to formulate linearized bifurcation buckling as an eigenvalue problem, in which it is assumed that prebuckling displacements are small, so that the prebuckling state can be determined from a linear analysis.

For specialized applications to idealized substructures (e.g., buckling of isolated plate or shell components), it is sometimes assumed for simplicity that the prebuckling state is a spatially constant pure membrane state.

The geometric stiffness presented above for the plane frame element is computed exactly, for the analytical formulation employed. It is therefore termed a *consistent* geometric stiffness, since the same shape functions are employed in both linear stiffness and geometric stiffness calculations. Note also that the consistent $\mathbf{k}_G$ is positive definite, which is relevant to its use in an eigenvalue formulation of linearized buckling. This consistent use of the same shape functions requires the integration of fourth-order polynomials for the consistent $\mathbf{k}_G$, whereas the exact calculation of the flexural stiffness terms in $\mathbf{k}_E$ requires only the integration of second-order polynomials.

The situation is similar in plate and shell applications. In a UDL formulation with material nonlinearity, it is a practical necessity to carry out these types of calculations numerically, usually by Gaussian integration. A consistent geometric stiffness would require more sampling points, at different sampling locations, than for the calculation of the linearized material tangent stiffness matrix.

As a result, the geometric stiffness has sometimes been calculated using various approximations, which can be viewed as using different, lower-order shape functions than for stiffness calculations. As an extreme example, a so-called "string stiffness" for a plane frame element has sometimes been used, based on a linear displacement shape,

$$v(x) = \left(\frac{1-x}{L}\right) v_1 + \left(\frac{x}{L}\right) v_2 \tag{10.72}$$

for the calculation of the 2D geometric stiffness, which leads to

$$\mathbf{k}_G \approx \frac{P}{L} \begin{bmatrix} 0 & 0 & 0 & 0 & 0 & 0 \\ 0 & 1 & 0 & 0 & -1 & 0 \\ 0 & 0 & 0 & 0 & 0 & 0 \\ 0 & 0 & 0 & 0 & 0 & 0 \\ 0 & -1 & 0 & 0 & 1 & 0 \\ 0 & 0 & 0 & 0 & 0 & 0 \end{bmatrix} \tag{10.73}$$

thereby eliminating the effect of nodal rotations on the geometric stiffness.

10.3.1.9 Elastic tangent stiffness matrices

Ideally, the simplest and perhaps most widely useful flat plate or shell element would be triangular with 6 DOF per node. Such an element based on the Kirchhoff hypothesis unfortunately cannot exist in a potential energy-based formulation. Irons and Draper (1965) showed in a simple and direct way that a plate bending element with only the desired minimum 3 DOF per node ($w, w_{,x}, w_{,y}$) cannot have unique values of curvatures ($w_{,xx}$, $w_{,yy}$, $w_{,xy}$) everywhere within the element.

Some of the alternatives that have subsequently been explored include (1) introduction of additional higher-order derivatives, for example, $w_{,xxy}$, as additional DOF (Bogner et al. 1965); (2) subdivision of the element into regions within which the displacements are (initially) interpolated independently; (3) relaxation of the Kirchhoff hypothesis and enforcement only at discrete locations; and (4) basing the development on the Mindlin plate theory, in which a simplified representation of transverse shear strains γ_{xz}, γ_{yz} is incorporated. Many of these approaches have undesirable side effects.

Mixed variational principles have also been exploited. A vast literature now exists regarding these finite element modeling issues, which arise in linear finite element analysis of plate and shell structures.

10.3.2 Mindlin plate theory

The Mindlin plate theory was originally developed for (moderately) thick plates in which transverse shear deformation is approximately accounted for. Figure 10.9 illustrates the deformation of a cross section x = constant according to the Mindlin theory. The midsurface rotation is $w_{,x}$, whereas the rotation (φ_x) of the normal differs from $w_{,x}$ due to transverse shear deformation, which we label as γ_x on a face x = constant. The kinematic constraints of the Mindlin theory are

$$\begin{cases} \bar{u} = u - z\,\varphi_x \\ \bar{v} = v - z\,\varphi_y \\ \bar{w} = w \end{cases} \tag{10.74}$$

Proceeding as in the Kirchhoff elements, we take the in-plane Green strains in the generic lamina of thickness dz located distance z above the midsurface as

$$\begin{cases} \bar{\varepsilon}_x = \bar{u}_{,x} + \dfrac{1}{2}\bar{w}_{,x}^2 \\ \bar{\varepsilon}_y = \bar{v}_{,y} + \dfrac{1}{2}\bar{w}_{,y}^2 \\ \bar{\gamma}_{xy} = \bar{u}_{,y} + \bar{v}_{,x} + \bar{w}_{,x}\bar{w}_{,y} \\ \bar{\gamma}_{xz} = \bar{w}_{,x} + \bar{u}_{,z} \\ \bar{\gamma}_{yz} = \bar{w}_{,y} + \bar{v}_{,z} \end{cases} \tag{10.75}$$

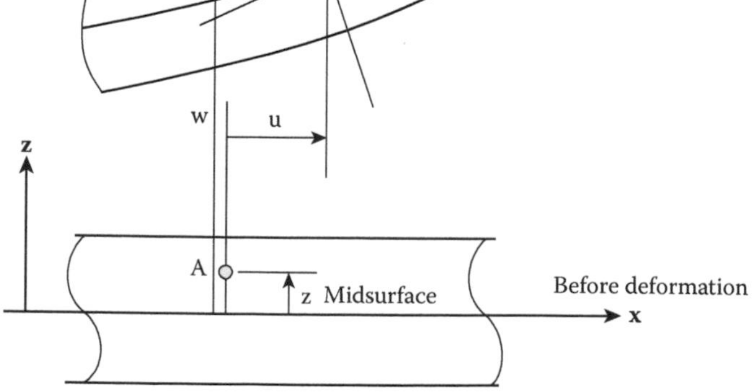

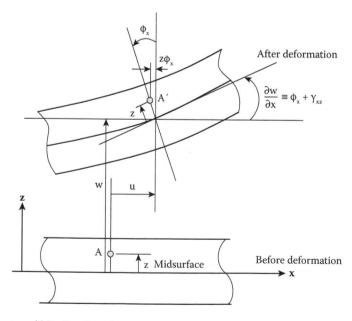

Figure 10.9 Kinematics of Mindlin plate theory.

where we have again assumed $\overline{u}_{,x}^2 \ll |\overline{u}_{,x}|$, and so forth. We first apply the constraint $\overline{w} = w$ to simplify the expressions for the lamina in-plane strains to

$$\begin{cases} \overline{\varepsilon}_x = \overline{u}_{,x} + \frac{1}{2} w_{,x}^2 \\ \overline{\varepsilon}_y = \overline{v}_{,y} + \frac{1}{2} w_{,y}^2 \\ \overline{\gamma}_{xy} = \overline{u}_{,y} + \overline{v}_{,x} + w_{,x} w_{,y} \end{cases} \tag{10.76}$$

and then find $\overline{u}_{,z} = -\varphi_x$ and $\overline{v}_{,z} = -\varphi_y$ from the first two constraints, which provides expressions for the two transverse shear strains as

$$\begin{cases} \overline{\gamma}_{xz} = \overline{w}_{,x} - \varphi_x \\ \overline{\gamma}_{yz} = \overline{w}_{,y} - \varphi_y \end{cases} \tag{10.77}$$

The lamina strains are then expressed as

$$\begin{cases} \overline{\varepsilon}_x = \varepsilon_x - z\varphi_{x,x} \\ \overline{\varepsilon}_y = \varepsilon_y - z\varphi_{y,y} \\ \overline{\gamma}_{xyx} = \gamma_{xy} - z(\varphi_{x,y} + \varphi_{y,x}) \\ \gamma_{xz} = w_{,x} - \varphi_x \\ \gamma_{yz} = w_{,y} - \varphi_y \end{cases} \tag{10.78}$$

where ε_x, ε_y,γ_{xy} are midsurface membrane strains containing nonlinear terms identical to those for the Kirchhoff elements (Equation 10.12).

The membrane forces **N** and bending moments **M** per unit length of midsurface are defined as usual by

$$\mathbf{N} \equiv \begin{Bmatrix} N_x \\ N_y \\ N_{xy} \end{Bmatrix} = \int_{-h/2}^{h/2} \begin{Bmatrix} \overline{\sigma}_x \\ \overline{\sigma}_y \\ \overline{\tau}_{xy} \end{Bmatrix} dz \tag{10.79}$$

and

$$\mathbf{M} \equiv \begin{Bmatrix} M_x \\ M_y \\ M_{xy} \end{Bmatrix} = \int_{-h/2}^{h/2} -z \begin{Bmatrix} \overline{\sigma}_x \\ \overline{\sigma}_y \\ \overline{\tau}_{xy} \end{Bmatrix} dz \tag{10.80}$$

leading, for plane stress isotropic elastic material behavior, to

$$\mathbf{N} = \mathbf{E}h \left(\begin{Bmatrix} \varepsilon_x \\ \varepsilon_y \\ \gamma_{xy} \end{Bmatrix}^{l} + \begin{Bmatrix} \varepsilon_x \\ \varepsilon_y \\ \gamma_{xy} \end{Bmatrix}^{nl} \right) \tag{10.81}$$

and

$$M = D\kappa \tag{10.82}$$

where **E** and **D** are the same as those given in Equations 10.10 and 10.16 for Kirchhoff elements. The linear and nonlinear membrane strains in Equation 10.81 are also identical to those given earlier for Kirchhoff elements. For Mindlin plates, the curvature vector $\boldsymbol{\kappa}$ is defined as

$$\boldsymbol{\kappa} = \left\{ \begin{array}{c} \varphi_{x,x} \\ \varphi_{y,y} \\ \varphi_{x,y} + \varphi_{y,x} \end{array} \right\} \tag{10.83}$$

The transverse shear forces Qx and Q_y per unit length of midsurface are

$$\left\{ \begin{array}{c} Q_x \\ Q_y \end{array} \right\} \equiv \int_{-h/2}^{h/2} \left\{ \begin{array}{c} \tau_{xz} \\ \tau_{yz} \end{array} \right\} dz = Gh \left\{ \begin{array}{c} \gamma_{xz} \\ \gamma_{yz} \end{array} \right\} \tag{10.84}$$

In contrast to Kirchhoff elements, in Mindlin plates transverse shear forces also store strain energy, which contributes to element stiffness.

10.3.3 Degeneration of isoparametric solid elements

We briefly describe here the process of degeneration of isoparametric continuum finite elements as a method of formulating isoparametric plate and shell finite elements. This is a simple process of imposing constraints of the Kirchhoff or Mindlin type on the through-thickness displacements of the continuum element. As a simple illustrative example, we describe the degeneration of a 2D Q9 Lagrangian element (described in Chapter 8) to form a one-dimensional (1D) Mindlin beam element.

In this section, we take the "thickness" dimension of the continuum element as 2h, for consistency with the parent element isoparametric coordinate system. Elsewhere, we label the thickness of plate and shell elements as h.

Figure 10.10 shows the nine-node (18 DOF) Lagrangian isoparametric element whose shape functions were listed in Chapter 8. The displacements of the 2D element are given by

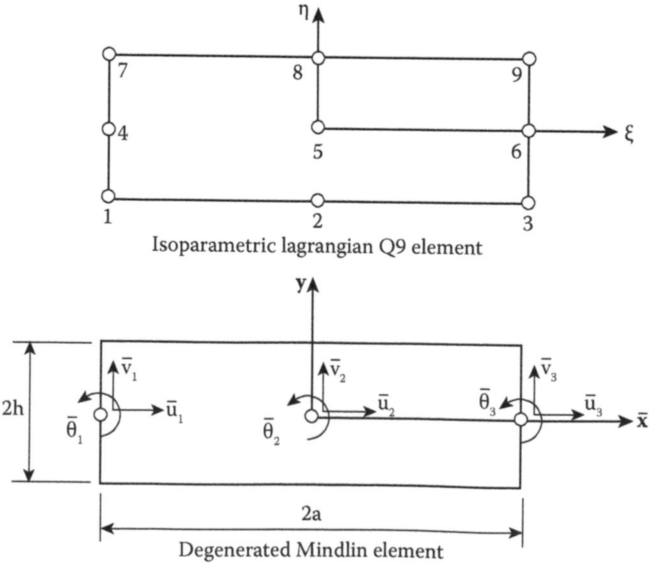

Figure 10.10 Degeneration of a continuum finite element.

$$\begin{cases} u(\xi,\eta) = \sum_{i=1}^{9} N_i(\xi,\eta)u_i \\ v(\xi,\eta) = \sum_{i=1}^{9} N_i(\xi,\eta)v_i \end{cases} \tag{10.85}$$

where the shape functions $N_i(\xi, \eta)$ are products of 1D quadratic Lagrangian shape functions. The 1D Lagrangian shape functions f_1, f_2, f_3 are defined as follows:

$$\begin{cases} f_1(\xi) \equiv -\frac{1}{2}\xi(1-\xi) \\ f_2(\xi) \equiv 1-\xi^2 \\ f_3(\xi) \equiv \frac{1}{2}\xi(1+\xi) \end{cases} \tag{10.86}$$

We may write the displacement $u(\xi,\eta)$ of the unconstrained continuum element explicitly in terms of the appropriate products of f_1,f_2,f_3 as

$$\begin{aligned} u(\xi,\eta) = {} & f_1(\xi)[f_1(\eta)u_1 + f_2(\eta)u_4 + f_3(\eta)u_7] \\ & + f_2(\xi)[f_1(\eta)u_2 + f_2(\eta)u_5 + f_3(\eta)u_8] \\ & + f_3(\xi)[f_1(\eta)u_3 + f_2(\eta)u_6 + f_3(\eta)u_9] \end{aligned} \tag{10.87}$$

We denote the displacements of the degenerated (constrained) element as $\bar{u}(\xi), \bar{v}(\xi), h\bar{\theta}(\xi)$, where $\bar{u}(\xi)$and $\bar{v}(\xi)$ are displacements of the centerline and $\bar{\theta}(\xi)$ is the rotation of the normal. We now constrain the transverse (η direction) variation of displacement. For example, the first bracketed term in Equation 10.87 becomes

$$\begin{aligned} & [f_1(\eta)(\bar{u}_1 + h\bar{\theta}_1) + f_2(\eta)\bar{u}_1 + f_3(\eta)(\bar{u}_1 - h\bar{\theta}_1)] = \\ & \{f_1(\eta) + f_2(\eta) + f_3(\eta)\}\bar{u}_1 + \{f_1(\eta) - f_3(\eta)\}h\bar{\theta}_1 = \\ & \bar{u}_1 - \eta h\bar{\theta}_1 \end{aligned} \tag{10.88}$$

in which we have used the relations

$$\begin{cases} f_1(\eta) + f_2(\eta) + f_3(\eta) = 1 \\ f_1(\eta) - f_3(\eta) = -\eta \end{cases} \tag{10.89}$$

Proceeding in the same way with the two remaining bracketed terms, we find the displacement $u(\xi, \eta)$ for the degenerated element,

$$u(\xi,\eta) = f_1(\xi)\bar{u}_1 + f_2(\xi)\bar{u}_2 + f_3(\xi)\bar{u}_3 - \eta\{f_1(\xi)h\bar{\theta}_1 + f_2(\xi)h\bar{\theta}_2 + f_3(\xi)h\bar{\theta}_3\} \tag{10.90}$$

which may be written compactly as

$$u(\xi,\eta) = \bar{u}(\xi) - \eta h\bar{\theta}(\xi) \tag{10.91}$$

The displacement $v(\xi, \eta)$ is constrained to be independent of η; that is, the normal remains unextended, and thus $v(\xi,\eta) = \bar{v}(\xi)$. We may summarize the displacements of the degenerated element as

$$\begin{Bmatrix} \bar{u} \\ \bar{v} \\ h\bar{\theta} \end{Bmatrix} = \sum_{i=1}^{3} f_i(\xi) \begin{Bmatrix} \bar{u}_i \\ \bar{v}_i \\ h\bar{\theta}_i \end{Bmatrix} \tag{10.92}$$

We note that the variables $\bar{u}(\xi), \bar{v}(\xi), h\bar{\theta}(\xi)$ are all interpolated with the same shape functions $f_i(\xi)$, which are in this case quadratic in ξ. This process extends directly to higher-order interpolations (additional nodes) and to degeneration of 3D continuum elements to form 2D plate or shell elements. In addition, isoperimetric mapping provides the capability to generate curvilinear shapes.

10.3.4 Mindlin plate and flat shell finite elements

We concentrate here on the modeling of flexural response using the Mindlin plate theory. This modeling approach produces a bending stiffness matrix $\mathbf{k}_b$ and, in addition, a shear stiffness matrix $\mathbf{k}_s$ that does not appear in Kirchhoff elements. The membrane stiffness $\mathbf{k}_m$ and geometric stiffness $\mathbf{k}_G$ are similar in Mindlin and Kirchhoff formulations and will be discussed later.

Finite elements based on the Mindlin plate theory use deflection w and normal rotations $\theta_x = \varphi_y$ and $\theta_y = -\varphi_x$ as independent nodal DOF. The interelement connectivity requirements are, as a result, much less onerous than for Kirchhoff elements. Only the independent normal rotations and the deflections, and not their derivatives, need be continuous across interelement boundaries for displacement continuity. This is usually termed C^0 continuity. Figure 10.11 shows a 12 DOF Mindlin plate bending parent element. Isoparametric mapping produces a general quadrilateral element in the (x, y) plane.

The deflection w is interpolated as

$$w = [N_1 \quad N_2 \quad N_3 \quad N_4] \begin{Bmatrix} w_1 \\ w_2 \\ w_3 \\ w_4 \end{Bmatrix} \tag{10.93}$$

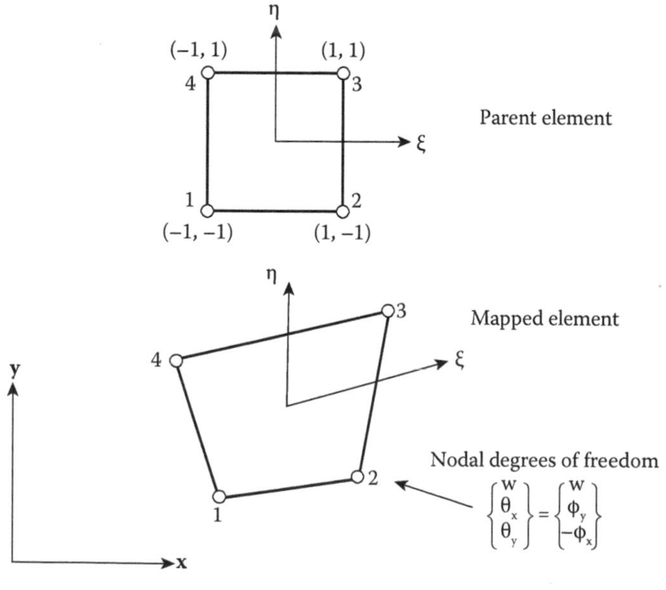

Figure 10.11 Quadrilateral Mindlin plate bending element.

The bilinear shape functions N_i are

$$\begin{cases} N_1 = \frac{1}{4}(1-\xi)(1-\eta) \\ N_2 = \frac{1}{4}(1+\xi)(1-\eta) \\ N_3 = \frac{1}{4}(1+\xi)(1+\eta) \\ N_4 = \frac{1}{4}(1-\xi)(1+\eta) \end{cases} \tag{10.94}$$

where $\xi \equiv x/a$ and $\eta \equiv y/b$.

The normal rotations φ_x and φ_y are interpolated using the same bilinear shape functions as the deflection w, that is,

$$\varphi_x = [N_1 \quad N_2 \quad N_3 \quad N_4]\begin{Bmatrix} \varphi_{x1} \\ \varphi_{x2} \\ \varphi_{x3} \\ \varphi_{x4} \end{Bmatrix}, \quad \varphi_y = [N_1 \quad N_2 \quad N_3 \quad N_4]\begin{Bmatrix} \varphi_{y1} \\ \varphi_{y2} \\ \varphi_{y3} \\ \varphi_{y4} \end{Bmatrix} \tag{10.95}$$

These shape functions ensure that w, φ_x, and φ_y are continuous between elements.

10.3.4.1 Bending stiffness

The curvature vector $\boldsymbol{\kappa}$ is determined from

$$\boldsymbol{\kappa} = \begin{Bmatrix} \varphi_{x,x} \\ \varphi_{y,y} \\ \varphi_{x,y} + \varphi_{y,x} \end{Bmatrix} = \mathbf{B}_b\mathbf{u} \tag{10.96}$$

where

$$\mathbf{B}_b = [\mathbf{B}_{b1} \quad \mathbf{B}_{b2} \quad \mathbf{B}_{b3} \quad \mathbf{B}_{b4}] \tag{10.97}$$

The block matrix $\mathbf{B}_{bi}$ for node i is

$$\mathbf{B}_{bi} = \begin{bmatrix} 0 & 0 & -N_{i,x} \\ 0 & N_{i,y} & 0 \\ 0 & N_{i,x} & -N_{i,y} \end{bmatrix} \tag{10.98}$$

The (12 × 1) nodal displacement vector is

$$\mathbf{u} = \begin{Bmatrix} \mathbf{u}_1 \\ \mathbf{u}_2 \\ \mathbf{u}_3 \\ \mathbf{u}_4 \end{Bmatrix} \tag{10.99}$$

and the (3 × 1) nodal displacement vector for node i is

$$\mathbf{u}_i = \begin{Bmatrix} w_i \\ \theta_{xi} \\ \theta_{yi} \end{Bmatrix} \tag{10.100}$$

The internal virtual work due to flexure is

$$\delta W_{int}^{b} = -\delta \mathbf{u}^T \mathbf{k}_b \mathbf{u} \tag{10.101}$$

where

$$\mathbf{k}_b = \int_A \mathbf{B}_b^T \mathbf{D}_b \mathbf{B}_b \, dA \tag{10.102}$$

For isotropic elastic material behavior, the matrix $\mathbf{D}_b$, as given previously for Kirchhoff elements, is

$$\mathbf{D}_b = \frac{Eh^3}{12(1-\nu^2)} \begin{bmatrix} 1 & \nu & 0 \\ \nu & 1 & 0 \\ 0 & 0 & (1-\nu)/2 \end{bmatrix} \tag{10.103}$$

In preparation for numerical integration, the integral in Equation 10.102 can be written as

$$\mathbf{k}_b = \int_{-1}^{1} \int_{-1}^{1} \mathbf{B}_b^T \mathbf{D}_b \mathbf{B}_b |\mathbf{J}| d\xi d\eta \tag{10.104}$$

Gaussian integration is described in Chapter 8; we briefly recall some of the details here. At each integration point (ξ_j, η_k) in turn, the integrand $\mathbf{B}_b^T \mathbf{D}_b \mathbf{B}_b$ and the determinant of the Jacobian matrix $\mathbf{J}$ are evaluated numerically as described below, multiplied by a weighting factor W_{jk}, and summed over all integration points. The shape function derivatives $N_{i,x}$ and $N_{i,y}$, which are required for $\mathbf{B}_b$, are calculated at the integration point from $N_{i,\xi}$ and $N_{i,\eta}$ via the simultaneous equations

$$\begin{bmatrix} \frac{\partial x}{\partial \xi} & \frac{\partial y}{\partial \xi} \\ \frac{\partial x}{\partial \eta} & \frac{\partial y}{\partial \eta} \end{bmatrix} \begin{Bmatrix} \frac{\partial N_i}{\partial x} \\ \frac{\partial N_i}{\partial y} \end{Bmatrix} = \begin{Bmatrix} \frac{\partial N_i}{\partial \xi} \\ \frac{\partial N_i}{\partial \eta} \end{Bmatrix} \tag{10.105}$$

where the (2 × 2) coefficient matrix in this equation is the Jacobian matrix $\mathbf{J}$ defined by the isoparametric mapping. The element of area $dA \equiv \det(\mathbf{J}) d\xi d\eta$. The transpose of the Jacobian matrix is determined from the nodal coordinates of the mapped element via

$$\begin{bmatrix} \frac{\partial x}{\partial \xi} & \frac{\partial y}{\partial \xi} \\ \frac{\partial x}{\partial \eta} & \frac{\partial y}{\partial \eta} \end{bmatrix} = \begin{bmatrix} \frac{\partial N_1}{\partial \xi} & \cdots & \frac{\partial N_n}{\partial \xi} \\ \frac{\partial N_1}{\partial \eta} & \cdots & \frac{\partial N_n}{\partial \eta} \end{bmatrix}_{(2\times n)} \begin{bmatrix} x_1 & y_1 \\ \vdots & \vdots \\ x_n & y_n \end{bmatrix}_{(n\times 2)} \tag{10.106}$$

In this equation, the number of nodes $n = 4$.

10.3.4.2 Shear stiffness

The shear stiffness matrix couples the deflection DOF w to the normal rotation DOF φ_x and φ_y. We write the transverse shear strains as a vector:

$$\gamma = \begin{Bmatrix} \gamma_{xz} \\ \gamma_{yz} \end{Bmatrix} = \begin{Bmatrix} w_{,x} - \varphi_x \\ w_{,y} - \varphi_y \end{Bmatrix} \tag{10.107}$$

The transverse shear strains are related to the nodal DOF by

$$\boldsymbol{\gamma} = \mathbf{B}_s\mathbf{u} \tag{10.108}$$

where

$$\mathbf{B}_s = [\mathbf{B}_{s1} \quad \mathbf{B}_{s2} \quad \mathbf{B}_{s3} \quad \mathbf{B}_{s4}] \tag{10.109}$$

The block matrix $\mathbf{B}_{si}$ for node i is

$$\mathbf{B}_{si} = \begin{bmatrix} N_{i,x} & 0 & N_i \\ N_{i,y} & -N_i & 0 \end{bmatrix} \tag{10.110}$$

The transverse shear forces, defined in terms of shear stresses in Equation 10.84, are written as a vector $\mathbf{Q}$,

$$Q = \begin{Bmatrix} Q_x \\ Q_y \end{Bmatrix} \tag{10.111}$$

which is related to the transverse shear strains by

$$Q = D_s\, \gamma \tag{10.112}$$

where

$$D_s = Gh \begin{bmatrix} 1 & 0 \\ 0 & 1 \end{bmatrix} \tag{10.113}$$

This relation emerges from the assumption that transverse shear strains, and therefore transverse shear stresses, are constant in the through-thickness direction, which is not realistic. An *ad hoc* shear correction factor $\alpha = 5/6$, based on a strain energy calculation assuming a parabolic distribution of shear stress, is often introduced.

The internal virtual work of the shear forces leads to the shear stiffness matrix

$$\delta W^s_{int} = -\delta\mathbf{u}^T\mathbf{k}_s\mathbf{u} \tag{10.114}$$

where

$$\mathbf{k}_s = \int_A \mathbf{B}_s^T \mathbf{D}_s \mathbf{B}_s \; dA \tag{10.115}$$

$$\mathbf{D}_s = \alpha Gh \begin{bmatrix} 1 & 0 \\ 0 & 1 \end{bmatrix} \tag{10.116}$$

The shear stiffness matrix $\mathbf{k}_s$ is of course also evaluated numerically via Gaussian integration.

10.3.4.3 In-plane response

The internal virtual work of the membrane stress resultants $\mathbf{N} = [N_x, N_y, N_{xy}]^T$ is

$$\delta W^m_{int} = -\int_A \delta\boldsymbol{\varepsilon}^T\mathbf{N}\, dA = -\int_A (\delta\boldsymbol{\varepsilon}^{l^T} + \delta\boldsymbol{\varepsilon}^{nl^T})\mathbf{N}\, dA \tag{10.117}$$

The expressions for the linear and nonlinear portions of the membrane strains were given in Equation 10.78 and are recalled here:

$$\boldsymbol{\varepsilon}^{l} \equiv \begin{Bmatrix} u_{,x} \\ v_{,y} \\ u_{,y} + v_{,x} \end{Bmatrix}, \boldsymbol{\varepsilon}^{nl} \equiv \frac{1}{2} \begin{Bmatrix} w_{,x}^{2} \\ w_{,y}^{2} \\ 2w_{,x}w_{,y} \end{Bmatrix} \qquad (10.118)$$

The internal virtual work of the membrane stress resultants acting through the linear portion of the membrane strains provides the membrane tangent stiffness $\mathbf{k}_m$, and that of the nonlinear portion of the membrane strains provides the geometric stiffness $\mathbf{k}_G$.

10.3.4.4 Geometric stiffness

We consider the internal virtual work of the membrane stress resultants acting through the nonlinear portion of the membrane strains, that is, the second term of Equation 10.117. Following the procedure developed earlier in this chapter (Equation 10.26), we write the nonlinear membrane strain as

$$\boldsymbol{\varepsilon}^{nl} = \frac{1}{2} \begin{bmatrix} w_{,x} & 0 \\ 0 & w_{,y} \\ w_{,y} & w_{,x} \end{bmatrix} \begin{Bmatrix} w_{,x} \\ w_{,y} \end{Bmatrix} \equiv \frac{1}{2}\mathbf{A}\boldsymbol{\psi} \qquad (10.119)$$

We use the symbol $\boldsymbol{\psi}$ (rather than $\boldsymbol{\theta}$) here to denote the displacement gradients to avoid confusion with the nodal rotation DOF θ_x and θ_y. The displacement gradients are

$$\boldsymbol{\psi} \equiv \begin{Bmatrix} w_{,x} \\ w_{,y} \end{Bmatrix} = [\mathbf{G}_1 \quad \mathbf{G}_2 \quad \mathbf{G}_3 \quad \mathbf{G}_4] \begin{Bmatrix} \mathbf{u}_1 \\ \mathbf{u}_2 \\ \mathbf{u}_3 \\ \mathbf{u}_4 \end{Bmatrix} \qquad (10.120)$$

where

$$\mathbf{G}_i = \begin{bmatrix} N_{i,x} & 0 & 0 \\ N_{i,y} & 0 & 0 \end{bmatrix} \qquad (10.121)$$

and the (3 × 1) nodal displacement vector $\mathbf{u}_i$ is defined in Equation 10.100. The block matrix $\mathbf{G}_i$ is filled with zeros in the second and third columns to reflect the fact that the displacement gradients $w_{,x}$ and $w_{,y}$ do not depend on the independent nodal rotations. As a result, the geometric stiffness matrix is quite sparse (depending only on the nodal displacements w_i), analogous to the geometric "string stiffness" for plane frame elements given in Equation 10.73.

The nonlinear membrane strain is

$$\boldsymbol{\varepsilon}^{nl} = \frac{1}{2}\mathbf{AGu} \qquad (10.122)$$

The associated internal virtual work expression is

$$\delta W_{int}^{G} = -\int_{A} \delta\boldsymbol{\varepsilon}^{nl,T}\mathbf{N}\ dA = -\delta\mathbf{u}^{T}\int_{A} \mathbf{G}^{T}\mathbf{A}^{T}\mathbf{N}\ dA \qquad (10.123)$$

in which the virtual nonlinear membrane strain is (Equation 10.29)

$$\delta\boldsymbol{\varepsilon}^{nl} = \mathbf{AG}\ \delta\mathbf{u} \qquad (10.124)$$

We rearrange the product A^TN as

$$A^TN = \tilde{N}\psi \tag{10.125}$$

where the membrane stress resultants are conveniently written as a (2 × 2) matrix $\tilde{N}$ rather than a (3 × 1) vector (namely, Equation 10.34):

$$\tilde{N} \equiv \begin{bmatrix} N_x & N_{xy} \\ N_{xy} & N_y \end{bmatrix} \tag{10.126}$$

Equation 10.123 then becomes

$$\delta W_{int}^G = -\delta \mathbf{u}^T \mathbf{k}_G \mathbf{u} \tag{10.127}$$

where the geometric stiffness $\mathbf{k}_G$ is

$$\mathbf{k}_G = \left[\int_A G^T \tilde{N} G \, dA \right] \tag{10.128}$$

10.3.4.5 Membrane stiffness

The linear membrane strains depend only on the in-plane midsurface displacements u(x, y), v(x, y) and not on the variables (w, φ_x, φ_y) governing the flexural response. As a result, the membrane stiffness is uncoupled from the bending, shear, and geometric stiffness matrices. The in-plane displacements u(x, y) and v(x, y)are related to the in-plane nodal DOF,

$$\begin{Bmatrix} u \\ v \end{Bmatrix} = \sum_{i=1}^{4} N_i(\xi, \eta) \begin{Bmatrix} u_i \\ v_i \end{Bmatrix} \tag{10.129}$$

using the same (in this case bilinear) shape functions. For the linear tangent stiffness, we use the linear relation

$$N = Eh\varepsilon^l \tag{10.130}$$

The detailed calculations for the membrane stiffness matrix exactly parallel those for the plane stress stiffness matrix of the isoparametric quadrilateral (see Chapter 8). We have only to multiply that result by the plate thickness h to convert from the (implied) unit thickness of the plane stress element. We expand the nodal displacement vector of the Mindlin flat shell quadrilateral element to accommodate the in-plane nodal DOF u_i,v_i. Finally, the element nodal displacement vector $\mathbf{u}_i$ has 5 DOF as follows:

$$\mathbf{u}_i = (u_i,\ v_i,\ w_i,\ \theta_{xi},\ \theta_{yi})^T \tag{10.131}$$

10.3.4.6 Performance of Mindlin elements

The Mindlin plate theory was originally formulated for application to moderately thick plates, taking into account the effect of transverse shear deformation. However, we think it fair to say that in developing Mindlin finite elements, thick plates have never been the primary interest. Rather, the intended application area has always been thin plates, and the motivation for using Mindlin theory in that context is the relative ease with which Mindlin finite elements can be formulated, compared with Kirchhoff finite elements.

There are, however, some well-known issues associated with Mindlin plate and shell elements that arise even in linear analysis. A succinct discussion is presented by Cook et al. (1989). We briefly summarize these issues below.

10.3.4.7 Shear locking, zero-energy modes, and hourglass control

It is well known that when Mindlin elements are used in thin-plate applications in which shear deformation should not be present, spurious transverse shear deformation, also known as "parasitic shear," sometimes develops. As the element thickness is reduced, the parasitic shear increases, thus artificially stiffening the element, eventually producing the phenomenon known as "shear locking."

Gaussian integration of an order sufficient to integrate the stiffness matrices "exactly" (at least when det $\mathbf{J}$ is constant) is termed *full integration*, lower-order integration is termed *reduced integration*, and lower-order integration of only the shear or membrane stiffness matrix is termed *selective reduced integration*. Although reduced integration eliminates parasitic shear, it can sometimes, depending on boundary condition restraints, introduce "zero-energy modes" (sometimes called "hourglassing" because of their characteristic shape), which are kinematic modes that generate zero strain energy in the underintegrated element. The same phenomenon appears in 2D plane stress–plane strain in-plane bending problems (e.g., modeling a deep cantilever beam), where it was first observed.

Standard remedies, variously termed "augmented stiffness," "hourglass control," or "stabilization," are often based on the concept of adding a small (arbitrary) stiffness to the rank-deficient stiffness matrix produced by reduced integration.

10.3.4.8 Membrane locking

A related phenomenon is the development of significant membrane strains in thin Mindlin curved shells in cases where the response should be essentially inextensional bending. As the shell thickness is reduced, the parasitic membrane strain energy increases and overwhelms the flexural strain energy, thereby stiffening and eventually "locking" the mesh (Parisch 1979). Belytschko et al. (1985) discuss the effectiveness of different remedies for membrane and shear locking in shell elements and compare the performance of several shell elements on three difficult test problems.

10.4 ISOPARAMETRIC CURVED SHELL ELEMENTS

A curved isoparametric shell element was first developed by Ahmad et al. (1970), starting from a conceptual model of the shell as a curved 3D isoparametric solid with top and bottom surface nodes used to define the shell geometry, a scheme later adopted by others, including Parisch (1978, 1981), Hughes and Liu (1981), Surana (1983), and Milford and Schnobrich (1986), who extended Ahmad's formulation to include geometric and material nonlinearities.

Figure 10.12 shows a schematic view of a nine-node Mindlin isoparametric curved shell element that has 5 DOF at each midsurface node: three Cartesian components of displacement and two in-plane rotation vectors tangent to the midsurface. The 3D element isoparametric coordinates (ξ, η, ζ) can be used to define a natural element coordinate system in which (ξ, η) are curvilinear coordinates of a point lying in the shell element midsurface.

As in the classic shell theory, the rotation about the surface normal vector is not included as an element DOF.

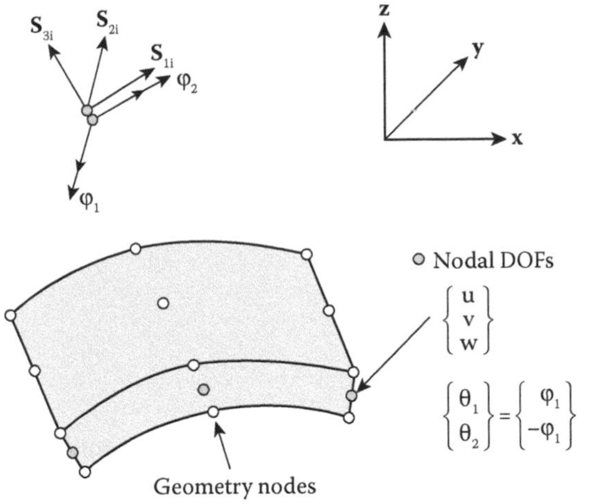

Figure 10.12 Ahmad shell element derived from a 3D isoparametric solid element.

We first elaborate on the normal rotation issue in classic thin-shell theory based on the Kirchhoff hypothesis.

10.4.1 Normal rotation in classic thin-shell theory

In shell theory formulated in lines of principal curvature coordinates (Seide 1975), the position vector of a generic point $\bar{\mathbf{x}}$ can be written as

$$\bar{\mathbf{x}}(\alpha, \beta, z) = \mathbf{x}(\alpha, \beta) + z\mathbf{n}(\alpha, \beta) \tag{10.132}$$

where (α, β) are surface "lines of curvature" coordinates and the unit vector **n** is strictly normal to the midsurface tangent plane at (α, β). The distance of point $\bar{\mathbf{x}}$ from the midsurface reference point **x**, measured along the normal vector, is z. The coordinates (α, β, z)define an *orthogonal* 3D coordinate system (Struik 1961).

We first summarize some background relations from differential geometry. Vectors tangent to the orthogonal coordinate lines (α, β) in the midsurface are $\mathbf{x}_{,\alpha}$ and $\mathbf{x}_{,\beta}$, respectively. A differential line element $d\mathbf{x}$ lying in the midsurface tangent plane is

$$d\mathbf{x} = \mathbf{x}_{,\alpha} d\alpha + \mathbf{x}_{,\beta} d\beta \tag{10.133}$$

Because the coordinates (α, β) are lines of principal curvature, and therefore orthogonal, the squared length $ds^2 \equiv d\mathbf{x} \cdot d\mathbf{x}$ may be written as

$$ds^2 = A^2 d\alpha^2 + B^2 d\beta^2 \tag{10.134}$$

where $A^2 \equiv (\mathbf{x}_{,\alpha} \cdot \mathbf{x}_{,\alpha})$ and $B^2 \equiv (\mathbf{x}_{,\beta} \cdot \mathbf{x}_{,\beta})$.

Thus, the element of midsurface area is da = AB dα dβ. Unit vectors tangent to the (α, β) coordinate lines in the midsurface are

$$\begin{aligned} \mathbf{e}_\alpha &= \mathbf{x}_{,\alpha}/A \\ \mathbf{e}_\beta &= \mathbf{x}_{,\beta}/B \end{aligned} \tag{10.135}$$

respectively, and the unit normal vector is $\mathbf{n} = \mathbf{e}_\alpha \times \mathbf{e}_\beta$. The triad ($\mathbf{e}_\alpha$, $\mathbf{e}_\beta$, **n**) constitutes a 3D orthonormal basis.

From Equation 10.132, we may find the tangent vectors lying in a generic surface parallel to the midsurface as

$$\begin{aligned} \overline{\mathbf{x}}_{,\alpha} &= \mathbf{x}_{,\alpha} + z\mathbf{n}_{,\alpha} \\ \overline{\mathbf{x}}_{,\beta} &= \mathbf{x}_{,\beta} + z\mathbf{n}_{,\beta} \end{aligned} \tag{10.136}$$

It is easily shown that the vectors $\mathbf{n}_{,\alpha}$ and $\mathbf{n}_{,\beta}$ lie in the midsurface (tangent plane) and can be expressed as

$$\begin{aligned} \mathbf{n}_{,\alpha} &= \frac{1}{R_\alpha}\mathbf{x}_{,\alpha} \\ \mathbf{n}_{,\beta} &= \frac{1}{R_\beta}\mathbf{x}_{,\beta} \end{aligned} \tag{10.137}$$

where R_α and R_β are the principal radii of curvature of the midsurface, which are given by

$$\frac{1}{R_\alpha} = \frac{(\mathbf{x}_{,\alpha\alpha} \cdot \mathbf{n})}{A^2} \quad \text{and} \quad \frac{1}{R_\beta} = \frac{(\mathbf{x}_{,\beta\beta} \cdot \mathbf{n})}{B^2} \tag{10.138}$$

Equation 10.137 is referred to by Struik (1961) as the "formula of Rodrigues."
Inserting Equations 10.137 in Equations 10.136, we find

$$\begin{aligned} \overline{\mathbf{x}}_{,\alpha} &= A\left(1 + \frac{z}{R_\alpha}\right)\mathbf{e}_\alpha \\ \overline{\mathbf{x}}_{,\beta} &= B\left(1 + \frac{z}{R_\beta}\right)\mathbf{e}_\beta \end{aligned} \tag{10.139}$$

Therefore, the differential arc length along a generic lamina of thickness dz located at a normal section α = constant illustrated in Figure 10.13 is $B(1 + \frac{z}{R_\beta})d\beta$. The figure is deceptive in that the thickness h is finite, whereas the arc length depicted is infinitesimal.

The membrane shear stress resultant and twisting moment (per unit length $Bd\beta$ of the midsurface) acting on the normal section α = constant are

$$\begin{cases} N_{\alpha\beta} = \displaystyle\int_{-h/2}^{h/2} \tau_{\alpha\beta}\left(1 + \frac{z}{R_\beta}\right)dz \\ M_{\alpha\beta} = \displaystyle\int_{-h/2}^{h/2} \tau_{\alpha\beta}z\left(1 + \frac{z}{R_\beta}\right)dz \end{cases} \tag{10.140}$$

On a section β = constant, we similarly find

$$\begin{cases} N_{\beta\alpha} = \displaystyle\int_{-h/2}^{h/2} \tau_{\beta\alpha}\left(1 + \frac{z}{R_\alpha}\right)dz \\ M_{\beta\alpha} = \displaystyle\int_{-h/2}^{h/2} \tau_{\beta\alpha}z\left(1 + \frac{z}{R_\alpha}\right)dz \end{cases} \tag{10.141}$$

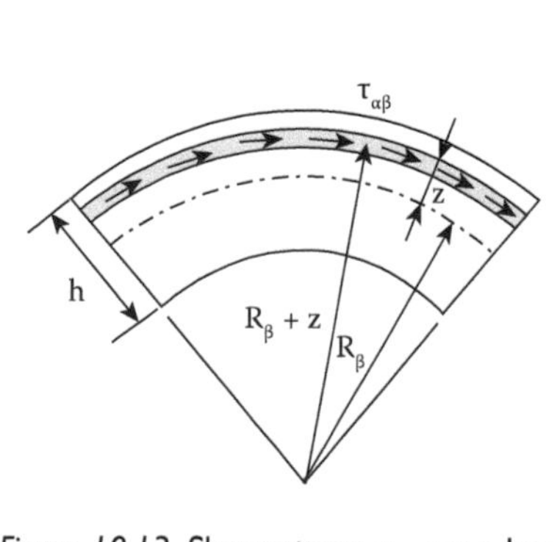

Figure 10.13 Shear stress $\tau_{\beta\alpha}$ on a lamina at section α = constant.

Therefore, even though the local shear stress $\tau_{\alpha\beta} = \tau_{\beta\alpha}$, in general, $N_{\alpha\beta} \neq N_{\beta\alpha}$ and $M_{\alpha\beta} \neq M_{\beta\alpha}$ in a curved shell (except for a sphere). Some shell theories assume $N_{\alpha\beta} \approx N_{\beta\alpha}$ and $M_{\alpha\beta} \approx M_{\beta\alpha}$ as an approximation, or introduce new variables $\overline{N}_{\alpha\beta} \equiv (N_{\alpha\beta} + N_{\beta\alpha})/2$ and $\overline{M}_{\alpha\beta} \equiv (M_{\alpha\beta} + M_{\beta\alpha})/2$.

Seide (1975) derives the six (partial differential) equilibrium equations in terms of stress resultants for a curved shell. Moment equilibrium about the shell normal **n** requires exactly

$$N_{\alpha\beta} - N_{\beta\alpha} + \frac{1}{R_\alpha} M_{\alpha\beta} - \frac{1}{R_\beta} M_{\beta\alpha} = 0 \tag{10.142}$$

(which is an algebraic relation, not a partial differential equation). With $\tau_{\alpha\beta} = \tau_{\beta\alpha}$, this equilibrium equation is an identity. We note that if the approximations $N_{\alpha\beta} \approx N_{\beta\alpha}$ and $M_{\alpha\beta} \approx M_{\beta\alpha}$ are adopted, then equilibrium of moments about the normal to the surface is, strictly speaking, not satisfied.

The normal rotation is undefined in classic thin-shell theory because it provides no additional useful information. This does not mean that the normal rotation is zero; rather, it is in a sense a DOF for which there is no corresponding generalized force, and it has been, in structural analysis terminology, "condensed out" of the theory. Thus, we see the difficulty with the introduction of an artificial normal rotational spring, together with an associated (problem-dependent) scalar stiffness parameter, as has sometimes been advocated in the past. We note that artificial springs have been used in the past as a convenient method of imposing *constraints* in finite element structural models. Here, we do not wish to constrain the normal rotation because it may be important in some cases, such as inextensional bending.

10.4.2 Isoparametric modeling of curved shell geometry

In Ahmad's formulation, the Cartesian coordinates $\overline{\mathbf{x}}$ of a generic point in the isoparametric shell element are defined by

$$\overline{\mathbf{x}} = \sum_{i=1}^{n} N_i(\xi,\eta)\left(\frac{1+\zeta}{2}\right)\mathbf{x}_i^{top} + \sum_{i=1}^{n} N_i(\xi,\eta)\left(\frac{1-\zeta}{2}\right)\mathbf{x}_i^{bot} \tag{10.143}$$

where $\mathbf{x}_i^{top}$ and $\mathbf{x}_i^{bot}$ are the Cartesian coordinates of the geometry nodes on the top and bottom surfaces of the shell. The isoparametric coordinate ζ has the range (−1, 1). We assume in this discussion that the shell thickness is constant and equal to h, although this is not essential.

The latter equation can be expressed in terms of the shell midsurface coordinates $\mathbf{x}_i$ and a unit vector $\mathbf{s}_{3i}$ as follows:

$$\overline{\mathbf{x}} = \sum_{i=1}^{n} N_i(\xi,\eta)\mathbf{x}_i + \sum_{i=1}^{n} N_i(\xi,\eta)\zeta\frac{h}{2}\mathbf{s}_{3i} \tag{10.144}$$

where $\mathbf{x}_i$ is the global position vector of midsurface node i defined by $\mathbf{x}_i \equiv (\mathbf{x}_i^{top} + \mathbf{x}_i^{bot})/2$. The nodal vector $\mathbf{s}_{3i}$, defined by

$$\mathbf{s}_{3i} = \frac{1}{h}(\mathbf{x}_i^{top} - \mathbf{x}_i^{bot}) \tag{10.145}$$

is used to establish a unique surface coordinate system at each global node i.

Equation 10.144 is the isoparametric interpolation of the equivalent pointwise mapping

$$\overline{\mathbf{x}}(\xi,\eta,\zeta) = \mathbf{x}(\xi,\eta) + \zeta\frac{h}{2}\mathbf{s}_3(\xi,\eta) \tag{10.146}$$

where $s_3(\xi, \eta)$ can be constructed from the input nodal geometry data (i.e., s_{3i}) as

$$s_3(\xi, \eta) \equiv \sum_{i=1}^{n} N_i(\xi, \eta) s_{3i} \tag{10.147}$$

A curved shell element can obviously model a smooth shell surface much more accurately than a faceted model built from flat shell elements. Nevertheless, even with curved shell elements, there will generally be midsurface slope discontinuities between adjacent elements, as a result of their independent local isoparametric coordinate systems.

10.4.3 Nodal surface coordinate system

A unique "surface" coordinate system must therefore be established at each node so that element midsurface rotations are in a consistent nodal coordinate system. The unit vector s_{3i} is the normal to the surface at node i. The orthogonal vectors (s_{1i}, s_{2i}), which span the surface tangent plane and complete the nodal surface triad (s_{1i}, s_{2i}, s_{3i}), are determined globally according to a predefined convention that is entirely element independent. This aspect of the geometry modeling has little to do with the details of the isoparametric shell finite element beyond the fact that it has only five nodal freedoms.

Figure 10.14 shows an extreme example: A curved shell modeled with a mesh of only two 4-node Mindlin isoparametric quadrilaterals, so that we can clearly see the distinction between the surface coordinate system to which the nodal rotations are referred and an element coordinate system based on the isoparametric coordinates (ξ, η, ζ), which are not orthogonal.

10.4.4 Jacobian matrix

The transformation from isoparametric coordinates (ξ, η, ζ) to global Cartesian coordinates $(\bar{x}, \bar{y}, \bar{z})$ is characterized by a Jacobian matrix, which follows from Equation 10.146.

This is of course also required for standard isoparametric solid elements, as described in Chapter 8.

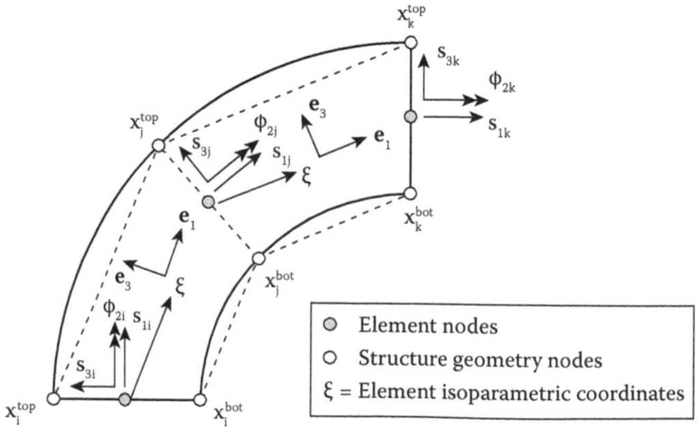

Figure 10.14 Nodal surface triad (s_1, s_2, s_3).

The Jacobian $\bar{J}$ is defined, as usual, by

$$\bar{J} = \begin{bmatrix} \frac{\partial \bar{x}}{\partial \xi} & \frac{\partial \bar{y}}{\partial \xi} & \frac{\partial \bar{z}}{\partial \xi} \\ \frac{\partial \bar{x}}{\partial \eta} & \frac{\partial \bar{y}}{\partial \eta} & \frac{\partial \bar{z}}{\partial \eta} \\ \frac{\partial \bar{x}}{\partial \zeta} & \frac{\partial \bar{y}}{\partial \zeta} & \frac{\partial \bar{z}}{\partial \zeta} \end{bmatrix} \tag{10.148}$$

From Equation 10.146, we find the transpose of the Jacobian matrix as

$$[\bar{\mathbf{x}}_{,\xi}, \bar{\mathbf{x}}_{,\eta}, \bar{\mathbf{x}}_{,\zeta}] = \left[\mathbf{x}_{,\xi}, \mathbf{x}_{,\eta}, \frac{h}{2}\mathbf{s}_3\right] + \zeta\frac{h}{2}\left[\mathbf{s}_{3,\xi}, \mathbf{s}_{3,\eta}, \mathbf{0}_3\right] \tag{10.149}$$

Here, $\mathbf{0}_3$ denotes a (3 × 1) zero vector. We note the similarity of this last equation to Equation 10.136 for the tangent vectors in shell principal axis coordinates. The linear variation of the Jacobian in the thickness direction ζ depends on the shell curvature (compare with Equations 10.137 and 10.139). Therefore, the Jacobian will in general be different at every Gaussian integration point (ξ_j, η_k, ζ_l), where ζ_l are the through-thickness sampling locations illustrated in Figure 10.15. The relatively large number of through-thickness integration points illustrated in the figure might be appropriate for modeling nonlinear material behavior, or for elastic multi-ply laminates. For thin (homogeneous) elastic shells, two through-thickness integration points are sufficient.

The first two rows of the midsurface Jacobian $\mathbf{J}(\zeta,\eta)$ can be determined from the nodal coordinates as

$$\begin{bmatrix} x_{,\xi} & y_{,\xi} & z_{,\xi} \\ x_{,\eta} & y_{,\eta} & z_{,\eta} \end{bmatrix} = \begin{bmatrix} N_{1,\xi} & \cdots & N_{n,\xi} \\ N_{1,\eta} & \cdots & N_{n,\eta} \end{bmatrix}_{(2\times n)} \begin{bmatrix} x_1 & y_1 & z_1 \\ \vdots & \vdots & \vdots \\ x_n & y_n & z_n \end{bmatrix}_{(n\times 3)} \tag{10.150}$$

and the third row from

$$\frac{h}{2}[s_{3x}, s_{3y}, s_{3z}] = \frac{h}{2}[N_1 \quad \cdots \quad N_n]_{(1\times n)} \begin{bmatrix} s^1_{3x} & s^1_{3y} & s^1_{3z} \\ \vdots & \vdots & \vdots \\ s^n_{3x} & s^n_{3y} & s^n_{3y} \end{bmatrix}_{(n\times 3)} \tag{10.151}$$

In Equation 10.151, $\mathbf{s}_3 = [s_{3x}, s_{3y}, s_{3z}]$ and a nodal value $\mathbf{s}_{3i}$ at node i is denoted by $[s^i_{3x}, s^i_{3y}, s^i_{3z}]^T$. The curvature terms that determine the variation of the Jacobian in the thickness direction can be determined from

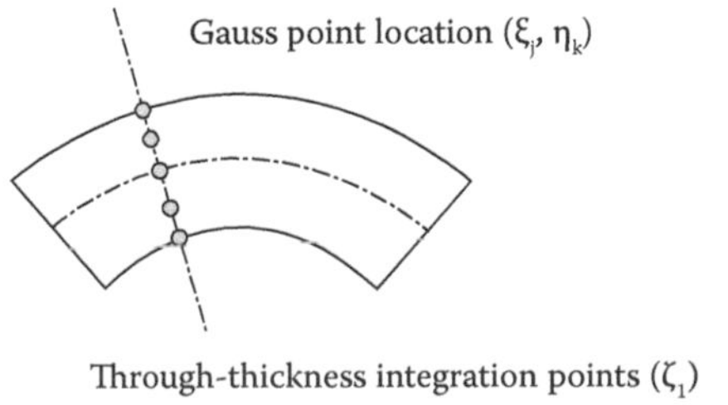

Figure 10.15 Numerical integration sampling points in the normal direction.

$$\begin{bmatrix} s_{3x,\xi} & s_{3y,\xi} & s_{3z,\xi} \\ s_{3x,\eta} & s_{3y,\eta} & s_{3z,\eta} \\ 0 & 0 & 0 \end{bmatrix} = \begin{bmatrix} N_{1,\xi} & \dots & N_{n,\xi} \\ N_{1,\eta} & \dots & N_{n,\eta} \\ 0 & \dots & 0 \end{bmatrix}_{(3\times n)} \begin{bmatrix} s_{3x}^1 & s_{3y}^1 & s_{3z}^1 \\ \vdots & \vdots & \vdots \\ s_{3x}^n & s_{3y}^n & s_{3z}^n \end{bmatrix}_{(n\times 3)} \tag{10.152}$$

Therefore, the Jacobian of the transformation from (ξ, η, ζ) isoparametric coordinates to global Cartesian coordinates $(\bar{x}, \bar{y}, \bar{z})$ has the general form

$$\bar{\mathbf{J}}(\xi, \eta, \zeta) = \mathbf{J}(\xi, \eta) + \zeta \mathbf{K}(\xi, \eta) \tag{10.153}$$

where $\mathbf{J}$ is the Jacobian at the midsurface, defined explicitly in Equations 10.150 and 10.151, and $\mathbf{K}$ is a measure of the shell curvature defined in Equation 10.152.

10.4.5 Lamina Cartesian coordinate system

We require an orthogonal element Cartesian coordinate system (e.g., x', y', z') in a generic lamina (i.e., at ζ = constant) in order to describe stresses, strains, and constitutive relations. This element coordinate system is constructed from the curvilinear isoparametric coordinates (ξ, η, ζ) as follows. Figure 10.16 shows the element isoparametric coordinate curves (ξ, η) at a specific through-thickness location ζ. A unit vector $\mathbf{e}_1$ tangent to the ζ coordinate curve is defined by

$$\mathbf{e}_1 \equiv \frac{\partial \bar{\mathbf{x}}}{\partial \xi} \bigg/ \left| \frac{\partial \bar{\mathbf{x}}}{\partial \xi} \right| \tag{10.154}$$

The unit normal vector $\mathbf{e}_3$ is then constructed by taking the cross product of $\mathbf{e}_1$ with $\partial \bar{\mathbf{x}} / \partial \eta$,

$$\mathbf{e}_3 = \left(\mathbf{e}_1 \times \frac{\partial \bar{\mathbf{x}}}{\partial \eta} \right) \bigg/ \left| \mathbf{e}_1 \times \frac{\partial \bar{\mathbf{x}}}{\partial \eta} \right| \tag{10.155}$$

and finally, $\mathbf{e}_2 = \mathbf{e}_3 \times \mathbf{e}_1$. Figure 10.14 illustrates the element Cartesian coordinate systems for the two simple quadrilateral Mindlin elements shown there. The orthogonal unit vectors $\mathbf{e}_1 = \mathbf{e}_2 \times \mathbf{e}_3$ define an orthogonal transformation matrix $\mathbf{E} = [\mathbf{e}_1, \mathbf{e}_2, \mathbf{e}_3]$ at every point (ξ, η, ζ) in the isoparametric element. We have suppressed overbars on the unit vectors $(\mathbf{e}_1, \mathbf{e}_2, \mathbf{e}_3)$ and the matrix $\mathbf{E}$ corresponding to the generic location $\bar{\mathbf{x}}$ (i.e., (ξ, η, ζ)) for simplicity. $\mathbf{E}$ can be built directly from the components of the Jacobian matrix $\bar{\mathbf{J}}$.

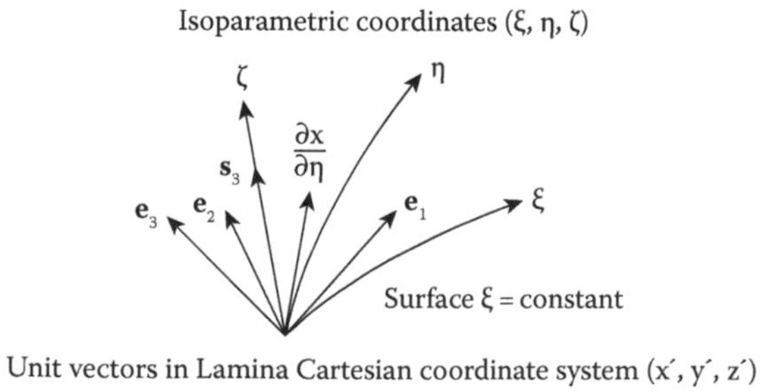

Figure 10.16 Lamina Cartesian coordinates.

10.4.6 Summary of coordinate systems and transformations

We have defined three coordinate systems (shown in Figure 10.17):

1. A global Cartesian coordinate system with base vectors (i_1, i_2, i_3) in which the coordinates of a generic point are ($\bar{x}, \bar{y}, \bar{z}$) and the coordinates of a midsurface point are (x, y, z)
2. A nodal (surface) coordinate system with orthonormal base vectors (s_1, s_2, s_3)
3. A local Cartesian or "lamina" coordinate system with orthonormal base vectors (e_1, e_2, e_3), in which the coordinates of a generic point are ($\overline{x'}, \overline{y'}, \overline{z'}$) and the coordinates of the corresponding midsurface point are (x′, y′, z′).

Rotation of a vector, say $\mathbf{y} \equiv [y_1, y_2, y_3]^T$, from the surface (nodal) coordinate system to the global Cartesian system is performed via the orthogonal matrix $S=[s_1, s_2, s_3]$, that is,

$$\mathbf{Y} = \mathbf{S}\mathbf{y} \tag{10.156}$$

where $\mathbf{Y} = [Y_1, Y_2, Y_3]^T$ is the representation of $\mathbf{y}$ in the basis (i_1, i_2, i_3) of the global Cartesian system.

The rotation of a vector $\mathbf{y}'$ in the lamina coordinate system into $\mathbf{Y}$ in the global Cartesian system is performed via the orthogonal matrix $\mathbf{E} = (e_1, e_2, e_3)$ as

$$\mathbf{Y} = \mathbf{E}\mathbf{y}' \tag{10.157}$$

The rotation of the vector $\mathbf{y}$ from the surface (nodal) coordinate system into the lamina coordinate system is therefore carried out via an orthogonal matrix $\mathbf{R}$, as indicated in Figure 10.17,

$$\mathbf{y}' = \mathbf{R}\mathbf{y} \tag{10.158}$$

where

$$\mathbf{R} = \mathbf{E}^T\mathbf{S} \tag{10.159}$$

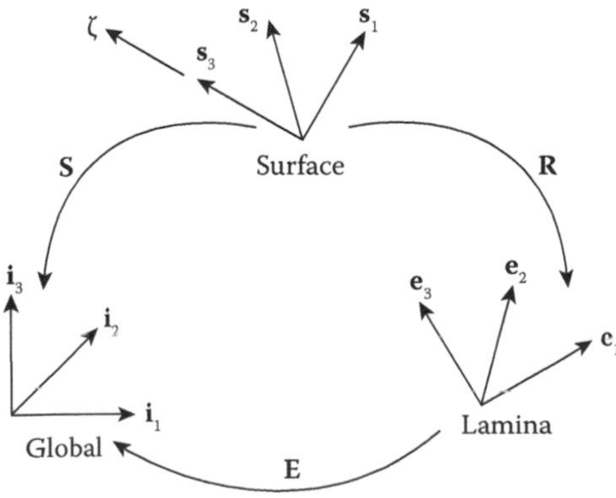

Figure 10.17 Vector transformations between coordinate systems.

The matrix **R** can be represented explicitly as

$$\mathbf{R} = \begin{bmatrix} e_1 \cdot s_1 & e_1 \cdot s_2 & e_1 \cdot s_3 \\ e_2 \cdot s_1 & e_2 \cdot s_2 & e_2 \cdot s_3 \\ e_3 \cdot s_1 & e_3 \cdot s_2 & e_3 \cdot s_3 \end{bmatrix} \tag{10.160}$$

10.4.7 Displacement interpolation

Element incremental displacements (referred to the global Cartesian coordinate system) are interpolated by

$$\Delta\bar{\mathbf{u}}(\xi,\eta,\zeta) = \sum_{i=1}^{n} N_i(\xi,\eta)\Delta\mathbf{u}_i - \sum_{i=1}^{n} N_i(\xi,\eta)\zeta\frac{h}{2}[s_{1i}, s_{2i}]\begin{Bmatrix} \Delta\varphi_1 \\ \Delta\varphi_2 \end{Bmatrix}_i \tag{10.161}$$

where $\Delta\varphi_1$ and $\Delta\varphi_2$ are the tangential (in-plane) components of the incremental rotation of the normal vector s_3, and $\Delta\mathbf{u}_i$ is the midsurface incremental displacement vector at node i. The five element incremental DOF at node i is

$$\Delta U_i = \begin{Bmatrix} \Delta u_i \\ \Delta v_i \\ \Delta w_i \\ \Delta\varphi_{1i} \\ \Delta\varphi_{2i} \end{Bmatrix} \tag{10.162}$$

The notation and sign convention for normal rotations is consistent with that adopted earlier in our discussion of Mindlin plate elements. We point out that $\Delta\varphi_1$ corresponds to a (vector) rotation about the $-s_2$ axis, which therefore produces an incremental displacement $-\zeta(h/2)\Delta\varphi_1$ in the direction of the s_1 axis. Similarly, $\Delta\varphi_2$ corresponds to a vector rotation about the s_1 axis, which produces an incremental displacement $-\zeta(h/2)\Delta\varphi_2$ in the direction of the s_2 axis.

To provide additional insight into the displacement constraints embedded in Equation 10.161, we consider a given midsurface location (ξ,η). The specified displacement variation through the thickness of the element may then be written, in global Cartesian coordinates, as

$$\Delta\bar{\mathbf{u}} = \Delta\mathbf{u} - \zeta\frac{h}{2}[s_1, s_2]\begin{Bmatrix} \Delta\varphi_1 \\ \Delta\varphi_2 \end{Bmatrix} \tag{10.163}$$

Transformation into the surface coordinate system (Equation 10.156) via premultiplication by S^T gives

$$\begin{Bmatrix} \Delta\bar{u} \\ \Delta\bar{v} \\ \Delta\bar{w} \end{Bmatrix}_S = \begin{Bmatrix} \Delta u \\ \Delta v \\ \Delta w \end{Bmatrix}_S - \zeta\frac{h}{2}\begin{bmatrix} s_1^T \\ s_2^T \\ s_3^T \end{bmatrix}[s_1, s_2]\begin{Bmatrix} \Delta\varphi_1 \\ \Delta\varphi_2 \end{Bmatrix} \tag{10.164}$$

where the subscript S denotes an incremental displacement vector referred to the surface coordinate system. Writing these three equations out explicitly, we find the familiar expressions

$$\begin{cases} \Delta\bar{u}_S = \Delta u_S - \zeta\dfrac{h}{2}\Delta\varphi_1 \\ \Delta\bar{v}_S = \Delta v_S - \zeta\dfrac{h}{2}\Delta\varphi_2 \\ \Delta\bar{w}_S = \Delta w_S \end{cases} \tag{10.165}$$

that we encountered in the Mindlin plate theory.

The element stresses, strains, and material properties are to be expressed in the lamina (x', y', z') orthogonal coordinate systems, whose orientations, strictly speaking, vary not only with midsurface location (ξ,η), but also with ζ because of the variation of the Jacobian with ζ(Equation 10.153).

10.4.8 Approximations

It is usually considered a reasonable approximation for thin shells to assume that the Jacobian is independent of ζ (or z), that is,

$$\bar{J}(\xi,\eta,\zeta) \simeq J(\xi,\eta) \qquad (10.166)$$

In classic thin-shell theory, this assumption would lead to, for example, dropping the terms $z/R\alpha$ and z/R_β from the integrands of Equations 10.140 and 10.141. We adopt this simplification here, because it allows integrations in the ζ direction to be carried out analytically, which greatly reduces the computational burden associated with the numerical integration of element matrices, especially when material nonlinearity is considered. An immediate consequence is that all lamina coordinate systems at a given midsurface sampling point (ξ_j, η_k) have the same orientation; that is, the orthogonal matrix **E** at a given sampling point does not vary with ζ. A further advantage is that the familiar bending moment and membrane stress resultants of shell theory then emerge explicitly.

We also adopt the approximation $s_3 \cdot e_3 \simeq 1$ to make the ensuing analysis tractable. Even though the shell normal vector s_3 is not exactly normal to the (e_1,e_2) plane of the lamina coordinate system, it should be a very good approximation for reasonable meshing of thin shells using curved Mindlin elements. In essence, the inextensibilty of a line in the e_3 direction (rather than in the s_3 direction) is assumed in the lamina kinematics, along with the plane stress approximation $\bar{\sigma}'_{33} \approx 0$.

10.4.9 Lamina kinematics

The displacements in a generic lamina are denoted $\bar{u}', \bar{v}', \bar{w}'$ in the (x', y', z') directions, respectively, that is, $\bar{\mathbf{u}}' = \bar{u}'e_1 + \bar{v}'e_2 + \bar{w}'e_3$. The corresponding midsurface displacements are denoted by u', v', w'.

Following the process used in the earlier development of the Mindlin flat shell element, we arrive at the following expressions for the total Green strains in the generic lamina in terms of midsurface displacements and rotations (in the lamina coordinate system at (ξ,η)).

$$\begin{cases} \bar{\varepsilon}_{x'} = u'_{,x'} + \dfrac{1}{2}w'^{2}_{,x'} - z'\varphi_{x',x'} \\ \bar{\varepsilon}_{y'} = v'_{,y'} + \dfrac{1}{2}w'^{2}_{,y'} - z'\varphi_{y',y'} \\ \bar{\gamma}_{x'y'} = u'_{,y'} + v'_{,x'} + w'_{,x'}w'_{,y'} - z'[\varphi_{x',y'} + \varphi_{y',x'}] \\ \bar{\gamma}_{x'z'} = w'_{,x'} - \varphi_{x'} \\ \bar{\gamma}_{y'z'} = w'_{,y'} - \varphi_{y'} \end{cases} \qquad (10.167)$$

In Equation 10.167, the notation "$(\cdot)_{,x'}$" denotes "$\partial(\cdot)/\partial x'$".

In these equations, the total Green strains in the lamina have been approximated by dropping terms such as $\bar{u}'^{2}_{,x'}$ in comparison with $|\bar{u}'_{,x'}|$ since strains are assumed to be small. In addition, we have assumed that (1) a line element in the direction of e_3 is inextensible (i.e., $\bar{\varepsilon}_{z'} = \overline{w'_{,z'}} = 0$), and (2) the transverse shear strains are constant in the through-thickness direction, that is, $\bar{\gamma}_{x'z'} = \gamma_{x'z'}$ and $\bar{\gamma}_{y'z'} = \gamma_{y'z'}$. This latter kinematic assumption (particular to the Mindlin theory) provides the necessary link between the normal rotations and the midsurface rotations ($w'_{,x'}$, $w'_{,y'}$) via the transverse shear strains.

The (5 × 1) strain vector $\bar{\boldsymbol{\varepsilon}}'$ in Equation 10.167 can be partitioned as

$$\bar{\boldsymbol{\varepsilon}}' \equiv \begin{Bmatrix} \bar{\boldsymbol{\varepsilon}}'_m \\ \bar{\boldsymbol{\gamma}}' \end{Bmatrix} = \begin{Bmatrix} \boldsymbol{\varepsilon}'_m \\ \boldsymbol{\gamma}' \end{Bmatrix} - z' \begin{Bmatrix} \boldsymbol{\kappa}' \\ 0 \end{Bmatrix} \tag{10.168}$$

where the midsurface quantities in this equation (i.e., the (3 × 1) membrane strain $\boldsymbol{\varepsilon}'_m$, the (2 × 1) transverse shear strain $\boldsymbol{\gamma}'$, and (3 × 1) curvature vector $\boldsymbol{\kappa}'$) are defined as

$$\boldsymbol{\varepsilon}'_m = \begin{Bmatrix} \varepsilon_{x'} \\ \varepsilon_{y'} \\ \gamma_{x'y'} \end{Bmatrix}, \quad \boldsymbol{\gamma}' = \begin{Bmatrix} \gamma_{x'z'} \\ \gamma_{y'z'} \end{Bmatrix} \quad \text{and} \quad \boldsymbol{\kappa}' = \begin{Bmatrix} \varphi_{x',x'} \\ \varphi_{y',y'} \\ \varphi_{x',y'} + \varphi_{y',x'} \end{Bmatrix} \tag{10.169}$$

The interpretation of these as "membrane" and "transverse" strains has physical significance only in a lamina coordinate system. The lamina midsurface membrane strain $\boldsymbol{\varepsilon}'_m$ contains both linear and nonlinear terms, as before.

The incremental strains in a generic lamina, determined from Equation 10.167, are

$$\begin{cases} \Delta\bar{\varepsilon}_{x'} = \Delta u'_{,x'} + w'_{,x'}\Delta w'_{,x'} - z'\Delta\varphi_{x',x'} \\ \Delta\bar{\varepsilon}_{y'} = \Delta v'_{,y'} + w'_{,y'}\Delta w'_{,y'} - z'\Delta\varphi_{y',y'} \\ \Delta\bar{\gamma}_{x'y'} = \Delta u'_{,y'} + \Delta v'_{,x'} + w'_{,y'}\Delta w'_{,x'} + w'_{,x'}\Delta w'_{,y'} - z'[\Delta\varphi_{x',y'} + \Delta\varphi_{y',x'}] \\ \Delta\bar{\gamma}_{x'z'} = \Delta w'_{,x'} - \Delta\varphi_{x'} \\ \Delta\bar{\gamma}_{y'z'} = \Delta w'_{,y'} - \Delta\varphi_{y'} \end{cases} \tag{10.170}$$

Identical relations pertain for virtual strains in terms of virtual displacements. These relations may be put in a form analogous to Equation 10.168:

$$\begin{Bmatrix} \Delta\bar{\boldsymbol{\varepsilon}}'_m \\ \Delta\bar{\boldsymbol{\gamma}}' \end{Bmatrix} = \begin{Bmatrix} \Delta\boldsymbol{\varepsilon}'_m \\ \Delta\boldsymbol{\gamma}' \end{Bmatrix} - z' \begin{Bmatrix} \Delta\boldsymbol{\kappa}' \\ 0 \end{Bmatrix} \tag{10.171}$$

Equation 10.171 can be further elaborated as

$$\begin{Bmatrix} \Delta\bar{\boldsymbol{\varepsilon}}'_m \\ \Delta\bar{\boldsymbol{\gamma}}' \end{Bmatrix} = \begin{Bmatrix} \Delta\boldsymbol{\varepsilon}'^{l}_m \\ \Delta\boldsymbol{\gamma}' \end{Bmatrix} + \begin{Bmatrix} \Delta\boldsymbol{\varepsilon}'^{nl}_m \\ 0 \end{Bmatrix} - z' \begin{Bmatrix} \Delta\boldsymbol{\kappa}' \\ 0 \end{Bmatrix} \tag{10.172}$$

by explicitly separating the midsurface incremental membrane strains into linear and nonlinear components, $\Delta\boldsymbol{\varepsilon}'^{l}_m$ and $\Delta\boldsymbol{\varepsilon}'^{nl}_m$, respectively.

10.4.10 Lamina stresses

In the updated current configuration, the second Piola–Kirchhoff (2PK) stress and Cauchy stress are momentarily identical, so we may write the stress in the lamina in partitioned vector form as

$$\bar{\boldsymbol{\sigma}}' = \begin{Bmatrix} \bar{\sigma}_{x'} \\ \bar{\sigma}_{y'} \\ \bar{\tau}_{x'y'} \\ \bar{\tau}_{x'z'} \\ \bar{\tau}_{y'z'} \end{Bmatrix} = \begin{Bmatrix} \bar{\boldsymbol{\sigma}}'_m \\ \boldsymbol{\tau}' \end{Bmatrix} \tag{10.173}$$

where the overbar denotes the current stress in a generic lamina. We omit the normal stress $\bar{\sigma}_{z'}$ from $\bar{\boldsymbol{\sigma}}'$ according to the standard plane stress assumption for thin plates and shells discussed previously in this chapter.

For the sake of simplicity, we assume a linear elastic stress–strain relation that incorporates the assumptions of the Mindlin theory, namely,

$$\begin{Bmatrix} \bar{\sigma}_{x'} \\ \bar{\sigma}_{y'} \\ \bar{\tau}_{x'y'} \\ \bar{\tau}_{x'z'} \\ \bar{\tau}_{y'z'} \end{Bmatrix} = \frac{E}{(1-\nu^2)} \begin{bmatrix} 1 & \nu & 0 & 0 & 0 \\ \nu & 1 & 0 & 0 & 0 \\ 0 & 0 & (1-\nu)/2 & 0 & 0 \\ 0 & 0 & 0 & \alpha(1-\nu)/2 & 0 \\ 0 & 0 & 0 & 0 & \alpha(1-\nu)/2 \end{bmatrix} \begin{Bmatrix} \bar{\varepsilon}_{x'} \\ \bar{\varepsilon}_{y'} \\ \bar{\gamma}_{x'y'} \\ \bar{\gamma}_{x'z'} \\ \bar{\gamma}_{y'z'} \end{Bmatrix} \tag{10.174}$$

which can be partitioned as

$$\begin{Bmatrix} \bar{\boldsymbol{\sigma}}'_m \\ \boldsymbol{\tau}' \end{Bmatrix} = \begin{bmatrix} \mathbf{D}_m & \mathbf{0} \\ \mathbf{0} & \mathbf{D}_s \end{bmatrix} \begin{Bmatrix} \bar{\boldsymbol{\varepsilon}}'_m \\ \bar{\boldsymbol{\gamma}}' \end{Bmatrix} \tag{10.175}$$

where

$$\mathbf{D}_m = \frac{E}{(1-\nu^2)} \begin{bmatrix} 1 & \nu & 0 \\ \nu & 1 & 0 \\ 0 & 0 & (1-\nu)/2 \end{bmatrix}, \quad \text{and} \quad \mathbf{D}_s = \alpha G \begin{bmatrix} 1 & 0 \\ 0 & 1 \end{bmatrix} \tag{10.176}$$

In these equations, the constant $\alpha = 5/6$, as usual in Mindlin finite elements. Incremental stresses are given by the linear relations assumed above as

$$\begin{Bmatrix} \Delta\bar{\boldsymbol{\sigma}}'_m \\ \Delta\bar{\boldsymbol{\tau}}' \end{Bmatrix} = \begin{bmatrix} \mathbf{D}_m & \mathbf{0} \\ \mathbf{0} & \mathbf{D}_s \end{bmatrix} \begin{Bmatrix} \Delta\bar{\boldsymbol{\varepsilon}}'_m \\ \Delta\bar{\boldsymbol{\gamma}}' \end{Bmatrix} \tag{10.177}$$

where the stress increments are actually increments of 2PK stress.

10.4.11 Virtual work density

The virtual work "density" δW^0_{int} (i.e., virtual work per unit midsurface area $|J|d\xi d\eta$) at a particular midsurface location (ξ, η) may be written, by integrating through the thickness of the shell, as

$$\delta W^0_{int} = -\int_{-h/2}^{h/2} [\delta\bar{\boldsymbol{\varepsilon}}'^T_m \quad \delta\bar{\boldsymbol{\gamma}}'^T] \begin{Bmatrix} \bar{\boldsymbol{\sigma}}'_m \\ \bar{\boldsymbol{\tau}}' \end{Bmatrix} dz' \tag{10.178}$$

Inserting the version of Equation 10.172 written for virtual strains rather than incremental strains, we obtain

$$\delta W^0_{int} = -[\delta\boldsymbol{\varepsilon}'^{lT}_m \mathbf{N}' + \delta\boldsymbol{\varepsilon}'^{nlT}_m \mathbf{N}' + \delta\boldsymbol{\gamma}'^T \mathbf{Q}' + \delta\boldsymbol{\kappa}'^T \mathbf{M}'] \tag{10.179}$$

where the stress resultants **N′**, **Q′**, **M′** are defined in terms of through-thickness integrals of stresses as for the flat Mindlin shell element. In Equation 10.179, we have now separated the membrane strain $\delta\boldsymbol{\varepsilon}'_m$ into its linear and nonlinear components. Either of these last two equations may be used to develop an expression for the element internal resisting force vector.

For the curved Mindlin shell element, we cannot (numerically) integrate this virtual work density over the midsurface of the element because the lamina coordinate systems vary, in general, with (ξ, η) on the midsurface. In order to proceed further, we must express the variables in the global Cartesian coordinate system.

To illustrate the general procedure, we first consider the organization of calculations to express the linear membrane strains $\boldsymbol{\varepsilon}'^{l}_{m}$ in the global (x, y, z) Cartesian coordinates.

The linear membrane strain vector can be formally written in terms of the lamina displacement gradients as

$$\boldsymbol{\varepsilon}'^{l}_{m} = \begin{bmatrix} 1 & 0 & 0 & 0 \\ 0 & 0 & 0 & 1 \\ 0 & 1 & 1 & 0 \end{bmatrix} \begin{bmatrix} u'_{,x'} \\ u'_{,y'} \\ v'_{,x'} \\ v'_{,y'} \end{bmatrix} \equiv \mathbf{L} \begin{bmatrix} u'_{,x'} \\ u'_{,y'} \\ v'_{,x'} \\ v'_{,y'} \end{bmatrix} \tag{10.180}$$

In general, displacement gradients in the lamina coordinate system are related to those in the global Cartesian coordinates by the transformation

$$\begin{bmatrix} u'_{,x'} & v'_{,x'} & w'_{,x'} \\ u'_{,y'} & v'_{,y'} & w'_{,y'} \\ u'_{,z'} & v'_{,z'} & w'_{,z'} \end{bmatrix} = \mathbf{E}^T \begin{bmatrix} u_{,x} & v_{,x} & w_{,x} \\ u_{,y} & v_{,y} & w_{,y} \\ u_{,z} & v_{,z} & w_{,z} \end{bmatrix} \mathbf{E} \tag{10.181}$$

The four lamina displacement gradients that are of interest for $\boldsymbol{\varepsilon}'^{l}_{m}$ can be related to Cartesian displacement gradients by a (4 × 9) matrix $\mathbf{H}_m$ that is built in a straightforward manner from Equation 10.181, that is,

$$\begin{bmatrix} u'_{,x'} \\ u'_{,y'} \\ v'_{,x'} \\ v'_{,y'} \end{bmatrix} = \mathbf{H}_m \begin{Bmatrix} u_{,x} \\ v_{,x} \\ w_{,x} \\ u_{,y} \\ v_{,y} \\ w_{,y} \\ u_{,z} \\ v_{,z} \\ w_{,z} \end{Bmatrix} \equiv \mathbf{H}_m \begin{Bmatrix} \boldsymbol{\theta}_x \\ \boldsymbol{\theta}_y \\ \boldsymbol{\theta}_z \end{Bmatrix} \tag{10.182}$$

where, for example, $\boldsymbol{\theta}_x \equiv [u_{,x}, v_{,x}, w_{,x}]^T$, and so forth. We note that expressions for the lamina displacement gradients $w'_{,x'}$ and $w'_{,y'}$ are also available from Equation 10.181.

The Cartesian displacement gradients in this last equation are related to the nodal displacement DOF via the isoparametric mapping

$$\begin{Bmatrix} \boldsymbol{\theta}_x \\ \boldsymbol{\theta}_y \\ \boldsymbol{\theta}_z \end{Bmatrix}_{(9\times 1)} = [\mathbf{G}_1 \quad \mathbf{G}_2 \quad \ldots \quad \mathbf{G}_n]_{(9\times 3n)} \begin{Bmatrix} \mathbf{u}_1 \\ \mathbf{u}_2 \\ \vdots \\ \mathbf{u}_n \end{Bmatrix}_{(3n\times 1)} \equiv \mathbf{G} \begin{Bmatrix} \mathbf{u}_1 \\ \mathbf{u}_2 \\ \vdots \\ \mathbf{u}_n \end{Bmatrix}_{(3n\times 1)} \tag{10.183}$$

where $\mathbf{u}_i \equiv [u_i, v_i, w_i]^T$ is the (3×1) nodal displacement vector at node i and the (9×3) nodal block matrix $\mathbf{G}_i$ is defined as

$$G_i = \begin{bmatrix} N_{i,x}I_3 \\ N_{i,y}I_3 \\ N_{i,z}I_3 \end{bmatrix} \tag{10.184}$$

The shape function derivatives in Equation 10.184 are found using the midsurface Jacobian J as

$$\begin{Bmatrix} N_{i,x} \\ N_{i,y} \\ N_{i,z} \end{Bmatrix} = J^{-1} \begin{Bmatrix} N_{i,\xi} \\ N_{i,\eta} \\ 0 \end{Bmatrix} \tag{10.185}$$

Therefore, the (linear) lamina membrane strains can be expressed in terms of nodal *displacement* DOF as

$$\boldsymbol{\varepsilon}'^{l}_{m} = LH_mG \begin{Bmatrix} u_1 \\ u_2 \\ \vdots \\ u_n \end{Bmatrix} \tag{10.186}$$

Incremental (or virtual) lamina membrane strains follow directly from this equation, for example,

$$\Delta\boldsymbol{\varepsilon}'^{l}_{m} = LH_mG \begin{Bmatrix} \Delta u_1 \\ \Delta u_2 \\ \vdots \\ \Delta u_n \end{Bmatrix} \tag{10.187}$$

10.4.12 Rotational freedoms

Both the transverse shear strain $\boldsymbol{\gamma}'$ and the curvature vector $\boldsymbol{\kappa}'$ are related to lamina normal rotations, which we now consider.

The lamina curvature vector $\boldsymbol{\kappa}'$ may be formally written as

$$\boldsymbol{\kappa}' = \begin{bmatrix} 1 & 0 & 0 & 0 \\ 0 & 0 & 0 & 1 \\ 0 & 1 & 1 & 0 \end{bmatrix} \begin{Bmatrix} \varphi_{x',x'} \\ \varphi_{x',y'} \\ \varphi_{y',x'} \\ \varphi_{y',y'} \end{Bmatrix} = L \begin{Bmatrix} \varphi_{x',x'} \\ \varphi_{x',y'} \\ \varphi_{y',x'} \\ \varphi_{y',y'} \end{Bmatrix} \tag{10.188}$$

similar to Equation 10.180 for $\boldsymbol{\varepsilon}'^{l}_{m}$.

A relation derived from that of Equation 10.181 can be used to determine the four partial derivatives of the lamina normal rotations that are of interest in Equation 10.188. The difference in the transformation relations in the current case arises from the fact that normal rotations are represented as vectors by $\varphi_y \mathbf{i}_1 - \varphi_x \mathbf{i}_2 + \varphi_z \mathbf{i}_3$ and $\varphi_{y'} \mathbf{e}_1 - \varphi_{x'} \mathbf{e}_2 + \varphi_{z'} \mathbf{e}_3$ in the (x, y, z) and (x′, y′, z′) coordinate systems, respectively, consistent with the representation of the nodal tangent vectors. The result is that the first and second rows of the matrices in Equation 10.181 have to be interchanged, with an appropriate sign change, as shown below.

$$\begin{bmatrix} \varphi_{y',x'} & \varphi_{y',y'} & \varphi_{y',z'} \\ -\varphi_{x',x'} & -\varphi_{x',y'} & -\varphi_{x',z'} \\ \varphi_{z',x'} & \varphi_{z',y'} & \varphi_{z',z'} \end{bmatrix} = \mathbf{E}^T \begin{bmatrix} \varphi_{y,x} & \varphi_{y,y} & \varphi_{y,z} \\ -\varphi_{x,x} & -\varphi_{x,y} & -\varphi_{x,z} \\ \varphi_{z,x} & \varphi_{z,y} & \varphi_{z,z} \end{bmatrix} \mathbf{E} \tag{10.189}$$

The four lamina rotation gradients can then be written as

$$\begin{Bmatrix} \varphi_{x',x'} \\ \varphi_{x',y'} \\ \varphi_{y',x'} \\ \varphi_{y',y'} \end{Bmatrix} = \mathbf{H}_b \begin{Bmatrix} \varphi_{x,x} \\ \varphi_{y,x} \\ \varphi_{z,x} \\ \varphi_{x,y} \\ \varphi_{y,y} \\ \varphi_{z,y} \\ \varphi_{x,z} \\ \varphi_{y,z} \\ \varphi_{z,z} \end{Bmatrix} = \mathbf{H}_b \begin{Bmatrix} \boldsymbol{\varphi}_x \\ \boldsymbol{\varphi}_y \\ \boldsymbol{\varphi}_z \end{Bmatrix} \tag{10.190}$$

The matrix $\mathbf{H}_b$ is again easily built, from Equation 10.189 this time. In this equation, $\boldsymbol{\varphi}_x = [\varphi_{x,x}, \varphi_{y,x}, \varphi_{z,x}]^T$, and so forth. We note that whereas $\varphi_{z'}$ is (assumed to be) zero in the lamina coordinate system, φ_z in the global Cartesian coordinate system will not in general be zero.

The normal rotations φ_x, φ_y, and φ_z must eventually be expressed in terms of the nodal tangential rotations in the surface coordinate system.

It is convenient to introduce the notation $\mathbf{V}_i$ for the nodal rotation vector

$$\mathbf{V}_i \equiv [\mathbf{s}_1 \quad \mathbf{s}_2]_i \begin{Bmatrix} \varphi_2 \\ -\varphi_1 \end{Bmatrix}_i = \bar{\mathbf{S}}_i \mathbf{v}_i \tag{10.191}$$

which is referred to global coordinates. The nodal vector $\mathbf{v}_i = [\varphi_2, -\varphi_1]_i^T$ lies in the surface coordinate system, and the (3 × 2) nodal matrix $\bar{\mathbf{S}}_i$ consists of the first two columns of $\mathbf{S}_i$.

The (3 × 1) nodal rotation vector $\bar{\mathbf{S}}_i \mathbf{v}_i$ lies in the surface tangent plane normal to $\mathbf{s}_3$. The interpolant $\mathbf{V}(\xi.\eta)$ is given by

$$\mathbf{V}(\xi, \eta) = \sum_{i=1}^{n} N_i(\xi, \eta) \bar{\mathbf{S}}_i \mathbf{v}_i \tag{10.192}$$

with Cartesian components $\mathbf{V}_i = [\varphi_y, -\varphi_x, \varphi_z]^T$. The partial derivatives $\mathbf{V}_{,x}$ in the global Cartesian system are

$$\mathbf{V}_{,x} = \sum_{i=1}^{n} N_{i,x}(\xi, \eta) \bar{\mathbf{S}}_i \mathbf{v}_i \tag{10.193}$$

where $\mathbf{V}_{,x} = [\varphi_{y,x}, -\varphi_{x,x}, \varphi_{z,x}]^T$.

The rotation gradient $\boldsymbol{\varphi}_x$ defined in Equation 10.190 can be formally represented as

$$\boldsymbol{\varphi}_x \equiv \begin{Bmatrix} \varphi_{x,x} \\ \varphi_{y,x} \\ \varphi_{z,x} \end{Bmatrix} = \begin{bmatrix} 0 & -1 & 0 \\ 1 & 0 & 0 \\ 0 & 0 & 1 \end{bmatrix} \begin{Bmatrix} \varphi_{y,x} \\ -\varphi_{x,x} \\ \varphi_{z,x} \end{Bmatrix} = \mathbf{W}\mathbf{V}_{,x} \tag{10.194}$$

Similarly, $\boldsymbol{\varphi}_Y \equiv \mathbf{W}\mathbf{V}_{,y}$ and $\boldsymbol{\varphi}_z \equiv \mathbf{W}\mathbf{V}_{,z}$.
Therefore, we may write

$$\begin{Bmatrix} \mathbf{V}_x \\ \mathbf{V}_y \\ \mathbf{V}_z \end{Bmatrix} = \mathbf{G} \begin{Bmatrix} \overline{\mathbf{S}}_1\mathbf{v}_1 \\ \overline{\mathbf{S}}_2\mathbf{v}_2 \\ \vdots \\ \overline{\mathbf{S}}_n\mathbf{v}_n \end{Bmatrix} \tag{10.195}$$

and

$$\begin{Bmatrix} \boldsymbol{\varphi}_x \\ \boldsymbol{\varphi}_y \\ \boldsymbol{\varphi}_z \end{Bmatrix} = \begin{bmatrix} \mathbf{W} & \mathbf{0}_3 & \mathbf{0}_3 \\ \mathbf{0}_3 & \mathbf{W} & \mathbf{0}_3 \\ \mathbf{0}_3 & \mathbf{0}_3 & \mathbf{W} \end{bmatrix} \mathbf{G} \begin{Bmatrix} \overline{\mathbf{S}}_1\mathbf{v}_1 \\ \overline{\mathbf{S}}_2\mathbf{v}_2 \\ \vdots \\ \overline{\mathbf{S}}_n\mathbf{v}_n \end{Bmatrix} \tag{10.196}$$

The matrix $\mathbf{G}$, which involves shape function derivatives, has been defined in Equation 10.184. Equation 10.196 may be written more compactly as

$$\begin{Bmatrix} \boldsymbol{\varphi}_x \\ \boldsymbol{\varphi}_y \\ \boldsymbol{\varphi}_z \end{Bmatrix} = \hat{\mathbf{G}} \begin{Bmatrix} \overline{\mathbf{S}}_1\mathbf{v}_1 \\ \overline{\mathbf{S}}_2\mathbf{v}_2 \\ \vdots \\ \overline{\mathbf{S}}_n\mathbf{v}_n \end{Bmatrix} \tag{10.197}$$

where the (9 × 3) block matrix $\hat{\mathbf{G}}_i$ is defined by

$$\hat{\mathbf{G}}_i = \begin{bmatrix} N_{i,x}\mathbf{W} \\ N_{i,y}\mathbf{W} \\ N_{i,z}\mathbf{W} \end{bmatrix} \tag{10.198}$$

Finally, we may absorb the nodal matrices $\overline{\mathbf{S}}_i$ into the (9 × 3) block matrices $\hat{\mathbf{G}}_i$ and relabel the nodal tangential rotations $\mathbf{v}_i$ as $\boldsymbol{\theta} \equiv [\theta_1,\theta_2]^T$, that is,

$$\begin{Bmatrix} \theta_1 \\ \theta_2 \end{Bmatrix} \equiv \begin{Bmatrix} \varphi_2 \\ -\varphi_1 \end{Bmatrix} \tag{10.199}$$

so that the nodal rotation vector $\boldsymbol{\theta} \equiv \theta_{1i}\,\mathbf{s}_{1i}+\theta_{2i}\mathbf{s}_{2i}$ has components $(\theta_{1i},\theta_{2i})$ directed along $\mathbf{s}_{1i}$ and $\mathbf{s}_{2i}$, respectively. The curvature vector may be written, in a form completely analogous to the membrane strain, as

$$\boldsymbol{\kappa}' = \mathbf{L}\mathbf{H}_b\overline{\mathbf{G}} \begin{Bmatrix} \boldsymbol{\theta}_1 \\ \boldsymbol{\theta}_2 \\ \vdots \\ \boldsymbol{\theta}_n \end{Bmatrix} \tag{10.200}$$

where the (9 × 2) nodal block matrices $\overline{G}_i$ that together constitute $\overline{G}$ are defined as

$$\overline{G}_i = \begin{bmatrix} N_{i,x}\mathbf{W} \\ N_{i,y}\mathbf{W} \\ N_{i,z}\mathbf{W} \end{bmatrix} \overline{S}_i \tag{10.201}$$

The incremental (and virtual) curvature vector follows directly from Equation 10.200 as

$$\Delta\boldsymbol{\kappa}' = LH_b\overline{G}\begin{Bmatrix} \Delta\boldsymbol{\theta}_1 \\ \Delta\boldsymbol{\theta}_2 \\ \vdots \\ \Delta\boldsymbol{\theta}_n \end{Bmatrix} \tag{10.202}$$

A matrix expression relating the transverse shear strain vector γ' to the nodal DOF is easily obtained from the relations just established. These formal relations can of course be constructed more efficiently for computation.

The relations for $\Delta\boldsymbol{\varepsilon}'^{l}_{m}$, $\Delta\boldsymbol{\kappa}'$, and $\Delta\boldsymbol{\gamma}'$ and their virtual equivalents $\delta\boldsymbol{\varepsilon}'^{l}_{m}$, $\delta\boldsymbol{\kappa}'$, and $\delta\boldsymbol{\gamma}'$ lead to membrane, bending, and shear (matrix) contributions to the linear tangent stiffness for UDL incremental analysis.

The lamina contribution to the element geometric stiffness is developed exactly as for the flat Mindlin shell element and is at that point related to increments of the midsurface rotations $\Delta w'_{,x'}$ and $\Delta w'_{,y'}$ in the lamina coordinate system. Displacement gradients in the lamina coordinate system are related to displacement gradients in the global Cartesian coordinate system via Equation 10.181, and in general involve spatial (x, y, z) gradients of all three midsurface displacements (u, v, w,). As a result, the geometric stiffness depends on all the nodal displacement increments $\Delta\mathbf{u}_i$.

10.4.13 Comments on the UDL formulation

In the UDL formulation outlined here, geometry updating (including the global coordinates of the element geometry nodes) must be done after each new equilibrium configuration is accepted. Even with linear material behavior, the tangent stiffness matrices, strictly speaking, evolve because of element shape changes, although this may not be a significant effect here. All this makes an inherently relatively expensive element even more expensive, which may be why much of the early published work utilized a TL formulation.

We here treat the tangential nodal rotation increments $\Delta\boldsymbol{\theta}_i$ as vectors that, although not strictly true, should be a reasonable approximation for small increments. We therefore avoid updating of the total tangential rotations as required in a TL formulation (Parisch 1978, 1981; Hughes and Liu 1981; Surana 1983; Milford and Schnobrich 1986). Updating the coordinates of the element geometry nodes implicitly incorporates information on rotations.

10.5 NONLINEAR MATERIAL BEHAVIOR

Nonlinear material behavior, predominantly due to in-plane flexural stresses, in the structural elements that have been discussed in this chapter can be handled in two different ways, which we now briefly discuss.

10.5.1 Through-thickness numerical integration

Element matrices are evaluated as usual via numerical integration, in the through-thickness direction first, followed by integration over the midsurface (or centerline in the case of space frame elements).

If only one material model (e.g., a von Mises plasticity model) is used uniformly throughout the element to characterize nonlinear behavior, then numerical integration in the through-thickness direction can be used, in conjunction with standard Gaussian integration over the midsurface. There are two issues to be considered here:

1. A plane stress form of the material model must be used since 3D (Cauchy) stresses are not in general available. For example, the plane stress version of the von Mises yield criterion is an ellipse in terms of principal Cauchy flexural stresses $(\bar{\sigma}_1, \bar{\sigma}_2)$ in planes parallel to the midsurface.
2. Since first yield normally occurs at the extreme fibers, it is important for the accurate detection of initial yield to have sampling points located there. Gaussian sampling points are of course not located at the end points of the integration interval. As a result, Lobatto (sometimes called Radau) integration is often employed in the through-thickness direction, in conjunction with Gaussian integration over the midsurface. The general form of 1D Lobatto integration is

$$\int_{-1}^{1} f(\xi)d\xi = W_1 f(-1) + \sum_{2}^{n-1} W_i f(\xi_i) + W_n f(1) \tag{10.203}$$

where the sampling points $\xi_i(i=2, n\text{-}1)$ and the weights $W_i(i=1, n)$ are chosen for optimality à la Gaussian integration, subject to symmetry constraints and $\sum_{i=1}^{n} W_i = 2$ (Irons 1966).

10.5.2 Layered models

When (elastic or inelastic) material properties vary in the through-thickness direction, a discretization of the thickness into distinct layers, each of which may have its own unique material model, may be utilized (Hand et al. 1973). The property variations may often be stress induced, and not present in the undeformed configuration. Each layer is again assumed to be in plane stress, and through-thickness integration is carried out by analytically integrating over each layer, assuming the material properties are constant within the layer. This results in general in coupled constitutive relations between the (incremental) stress resultants ($\Delta\mathbf{M}$, $\Delta\mathbf{N}$) and the (incremental) midsurface deformation measures ($\Delta\boldsymbol{\varepsilon}_m$, $\Delta\boldsymbol{\kappa}$).

Suppose the plate or shell thickness is discretized into n layers, as illustrated in Figure 10.18.

We assume only that the material properties within a generic layer do not vary in the normal (z) direction. Therefore, in layer k, which has thickness $h_k \equiv z_{k+1} - z_k$, we may write the tangential stress–strain relation as

$$\Delta\boldsymbol{\sigma}_k = \mathbf{E}_k \Delta\boldsymbol{\varepsilon}_k \tag{10.204}$$

The variation of the incremental membrane strain $\Delta\boldsymbol{\varepsilon}$ in the z direction is

$$\Delta\boldsymbol{\varepsilon} = \Delta\boldsymbol{\varepsilon}_m - z\Delta\boldsymbol{\kappa} \tag{10.205}$$

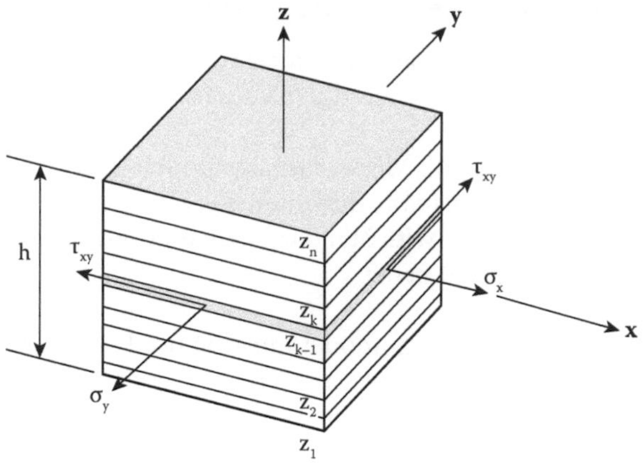

Figure 10.18 Layer discretization for material nonlinearity.

where $\Delta\boldsymbol{\varepsilon}_m$ is the incremental midsurface membrane strain and $\Delta\boldsymbol{\kappa}$ is the curvature vector. The incremental stress in layer k is then

$$\Delta\boldsymbol{\sigma}_k = E_k(\Delta\boldsymbol{\varepsilon}_m - z\Delta\boldsymbol{\kappa}) \tag{10.206}$$

and the section stress resultants are found as

$$\begin{aligned}\Delta N &= \int_{-h/2}^{h/2} \Delta\boldsymbol{\sigma} dz = \sum_{k=1}^{n}\left\{\int_{z_k}^{z_{k+1}} E_k(\Delta\boldsymbol{\varepsilon}_m - z\Delta\boldsymbol{\kappa})dz\right\} \\ &= \left[\sum_{k=1}^{n} E_k h_k\right]\Delta\boldsymbol{\varepsilon}_m - \left[\sum_{k=1}^{n} E_k \frac{1}{2}(z_{k+1}^2 - z_k^2)\right]\Delta\boldsymbol{\kappa}\end{aligned} \tag{10.207}$$

and

$$\begin{aligned}\Delta M &= -\int_{-h/2}^{h/2} \Delta\boldsymbol{\sigma} z dz = -\sum_{k=1}^{n}\left\{\int_{z_k}^{z_{k+1}} E_k(\Delta\boldsymbol{\varepsilon}_m - z\Delta\boldsymbol{\kappa})z dz\right\} \\ &= -\left[\sum_{k=1}^{n} E_k \frac{1}{2}(z_{k+1}^2 - z_k^2)\right]\Delta\boldsymbol{\varepsilon}_m + \left[\sum_{k=1}^{n} E_k \frac{1}{3}(z_{k+1}^3 - z_k^3)\right]\Delta\boldsymbol{\kappa}\end{aligned} \tag{10.208}$$

These relations may be compactly written as

$$\begin{Bmatrix}\Delta N \\ \Delta M\end{Bmatrix} = \begin{bmatrix}\mathbf{D}_{11} & \mathbf{D}_{12} \\ \mathbf{D}_{21} & \mathbf{D}_{22}\end{bmatrix}\begin{Bmatrix}\Delta\boldsymbol{\varepsilon}_m \\ \Delta\boldsymbol{\kappa}\end{Bmatrix} \tag{10.209}$$

where

$$D_{11} = \left[\sum_{k=1}^{n} E_k h_k\right]$$

$$D_{12} = D_{21}^{T} = -\left[\sum_{k=1}^{n} E_k \frac{1}{2}(z_{k+1}^{2} - z_k^{2})\right] \tag{10.210}$$

$$D_{22} = \left[\sum_{k=1}^{n} E_k \frac{1}{3}(z_{k+1}^{3} - z_k^{3})\right]$$

Note that we do not assume that the stresses (or strains) are constant (with respect to z) within a given layer. We also observe that if the material properties E_k (whether elastic or inelastic) and layer thicknesses h_k are asymmetrically disposed with respect to the midsurface, there will be coupling between the membrane and bending stress resultants. In beams exhibiting material nonlinearity, this is sometimes described as a "shift of the neutral axis," although the concept does not immediately extend to plates and shells.

Chapter 11

Solution methods

11.1 INTRODUCTION

The numerical solution of nonlinear structural problems is generally carried out by means of an incremental-iterative solution procedure in which the external load is gradually incremented and equilibrium is sought at each load level via an iterative process. We therefore consider a sequence of external load vectors ($\mathbf{P}_1, \mathbf{P}_2, \ldots, \mathbf{P}_n, \mathbf{P}_{n+1}, \ldots$) applied to the structure and seek the corresponding equilibrium configurations ($\mathbf{U}_1, \mathbf{U}_2, \ldots, \mathbf{U}_n, \mathbf{U}_{n+1}, \ldots$). At load-level $\mathbf{P}_n$, equilibrium is expressed as $\mathbf{P}_n - \mathbf{I}_n = 0$, where $\mathbf{I}_n$ is the internal resisting force vector. In practice, an equilibrium configuration is accepted when equilibrium is satisfied to within a specified tolerance. The residual load vector $\mathbf{R}_n$, defined as

$$\mathbf{R}_n = \mathbf{P}_n - \mathbf{I}_n \tag{11.1}$$

is a measure of the equilibrium "error." The iterative process seeks to drive the residual to zero, that is, $\mathbf{R}_n \to 0$, and equilibrium is accepted when, for example, $|\mathbf{R}_n| < \varepsilon |\mathbf{P}_n|$, where ε is a specified tolerance.

In the following sections, we briefly discuss the

1. Pure incremental (Euler–Cauchy) method
2. Incremental method with equilibrium corrections
3. Newton–Raphson incremental-iterative methods
4. Arc-length methods

11.2 PURE INCREMENTAL (EULER–CAUCHY) METHOD

If one multiplies the rate form of the nodal equilibrium equations $\mathbf{K}_t(\mathbf{U})\dot{\mathbf{U}} = \dot{\mathbf{P}}$ by a small but finite pseudo–time step Δt (e.g., setting $\dot{\mathbf{U}}\Delta t = \Delta \mathbf{U}$), the rate form is converted directly to an incremental system of equilibrium equations

$$\mathbf{K}_t(\mathbf{U}_n)\Delta \mathbf{U}_{n+1} = \Delta \mathbf{P}_{n+1} \tag{11.2}$$

where now $\mathbf{K}_t$ is the tangent stiffness at configuration $\mathbf{U}_n$, the beginning of the increment. We note that the specific form of the tangent stiffness depends on the underlying formulation (total Lagrangian [TL], updated Lagrangian [UDL], or corotational [CR]), but in general we may say that it is a function of the current configuration and stresses, and may also include other contributions, such as pressure stiffness.

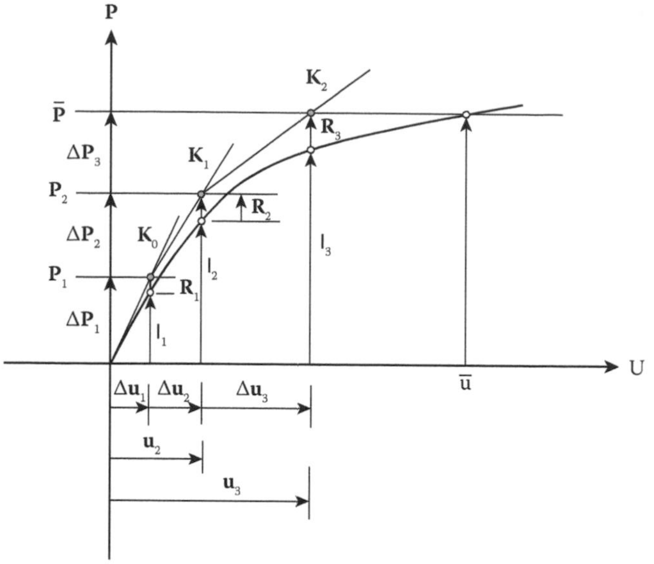

Figure 11.1 Pure incremental method (schematic).

If we simply apply this equation sequentially for a series of load increments ($\mathbf{\Delta P}_1, \mathbf{\Delta P}_2, \ldots, \mathbf{\Delta P}_n$) where

$$\mathbf{P}_n = \sum_{k=1}^{n} \mathbf{\Delta P}_k \tag{11.3}$$

and we update the tangent stiffness at the beginning of each new increment, we obtain a computed equilibrium path like that shown schematically by the discrete points in Figure 11.1 (which, strictly speaking, applies to a single degree of freedom case).

We note that there is a significant accumulating drift of the computed solution points away from the actual solution. In practice, it is found that even for very small load increments, the error accumulation is very often unacceptable. A significant improvement can be made, still without carrying out iterations, by means of a simple equilibrium correction.

11.3 INCREMENTAL METHOD WITH EQUILIBRIUM CORRECTION

In view of the numerical error accumulation that occurs in the purely incremental (Euler–Cauchy) method, an equilibrium correction scheme is clearly required in order to render the incremental approach viable for practical usage. The equilibrium correction compensates *at the next step* (n, n + 1) for the unbalanced force $\mathbf{R}_n$ existing at the end of increment (n – 1, n). Therefore, to advance the solution from $\mathbf{U}_n$ to $\mathbf{U}_{n+1}$, a load increment $\mathbf{P}_{n+1} - \mathbf{I}_n$ ($\equiv \mathbf{\Delta P}_{n+1} + \mathbf{R}_n$) is applied as illustrated in Figure 11.2, rather than just $\mathbf{\Delta P}_{n+1}$.

The incremental system of equilibrium equations becomes

$$\mathbf{K}_t(\mathbf{U}_n)\mathbf{\Delta U}_{n+1} = \mathbf{P}_{n+1} - \mathbf{I}_n \tag{11.4}$$

Note that the tangent stiffness for the current increment is evaluated at configuration $\mathbf{U}_n$ as previously, and that the only change from the pure incremental method is the inclusion of the equilibrium correction. Each computed solution point will now coincide more closely with the

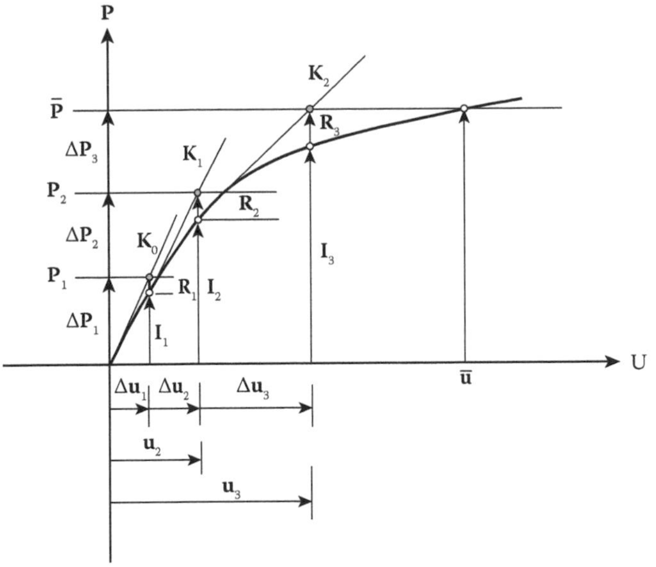

Figure 11.2 Incremental method with equilibrium correction (schematic).

actual solution, but will still contain a measurable residual equilibrium error that is simply accepted, since there is no mechanism in this method by which it can be reduced at the current step. Improvement over the pure incremental method is due solely to preventing the equilibrium error from accumulating. If the computed solution is deemed to be not sufficiently accurate (perhaps judged by the magnitude of $\|\mathbf{R}_n\|/\|\mathbf{P}_n\|$), the recourse would then be to repeat the solution using smaller load increments.

The equilibrium correction entails some additional cost, as it requires the computation of the internal resisting force vector, which we can represent as

$$\mathbf{I}_n = \sum_e \int_{V_e} \mathbf{B}_n^T \boldsymbol{\sigma}_n dV_e \tag{11.5}$$

for solid elements in a TL formulation, for example. The symbol $\sum_e$ denotes direct stiffness assembly, as usual. In this chapter we dispense with notation $\{\sigma\}$ to indicate the vector form of the tensor $\{\sigma\}$ and write the vector form simply as $\boldsymbol{\sigma}$.

We remark here that while schematic plots, such as Figures 11.1 and 11.2, for single degree of freedom structures are very useful for illustrating concepts, they can sometimes be misleading. As an example, it might initially be thought from an examination of Figure 11.1 or Figure 11.2 that a point corresponding to the internal resisting force vector $\mathbf{I}_n$ at configuration $\mathbf{U}_n$ lies on the true solution path. It is true that the internal resisting force vector always represents an equilibrium configuration, but the external load that is equilibrated at that configuration is not $\mathbf{P}_n$ as required, but rather $\mathbf{P}_n^* (= \mathbf{P}_n - \mathbf{R}_n)$. Only if $\mathbf{R}_n$ happened to be proportional to the load vector $\mathbf{P}_n$, that is, $\mathbf{R}_n = \gamma \mathbf{P}_n$, could $\mathbf{I}_n$ be considered a solution point, and in that case it would correspond to a scaled external load $\mathbf{P}_n^* = (1-\gamma)\mathbf{P}_n$. This scaling of the external load magnitude can always be done for a single degree of freedom system, as implied by Figure 11.2, but essentially "never" for a multi-degree-of-freedom system. Another way of saying this is that the internal resisting force vector will not generally be proportional to the external load pattern.

In order to reduce the equilibrium error $\mathbf{R}_{n+1}$ at $\mathbf{P}_{n+1}$, we must solve Equation 11.4 iteratively, recognizing that the incremental equilibrium equations are inherently nonlinear, that is, that the actual tangent stiffness does not remain constant, but varies during the increment.

Although other solution methods for nonlinear equations can be used, we focus our attention here on the well-known Newton–Raphson method and its variants.

11.4 INCREMENTAL-ITERATIVE (NEWTON–RAPHSON) METHOD

The Newton–Raphson method, sometimes simply called Newton's method, is a classic iterative root-finding method for nonlinear algebraic equations.

It is derived from a truncated Taylor series expansion of the function, say f(x), whose root $\overline{x} \Rightarrow f(\overline{x}) = 0$ is sought. The single-variable method starts with an initial estimate $x = x_1$ of the root. If $f(x_1) \neq 0$, then a Taylor series expansion of f(x) about $x = x_1$ gives

$$f(x - x_1) = f(x_1) + (x - x_1)f'(x_1) + R_1(x) \tag{11.6}$$

where $R_1(x)$ is the remainder, which may be expressed analytically in different forms. An improved estimate x_2 of the root is obtained from Equation 11.6 by setting $f(x_2 - x_1) = 0$ with $R_1 = 0$,

$$x_2 = x_1 - \frac{f(x_1)}{f'(x_1)} \tag{11.7}$$

and in general,

$$x_{i+1} = x_i - \frac{f(x_i)}{f'(x_i)}, \quad i = 1, 2, \ldots \tag{11.8}$$

as illustrated in Figure 11.3. We note that the tangent $f'(x_i)$ in equation 11.8 is evaluated at each new approximation $x = x_i$.

It can be proved that the method exhibits quadratic convergence to the solution $f(\overline{x}) = 0$ under certain favorable conditions. The most obvious of these conditions, evident from Figure 11.3, is that the tangent $f'(x_i) \neq 0$ in the interval where iterations are carried out. In addition, the function must be smooth [i.e., $f''(x)$ continuous in the interval] and the initial trial solution should be "close" to the root $x = \overline{x}$.

To apply Newton–Raphson iteration to reduce the residual $\mathbf{R}_{n+1}$ to a prescribed level, we write Equation 11.4 in the form

$$\mathbf{K}_{tn+1}^{i} \, \Delta\mathbf{U}_{n+1}^{i+1} = \mathbf{P}_{n+1} - \mathbf{I}_{n+1}^{i} \tag{11.9}$$

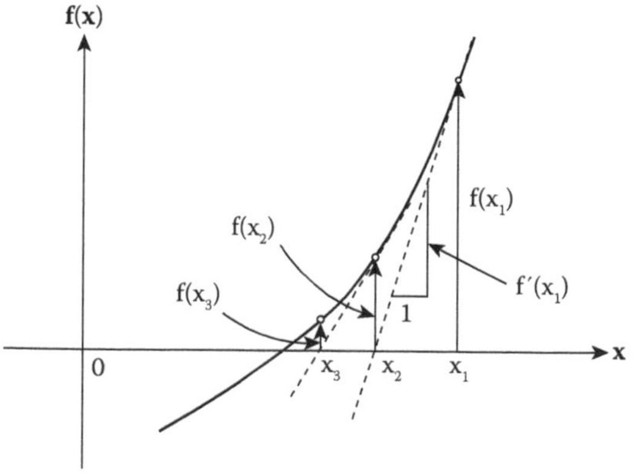

Figure 11.3 Newton–Raphson iteration in one variable.

where the superscript i denotes the iteration number within the current increment (n, n + 1). The external load $\mathbf{P}_{n+1}$ remains constant during the increment, but the tangent stiffness and internal resisting force vector are updated at the conclusion of each iteration i. At the beginning of the increment (n, n + 1), when i = 0,

$$\begin{cases} \mathbf{K}^{0}_{t\,n+1} \equiv \mathbf{K}_{t\ n} \\ \mathbf{I}^{0}_{n+1} \equiv \mathbf{I}_{n} \\ \mathbf{R}^{0}_{n+1} \equiv \mathbf{R}_{n} \end{cases} \tag{11.10}$$

are the final updated tangent stiffness, internal resisting force vector, and residual force vector, respectively, at the previous converged equilibrium configuration $\mathbf{U}_n$.

At each iteration, the system of linear equations in Equation 11.9 can be solved by using either a direct solution method, such as Gauss elimination, or a semi-iterative method, such as conjugate gradient or preconditioned conjugate gradient. In the preconditioned conjugate gradient method, the tangent stiffness matrix is not explicitly formed, which makes it well suited for very large finite element systems. Conjugate gradient algorithms are described in Ghaboussi and Wu (2016).

Iterations are terminated and an equilibrium configuration is accepted when a preselected convergence criterion is met. Various criteria may be used, based, for example, on the residual load, that is,

$$\|\mathbf{R}_{n+1}\| < \varepsilon\|\mathbf{P}_{n+1}\| \tag{11.11}$$

where ε is a specified tolerance, or perhaps a norm of the incremental displacement.

Figure 11.4 illustrates the expected effect of iterating at a fixed external load level, namely, the gradual reduction of the residual load vector and the consequent improvement of the computed solution point versus the true solution.

This is typically a computationally expensive procedure, particularly for large finite element systems, because it requires that the tangent stiffness matrix be updated and triangulated at each iteration within the increment.

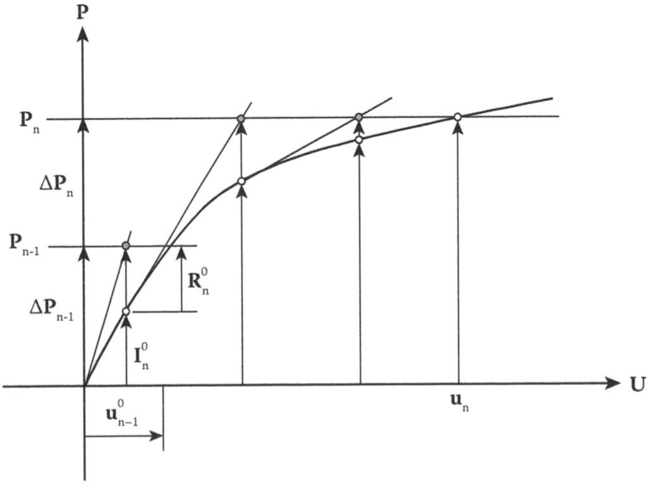

Figure 11.4 Incremental-iterative Newton–Raphson method (schematic).

11.5 MODIFIED NEWTON–RAPHSON METHOD

The modified Newton–Raphson (MNR) method, characterized by less frequent updates of the tangent stiffness matrix, is often a more suitable strategy for cases involving a moderate degree of structural nonlinearity.

This is a type of so-called quasi-Newton method in which the tangent stiffness is updated and decomposed (triangularized) only a few times within each load step, that is,

$$\overline{\mathbf{K}}_{tn+1}\ \Delta\mathbf{U}_{n+1}^{i+1} = \mathbf{P}_{n+1} - \mathbf{I}_{n+1}^{i} \tag{11.12}$$

where $\overline{\mathbf{K}}_{tn+1}$ represents the infrequently updated tangent stiffness. We note that the quadratic rate of convergence associated with Newton–Raphson iteration is thereby lost. This is a typical trade-off: additional cheap iterations versus fewer expensive iterations to achieve convergence. The preferred choice is primarily dependent on the degree and character of the structural nonlinearity.

Many variations of this method are possible. Figure 11.5 illustrates the most common version of the MNR procedure, in which a tangent stiffness update is made only at the start of the increment.

The decision of whether to update at a particular point in the iterative solution process can be based on, for example, monitoring the rate of convergence of successive iterative corrections, such as comparison of $\|\Delta\mathbf{U}_{n+1}^{i}\|$ versus $\|\Delta\mathbf{U}_{n+1}^{i-1}\|$ or comparison of $\|\Delta\mathbf{U}_{n+1}^{i}\|$ with the total incremental displacement $\|\Delta\overline{\mathbf{U}}_{n+1}\|$ thus far accumulated for the step, or it may be based simply on reaching a specified maximum number of iterations for the increment.

We note that in structural problems, the structural displacement vector **U** generally contains degrees of freedom of differing dimensions, that is, displacements and rotations, and the load vectors **P**, **I**, and **R** therefore contain components with differing (corresponding) dimensions, that is, forces and moments. The Euclidian norm of one of these vectors would therefore depend on the units chosen for the solution, an obviously undesirable situation. As a result, a decision must be made on how to measure the magnitudes of these vectors for use in assessing convergence and rate of convergence of the computed solution. For example, one procedure that has been used is to introduce a length scale by which rotations are multiplied

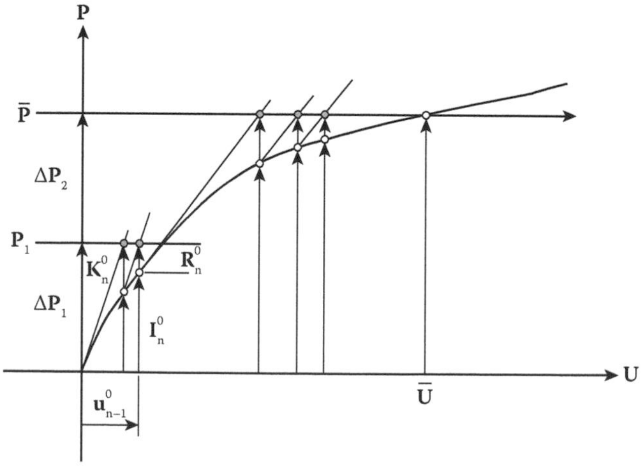

Figure 11.5 Modified Newton–Raphson method (schematic).

and moments are divided, for evaluating norms. Another procedure that is sometimes used is to simply exclude rotations from computation of displacement norms and moments from computation of force norms.

In practice, in order to achieve computational efficiency, many parameters that directly affect the solution process have to be selected. For example, very often we are required to prescribe the number of load steps or the load step size at the beginning of the solution process. Within each load step, we may need to prescribe the maximum number of Newton–Raphson iterations that are to be allowed or prescribe the frequency of tangent stiffness updates. All this involves trying to balance the number of iterations required for convergence against the computational cost of each iteration.

Adaptive solution schemes in which load increments are automatically selected based on the nonlinearity of the computed solution path can improve computational efficiency. In addition, the detection of points on the solution path at which the tangent stiffness becomes singular is essential so that the solution can be continued past those points if required.

11.6 CRITICAL POINTS ON THE EQUILIBRIUM PATH

A state of structural stability is associated with a positive-definite tangent stiffness matrix. If a so-called "critical" point is reached on the equilibrium path at which the tangent stiffness matrix becomes singular (i.e., it is rank deficient), a solution $\dot{\mathbf{U}}$ to the linear system of rate equations $\mathbf{K}_t\dot{\mathbf{U}} = \dot{\mathbf{P}}$ either does not exist or, if it does exist, is not unique. Of course, it is extremely unlikely that such a critical point will be "precisely" found in the normal course of an incremental numerical solution. However, slower convergence of the iterative solution process is often experienced "near" a critical point due to ill conditioning of the tangent stiffness matrix. This is often taken as an empirical indication that a critical point is nearby.

A critical point is classified either as a *limit point* or a *bifurcation*. An example of a limit point instability is shown in Figure 11.6, in which a shallow symmetric elastic arch is loaded at its centerline by a vertical concentrated load P. Due to the mirror symmetry of both the arch and the external loading about a vertical plane through the centerline, the primary equilibrium path (shown in Figure 11.7) lies in the P–U plane, consisting of a vertical displacement U (with the horizontal displacement V = 0). At the limit point, the equilibrium path reaches a local maximum at which dP/dU = 0. As the vertical displacement increases past the limit point, the primary path is unstable and the vertical load P must be decreased in order to satisfy equilibrium. The structure would like to "snap" or jump dynamically to a point on the stable rising equilibrium path on the right-hand side of the figure, where dP/dU > 0.

This so-called "snap-through" of the arch is an instance of elastic instability due to geometric nonlinearity, but limit point behavior can also occur purely as a result of material

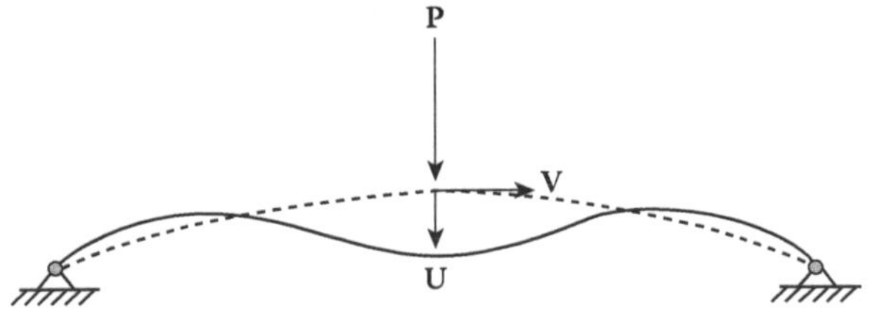

Figure 11.6 Snap-through of a shallow elastic arch.

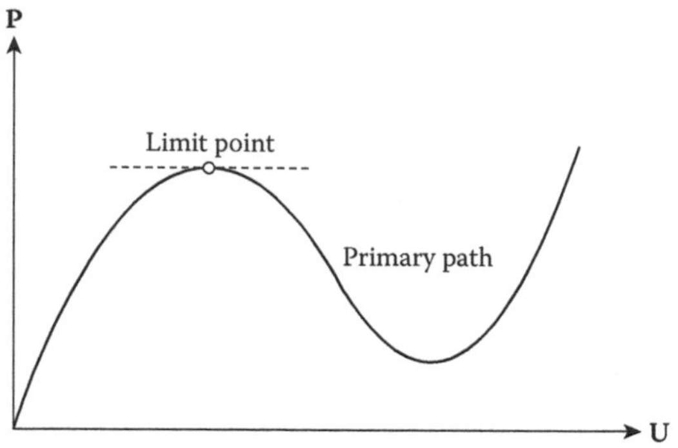

Figure 11.7 Limit point instability—shallow arch.

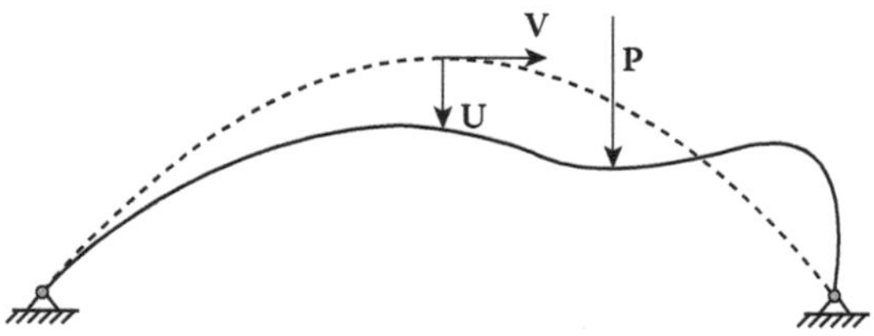

Figure 11.8 Bifurcation buckling of a deep arch.

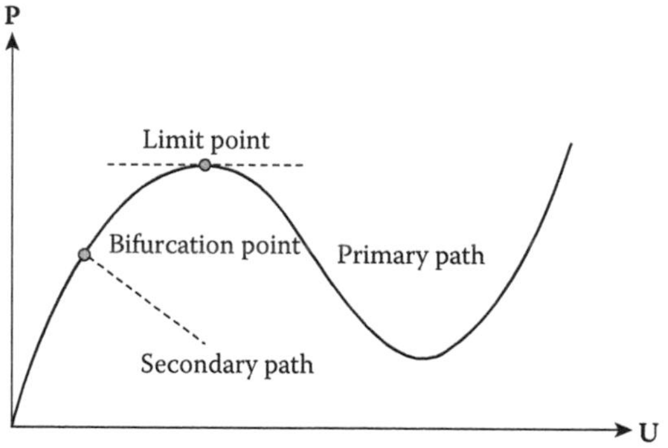

Figure 11.9 Equilibrium paths of a deep arch.

softening (e.g., metal plasticity), in which case the subsequent elastic stiffening shown in Figure 11.7 would not occur.

Bifurcation is the term used to describe the appearance of multiple equilibrium paths that intersect at, or branch from, a critical point. Figure 11.8 illustrates the bifurcation (in this case, side-sway) buckling of a deep elastic arch. Figure 11.9 illustrates the load–deflection (P vs. U) response on the primary path, and also indicates the location of a bifurcation point and the secondary path that emanates from it (shown as a dotted line). We define the "primary path" as the equilibrium path emanating from the unloaded configuration. Figure 11.9 shows the *projection* of the secondary path on the P-U plane. However, the secondary path is actually a space curve in the P-U-V space. Figure 11.10 shows its projection on the P-V plane.

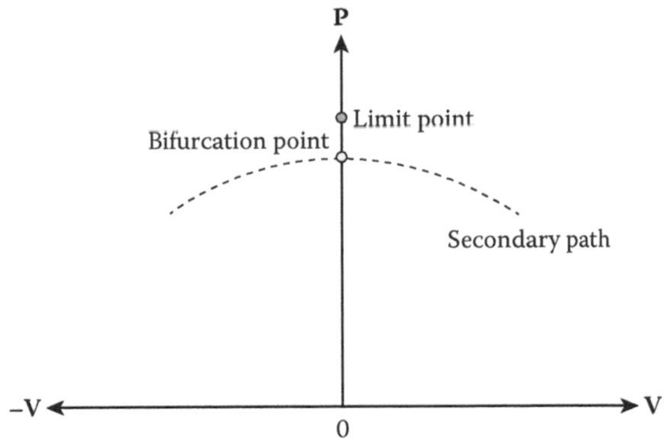

Figure 11.10 Secondary equilibrium path of a deep arch.

11.6.1 Characterization of critical points

In order to explore the analytical distinction between limit point and bifurcation instabilities, we consider the rate form of the equilibrium equations,

$$\mathbf{K}_t\dot{\mathbf{U}} = \dot{\lambda}\hat{\mathbf{P}} \tag{11.13}$$

for a proportional loading $\mathbf{P} = \lambda\hat{\mathbf{P}}$, where $\hat{\mathbf{P}}$ is a reference load pattern that specifies the spatial distribution of external loading, and λ is a scalar load factor that measures its magnitude. This equation can be decomposed using the spectral decomposition of the tangent stiffness matrix (Pecknold et al. 1985). At a particular configuration $\mathbf{U}$, the spectral decomposition can be written as

$$\mathbf{K}_t = \sum_{k=1}^{N} \omega_k \boldsymbol{\varphi}_k \boldsymbol{\varphi}_k^T \tag{11.14}$$

where ω_k, $k = 1,\ldots,N$ are the eigenvalues of $\mathbf{K}_t$ arranged in ascending order, and $\boldsymbol{\varphi}_k$, $k = 1,\ldots,N$ are its orthonormal eigenvectors.

The reference load vector $\hat{\mathbf{P}}$ and the displacement rate vector $\dot{\mathbf{U}}$ can be expanded in the eigenvector basis of $\mathbf{K}_t$ as

$$\begin{cases} \hat{\mathbf{P}} = \sum_{k=1}^{N} \hat{p}_k \boldsymbol{\varphi}_k \\ \dot{\mathbf{U}} = \sum_{k=1}^{N} \dot{u}_k \boldsymbol{\varphi}_k \end{cases} \tag{11.15}$$

Substituting these expressions into the rate equilibrium equation $\mathbf{K}_t\dot{\mathbf{U}} = \dot{\lambda}\hat{\mathbf{P}}$ and using the orthonormality of the eigenvectors ($\boldsymbol{\varphi}_i^T\boldsymbol{\varphi}_j = \delta_{ij}$), we obtain the following uncoupled set of relations:

$$\omega_k \dot{u}_k = \dot{\lambda}\ \hat{p}_k, \quad k = 1, \ldots, N \quad \text{(no sum)} \tag{11.16}$$

For the smallest eigenvalue (k = 1), we have

$$\omega_1 \dot{u}_1 = \dot{\lambda}\hat{p}_1 \equiv \dot{\lambda}\boldsymbol{\varphi}_1^T\hat{\mathbf{P}} \tag{11.17}$$

At a critical point where the tangent stiffness matrix is singular,

$$\omega_1 \to 0 \Rightarrow \begin{cases} \dot{\lambda} = 0 & \text{(limit point)} \\ \boldsymbol{\varphi}_1^T \hat{\mathbf{P}} = 0 & \text{(bifurcation)} \end{cases} \tag{11.18}$$

We observe that the displacement (rate) on the secondary path is initially in a direction $\boldsymbol{\varphi}_1$ orthogonal to the reference load vector $\hat{\mathbf{P}}$. We remark that it is possible, perhaps even likely, that more than one eigenvalue goes to zero at a bifurcation point in a structure with a high initial degree of symmetry. An example is discussed later in this chapter.

We know that a Newton–Raphson iteration will experience convergence difficulties in the close neighborhood of critical points due to ill conditioning of the tangent stiffness matrix. In order to continue an incremental solution past a limit point, the load must obviously be decreased, and stiffness updates may also have to be temporarily suspended. It is therefore important to have a means of determining when those measures should be taken.

In the case of a bifurcation point, especially one on an ascending primary equilibrium path, such as illustrated in Figure 11.9, we want to determine its approximate location, since that information is of intrinsic importance for the discovery of potential structural failure modes. Beyond that, we wish to develop effective strategies for bypassing the bifurcation point and continuing on the primary equilibrium path to determine the limit point failure load.

11.6.2 Monitoring the incremental solution on the primary path

Ideally, computation of the (say 10 or so) lowest eigenvalues and corresponding eigenvectors of the tangent stiffness matrix at converged equilibrium points on the solution path would provide the essential information required for detecting and locating both limit points and bifurcation points. However, this may sometimes be computationally too expensive for large finite element systems. Inverse power iteration can be used to find the single lowest eigenvalue and associated eigenvector. Subspace iteration (Bathe and Wilson 1973; Ghaboussi and Wu 2016) is more suitable, as it allows multiple eigenpairs to be computed simultaneously.

11.6.3 Determinant of the tangent stiffness

An alternative that provides some of the same information, but with no added computational expense, is to monitor the determinant of the tangent stiffness matrix. In general, this is easily accomplished, as described subsequently.

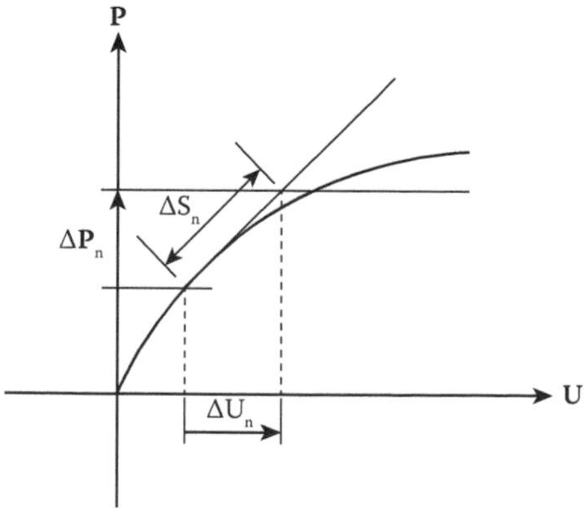

Figure 11.11 Parameters defined in the arc-length method.

The determinant of the tangent stiffness is useful because it can be expressed as the product of its eigenvalues, that is,

$$\det(\mathbf{K}_t) = \prod_{k=1}^{N} \omega_k \tag{11.19}$$

Consider, for example, an incremental MNR solution in which the tangent stiffness is updated only at the beginning of each increment. A sign change of the determinant from the beginning of one increment to the beginning of the next increment shows that an odd number of eigenvalues have changed sign during the increment. This indicates that a critical point has been passed, and isolates it within a specific load range $(\lambda_n, \lambda_{n+1})$. However, this information by itself does not reveal whether the critical point is a bifurcation or a limit point.

The determinant of the tangent stiffness matrix can be computed as a by-product during the regular solution process. For example, when we use a direct solution method, such as Gauss elimination, we perform a triangular decomposition of the tangent stiffness matrix.

$$\mathbf{K}_t = \mathbf{L}\overline{\mathbf{U}} \tag{11.20}$$

where $\mathbf{L}$ is the unit lower triangular (i.e., with unit diagonal elements) and $\overline{\mathbf{U}}$ is the upper triangular. The determinant of $\mathbf{K}_t$ can then be computed directly from the diagonal elements of $\overline{\mathbf{U}}$:

$$\begin{aligned}\det(\mathbf{K}_t) &\equiv \det(\mathbf{L})\det(\overline{\mathbf{U}}) \\ &= \det(\overline{\mathbf{U}}) = \prod_{k=1}^{N} \overline{\mathrm{U}}_{kk} \quad \text{(no sum on k)}\end{aligned} \tag{11.21}$$

Since the triangular decomposition is performed as a routine part of the regular solution process, the computation of $\det(\mathbf{K}_t)$ does not incur any additional computational expense.

11.6.4 Current stiffness parameter

Bergan et al. (1978) proposed the use of a "current stiffness parameter" that can be computed from information already available from the incremental solution. It can be effectively used to adaptively select load increments (Bergan 1980) as the incremental solution approaches a limit point, and to estimate its location. For proportional loadings $\mathbf{P} = \lambda\hat{\mathbf{P}}$, the current stiffness parameter (SP_n) is defined in rate form as

$$SP_n = \frac{\dot{\mathbf{U}}_0^T\hat{\mathbf{P}}}{\dot{\mathbf{U}}_n^T\hat{\mathbf{P}}} \tag{11.22}$$

where $\dot{\mathbf{U}}_0$ is the initial displacement rate at $\lambda = 0$, and $\dot{\mathbf{U}}_n$ is the displacement rate at $\lambda = \lambda_n$. A displacement rate $\dot{\mathbf{U}}$ is interpreted as $\dot{\mathbf{U}} = d\mathbf{U}/d\lambda$, which is essentially an inverse measure of stiffness *along the solution path*. That is, as the structure softens, a given load increment will produce progressively larger displacements, and thus $\dot{\mathbf{U}}_n^T\hat{\mathbf{P}}$ will increase relative to the reference value $\dot{\mathbf{U}}_0^T\hat{\mathbf{P}}$, reducing the current stiffness parameter accordingly.

When the rates are replaced by incremental changes, that is, $\dot{U} = \Delta U/\Delta\lambda$, the current stiffness parameter is expressed as

$$SP_n = \frac{\Delta\lambda_n}{\Delta\lambda_0}\frac{\Delta U_0^T \hat{P}}{\Delta U_n^T \hat{P}} \tag{11.23}$$

and $SP_n \to 0$ indicates a limit point. We note that the current stiffness parameter does not detect bifurcations on the solution path.

Basically, we need two types of special numerical schemes. First, we need diagnostic methods that can identify an approaching limit point or bifurcation point. Within the neighborhood of a limit point or bifurcation point, we may then need to adjust load step size in order to bypass the critical (limit or bifurcation) point.

Second, once a bifurcation point is identified, we may wish to proceed onto a secondary solution path emanating from the bifurcation point. Some special numerical techniques have been developed and used in nonlinear finite element analysis of problems involving limit points and bifurcation points.

11.7 ARC-LENGTH SOLUTION METHODS

Riks (1972, 1979) presented a nonlinear analysis formulation for multi-degree-of-freedom systems that can effectively handle limit point instabilities. In Riks's method, an N degree of freedom system subjected to a proportional loading $\mathbf{P} = \lambda\hat{\mathbf{P}}$ is treated in an (N + 1)-dimensional space (**U**, λ). The incremental equilibrium equations

$$\mathbf{K}_{t\,n}\ \Delta\mathbf{U}_{n+1} = \lambda_{n+1}\hat{\mathbf{P}} - \mathbf{I}_n \tag{11.24}$$

can be written in the alternative form

$$\mathbf{K}_{t\,n}\ \Delta\mathbf{U}_{n+1} - \hat{\mathbf{P}}\Delta\lambda_{n+1} = \mathbf{R}_n \tag{11.25}$$

in which we have used $\lambda_{n+1} = \Delta\lambda_{n+1} + \lambda_n$ and $\lambda_n\hat{\mathbf{P}} - \mathbf{I}_n \equiv \mathbf{R}_n$. $\hat{\mathbf{P}}$ is a constant load pattern, and the load parameter λ is now treated as an additional variable.

The solution path in the (N + 1)-dimensional space is a space curve with a differential arc length dS, namely,

$$d\mathbf{U}^T d\mathbf{U} + \alpha\ d\lambda^2\hat{\mathbf{P}}^T\hat{\mathbf{P}} = dS^2 \tag{11.26}$$

and instead of load control in which the sequence of external loads ($\mathbf{P}_1, \mathbf{P}_2, \ldots, \mathbf{P}_n, \mathbf{P}_{n+1}, \ldots$) is specified, the solution is controlled by specifying the arc length. The scale factor α, which may be chosen in different ways, is introduced so that the arc length is defined consistently in units of length. This last equation can be written in incremental form as

$$\Delta\mathbf{U}_{n+1}^T\Delta\mathbf{U}_{n+1} + \Delta\lambda_{n+1}^2\ \alpha\hat{\mathbf{P}}^T\hat{\mathbf{P}} = \Delta S_{n+1}^2 \tag{11.27}$$

where ΔS_{n+1} is the specified arc length for the increment (n, n + 1) (Figure 11.11). Equations 11.25 and 11.27 constitute an (N + 1) system of incremental equations in terms of the unknowns ($\Delta\mathbf{U}_{n+1}$, $\Delta\lambda_{n+1}$).

The arc-length constraint Equation 11.27 is nonlinear in the incremental unknowns ($\Delta\mathbf{U}_{n+1}$, $\Delta\lambda_{n+1}$). For a pure incremental solution, it may be linearized as

$$\Delta\mathbf{U}_n^T\Delta\mathbf{U}_{n+1} + \Delta\lambda_n\Delta\lambda_{n+1}\alpha\hat{\mathbf{P}}^T\hat{\mathbf{P}} = \Delta S_{n+1}^2 \tag{11.28}$$

by introducing the approximations $\Delta\mathbf{U}_{n+1}^T\ \Delta\mathbf{U}_{n+1} \approx \Delta\mathbf{U}_n^T\ \Delta\mathbf{U}_{n+1}$ and $\Delta\lambda_{n+1}^2 \approx \Delta\lambda_n\ \Delta\lambda_{n+1}$.

Equations 11.25 and 11.28 would then together provide the linear system of incremental equations

$$\begin{bmatrix} \mathbf{K}_{tn} & -\hat{\mathbf{P}} \\ \Delta\mathbf{U}_n^T & \Delta\lambda_n \alpha \hat{\mathbf{P}}^T \hat{\mathbf{P}} \end{bmatrix} \begin{Bmatrix} \Delta\mathbf{U}_{n+1} \\ \Delta\lambda_{n+1} \end{Bmatrix} = \begin{Bmatrix} \mathbf{R}_n \\ \Delta S_{n+1}^2 \end{Bmatrix} \tag{11.29}$$

which is nonsingular at a *limit point* (Riks 1979). It does not, however, remove singularities at bifurcation points, which must still be dealt with by other means.

We can see that there would be some serious computational disadvantages to this particular arc-length formulation: the inflated "tangent stiffness" is obviously no longer symmetric, and it would not in general have a narrow-band structure.

Crisfield (1981) and Ramm (1981, 1982) proposed incremental-iterative MNR arc-length methods that do not have these undesirable features and are therefore practical for large finite element systems.

Crisfield and Ramm both simplify the arc-length constraint by choosing the parameter $\alpha = 0$ in Equation 11.27. These approaches are sometimes termed "modified arc length" methods. In this case, the arc-length constraint simplifies to

$$\Delta\mathbf{U}_{n+1}^T \Delta\mathbf{U}_{n+1} = \Delta S_{n+1}^2 \tag{11.30}$$

This is now essentially a generalized displacement control method, in which the Euclidean norm of the incremental displacement is specified, that is, $\|\Delta\mathbf{U}_{n+1}\| = \Delta S_{n+1}$, in contrast to classic displacement control in which a single displacement degree of freedom is incremented. We note that this latter approach, which is not discussed herein, has very limited applicability.

Both methods involve iterative corrections to the load factor increment $\Delta\lambda_{n+1}$ and the displacement increment $\Delta\mathbf{U}_{n+1}$. We denote these iterative corrections as $\delta\lambda_{n+1}^{i+1}$ and $\delta\mathbf{U}_{n+1}^{i+1}$, respectively, defined as follows:

$$\begin{cases} \Delta\lambda_{n+1}^{i+1} = \Delta\lambda_{n+1}^{i} + \delta\,\lambda_{n+1}^{i+1} \\ \Delta\mathbf{U}_{n+1}^{i+1} = \Delta\mathbf{U}_{n+1}^{i} + \delta\mathbf{U}_{n+1}^{i+1} \end{cases} \tag{11.31}$$

where $\Delta\lambda_{n+1}^0 \equiv 0$ and $\Delta\mathbf{U}_{n+1}^0 \equiv 0$ and the superscript $(i \geq 0)$ denotes an iteration number within the increment (n, n + 1).

MNR iterations on Equation 11.24 at load step (n, n + 1) are given by

$$\mathbf{K}_{tn} \delta\mathbf{U}_{n+1}^{i+1} = \lambda_{n+1}^{i+1} \hat{\mathbf{P}} - \mathbf{I}_{n+1}^{i} \tag{11.32}$$

which can be written as

$$\mathbf{K}_{tn} \delta\mathbf{U}_{n+1}^{i+1} = \delta\lambda_{n+1}^{i+1} \hat{\mathbf{P}} + \mathbf{R}_{n+1}^{i} \tag{11.33}$$

in which we have used the relations

$$\begin{cases} \lambda_{n+1}^{i+1} = \lambda_{n+1}^{i} + \delta\lambda_{n+1}^{i+1} \\ \lambda_{n+1}^{i} \hat{\mathbf{P}} - \mathbf{I}_{n+1}^{i} = \mathbf{R}_{n+1}^{i} \end{cases} \tag{11.34}$$

The iterative correction $\delta\mathbf{U}_{n+1}^{i+1}$ is constructed as a linear combination of independent solutions of Equation 11.33 for the two load vectors $\hat{\mathbf{P}}$ and $\mathbf{R}_{n+1}^{i}$ using the tangent stiffness $\mathbf{K}_{tn}$ (thus reintroducing the limit point singularity). Therefore,

$$\delta\mathbf{U}_{n+1}^{i+1} = \delta\lambda_{n+1}^{i+1} \mathbf{v}_{n+1} + \mathbf{w}_{n+1}^{i+1} \tag{11.35}$$

where

$$\begin{cases} \mathbf{K}_{tn}\ \mathbf{v}_{n+1} = \hat{\mathbf{P}} \\ \mathbf{K}_{tn}\ \mathbf{w}_{n+1}^{i+1} = \mathbf{R}_{n+1}^{i} \end{cases} \tag{11.36}$$

We note that the solution $\mathbf{v}_{n+1}$ is obtained only at the beginning of the increment for MNR iterations, since $\mathbf{K}_{tn}$ remains constant during the imcrement.

The iterative load correction $\delta\lambda_{n+1}^{i+1}$ is determined using Equation 11.35 in the arc-length constraint. The way that the constraint is applied differentiates the Crisfield and Ramm methods, as described in the following two sections.

It is convenient to determine the arc length ΔS_1 for the first increment indirectly by specifying the load increment $\Delta\lambda_1$, that is,

$$\begin{aligned} &\mathbf{K}_{t0}\mathbf{v}_1 = \hat{\mathbf{P}} \\ &\Delta\mathbf{U}_1 = \Delta\lambda_1\mathbf{v}_1 \\ &\Delta S_1 = \Delta\lambda_1\sqrt{\mathbf{v}_1^T\mathbf{v}_1} \end{aligned} \tag{11.37}$$

The arc length ΔS_{n+1} for subsequent load steps may be taken as constant ($\Delta S_{n+1} = \Delta S_1$), or it may be varied using, for example, the current stiffness parameter (Bergan 1980). In Crisfield's method, the arc length is constant during iterations for a given increment (n, n + 1). In Ramm's method, ΔS_{n+1} can be regarded as an initial value, but it will in general vary thereafter in the course of iterations.

11.7.1 Crisfield's spherical method

Crisfield's modified arc-length method (Crisfield 1981) is often termed the "spherical method" since it seeks to find an intersection of the equilibrium path with a hypersphere of radius ΔS_{n+1} in (N + 1)-dimensional Euclidian space.

In Crisfield's method, the arc-length constraint equation is not linearized. It yields a quadratic equation for $\delta\lambda_{n+1}^{i+1}$ (which we here denote simply as $\delta\lambda$ to simplify the notation), that is,

$$a\ \delta\lambda^2 + 2b\ \delta\lambda + c = 0 \tag{11.38}$$

The coefficients in this quadratic equation are

$$\begin{cases} a = \mathbf{v}_{n+1}^T\mathbf{v}_{n+1} \\ b = \mathbf{v}_{n+1}^T(\Delta\mathbf{U}_{n+1}^i + \mathbf{w}_{n+1}^{i+1}) \\ c = (\Delta\mathbf{U}_{n+1}^i + \mathbf{w}_{n+1}^{i+1})^T(\Delta\mathbf{U}_{n+1}^i + \mathbf{w}_{n+1}^{i+1}) - \Delta S_{n+1}^2 \end{cases} \tag{11.39}$$

The two roots of Equation 11.38 are

$$\delta\lambda_{1,2} = \frac{1}{a}\left[-b \pm \sqrt{b^2 - ac}\right] \tag{11.40}$$

One of the two values of $\delta\lambda$ provided by Equation 11.40 must be selected for $\delta\lambda_{n+1}^{i+1}$. The main objective is to prevent reversal (i.e., "backtracking") of the incremental solution.

One approach is to choose the value of $\delta\lambda$ that yields a displacement correction δU_{n+1}^{i+1} that is in the direction closest to the accumulated displacement increment ΔU_{n+1}^{i}. That is, choose the $\delta\lambda_i$ that provides the smallest positive value of

$$\cos\theta \equiv (\Delta U_{n+1}^{i})^T \delta U_{n+1}^{i+1} \tag{11.41}$$

with $\Delta\lambda_1$ or $\Delta\lambda_2$ inserted in δU_{n+1}^{i+1}, as given by Equation 11.35.

However, this approach for selecting $\delta\lambda_i$ appears not to be entirely foolproof, and additional criteria may be needed in some cases.

11.7.2 Ramm's normal plane method

Ramm (1981, 1982) proposed a predictor–corrector method that seeks to find the intersection of the equilibrium path with a hyperplane defined by a linearized arc-length constraint. In this approach, the arc–length constraint is imposed for only the first iteration of each increment and is allowed to vary with subsequent iterations.

At the beginning of each load increment (n, n + 1), the auxiliary vector $\mathbf{v}_{n+1}$ corresponding to the load pattern is computed (Equation 11.36) and used to produce the predictor δU_{n+1}^{0} from the following steps:

$$\begin{cases} \mathbf{K}_{tn}\mathbf{v}_{n+1} = \hat{\mathbf{P}} \\ \Delta\lambda_{n+1}^{0} = \Delta S_{n+1} \Big/ \sqrt{\mathbf{v}_{n+1}^{T}\mathbf{v}_{n+1}} \\ \Delta U_{n+1}^{0} = \Delta\lambda_{n+1}^{0}\mathbf{v}_{n+1} \end{cases} \tag{11.42}$$

At each iteration $i \geq 0$, the auxiliary solution $\mathbf{w}_{n+1}^{i+1}$ determined from the residual $\mathbf{R}_{n+1}^{i}$ (Equation 11.36) is used along with $\mathbf{v}_{n+1}$ to construct iterative correctors from linearized arc-length constraints,

$$\Delta\overline{U}_{n+1}^{T}\delta U_{n+1}^{i+1} = 0, \quad i \geq 0 \tag{11.43}$$

where $\delta U_{n+1}^{i+1} = \delta\lambda_{n+1}^{i+1}\mathbf{v}_{n+1} + \mathbf{w}_{n+1}^{i+1}$. This leads to the iterative load factor correction

$$\delta\lambda_{n+1}^{i+1} = -\frac{\Delta\overline{U}_{n+1}^{T}\mathbf{w}_{n+1}^{i+1}}{\Delta\overline{U}_{n+1}^{T}\mathbf{v}_{n+1}} \tag{11.44}$$

$\Delta\overline{U}_{n+1}$ in these last two equations may be taken as either the predictor ΔU_{n+1}^{0}, in which case we obtain Ramm's normal plane method, or the accumulated incremental correction ΔU_{n+1}^{i}, in which case we obtain Ramm's updated normal plane method.

$$\begin{cases} \Delta\overline{U}_{n+1} = \Delta U_{n+1}^{0}, & \text{Normal Plane} \\ \Delta\overline{U}_{n+1} = \Delta U_{n+1}^{i}, & \text{Updated Normal Plane} \end{cases} \tag{11.45}$$

Figure 11.12 schematically illustrates the normal plane method in which the plane appears as a straight line in edge view. It shows a gradual decrease of the accumulated load factor increment $\Delta\lambda_{n+1}^{i}$ with each succeeding iteration. A similar illustration for the updated normal plane method would show a series of planes in edge view (i.e., lines) with increasing negative slopes.

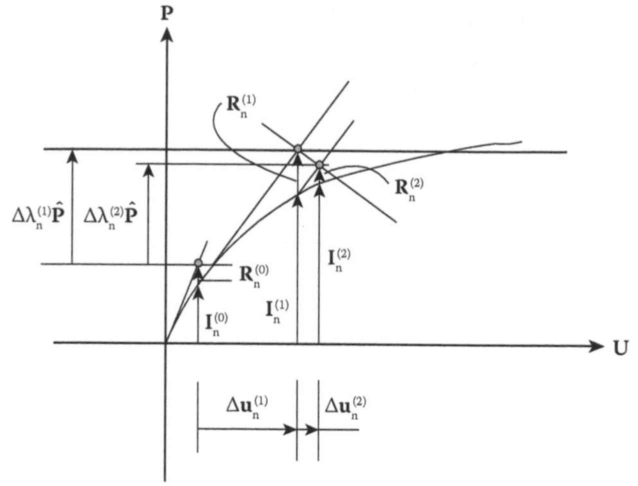

Figure 11.12 Ramm's normal plane arc-length method (schematic).

11.8 LATTICE DOME EXAMPLE

We now consider as an example the nonlinear analysis of the shallow 3D dome shown in Figure 11.13 (Healey 1988). The dome structure is a latticework of slender hyperelastic bar elements interconnected by frictionless hinges that are capable of transmitting only axial force. The dome is supported at nodes labeled H through M in the figure, at each of which all degrees of freedom are constrained. The apex node A and nodes B through G, which are arranged in a regular hexagonal pattern around the apex A, are free to displace in the global (x, y, z) directions.

We specify an external loading consisting of equal-magnitude vertical concentrated loads at the free nodes A through G. As a result, the structure and its loading are both obviously highly symmetric; for example, rotations $\theta_n = \frac{n\pi}{3}, n = 1, 2, 3, \ldots$ about a vertical axis through the apex A leave both the structure and its external loading unchanged.

The finite element model of the dome structure consists of 24 hyperelastic bar elements, 13 nodes, and 21 degrees of freedom, that is, displacements (u, v, w) in directions (x, y, z), respectively, at each of the 7 free nodes A through G.

We employ a TL formulation for the nonlinear analysis. The TL formulation for the hyperelastic bar element (shown again in Figure 11.14) was described in detail in Chapter 8.

The (21 × 1) external load pattern $\hat{P}$ acting on the structuure is defined as

$$\hat{\mathbf{P}} = [\mathbf{P}_j^T \;\; \mathbf{P}_j^T \;\; \mathbf{P}_j^T \;\; \mathbf{P}_j^T \;\; \mathbf{P}_j^T \;\; \mathbf{P}_j^T \;\; \mathbf{P}_j^T]^T$$
$$\text{where} \quad \mathbf{P}_j \equiv \begin{Bmatrix} 0 \\ 0 \\ -1 \end{Bmatrix} \tag{11.46}$$

The *relative* dimensions of the finite element model are shown in Figure 11.13. In order to present the results in nondimensional form, we introduce a geometric scale factor L_s for bar lengths. The corresponding scale factor for loads is the product AE, where A is the cross-sectional area of the hyperelastic bars and E is their (initial) elastic modulus. The numerical solution is of course carried out using specific values for these scale factors, and then the

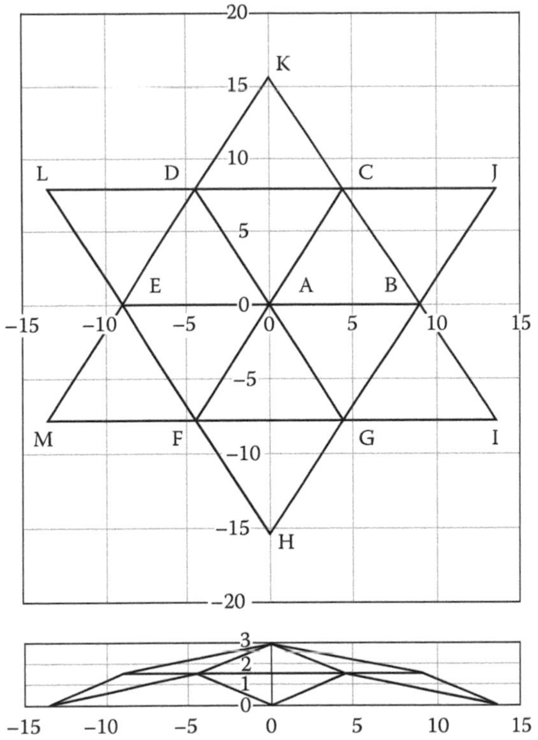

Figure 11.13 3D hexagonal lattice dome.

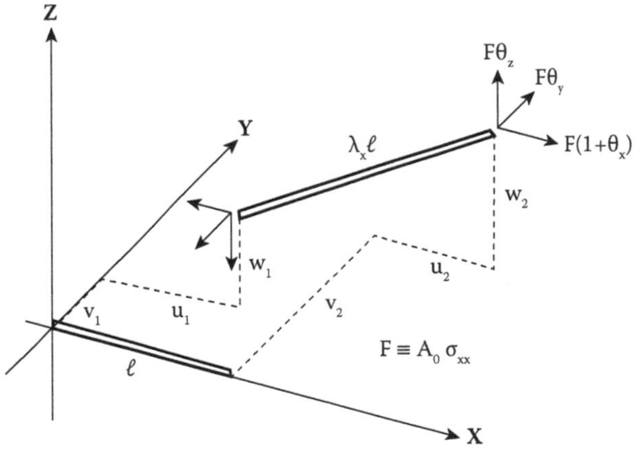

Figure 11.14 Hyperelastic bar element in 3D.

results are normalized accordingly. Numerical parameter values chosen for the numerical analysis are

Geometric scale factor = 12
Force scale factor AE = 30,000
Incremental load factor (unscaled) for first load increment $\Delta\lambda_1 = 1.0$
Convergence criterion $\|\mathbf{R}_{n+1}\| < \varepsilon\|\mathbf{P}_{n+1}\|$
Tolerance $\varepsilon = 0.0001$

The computational steps are

Arc-length calculation $\mathbf{K}_{t0}\mathbf{v}_1 = \hat{\mathbf{P}}$, $\Delta S = \Delta\lambda_1\sqrt{\mathbf{v}_1^T\mathbf{v}_1}$

For $n \geq 1$, $\mathbf{K}_{tn}\mathbf{v}_{n+1} = \hat{\mathbf{P}}$

Trial incremental load factor $\Delta\lambda = \Delta S \Big/ \sqrt{\mathbf{v}_{n+1}^T\mathbf{v}_{n+1}}$

If $\mathbf{v}_{n+1}^T\mathbf{K}_{tn}\mathbf{v}_{n+1} > 0$, then $\Delta\lambda_{n+1}^{(1)} = \Delta\lambda$; else $\Delta\lambda_{n+1}^{(1)} = -\Delta\lambda$ (load reversal)
Iterate until convergence $\|\mathbf{R}_{n+1}\| < \varepsilon\|\mathbf{P}_{n+1}\|$ (normal plane method)

11.8.1 Load–deflection response of hexagonal dome

The computed primary equilibrium path of the hexagonal dome is shown in Figure 11.15. The nondimensional load $\bar{\lambda} \equiv \lambda(10^4)/AE$ is plotted on the vertical axis, and the nondimensional displacement norm $\|U\|/L_S$ is plotted on the horizontal axis.

A limit point was reached at $\bar{\lambda}_{LP} \approx 10.25$. No particular convergence problems were noted on the rising portion of the primary path, although it is evident from the figure that the incremental load parameter $\Delta\lambda_n$ decreases significantly along the solution path ($\Delta\lambda_1 = 1.0$).

We have discussed monitoring the solution via the eigenvalues ω_i or the determinant of the tangent stiffness $\mathbf{K}_{tn}$ in order to detect bifurcation points. Figure 11.16 shows the computed value of $\det(\mathbf{K}_t)$ versus scaled load factor $\bar{\lambda}$ along the primary solution path. In this figure, the determinant is normalized by its initial value $\det(\mathbf{K}_{t0})$.

The curve cuts the $\bar{\lambda}$ axis at $\bar{\lambda} \approx 3.42$, indicating a bifurcation point there. The eigenvector $\boldsymbol{\varphi}_1$ is the corresponding buckling mode shape, which is "symmetry breaking," a term often used to describe bifurcation buckling in general (since $\boldsymbol{\varphi}_1^T\hat{\mathbf{P}} = 0$, $\boldsymbol{\varphi}_1$ cannot have the same rotational symmetry as $\hat{\mathbf{P}}$).

At $\bar{\lambda} \approx 5.33$ and $\bar{\lambda} \approx 9.49$, where again $\det(\mathbf{K}_t) = 0$, we observe that the $\det(\mathbf{K}_t)$ versus $\bar{\lambda}$ curve is also tangent to the $\bar{\lambda}$ axis. Therefore, on the primary path leading up to the limit point, there are three critical points (bifurcations). The structure is unstable on the primary path for $\bar{\lambda} \geq 3.42$ (i.e., beyond the first bifurcation).

Accurate locations of critical points have to be obtained by interpolation from discrete computed values near a critical point. Computed values of several of the lowest eigenvalues of $\mathbf{K}_t$ are much better suited to this purpose than $\det(\mathbf{K}_t)$. Figure 11.17 shows the computed value of ω_1, the lowest eigenvalue of $\mathbf{K}_t$ on the primary solution path. Linear interpolation

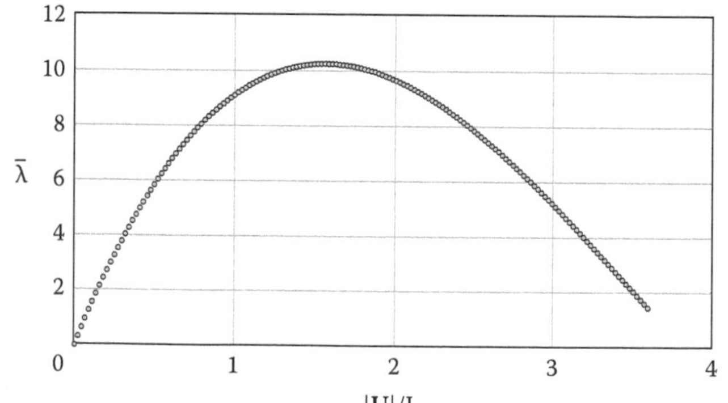

Figure 11.15 Load–deflection response of hexagonal dome on primary path.

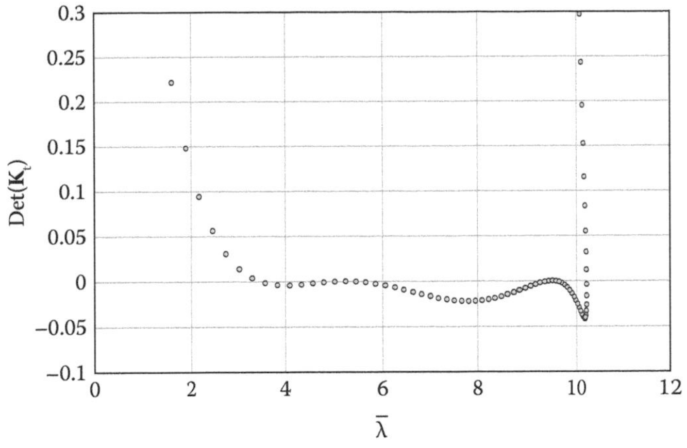

Figure 11.16 Det($\mathbf{K}_t$) on primary solution path.

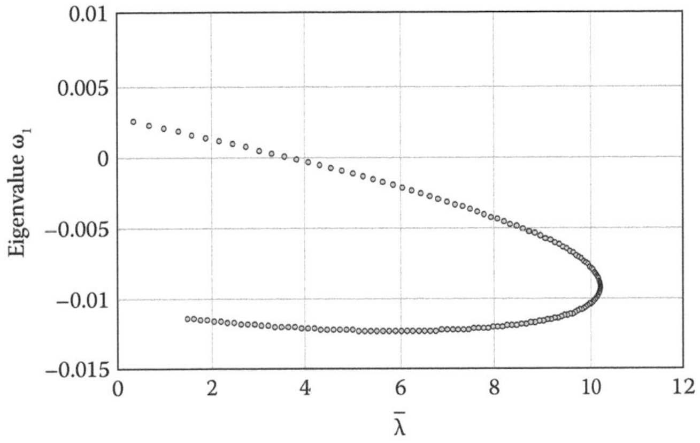

Figure 11.17 Lowest eigenvalue ω_1 of $\mathbf{K}_t$ on primary path.

between two computed points on either side of the bifurcation point leads to the value $\bar{\lambda} \approx 3.42$ reported above. At the limit point ($\bar{\lambda}_{LP} \approx 10.25$), the ω_1 versus λ curve has a vertical tangent corresponding to $d\bar{\lambda}/d\omega = 0$.

Figure 11.18 shows the computed values of the repeated eigenvalues $\omega_2 \equiv \omega_3$ on the primary solution path. Linear interpolation yields $\bar{\lambda} \approx 5.33$ for the location of this double root. We observe that the ω versus λ curves cut the λ axis at an angle, so that the double root can be accurately located via linear interpolation in contrast to the det(K_t) versus $\bar{\lambda}$ curve shown in Figure 11.16.

Figure 11.19 shows the computed value of the eigenvalue ω_4 on the primary solution path. The bifurcation corresponding to this root is at $\bar{\lambda} \approx 9.49$. Figure 11.20 shows the computed value of the eigenvalue ω_5 on the primary solution path. It cuts the $\bar{\lambda}$ axis at the limit point $\bar{\lambda}_{LP} \approx 10.25$, where it also has a vertical tangent $d\lambda/d\omega = 0$.

We note that our numerical values differ slightly from those reported by Healey, whose detailed analysis of the hexagonal dome included a thorough exploration of numerous secondary branches emanating from bifurcations (Healey 1988).

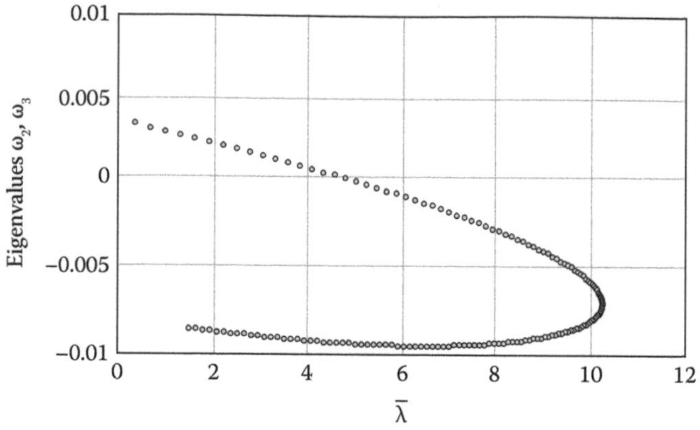

Figure 11.18 Eigenvalues $\omega_2 \equiv \omega_3$ of $\mathbf{K}_t$ on primary path.

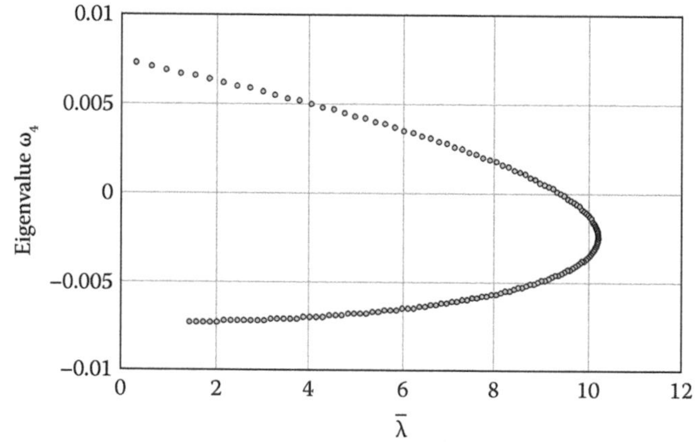

Figure 11.19 Eigenvalue ω_4 of $\mathbf{K}_t$ on primary path.

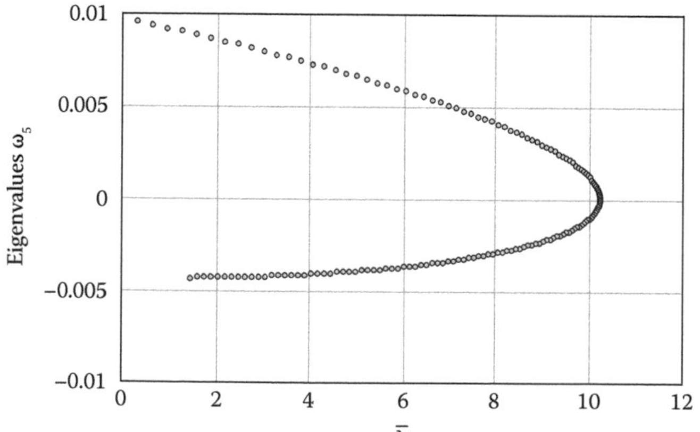

Figure 11.20 Eigenvalue ω_5 of $\mathbf{K}_t$ on primary path.

11.9 SECONDARY SOLUTION PATHS

Secondary solution paths emanating from a bifurcation point may sometimes be of considerable practical interest, at least insofar as determining whether the load multiplier λ is initially increasing or decreasing along the secondary equilibrium path. This information can provide an indication of the degree of imperfection sensitivity of a structure. The buckling of axially compressed thin cylindrical shells (Bushnell 1985) is a well-known example of extreme sensitivity to geometric imperfections.

Koiter's (1945) seminal work utilizing perturbation theory laid the foundation for analytically determining the initial portion of secondary branches, thereby enabling the assessment of imperfection sensitivity. At that time, Koiter's approach could only be applied for cases in which the primary path and bifurcation point could be calculated analytically. Haftka et al. (1971) later developed a Koiter-type finite element procedure, based on the nonlinear finite element formulation of Mallett and Marcal (1968), which is particularly convenient for that purpose (see also Rajasekaran and Murray 1973). Thurston et al. (1986) proposed an incremental-iterative finite element approach that is well suited for general-purpose finite element systems, using for the most part standard direct stiffness algorithms that would already be available in such systems. The essence of that method is a change of variables near a bifurcation point: The "tangent" stiffness is modified using one or more computed eigenvectors corresponding to the lowest eigenvalues. The change of variables introduces one "modal" degree of freedom corresponding to each eigenvector that is included in the modified tangent stiffness. Healey (1988) presented a group theoretic approach for symmetric structures that greatly simplifies the computation of secondary equilibrium paths.

In all these approaches, initiating the computation of structural response along a secondary solution path requires tracking a small number of the lowest eigenvalues and corresponding eigenvectors of $\mathbf{K}_t$ along the primary solution path approaching the bifurcation point.

A simple practical approach that does not require special algorithms and that can be used to assess imperfection sensitivity is described in the next section.

11.10 STRUCTURAL IMPERFECTIONS

A finite element model is often an idealized representation of actual geometry. In addition, load distributions are often simplified for analysis. Even when a finite element model is a faithful representation of structure geometry, the physical realization will undoubtedly be imperfect because of (unknown) variations in actual dimensions and material properties. As a result, it is unlikely for a real structure to (ideally) bifurcate, instead undergoing a limit point–type instability.

A simple example of geometric imperfection is the pin-ended column shown in Figure 11.21. A perfectly straight pin-ended column first bifurcates at the classic Euler load

$$P_E = \pi^2 \frac{EI}{L^2} \tag{11.47}$$

in the buckling mode shape (i.e., eigenvector) sin $\pi x/L$, where EI is the flexural stiffness of the column.

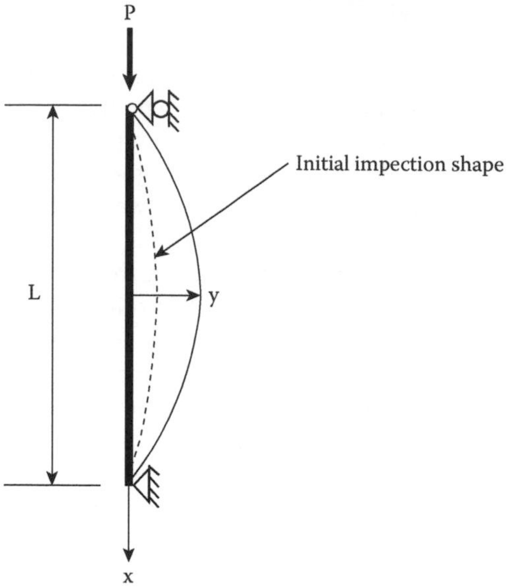

Figure 11.21 Buckling of a pin-ended column.

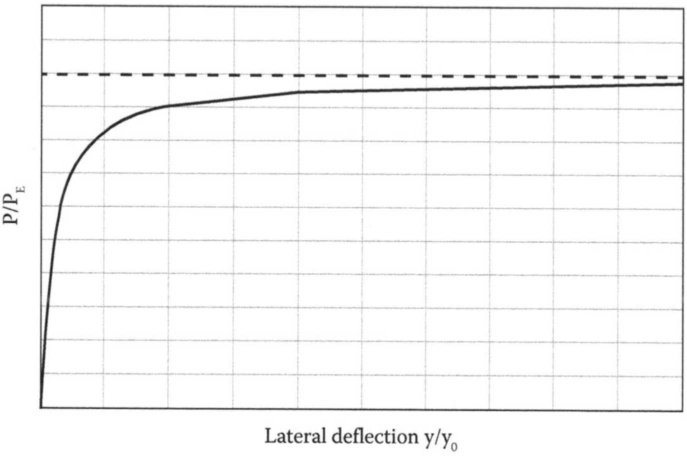

Figure 11.22 Limit point instability of geometrically imperfect column.

If the column is assumed to have an initial lateral out-of-straightness $y_i(x)$ in the shape of the lowest buckling mode

$$y_i(x) = y_0 \sin \pi \frac{x}{L} \tag{11.48}$$

the lateral response y(x) of the imperfect column is

$$\frac{y(x)}{y_0} = \frac{1}{1 - \dfrac{P}{P_E}} \sin \pi \frac{x}{L} \tag{11.49}$$

shown in nondimensional form in Figure 11.22. The initial geometric imperfection transforms the bifurcation buckling into a limit point instability: $P \rightarrow P_E$ as $y(x)/y_0 \rightarrow \infty$.

Qualitatively, the same effect is seen if instead we were to assume that a small lateral load (imperfection) acts on the geometrically perfect column in addition to the axial load P.

Pecknold et al. (1985) show the equilibrium paths of a simple two-bar truss subjected to a load imperfection.

Here, we employ the more convenient load imperfection approach to obtain information about secondary solution paths. We suppose that an initial analysis has first been performed to determine the primary equilibrium path under the ideal loading $\mathbf{P}_n = \lambda_n \hat{\mathbf{P}}$, and that eigenvalues and eigenvectors have been monitored to locate critical points. We suppose further that a bifurcation point defined by $\omega_1 = 0$, $\boldsymbol{\varphi}_1^T \hat{\mathbf{P}} = 0$ has been (approximately) located on the primary path where the tangent stiffness $\mathbf{K}_t$ becomes singular. We then initiate a new analysis, again along the primary path, and as the incremental-iterative numerical solution approaches the bifurcation point, we augment the applied loading rate with a load rate imperfection $\bar{\varepsilon}\dot{\lambda}\hat{\mathbf{P}}_\varepsilon$, where $\bar{\varepsilon}$ is a small numerical constant and $\hat{\mathbf{P}}_\varepsilon$ is a load imperfection pattern that is chosen to contain some of the eigenvector $\boldsymbol{\varphi}_1$, that is, $\boldsymbol{\varphi}_1^T \hat{\mathbf{P}}_\varepsilon \neq 0$. At this point, the external loading rate is therefore

$$\dot{\mathbf{P}} = \dot{\lambda}(\hat{\mathbf{P}} + \bar{\varepsilon}\hat{\mathbf{P}}_\varepsilon) \tag{11.50}$$

Since $\omega_1 \to 0$ (approximately), because $\boldsymbol{\varphi}_1^T(\hat{\mathbf{P}} + \bar{\varepsilon}\hat{\mathbf{P}}_\varepsilon) = \bar{\varepsilon}\boldsymbol{\varphi}_1^T \hat{\mathbf{P}}_\varepsilon \neq 0$, $\dot{\lambda} \to 0$ and the bifurcation point is essentially transformed into a limit point.

In practice, we apply the augmented load pattern (i.e., including the load imperfection) from the beginning of the reanalysis. The load imperfection pattern $\hat{\mathbf{P}}_\varepsilon$ is easily selected since the eigenvector $\boldsymbol{\varphi}_1$ is known.

11.10.1 Application to hexagonal dome example

We now apply this simple procedure to the hexagonal dome structure that was discussed in Section 11.8 to explore the secondary equilibrium path emanating from the bifurcation point at $\bar{\lambda} \approx 3.42$.

The "perfect" load pattern $\hat{\mathbf{P}}$ that was defined in Equation 11.46 is recalled here:

$$\hat{\mathbf{P}} = [\mathbf{P}_j^T \;\; \mathbf{P}_j^T \;\; \mathbf{P}_j^T \;\; \mathbf{P}_j^T \;\; \mathbf{P}_j^T \;\; \mathbf{P}_j^T \;\; \mathbf{P}_j^T]^T \qquad \text{where} \quad \mathbf{P}_j \equiv \begin{Bmatrix} 0 \\ 0 \\ -1 \end{Bmatrix} \tag{11.51}$$

The seven free nodes A through G are loaded with equal loads in the –Z direction. The load imperfection pattern $\hat{\mathbf{P}}_\varepsilon$ is selected as

$$\hat{\mathbf{P}}_\varepsilon = [\mathbf{0}_3^T \;\; \mathbf{P}_j^T \;\; -\mathbf{P}_j^T \;\; \mathbf{P}_j^T \;\; -\mathbf{P}_j^T \;\; \mathbf{P}_j^T \;\; -\mathbf{P}_j^T]^T \qquad \text{where} \quad \mathbf{0}_3 \equiv \begin{Bmatrix} 0 \\ 0 \\ 0 \end{Bmatrix} \tag{11.52}$$

That is, for the load imperfection pattern, the apex (node A) is not loaded and the remaining six free nodes B through G have Z direction loads that alternate in sign, proceeding around the hexagon. We select the imperfection parameter $\bar{\varepsilon} = 0.01$ and repeat the arc-length solution as described previously in Section 11.7.

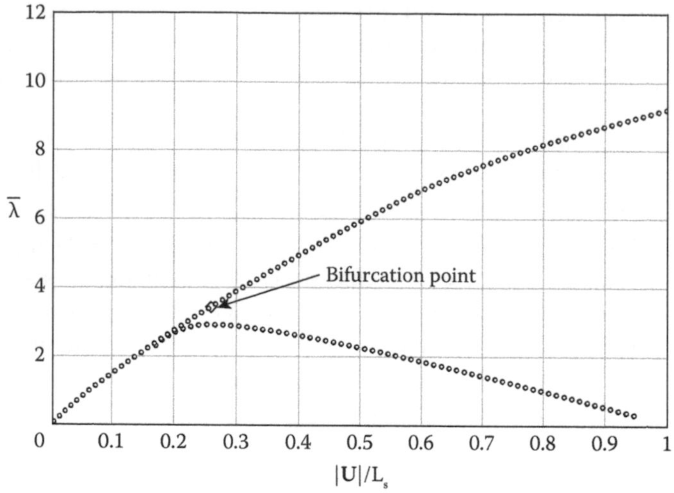

Figure 11.23 Hexagonal dome with load imperfection.

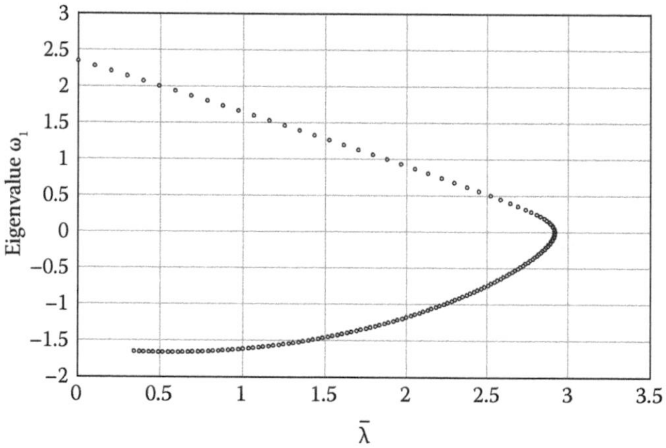

Figure 11.24 Lowest eigenvalue ω_1 of K_t on imperfect primary path.

Figure 11.23 shows the computed response for the *augmented load pattern* $\hat{\mathbf{P}} + \varepsilon\hat{\mathbf{P}}_\varepsilon$, superimposed on the initial portion of the primary equilibrium path (previously shown in Figure 11.15). A limit point is observed at $\bar{\lambda} \approx 2.92$, and it appears that the computed λ versus $|\mathbf{U}|$ response asymptotically approaches a straight line passing through the bifurcation point on the primary equilibrium path.

Figure 11.24 shows the lowest eigenvalue of the tangent stiffness $\mathbf{K}_t$ on what is now actually the *primary* equilibrium path for the structure under the *augmented* loading, and is also an approximation to the secondary equilibrium path of the perfectly loaded structure. The limit point is located where the ω versus $\bar{\lambda}$ curve cuts the $\bar{\lambda}$ axis with a vertical tangent $d\bar{\lambda}/d\omega = 0$ (i.e., at $\bar{\lambda} \approx 2.92$).

Chapter 12

Soft computing in computational mechanics

12.1 INTRODUCTION

Soft computing methods are biologically inspired, and as such, they are fundamentally different than the computational mechanics that has been covered in previous chapters in this book. We refer to computational mechanics as *hard computing*. We discuss the fundamental differences between hard computing and soft computing later in this chapter. The soft computing methods that we consider in this chapter are based on adaptive learning systems, specifically neural networks. Soft computing in computational mechanics was initiated in the late 1980s and early 1990s, and extensive research has been conducted to address some of the fundamental questions (Ghaboussi et al. 1990, 1991; Ghaboussi and Wu 1998).

Neural networks are inspired by the massively parallel structure of the human brain. Another soft computing method that also has applications in computational mechanics is genetic algorithms that are based on natural evolution. We should also point out that there are other soft computing methods, such as fuzzy logic, that are based on languages and swarm methods that are inspired by ant colonies. In this chapter, we explore the application of neural networks in material modeling. Additional information on soft computing and its engineering applications is presented by the first author in a forthcoming book (Ghaboussi 2017).

We should point out that although neural networks share some common features with the human brain, they are not actually intended to be models of the brain. However, they are massively parallel systems for knowledge representation that contain a highly simplified form of neurons and synapses. Our brains are highly complex systems, and we have at present only a limited understanding of how they work. We do know that human brains perform highly complex functions in a robust and fault-tolerant way. Thus, we may surmise that as our understanding of the human brain increases, the architecture of artificial neural networks will evolve. This implies that there is room for almost unlimited improvement in neural networks and their applications in soft computational mechanics. Therefore, we present in this chapter what should be considered the early stages of soft computational mechanics.

The first application of neural networks was in material modeling (Ghaboussi et al. 1990, 1991; Wu 1991) and structural damage detection and monitoring (Wu et al. 1992). Although neural networks and genetic algorithms have been used in the modeling, design, and condition monitoring of structural systems, the application of neural networks in material modeling is potentially an important development in computational mechanics. In this chapter, we concentrate on neural networks for the modeling of the constitutive behavior of materials.

12.2 MULTILAYER NEURAL NETWORKS

We consider neural networks that are mainly used to model complex relationships (contained in data) that are not amenable to analytical formulations. As illustrated in Figure 12.1,

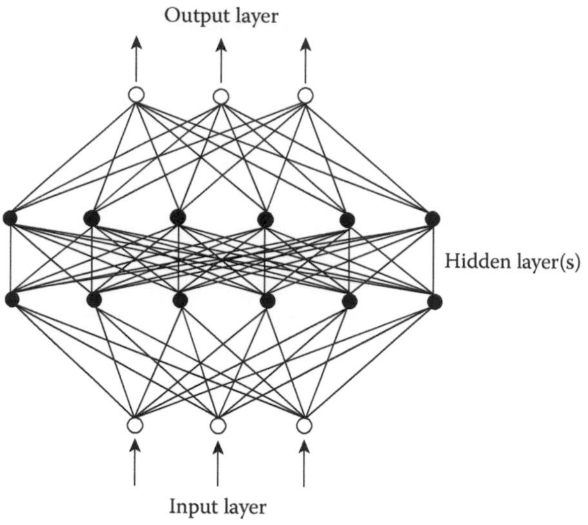

Figure 12.1 Multilayer neural network.

these neural networks have several layers of neurons (or nodes). The bottom layer in the figure contains the input nodes, and the top layer contains the output nodes. The layers in between are called *hidden layers*.

Each node (*neuron*) in a layer is connected to all the nodes in the layer above it. These connections are the *synapses*, and each synapse has a numerical weight. Signals propagate along these synapses from the input layer, through the hidden layers, to the output layer. Each node receives signals from the nodes in the layer below it. Each node computes a weighted sum of the incoming signals (using the weights of the incoming connections) and passes that weighted sum through an activation function to determine its output, which is sent through the outgoing connections to the nodes in the layer above it. When we provide input values to the input layer nodes, they initiate the passing of the signals through the connections from one layer to the next, until they reach the output layer. The nodes in the output layer provide the neural network response to the input data.

12.2.1 Training of neural networks

From this brief description, we see that the information (knowledge) acquired by the neural network is stored in its connection weights. Connection weights, initially assigned random values, are determined during a *training* process in which the neural network is presented with a *comprehensive* training set consisting of a collection of input–output pairs. It is important that the training data contain all the information that the neural network is desired to learn.

There are as yet no general rules for determining whether a training set is sufficiently comprehensive; judgment must be used. This is but one example of the difference between soft and hard computing methods.

During training (sometimes termed *supervised learning*), each individual input–output data pair is passed through the neural network in a *feed-forward* process. The output of a partially trained neural network will of course be different than the desired (correct) output. The error is *back-propagated* through the neural network, and its connection weights are adjusted to reduce the output error.

This feed-forward and back-propagation process for the complete set of training data constitutes a training *epoch*, and it is repeated for a number of epochs until the output error is reduced to a prespecified tolerance level. At that point, the neural network is considered to be trained.

A properly trained neural network will not only learn the information in the training data, but also be able to generalize that information. This is somewhat akin to data interpolation and extrapolation in classical data analysis. However, this is not a complete analogy. A properly trained neural network is robust, noise and fault tolerant, self-organizing, and adaptive, in addition to its generalization capability.

As an example, we may use the neural network to model the stress–strain relationship for a material. In this case, we may choose to use strains as input and stresses as output. The training data can be the stress–strain relationships measured from material tests. At this point, we can see that many important questions arise:

- Intuitively, the distributed patterns of connection weights are an indication of capacity of a neural network. How large a neural network (i.e., number of hidden layers and nodes per layer) is needed to capture the information in the data?
- How many epochs of training are required? What output error tolerance should be selected?
- What type of material tests are needed to generate the data to train the neural network? How do we know if the training data contain sufficient information for the neural network to learn the complete material behavior?
- How can we account for objectivity, frame indifference, history dependence, and other features of material behavior that we discussed in Chapters 5 through 7?
- How can we employ a neural network material model in finite element analysis?

These are some of the important questions that we need to address. Similar questions arise when neural networks are used in applications other than material modeling. Before addressing these specific questions, we describe some of the ways in which soft computing methods differ fundamentally from hard computing methods. This will help us address the questions raised above.

12.3 HARD COMPUTING VERSUS SOFT COMPUTING

Hard computing methods are generally mathematically based; they inherit their characteristic properties from the mathematics. These properties include precision, universality, and functional uniqueness. Exploring these properties will help us understand the fundamental differences between hard and soft computing methods.

All hard computing methods are inherently precise to within the round-off error. Of course, round-off error can lead to very large errors in some cases, resulting in a loss of accuracy, which measures the fidelity between a mathematical model and its physical counterpart. Nevertheless, hard computing methods are intended to be precise (or exact) because their underlying theoretical bases are mathematical. The input–output in any problem in computational mechanics is intended to be precise. The physical problems, with their inherent random variability, that are modeled and solved via computational mechanics are not precise and exact by nature, and of course, the level of precision varies in different problems. For example, in any finite element analysis, the geometry and the material properties are assumed to be precise and exact, although in reality they are not. We often take

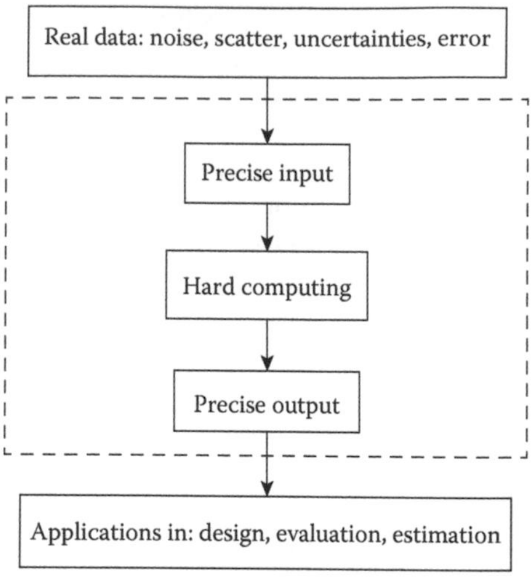

Figure 12.2 Illustration of the role of precision in hard computing methods in actual applications.

approximate characteristic values, such as the mean or lower bound, and input them as exact values to the finite element analysis and receive exact results in applications that may only require approximate values to some degree. This process of using hard computing methods in applications to real-world problems is schematically illustrated in Figure 12.2. Of course, in some cases we use stochastic computational methods to account for inherent variability, in which the variability is characterized by probability distributions that are again described in exact mathematics.

Universality is another feature of the mathematical functions that are incorporated in hard computational mechanics. Mathematical functions are valid for all possible values of the variables, whereas universality does not exist in real-world problems. For example, when we assume a material to be linear elastic, the associated mathematically based material model is assumed to respond in a linear elastic manner for all possible values of strains. In the real world, however, linear elasticity generally approximates material behavior only over a very small range of strains near the origin.

Functional uniqueness is another feature of mathematically based methods that affects hard computing in computational mechanics. For example, there is only one sin(x) function that is universally valid for x from $-\infty$ to $+\infty$.

Soft computing methods such as neural networks, genetic algorithms, fuzzy logic, and swarm intelligence are biologically inspired. They are based on the problem solving in nature that takes place without relying on exact mathematics. The vast majority of problems in nature are inverse problems, and they are solved with massively parallel computational systems, such as brains, or through random search and reduction in disorder in natural evolution. Precision is not important in problem solving in nature. Soft computing methods are imprecision tolerant, nonuniversal, and functionally nonunique. We next explore these properties of soft computing in the context of neural networks.

We mentioned earlier that neural networks are inherently imprecision tolerant. We can train a neural network to learn the information in the training data to within a specified output error tolerance. We can never expect an exact or precise output from a neural network. If we are training a neural network with actual measured data, there is always inherent noise and scatter in the data to start with and the neural network may ideally learn to approximate the data with scatter.

Soft computing methods are also inherently nonuniversal. For example, if we train a neural network on the linear elastic property of a material, it will only learn to approximate the linear elastic behavior within the range of strains that are present in the data. Outside that range, the response of the neural network is inherently nonlinear and will not represent the behavior of the actual material. It is important to remember that neural networks always learn the information present in the training data only over the limited range of that data. In that sense, modeling any phenomenon with neural networks is very different than modeling the same phenomenon with a mathematical function. If the range of interest increases, we need to generate new data with information about the additional range and retrain the neural network.

Functional nonuniqueness is the third fundamental difference between soft computing and hard computing. While a mathematical function may be unique, different neural networks with different internal structures can approximate that function in a range of its variables to within the same specified output error tolerance.

In summary, hard computing methods are based on mathematics and they are therefore precise, universal, and functionally unique. Soft computing methods are based on problem-solving methods in nature, and they are imprecision tolerant, nonuniversal, and functionally nonunique. Importantly, hard computing takes place in a sequential manner in our computers. Hard computing methods are suitable mainly for solving forward problems. Soft computing methods are based on massively parallel systems that are more suitable for solving inverse problems.

We next discuss the application of neural networks in modeling constitutive behavior of materials.

12.4 NEURAL NETWORKS IN MATERIAL MODELING

Modeling of constitutive behavior of materials is inherently an inverse problem. We apply stresses to a sample of the material and measure the strains (or vice versa). If we consider the material as a system, stresses are the input to that system and strains are the output. That is, we measure the input and the output and wish to identify the system—this is a classic inverse problem. After the system is identified and we have the material model, we use it in solving forward problems, where the input and system model are known and we determine the output.

The inverse problem of material modeling is usually approached by developing a mathematical model. When we use neural networks, we train a neural network to learn the material behavior. In its simplest form, a neural network material model that can be used in a finite element analysis takes strains as input and gives the stresses as output. In the earliest applications, neural networks were used in this form. In this case, we simply replace the mathematical model with a neural network. However, neural networks can be used in far more effective ways. Any new paradigm is initially used in a way similar to that of previous applications; later, its potential is often recognized and it is then used in more effective ways. We summarize below some of the early applications of neural networks in material modeling.

In the first applications in computational mechanics (Ghaboussi et al. 1990, 1991; Wu 1991), neural networks were used to model the behavior of plain concrete in a two-dimensional (2D) plane stress condition under monotonic loading (Figure 12.3). Existing experimental results were used to train a neural network. Because of the nonlinearity and path dependence, the current stresses and strains and the strain increments were used as input to the neural network, with the corresponding stress increments as the output. The notation that was developed later can be used to describe this neural network in an equation form.

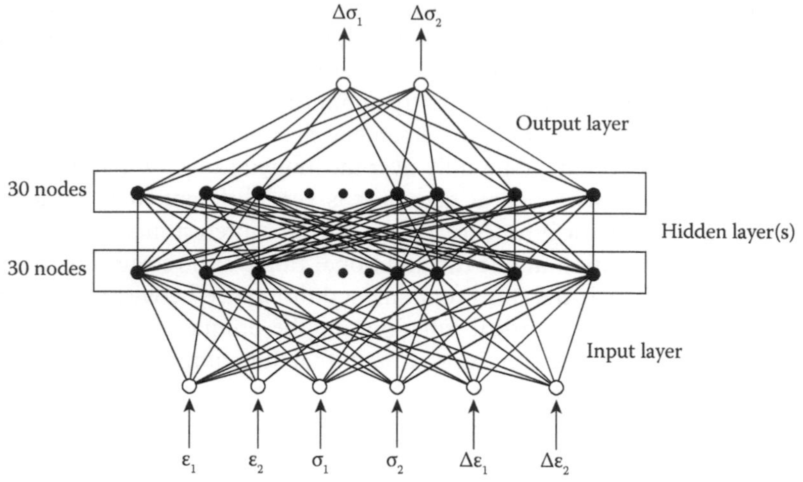

Figure 12.3 First application of neural network in modeling the behavior of plain concrete in 2D stress–strain states.

We can see that the terms in the brackets define the input to the neural network, and the numbers indicate the number of nodes in each layer, starting with the input layer and ending with the output layer. This type of neural network material model was described as a "one-point scheme"; it uses the current state of stresses and strains. This neural network was able to learn the material behavior in monotonic loading experiments. Typically, a neural network is trained on some experimental results and then tested on completely different experimental results to verify that it has learned the underlying material behavior with sufficient accuracy, and that it can generalize to other stress paths that were not present in the training data.

$$\Delta\boldsymbol{\sigma} = \mathrm{NN}[\boldsymbol{\sigma}, \boldsymbol{\varepsilon}, \Delta\boldsymbol{\varepsilon} : 6, 30, 30, 2] \tag{12.1}$$

The input to this particular neural network is not sufficient to describe the material behavior in loading and unloading. Materials like metals, concrete, rocks, and geomaterials undergo microstructural changes, and their loading and unloading behavior can be significantly different. For the neural network to learn the loading and unloading behavior of the material, three history points were provided as input. This "three-point scheme," described by the following equation, was first used for modeling plain concrete material behavior under uniaxial cyclic loading (Ghaboussi et al. 1991; Wu 1991).

$$\Delta\boldsymbol{\sigma}_{j+1} = \mathrm{NN}[\boldsymbol{\sigma}_{j-2}, \boldsymbol{\varepsilon}_{j-2}, \boldsymbol{\sigma}_{j-1}, \boldsymbol{\varepsilon}_{j-1}, \boldsymbol{\sigma}_{j}, \boldsymbol{\varepsilon}_{j}, \Delta\boldsymbol{\varepsilon}_{j+1} :] \tag{12.2}$$

Since this equation does not represent a specific neural network, the number of nodes in layers is not given in this equation.

12.5 NESTED ADAPTIVE NEURAL NETWORKS

In the first applications of neural networks in material modeling, the number of nodes in the hidden layers was specified before the start of training. The size of the neural networks (i.e., the number of nodes in the hidden layers) determines the capacity of the neural network in acquiring and storing the information in the training data. Neural networks that are either too large or too small are problematic. Obviously, if the neural network is too small, it may not have sufficient capacity to adequately learn the information in the training data. On the other hand, if the capacity of a large neural network is much larger than needed for extracting and storing the information in the training data, it may learn the details of the data, including the noise and scatter that are present in any measured data. When a too large neural network learns the information in the training data with a high degree of accuracy, it loses the capability to generalize beyond the training data. In general, there are no rule-based methods for determining the appropriate size of a neural network. *Adaptive neural networks* were developed to address this problem—the size of the neural network gradually evolves during the training (Wu 1991; Ghaboussi et al. 1997; Ghaboussi and Sidarta 1998).

Adaptive neural network training starts with a small number of nodes in the hidden layers. As the training proceeds, if the capacity of the neural network is insufficient for learning the information in the training data, the output error reaches a plateau and will not decrease further. Monitoring the output error allows us to determine if the capacity of the neural network has been reached. If so, new nodes are added to the hidden layers. In a first application of adaptive training through node generation, weights of the new connections to the new nodes were assigned random values and the training of the whole network was continued (Wu 1991). Although this approach was useful in adjusting the size of hidden layers, it was observed to be not efficient. Later, this approach was enhanced (Ghaboussi et al. 1997) by freezing the connection weights of the existing nodes and continuing the training of only the connection weights of the new nodes. The rationale was that we wanted the connection weights of the new nodes to learn the information that the existing neural network had not learned. After a few training epochs, the connection weights of the original neural network are unfrozen and training is continued for all the connection weights. This process of adaptively adding new nodes to the hidden layers is continued until the output error reaches an acceptable level. One step of this process is illustrated in Figure 12.4.

While the "adaptive" part of nested adaptive neural networks (NANNs) addresses the extent or complexity of the information in the training data, the "nested" part deals with the structure of the information in the data. In some cases, there are subsets of the data that have a clear hierarchical nested structure (Ghaboussi et al. 1997; Ghaboussi and Sidarta 1998).

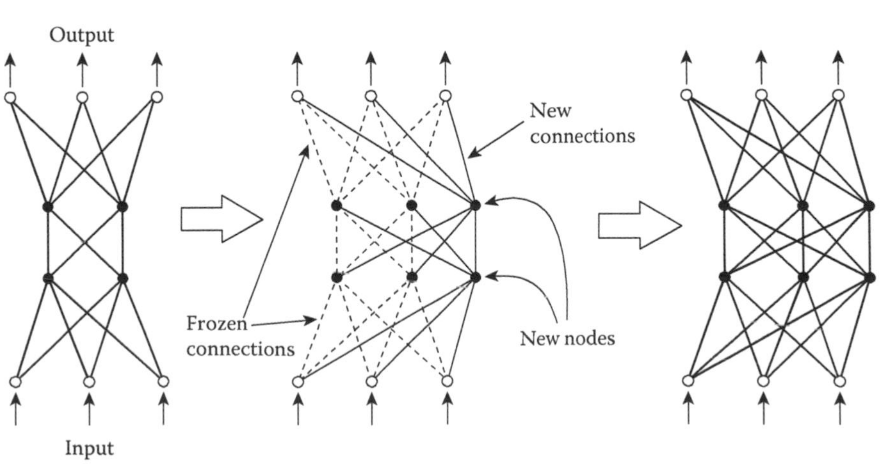

Figure 12.4 Process of adaptive node generation during the training of neural networks.

One type of data hierarchy in the constitutive behavior of materials arises from dimensionality. The data from one-dimensional (1D) constitutive material behavior are a subset of the data from 2D constitutive material behavior that, in turn, are a subset of the data from three-dimensional (3D) material behavior. The same nested structure is present in the information contained in the training data. If we assume that the functions $\mathbf{f}_j$ describe the material behavior in one, two, and three dimensions, they belong to 1D, 3D, and six-dimensional function spaces $\mathbf{F}_j$.

$$\dot{\boldsymbol{\sigma}}_j = \mathbf{f}_j(\dot{\boldsymbol{\varepsilon}}_j) \quad \text{for} \quad j = 1, 3, 6 \tag{12.3}$$

$$\mathbf{f}_j \in \mathbf{F}_j \tag{12.4}$$

These function spaces have a nested structure described by the following equation:

$$\mathbf{F}_1 \subset \mathbf{F}_3 \subset \mathbf{F}_6 \tag{12.5}$$

Neural network material models also use incremental formulations, similar to mathematical material models. In some cases, such as nonlinear elastic materials, path dependence is not important. In these cases, the input to the neural network consists of stresses and strains at the beginning of the increment and the strains at the end of the increment, leading to the output of stresses at the end of the time step. This is the one-point model described earlier. This information is not sufficient if path dependency is important, for instance, when unloading and reloading occurs. In this case, the material behavior is influenced by the past history. In the first application of neural networks in modeling of the path dependency in loading–unloading, three history points were used (Ghaboussi et al. 1991; Wu 1991). This was called the "three-point model." Modeling the path dependency of material behavior with history points is another example of the nested structure of information in the data.

The following equation describes the neural network architecture for a path-dependent material model with k history points. Each k-point function $\mathbf{f}_k$ belongs to a k-point function space $\mathbf{F}_k$.

$$\dot{\boldsymbol{\sigma}}_n = \mathbf{f}_k(\boldsymbol{\sigma}_{n-k}, \boldsymbol{\varepsilon}_{n-k}, \cdots, \boldsymbol{\sigma}_{n-1}, \boldsymbol{\varepsilon}_{n-1}, \boldsymbol{\sigma}_n, \boldsymbol{\varepsilon}_n, \dot{\boldsymbol{\varepsilon}}_n) \quad \text{for} \quad k = 0, 1, 2, \cdots \tag{12.6}$$

$$\mathbf{f}_k \in \mathbf{F}_k \tag{12.7}$$

The function spaces have a nested structure described by the following equation:

$$\mathbf{F}_k \subset \mathbf{F}_{k+1} \tag{12.8}$$

A typical NANN of a 1D material model is shown in Figure 12.5. First, the base (one-point) module is trained adaptively, as shown in Figure 12.5a. The hidden layers begin with two nodes and end with four nodes each. Next, the first history point module is added and trained adaptively, as shown in Figure 12.5b. We note that there are only one-way connections between the history point module and the base module; there are no connections from the base module to the history point module. This reflects the nested structure of the training data. We also note that the one-way connections reflect the fact that the past affects the present, but the present has no effect on the past. Thus, the one-way connections go from history point to the base module. Then a second history point is added and trained adaptively, as shown in Figure 12.5c.

A NANN was first applied to 1D cyclic behavior of plain concrete (Zhang 1996; Ghaboussi 2009). NANNs were trained using the data from one experiment, and the trained neural network was tested on the results of a different experiment. Three NANNs with one, two, and

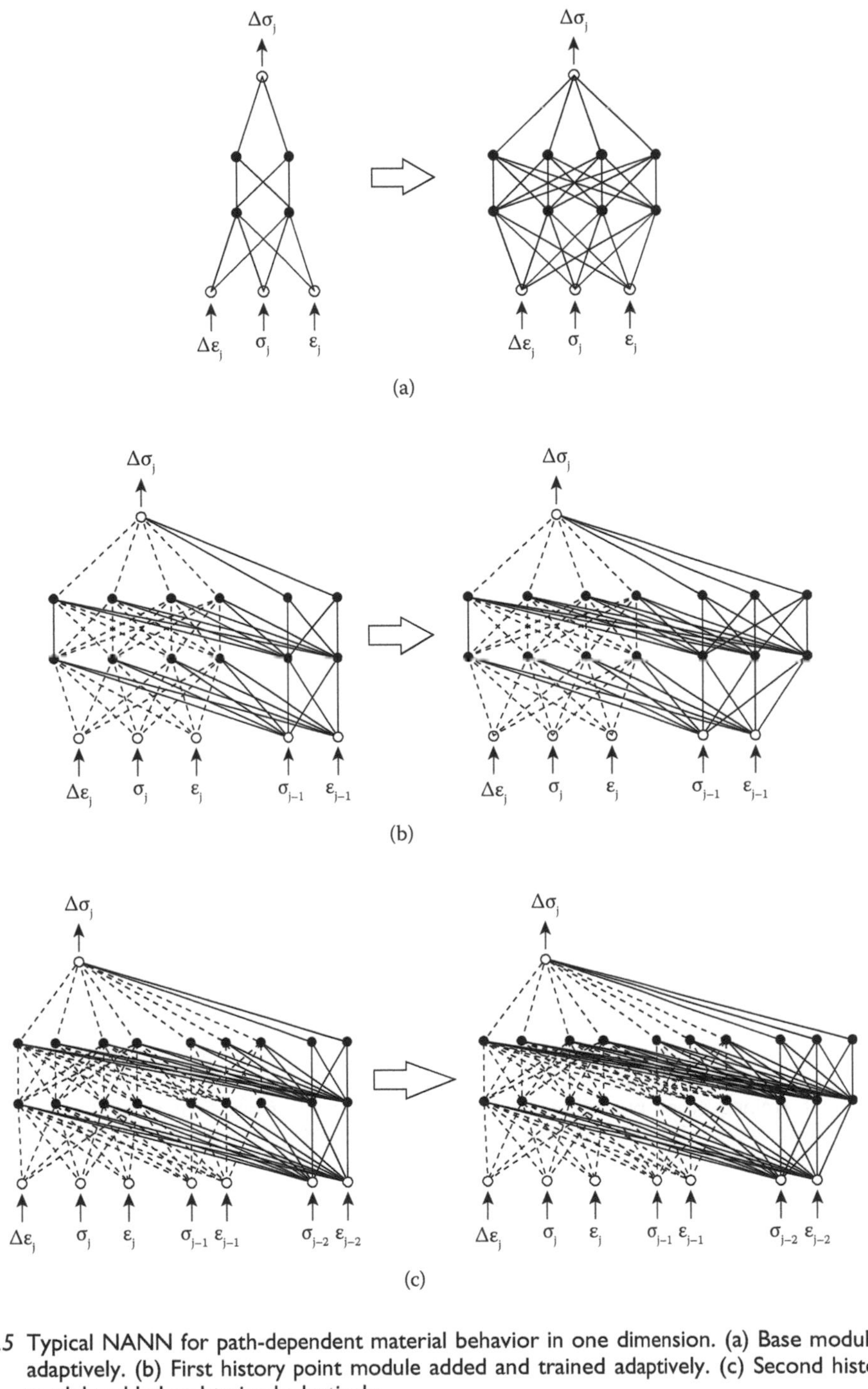

Figure 12.5 Typical NANN for path-dependent material behavior in one dimension. (a) Base module trained adaptively. (b) First history point module added and trained adaptively. (c) Second history point module added and trained adaptively.

three history points, respectively, were trained and tested. The results are shown in Figure 12.6. On the left, the response of the trained neural network is compared with its training data. On the right, the response *predictions* of the trained NANNs are compared with the results of a *novel* 1D cyclic test (i.e., a new test "not seen" previously by the NANN). It is evident that additional history points improve the predictions.

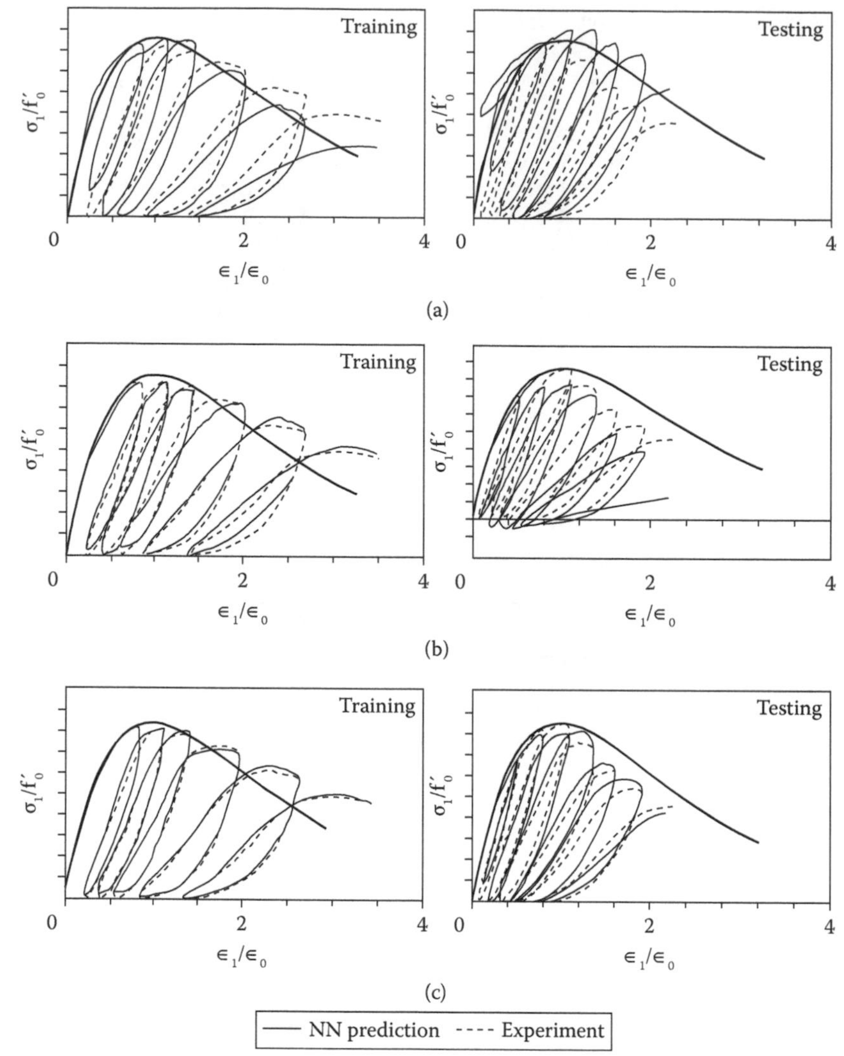

Figure 12.6 Performance of NANN with (a) one, (b) two, and (c) three history points. (From Zhang, M., Determination of Neural Network Material Models from Structural Tests, PhD thesis, Department of Civil and Environmental Engineering, University of Illinois at Urbana-Champaign, 1996; experimental results from Karsan, I.D., and Jirsa, J.O., *J. Struct. Div.*, 95, 2543–2563, 1969.)

For the results shown in Figure 12.6, the base module was first trained adaptively beginning with two nodes and ending with four nodes in each of the hidden layers. This is described in the following equation.

$$\Delta\sigma_{n+1} = \mathbf{NN}_0[\Delta\varepsilon_{n+1} : 1, 2-4, 2-4, 1] \tag{12.9}$$

Then the history point modules were added and trained adaptively. The resulting NANNs are described by the following equations:

One history point

$$\Delta\sigma_{n+1} = \mathbf{NN}_1[\sigma_n, \varepsilon_n, \Delta\varepsilon_{n+1} : (1,2), (2-4, 2-12), (2-4, 2-12), (1)] \tag{12.10}$$

Two history points

$$\Delta\sigma_{n+1} = NN_2[\sigma_{n-1}, \varepsilon_{n-1}, \sigma_n, \varepsilon_n, \Delta\varepsilon_{n+1} : (1,2,2), (2-4, 2-12, 2-10), (2-4, 2-12, 2-10), (1)] \quad (12.11)$$

Three history points

$$\Delta\sigma_{n+1} = NN_3[\sigma_{n-2}, \varepsilon_{n-2}, \sigma_{n-1}, \varepsilon_{n-1}, \sigma_n, \varepsilon_n, \Delta\varepsilon_{n+1} : (1,2,2,2), (2-4, 2-12, 2-10, 2-9), (2-4, 2-12, 2-10, 2-9), (1)] \quad (12.12)$$

The cyclic behavior of plain concrete in one dimension shown in Figure 12.6 clearly shows a strong path and history dependence. History points are needed in order to capture this history dependence, as clearly indicated by the fact that the performance of the NANN improves with the addition of the new history points. The neural network NN_3 with three history points appears to have learned the cyclic material behavior with a reasonable level of accuracy. This figure also demonstrates the generalization capability of the NANN. That is, the trained neural network has learned the underlying material behavior, not just a specific response curve. The trained neural network is able to predict the results of a different experiment that was not included in the training data.

Neural network material models with history points have been successfully used in a number of applications in modeling the multidimensional behavior of materials. Most of these applications employed the autoprogressive algorithm for training of neural networks (see Section 12.9 for a description). We next describe another method for modeling of the hysteretic behavior of materials.

12.6 NEURAL NETWORK MODELING OF HYSTERETIC BEHAVIOR OF MATERIALS

In the previous section, we described the use of history points in the three-point method for modeling of the behavior of materials. In this section, we describe another method for modeling hysteretic behavior of materials (Yun 2006; Yun et al. 2008a). This method has been successfully applied in the modeling of dynamics of beam–column connections in structural frames.

The basic idea is to introduce two variables, ξ_n and η_n, at the input to the one-point neural network material model. These two variables define current positions ε_{n-1} and σ_{n-1} in the stress–strain space and the direction of the movement in that space. These variables, illustrated in Figure 12.7, are given in the following equations. The stresses and strains are in vector form.

$$\xi_n = \boldsymbol{\sigma}_{n-1}^T \boldsymbol{\varepsilon}_{n-1} \quad (12.13)$$

$$\begin{aligned} \eta_n &= \boldsymbol{\sigma}_{n-1}^T \boldsymbol{\varepsilon}_n \\ &= \xi_n + \Delta\eta_n \end{aligned} \quad (12.14)$$

$$\Delta\eta_n = \boldsymbol{\sigma}_{n-1}^T \Delta\boldsymbol{\varepsilon}_n \quad (12.15)$$

The neural network material model in this case is given by the following equation:

$$\boldsymbol{\sigma}_n = NN[\boldsymbol{\sigma}_{n-1}, \boldsymbol{\varepsilon}_{n-1}, \boldsymbol{\varepsilon}_n, \xi_n, \Delta\eta_n:] \quad (12.16)$$

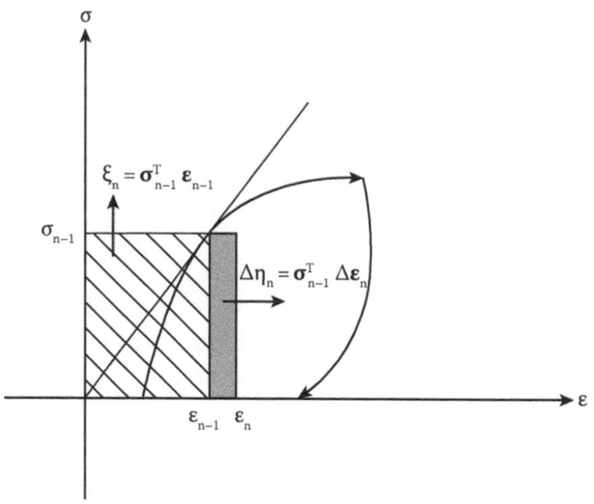

Figure 12.7 Hysteretic variables illustrated on a 1D cyclic stress–strain diagram. (From Yun, G.-J., et al.: A new neural network based model for hysteretic behavior of materials. *Int. J. Numer. Methods Eng.* 2008. 73. 447–469. Copyright Wiley-VCH Verlag GmbH & Co. KGaA. Reproduced with permission.)

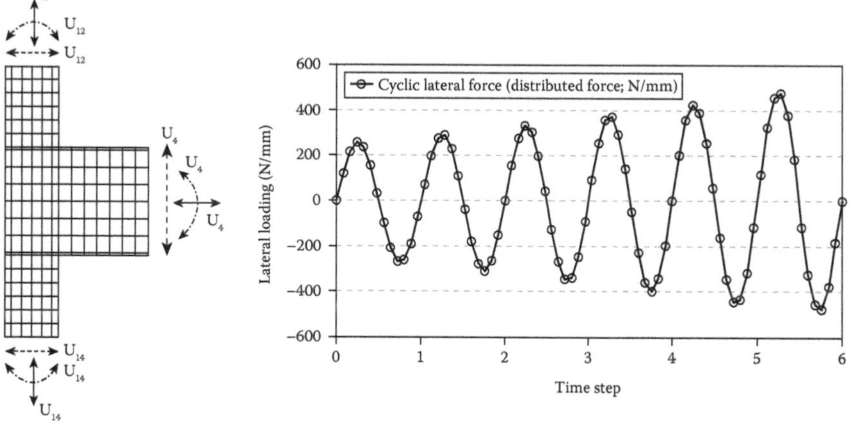

Figure 12.8 Finite element model of beam column connection in a three-story building subjected to cyclic lateral force. (From Yun, G.-J., et al.: A new neural network based model for hysteretic behavior of materials.*Int. J. Numer. Methods Eng.* 2008. 73. 447–469. Copyright Wiley-VCH Verlag GmbH & Co. KGaA. Reproduced with permission.)

The architecture of the neural network is not specified in this equation. We should point out that since there is no obvious hierarchical structure in the training data, a nested neural network architecture was not employed.

The performance of the neural network was evaluated on a finite element *simulation* of a beam column connection subjected to cyclic force, as illustrated in Figure 12.8. The material in the finite element model was represented with an elastoplastic model with kinematic hardening. Stresses and strains from the finite element analysis were collected at a number of points and used to train a neural network with hysteretic variables. The trained neural network was then tested with the stresses and strains at some other elements that were not included in the training data. A typical set of results is shown in Figure 12.9. This figure

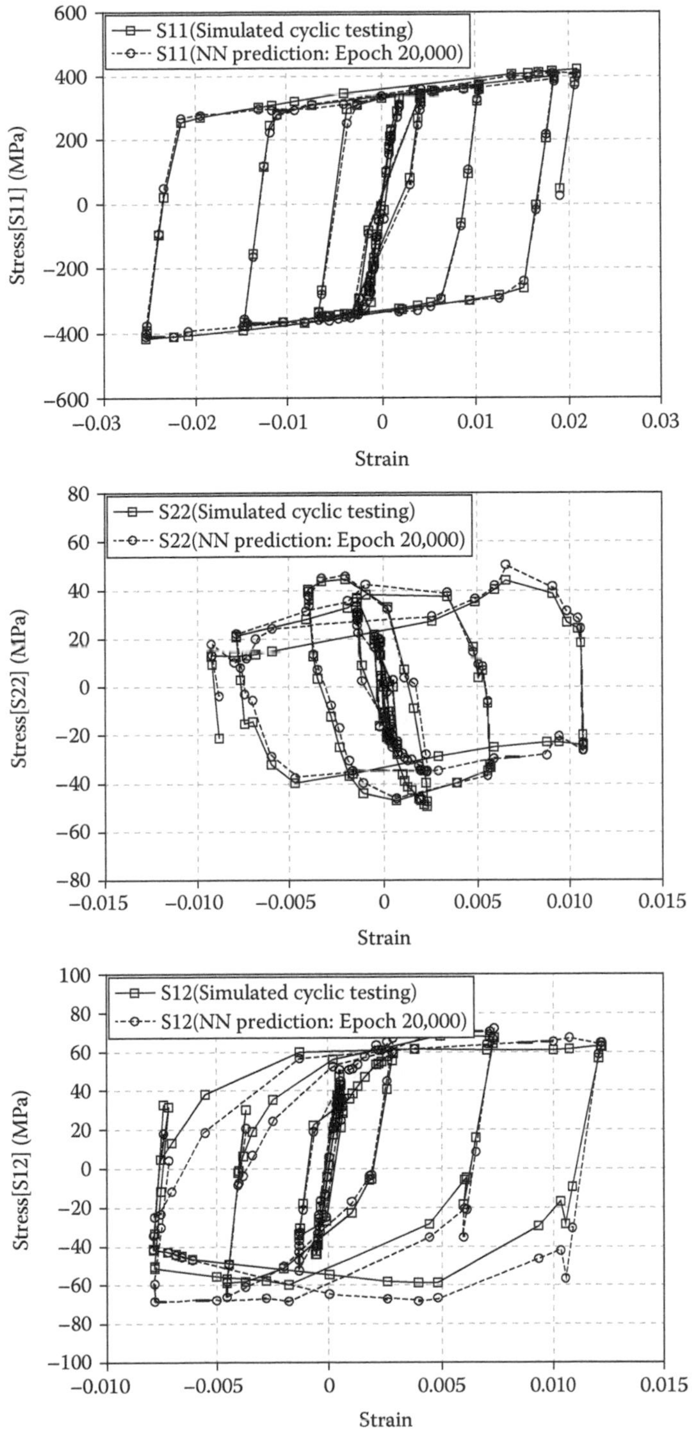

Figure 12.9 Performance of neural network material model with hysteretic variables for stresses at one point in the finite element mesh. (From Yun, G.-J., et al.: A new neural network based model for hysteretic behavior of materials. *Int. J. Numer. Methods Eng.* 2008. 73. 447–469. Copyright Wiley-VCH Verlag GmbH & Co. KGaA. Reproduced with permission.)

shows that the hysteretic variables are quite effective in modeling the hysteretic behavior of materials.

12.7 ACQUISITION OF TRAINING DATA FOR NEURAL NETWORK MATERIAL MODELS

Before discussing further the training of neural network material models, we examine some of the fundamentals of constitutive modeling. Constitutive modeling from material tests is inherently an inverse problem; the input (applied stresses) and output (measured strains) are known (measured), and the system that is the constitutive model needs to be determined. Rather than directly determining the model for the material behavior, we normally use a combination of principles (i.e., conservation laws, objectivity, and frame indifference) and assumptions or idealizations (e.g., isotropy, elasticity, plasticity, yield surface, normality rule, and hardening) and so on. Using these principles and assumptions reduces the objectives of material testing to just determining the parameters of the mathematical material model. Almost all material tests in practice are performed in order to determine the parameters of a specific material model that we have already chosen, and which satisfies the fundamental principles and our assumptions.

This places some restrictions on material tests. Since the material model represents the behavior of a generic material point, the test specimen, or a subregion within it, must be subjected to an essentially uniform state of stress and strain.

The situation with developing training data for neural network material models is different. The results of uniform material tests intended for determining the parameters of analytical material models are not sufficient for neural network training. In the early stages of development of neural network material models, the results of uniform material tests were used in order to explore and evaluate the capability of neural networks to learn constitutive behavior. For example, a neural network was trained to learn the cyclic behavior of plain concrete from 1D tests, as demonstrated in Figure 12.6. The 1D cyclic tests reported in the literature (Karsan and Jirsa 1969) were performed to explore the cyclic behavior of plain concrete and were not, of course, intended for neural network training. However, the 1D neural network trained on these data obviously cannot be used in finite element simulations involving 2D or 3D states of stress.

A neural network material model that can be used in a finite element simulation has to have comprehensive knowledge of the material behavior in the range of interest, which means that it should be trained on a comprehensive training data set that contains the information about material behavior along many stress paths in the region of interest in stress space. Uniform material tests generally follow just one specific stress path. In order to generate comprehensive training data, we need to perform many material tests, each subjected to a different stress path. This appears to require a voluminous amount of test data for training, which may seem to be impractical.

Everything that a neural network material model learns is necessarily contained in the training data. Conservation laws are satisfied since the training data originate from the response of real materials that presumably do satisfy them. We can also rotate the reference axes, thereby generating additional training data, to ensure that the neural network learns about frame indifference. However, we make no assumptions about the material behavior itself; the neural network must learn all aspects of the material behavior from the training data. That is why the training data have to be comprehensive. In order to generate that type of data, the alternative to performing many uniform material tests subjected to a variety of different stress paths is to perform *nonuniform* material tests. The state of stress in the sample has to be spatially as nonuniform as possible. This is directly opposite to previous conventional thinking about material tests.

In a nonuniform material test, different points in the sample follow different stress paths, thus generating information about material behavior along many, but not all the possible, stress paths. The question is therefore, how can we extract that information from the results of a nonuniform test to train a neural network?

The only way to extract information from the results of a nonuniform test is to simulate that test in a finite element analysis. But the dilemma is that to do the finite element analysis, we need a trained neural network material model. So we need a trained neural network material model to perform the finite element simulation in order to generate the training data needed to train the neural network material model. This does not appear to be logical. However, there is a way to accomplish this; we can iteratively combine the two tasks of finite element simulation to generate the data and to use that data to train the neural network material model. This is done by means of the autoprogressive algorithm, which is described next. But, first we need to discuss how neural network material models can be used as constitutive models directly in finite element analysis. In the autoprogressive algorithm, we perform finite element analysis with neural network material models.

12.8 NEURAL NETWORK MATERIAL MODELS IN FINITE ELEMENT ANALYSIS

In order to use the trained neural network material model in a nonlinear finite element analysis, we need to develop the incremental stress–strain relations.

$$\Delta\sigma = \mathbf{D}\ \Delta\varepsilon \tag{12.17}$$

There are two possible ways of achieving this objective.

The most obvious method is to probe the neural network material model to determine the terms of the tangent stress–strain matrix **D**. For example, in a 2D problem we provide the following strain increment input vectors to the trained neural network and determine the output stress increment vectors:

$$\begin{Bmatrix} \Delta\varepsilon_{11} \\ 0 \\ 0 \end{Bmatrix}; \quad \begin{Bmatrix} 0 \\ \Delta\varepsilon_{22} \\ 0 \end{Bmatrix}; \quad \begin{Bmatrix} 0 \\ 0 \\ \Delta\varepsilon_{12} \end{Bmatrix} \tag{12.18}$$

The terms of the stress–strain matrix **D** can be directly determined by dividing the output stress increments by the input strain increments. Of course, in probing with the trained neural network, we must also provide any other input, such as history points, in addition to the above strain increments.

A direct method of determining the stress–strain matrix **D** from the connection weights of the neural network was presented in Hashash et al. (2004) and is discussed in detail in a forthcoming book on soft computing (Ghaboussi 2017). This method is based on the fact that in a feed-forward neural network material model, we apply input strain increments (and history points, if needed) at the input layer and propagate the signal from one layer to the next, until we arrive at the stress increments at the output layer. If we reverse this process (i.e., starting from the output layer and proceeding to the input layer), we can determine the derivatives of the stress increments with respect to the strain increments in terms of the connection weights; these are the terms of the incremental stress–strain matrix **D**.

The neural network material model then simply replaces the conventional analytical material model in a standard incremental-iterative nonlinear finite element analysis, or in any other nonlinear analysis procedure.

12.9 AUTOPROGRESSIVE ALGORITHM

In addition to the information-rich nonuniform material tests mentioned earlier, there are many other potential sources of data that contain information on material behavior. The measured response of a structural or mechanical system subjected to a known (dynamic or quasi-static) excitation contains information about the constitutive behavior of the material (or materials) in that system. As an example, we consider a physical experiment on a structural system; suppose we apply forces and measure displacements at a number of selected locations. The known forces and measured displacements contain information about the constitutive properties of the materials within the structural system. This is again an inverse problem; the forces and displacements are the known input and output of the system, and the constitutive properties of the materials within the structural system are to be determined. A nonuniform material test can also be considered as a miniature structural system. Unlike uniform material tests, which ideally induce a uniform state of stress within the sample, structural tests usually induce nonuniform states of stress and strain within the structure. Since points in the structure follow different stress paths, a single structural test potentially has sufficient information on the material behavior to train a neural network material model. The autoprogressive algorithm is a method for training a neural network to learn the constitutive properties of materials from the results of structural tests (Ghaboussi et al. 1998).

In the autoprogressive algorithm, we begin with a finite element model of the structure being tested and use a neural network (or several neural networks, as needed) to represent the constitutive model of the material (or materials). The neural network material model is trained via incremental-iterative finite element analyses. Initially, the neural network material model is pretrained with an idealized, but necessarily inaccurate, material behavior. Linearly elastic material properties are often used for pretraining. Occasionally, it may be useful to use a previously trained neural network from another problem as an initial material model. The pretrained or previously trained neural network is then initially used to represent the material behavior in order to begin the inverse analysis.

The autoprogressive algorithm simulates the structural test through a series of incremental steps. Two analyses are performed in each incremental step using the current trained neural network material model. The first analysis (FEA-A) is a standard nonlinear finite element analysis, where the *known* load increments are applied and the corresponding displacement increments are computed. In the second analysis (FEA-B), the *measured* displacement increments are applied at the preselected measurement locations. The stresses from the first analysis and the strains from the second analysis are used to continue the training of the neural network material model. This process is repeated for several iterations at the same load increment; in each iteration, the newly updated trained neural network is used, until the differences between the results from FEA-A and FEA-B fall below a certain specified tolerance level. As the iterations and increments are continued, the neural network material model gradually learns the constitutive behavior of the material so that FEA-A and FEA-B produce the same response. The displacements computed from FEA-A match the measured displacements, and forces computed in FEA-B match the applied forces. The continuation of this process until all incremental steps are completed is termed a *pass*. Several passes may be needed for the neural network to learn the material behavior. At that point, the two finite element analyses with the trained neural network will produce the same results, to within a numerical tolerance.

The fact that in every finite element analysis both equilibrium and compatibility are satisfied plays an important role in the autoprogressive algorithm. Equilibrium is satisfied in relating the applied nodal forces to computed material point stresses. Similarly, compatibility is satisfied in relating the measured nodal displacements to computed material point strains. The known forces are applied in FEA-A, and therefore the computed stresses are (expected to be) closer to the true stresses in the structural system because of the equilibrium constraint. Similarly, in FEA-B the measured displacements are applied, and therefore the computed strains are (expected to be) closer to the true strains in the structural system because of the compatibility constraint. This is the underlying rationale for using the stresses from FEA-A and the strains from FEA-B to continue training the neural network material model. When the retrained neural network material model is used in the FEA-A and FEA-B in the next iteration, the computed stresses and strains will be closer to the true values in the structural system. This is why the trained neural network gradually approaches the true material behavior from one iteration to the next and from one increment to the next, eventually learning the true material behavior.

12.9.1 Autoprogressive algorithm in modeling composite materials

One of the first applications of the autoprogressive algorithm was in modeling damage and failure in laminated graphite–epoxy composite plates (Ghaboussi et al. 1998). Laminated composites can display a number of complex damage modes under increasing load (e.g., transverse cracking, matrix shearing or crushing, breaking or microbuckling of fibers followed by kink-band formation, fiber–matrix debonding, and interply delamination) that are difficult to characterize using conventional material modeling approaches.

Lessard and Chang (1991) reported tests on a series of laminated graphite–epoxy structural plates containing a central open hole, as illustrated in Figure 12.10.

Plates with different layups (e.g., cross-ply, angle-ply, and quasi-isotropic-ply orientations) were made from thin unidirectional graphite–epoxy laminae. Two angle-ply laminates (at ±45° and ±30° to the loading direction, respectively), each with 24 individual plies arranged symmetrically with respect to the midplane, were selected. The ±45° laminated plate (designated $[\pm 45^\circ_6]_S$ in standard notation) is used for training the neural network, and the ±30° laminated plate ($[\pm 30^\circ_6]_S$) is used for testing the trained neural network. Here, the "material" is considered to be the thin unidirectional graphite–epoxy lamina. A finite element model of a composite plate with any layup can easily be developed if the 2D in-plane properties of a lamina are known.

Figure 12.11 illustrates the hierarchical modeling procedure for building the finite element model of the test specimen.

The autoprogressive algorithm was used to solve the inverse problem of determining the neural network material model of the unidirectional lamina from the results of the compression test on the $[\pm 45_6]_S$ plate with the open hole. The presence of the open hole of course provides the necessary (spatial) variation in stress–strain paths.

Two passes were made through the experimental load deflection data using the autoprogressive algorithm in order to train the neural network material model. After each pass, the trained neural network was used in a simulation of the experiment. The results of these simulations are shown in Figure 12.12.

Also shown in Figure 12.12 are the results of a forward analysis of the $[\pm 30^\circ_6]_S$ plate using the trained neural network material model. This is a different experiment on a different plate with different macroscopic properties. It demonstrates that the neural network has adequately learned the material behavior of the lamina, and could then be used to predict the response of other laminated structural components with different layups made from the same unidirectional graphite–epoxy plies.

We now look a little further into the learned characteristics of the lamina neural network material model itself. It was initially pretrained with linearly elastic data. Most of the nonlinearity is expected to occur in the shear stress–strain response, which is contributed almost exclusively by the epoxy matrix. The evolution of the lamina shear stress–strain relation in

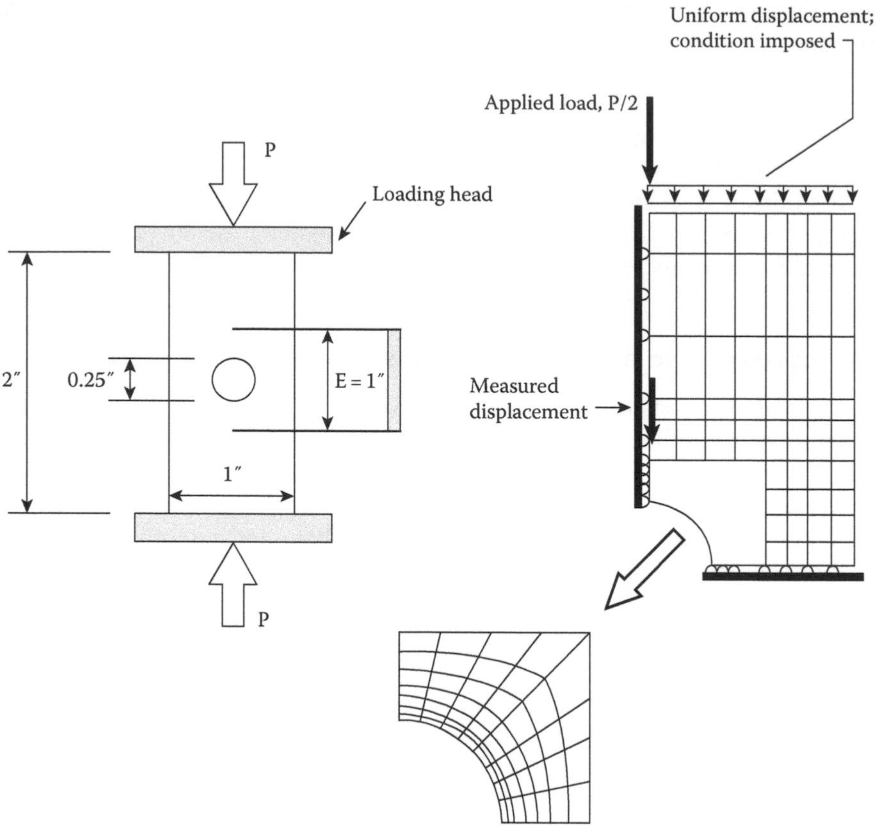

Figure 12.10 Test setup and finite element model for the composite plate experiment. (Test setup from Ghaboussi, J., et al.: Autoprogressive training of neural network constitutive models. *Int. J. Numer. Methods Eng.* 1998. 42. 105–126. Copyright Wiley-VCH Verlag GmbH & Co. KGaA. Reproduced with permission.)

the neural network during autoprogressive training is shown in Figure 12.13 in three load steps in each of the two passes. We can see that the lamina material behavior in load step 7 in pass 1 is still almost linearly elastic. As we progress through the load steps and passes, we can see that the neural network learns the nonlinear shear stress–strain behavior of the lamina. We remark that it approaches the analytical shear stress–strain relationship proposed by Chang and Lessard (1991).

More details of this application of the autoprogressive algorithm to the modeling of composite plates can be found in Ghaboussi et al. (1998).

12.9.2 Autoprogressive algorithm in structural mechanics and in geomechanics

The autoprogressive algorithm has been successfully used in a number of additional engineering applications. The following are some examples of the application of the autoprogressive algorithm by the first author and his colleagues and former doctoral students:

- Deep excavations in urban areas (Hashash et al. 2003, 2006)
- Modeling of the behavior of geomaterials from nonuniform tests (Sidarta and Ghaboussi 1998; Fu et al. 2007; Hashash et al. 2009)

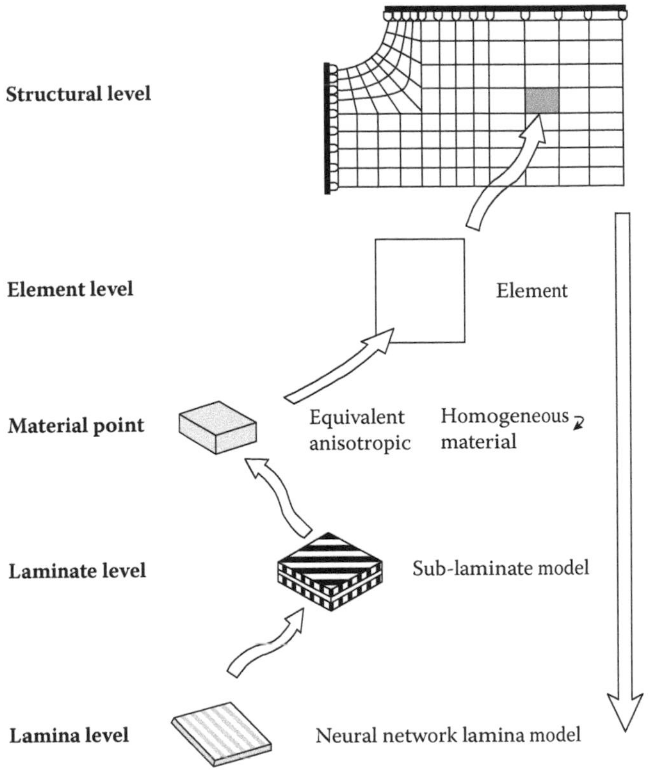

Figure 12.11 Finite element procedure for laminated composite modeling. (From Ghaboussi, J., et al.: Autoprogressive training of neural network constitutive models. *Int. J. Numer. Methods Eng.* 1998. 42. 105–126. Copyright Wiley-VCH Verlag GmbH & Co. KGaA. Reproduced with permission.)

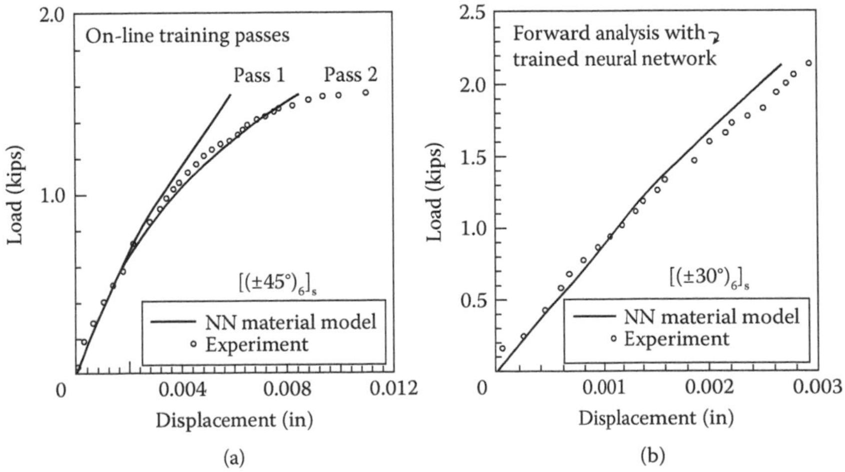

Figure 12.12 (a, b) Load–displacement results from forward analyses, using the neural network lamina material model trained on $[(\pm 45)_6]_s$. (Data from Ghaboussi, J., et al.: Autoprogressive training of neural network constitutive models. *Int. J. Numer. Methods Eng.* 1998. 42. 105–126. Copyright Wiley-VCH Verlag GmbH & Co. KGaA. Reproduced with permission.)

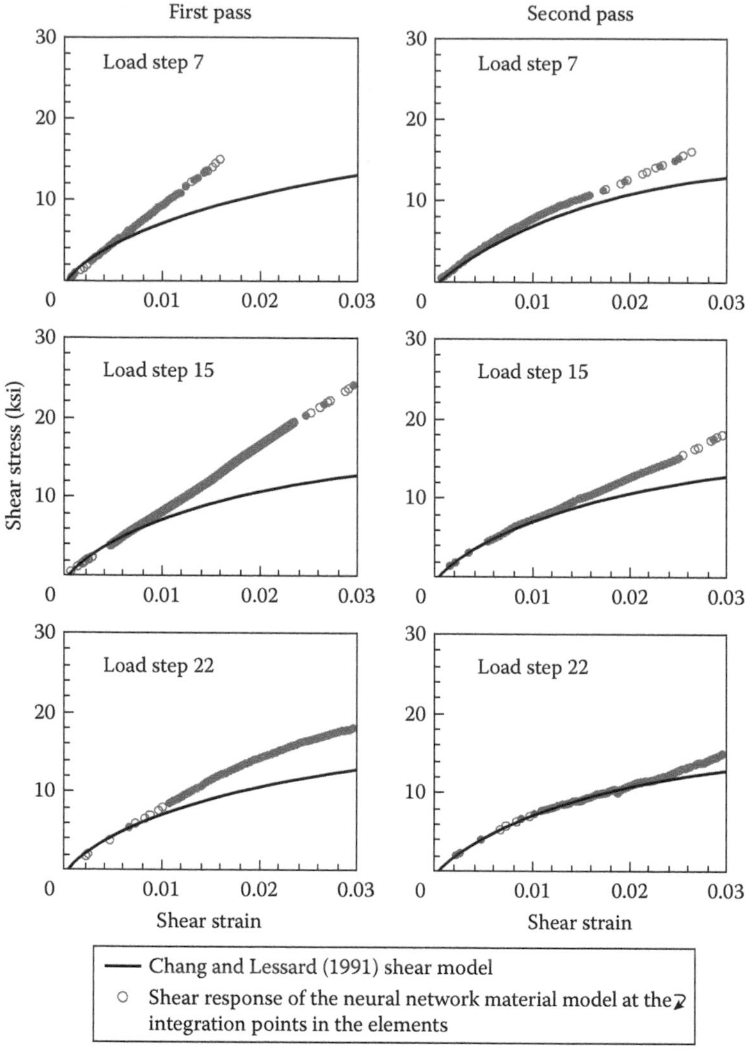

Figure 12.13 Evolution of the shear stress–strain relationship in the neural network lamina material model. (From Ghaboussi, J., et al.: Autoprogressive training of neural network constitutive models. *Int. J. Numer. Methods Eng.* 1998. 42. 105–126. Copyright Wiley-VCH Verlag GmbH & Co. KGaA. Reproduced with permission.)

- Time-dependent behavior of concrete in segmented long-span bridges (Jung et al. 2007)
- Modeling of hysteretic behavior of beam–column connections from the results of dynamic tests on structural frames (Yun et al. 2008b; Kim et al. 2012).

12.9.3 Autoprogressive algorithm in biomedicine

Biomedicine and biomedical imaging comprise another important area of application for self-learning simulation with the autoprogressive algorithm. The autoprogressive algorithm has great potential for developing noninvasive methods for *in vivo* determination of constitutive properties of soft tissue.

A successful application of the autoprogressive algorithm in determining the constitutive properties of the human cornea has been reported in Kwon (2006) and Kwon et al. (2008).

Corneal constitutive properties are significant in virtual laser surgery to individualize and optimize laser surgery procedures. More accurate clinical determination of intraocular pressure (IOP) is made possible by accounting for the influence of corneal biomechanical properties on standard IOP measurement methods (Ghaboussi et al. 2009; Kwon et al. 2010).

Another application of the autoprogressive algorithm is in elastography, imaging the constitutive properties of soft tissue (Hoerig et al. 2015, 2016, 2017), with potential applications in diagnostics and monitoring of disease progression.

References

Ahmad, S., Irons, B.M. and Zienkiewicz, O.C., (1970), Analysis of Thick and Thin Shell Structures by Curved Finite Elements, *International Journal for Numerical Methods in Engineering*, 2, 419–451.

Alexander, H., (1971), Tensile Instability of Initially Spherical Balloons, *International Journal of Engineering Science*, 9, 151–162.

Allman, D.J., (1984), A Compatible Triangular Element Including Vertex Rotations for Plane Elasticity Analysis, *Computers and Structures*, 19, 1–8.

Argyris, J., (1982), An Excursion into Large Rotations, *Computer Methods in Applied Mechanics and Engineering*, 32, 85–155.

Argyris, J.H., Fried, I. and Scharpf, D.W., (1968), The TUBA Family of Plate Elements for the Matrix Displacement Method, *Aeronautical Journal of the Royal Aeronautical Society*, 72, 702–709.

Bathe, K.J. and Wilson E.L., (1973), Solution Methods for Eigenvalue Problems in Structural Mechanics, *International Journal for Numerical Methods in Engineering*, 6, 213–226.

Belytschko, T. and Hsieh, B.J., (1973), Non-Linear Transient Finite Element Analysis with Convected Co-Ordinates, *International Journal for Numerical Methods in Engineering*, 7, 255–271.

Belytschko, T., Stolarski, H., Liu, W.K., Carpenter, N. and Ong, J.S.-J., (1985), Stress Projection for Membrane and Shear Locking in Shell Finite Elements, *Computer Methods in Applied Mechanics and Engineering*, 51, 221–258.

Bergan, P.G., (1980), Solution Algorithms for Nonlinear Structural Problems, *Computers and Structures*, 12, 497–509.

Bergan, P.G. and Felippa, C.A., (1985), A Triangular Membrane Element with Rotational Degrees of Freedom, *Computer Methods in Applied Mechanics and Engineering*, 50, 25–69.

Bergan, P.G., Horrigmoe, G., Krakeland, B. and Soreide, T.H., (1978), Solution Techniques for Non-Linear Finite Element Problems, *International Journal for Numerical Methods in Engineering*, 12, 1677–1696.

Bogner, F.K., Fox, R.L. and Schmidt Jr., L.A., (1965), The Generation of Interelement Compatible Stiffness and Mass Matrices by the Use of Interpolation Formulas, *Proceedings of the Conference on Matrix Methods in Structural Mechanics*, AFFDL-TR-66-80, 397–443.

Bridgman, P.W., (1923), Compressibility of Thirty Metals as a Function of Temperature and Pressure, *Proceedings of the American Academy of Arts and Sciences*, 58, 163–242.

Bridgman, P.W., (1952), *Studies in Large Plastic Flow and Fracture*, McGraw-Hill, New York.

Bushnell, D., (1985), *Computerized Buckling Analysis of Shells*, Martinus Nijhoff Publishers, Dordrecht, the Netherlands.

Cayley, A., (1843), On the Motion of Rotation of a Solid Body, *Cambridge Mathematical Journal*, 3, 224–232.

Cayley, A., (1845), On Certain Results Relating to Quaternions, *Philosophical Magazine*, 26, 141–145.

Cayley, A., (1846), Sur Quelques Propriertes des Determinants Gauches, *Journal fur die Reine und Angewandte Mathematik*, XXII, 119–123.

Cayley, A., (1848a), Sur Les Determinants Gauches, *Journal fur die Reine und Angewandte Mathematik*, XXXVIII, 93–96.

Cayley, A., (1848b), On the Application of Quaternions to the Theory of Rotation, *Philosophical Magazine*, XXXIII, 196–200.

Chang, F. and Lessard, L., (1991), Damage Tolerance of Laminated Composites Containing an Open Hole and Subjected to Compressive Loadings. Part I. Analysis, *Journal of Composite Materials*, 25, 2–43.

Chasles, M., (1830), Note sur les Proprietes Generales du Systeme de Duex Semblables Entr'eux, *Bulletin des Sciences Mathematiques, Astronomiques, Physiques et Chemiques*, 14, 321–326.

Cheng, H. and Gupta, K.C., (1989), An Historical Note on Finite Rotations, *Journal of Applied Mechanics*, 56, 139–145.

Chuong, C.J. and Fung, Y.C., (1983), Three-Dimensional Stress Distribution in Arteries, *Journal of Biomechanical Engineering*, 105, 269–274.

Cook, R.D., Malkus, D.S. and Plesha, M.E., (1989), *Concepts and Applications of Finite Element Analysis*, 3rd ed., John Wiley & Sons, New York.

Cowper, G.R., (1973), Gaussian Quadrature Formulas for Triangles, *International Journal for Numerical Methods in Engineering*, 405–408.

Crisfield, M.A., (1981), A Fast Incremental/Iterative Solution Procedure that Handles "Snap-Through," *Computers and Structures*, 13, 55–62.

Crisfield, M.A., (1990), A Consistent Corotational Formulation for Non-Linear, Three-Dimensional, Beam-Elements, *Computer Methods in Applied Mechanics and Engineering*, 81, 131–150.

de Borst, R., (1991), The Zero-Normal-Stress Condition in Plane-Stress and Shell Elastoplasticity, *Communications in Applied Numerical Methods*, 7, 29–33.

de Borst, R. and Groen, A.E., (1994), A Note on the Calculation of Consistent Tangent Operators for von Mises and Drucker–Prager Plasticity, *Communications in Numerical Methods in Engineering*, 10, 1021–1025.

Dhatt, G., Marcotte, L. and Matte, Y., (1986), A New Triangular Discrete-Kirchhoff Plate/Shell Element, *International Journal for Numerical Methods in Engineering*, 23, 453–470.

Dienes, J.K., (1979), On the Analysis of Rotation and Stress Rate in Deforming Bodies, *Acta Mechanica*, 32, 217–232.

Drucker, D.C. and Prager, W., (1952), Soil Mechanics and Plastic Analysis or Limit Design, *Quarterly of Applied Mathematics*, 10, 157–165.

Felippa, C.A. and Haugen, B., (2005), A Unified Formulation of Small-Strain Co-Rotational Finite Elements. I. Theory, *Computer Methods in Applied Mechanics and Engineering*, 194, 2285–2335.

Fu, Q., Hashash, Y.M.A., Jung, S. and Ghaboussi, J., (2007), Integration of Laboratory Testing and Constitutive Modeling of Soils, *Computer and Geotechnics*, 34(5), 330–345.

Fung, Y.C., (1993), *Biomechanics: Mechanical Properties of Living Tissues*, 2nd ed., Springer-Verlag, New York.

Gent, A.N., (1996), A New Constitutive Relation for Rubber, *Rubber Chemistry and Technology*, 69, 59–61.

Gent, A.N., (1999), Elastic Instabilities of Inflated Rubber Shells, *Rubber Chemistry and Technology*, 72, 263–268.

Ghaboussi, J., (2009), Advances in Neural Networks in Computational Mechanics and Engineering, in *Advances of Soft Computing in Engineering*, ed. Z. Waszczyszyn, Springer, New York, pp. 191–236.

Ghaboussi, J., (2017), *Biologically Inspired Soft Computing in Engineering*, CRC Press, Taylor & Francis Group, Boca Raton, FL.

Ghaboussi, J., Garrett, J.H. and Wu, X., (1990), Material Modeling with Neural Networks, *Proceedings of International Conference on Numerical Methods in Engineering: Theory and Applications*, Swansea, UK, 701–717.

Ghaboussi, J., Garrett, J.H. and Wu, X., (1991), Knowledge-Based Modeling of Material Behavior with Neural Networks, *Journal of Engineering Mechanics Division*, 117(1), 132–153.

Ghaboussi, J., Kwon, T.-H., Pecknold, D.A. and Hashash, Y.M.A., (2009), Accurate Intraocular Pressure Prediction from Applanation Response Data Using Genetic Algorithm and Neural Networks, *Journal of Biomechanics*, 42(14), 2301–2306.

Ghaboussi, J., Pecknold, D.A., Zhang, M. and Haj-Ali, R., (1998), Autoprogressive Training of Neural Network Constitutive Models, *International Journal for Numerical Methods in Engineering*, 42, 105–126.

Ghaboussi, J. and Sidarta, D.E., (1998), A New Nested Adaptive Neural Network for Modeling of Constitutive Behavior of Materials, *International Journal of Computer and Geotechnics*, 22(1), 29–51.

Ghaboussi, J. and Wu, X., (1998), Soft Computing with Neural Networks for Engineering Applications: Fundamental Issues and Adaptive Approaches, *Journal of Structural Engineering and Mechanics*, 6(8), 955–969.

Ghaboussi, J. and Wu, X., (2016), *Numerical Methods in Computational Mechanics*, CRC Press, Taylor & Francis Group, Boca Raton, FL.

Ghaboussi, J., Zhang, M., Wu, X. and Pecknold, D.A., (1997), Nested Adaptive Neural Networks: A New Architecture, *Proceedings of International Conference on Artificial Neural Networks in Engineering, ANNIE*, St. Louis, MO.

Goldstein, H., (1980), *Classical Mechanics*, 2nd ed., Addison-Wesley, Reading, MA.

Gurson, A.L., (1975), *Plastic Flow and Fracture Behavior of Ductile Materials Incorporating Void Nucleation, Growth, and Interaction*, PhD thesis, Brown University, Providence, RI.

Gurson, A.L., (1977), Continuum Theory of Ductile Rupture by Void Nucleation and Growth. Part I. Yield Criteria and Flow Rules for Porous Ductile Media, *Journal of Engineering Materials and Technology*, 99, 2–15.

Haftka, R.T., Mallett, R.H. and Nachbar, W., (1971), Adaption of Koiter's Method to Finite Element Analysis of Snap-Through Buckling Behavior, *International Journal of Solids and Structures*, 7, 1427–1445.

Hamilton, W.R., (1844), On Quaternions; or on a New System of Imaginaries in Algebra, *Philosophical Magazine*, 25, 489–495.

Hand, F.R., Pecknold, D.A. and Schnobrich, W.C., (1973), Nonlinear Layered Analysis of RC Plates and Shells, *Journal of the Structural Division*, 99, 1491–1505.

Hashash, Y.M.A., Fu, Q.-W., Ghaboussi, J., Lade, P.V. and Saucier, C., (2009), Inverse Analysis Based Interpretation of Sand Behavior from Triaxial Shear Tests Subjected to Full End Restraint, *Canadian Geotechnical Journal*, 46(7), 768–791.

Hashash, Y.M.A., Jung, S. and Ghaboussi, J., (2004), Numerical Implementation of a Neural Networks Based Material Model in Finite Element, *International Journal for Numerical Methods in Engineering*, 59, 989–1005.

Hashash, Y.M.A., Marulanda, C., Ghaboussi, J. and Jung, S., (2003), Systematic Update of a Deep Excavation Model Using Field Performance Data, *Computers and Geotechnics*, 30, 477–488.

Hashash, Y.M.A., Marulando, C., Ghaboussi, J. and Jung, S., (2006), Novel Approach to Integration of Numerical Modeling and Field Observation for Deep Excavations, *Journal of Geotechnical and Geo-Environmental Engineering*, 132(8), 1019–1031.

Haugen, B., (1994), Buckling and Stability Problems for Thin Shell Structures Using High Performance Finite Elements, PhD thesis, University of Colorado, Boulder.

Healey, T.J., (1988), A Group-Theoretic Approach to Bifurcation Problems with Symmetry, *Computer Methods in Applied Mechanics and Engineering*, 67, 257–295.

Hill, R., (1950), *The Mathematical Theory of Plasticity*, Oxford University Press, Oxford.

Hoerig, C., Ghaboussi, J., Fatemi, M. and Insana, M.F., (2016), A New Approach to Ultrasonic Elasticity Imaging, *Proceedings of SPIE Medical Imaging: Ultrasonic Imaging and Tomography*, 9790, G1–G9.

Hoerig, C., Ghaboussi, J. and Insana, M.F., (2015), Informational Modeling of Tissue-Like Materials Using Ultrasound, *Proceedings of 12th IEEE International Symposium on Biomedical Imaging, ISBI 2015*, 239–242.

Hoerig, C., Ghaboussi, J. and Insana, M.F., (2017), An Information-Based Machine Learning Approach to Elasticity Imaging, *Journal of Biomechanics and Modeling in Mechanobiology*, 1–18, doi: 10.1007/s10237-016-0854-6.

Hughes, T.J.R. and Liu, W.K., (1981), Nonlinear Finite Element Analysis of Shells. Part I. Three-Dimensional Shells, *Computer Methods in Applied Mechanics and Engineering*, 26, 331–362.

Hughes, T.J.R. and Winget, J., (1980), Finite Rotation Effects in Numerical Integration of Rate Constitutive Equations Arising in Large-Deformation Analysis, *International Journal for Numerical Methods in Engineering*, 15, 1862–1867.

Irons, B.M., (1966), Engineering Applications of Numerical Integration in Stiffness Methods, *Journal of the American Institute of Aeronautics and Astronautics*, 4, 2035–2037.

Irons, B.M. and Draper, K.J., (1965), Inadequacy of Nodal Connections in a Stiffness Solution for Plate Bending, *Journal of the American Institute of Aeronautics and Astronautics*, 3, 961.

Johnson, G.C. and Bammann, D.J., (1984), A Discussion of Stress Rates in Finite Deformation Problems, *International Journal of Solids and Structures*, 20, 725–737.

Jung, S.-M., Ghaboussi, J. and Marulanda, C., (2007), Field Calibration of Time-Dependent Behavior in Segmental Bridges Using Self-Learning Simulations, *Engineering Structures*, 29, 2692–2700.

Karsan, I.D. and Jirsa, J.O., (1969), Behavior of Concrete under Compressive Loading, *Journal of the Structural Division*, 95, 2543–2563.

Kim, J.-H., Ghaboussi, J. and Elnashai, A.S., (2012), Hysteretic Mechanical-Informational Modeling of Bolted Steel Frame Connections, *Engineering Structures*, 45(1), 1–11.

Koiter, W.T., (1945), On the Stability of Elastic Equilibrium, PhD thesis, Polytechnic Institute, Delft, the Netherlands. NASA Technical Translation TT F-10 (1967).

Kojic, M. and Bathe, K.-J., (1987), Studies of Finite Element Procedures—Stress Solution of a Closed Elastic Strain Path with Stretching and Shearing Using the Updated Lagrangian Jaumann Formulation, *Computers and Structures*, 26, 175–179.

Kupfer, H.K., Hilsdorf, H. and Rusch, H., (1969), Behavior of Concrete under Biaxial Stresses, *Journal of the American Concrete Institute*, 66, 656–666.

Kwon, T.-H., (2006), Minimally Invasive Characterization and Intraocular Pressure Measurement via Numerical Simulation of Human Cornea, PhD thesis, Department of Civil and Environmental Engineering, University of Illinois at Urbana–Champaign.

Kwon, T.-H., Ghaboussi, J., Pecknold, D.A. and Hashash, Y.M.A., (2008), Effect of Cornea Material Stiffness on Measured Intraocular Pressure, *Journal of Biomechanics*, 41(8), 1707–1713.

Kwon, T.-H., Ghaboussi, J., Pecknold, D.A. and Hashash, Y.M.A., (2010), Role of Cornea Biomechanical Properties in Applanation Tonometry Measurements, *Journal of Refractive Surgery*, 26 (7), 512–519.

Lade, P.V., (1977), Experimental Observations of Stability, Instability and Shear Planes in Granular Materials, *Ingenieur-Archiv*, 59, 114–123.

Lade, P.V., (2014), Estimating Parameters from a Single Test for the Three-Dimensional Failure Criterion for Frictional Materials, *Journal of Geotechnical and Geoenvironmental Engineering*, 140(8), 140–144.

Lee, E.H., (1981), Some Comments on Elastic-Plastic Analysis, *International Journal of Solids and Structures*, 17, 859–872.

Lessard, L. and Chang, F.-K., (1991), Damage Tolerance of Laminated Composites Containing an Open Hole and Subjected to Compressive Loadings. Part II. Experiment, *Journal of Composite Materials*, 25, 44–64.

Mallett, R.H. and Marcal, P.V., (1968), Finite Element Analysis of Nonlinear Structures, *Journal of the Structural Division*, 94, 2081–2105.

Martin, J.B., (1975), *Plasticity: Fundamentals and General Results*, MIT Press, Cambridge, MA.

Mendelson, A., (1968), *Plasticity: Theory and Application*, Macmillan, New York.

Milford, R.V. and Schnobrich, W.C., (1986), Degenerated Isoparametric Finite Elements Using Explicit Integration, *International Journal for Numerical Methods in Engineering*, 23, 133–154.

Mroz, Z., (1963), Non-Associated Flow Laws in Plasticity, *Journal de Mecanique*, 2, 21–42.

Nagtegaal, J.C., (1982), On the Implementation of Inelastic Constitutive Equations with Special Reference to Large Deformation Problems, *Computer Methods in Applied Mechanics and Engineering*, 33, 469–484.

Nayak, G.C. and Zienkiewicz, O.C., (1972), Convenient Form of Stress Invariants for Plasticity, *Journal of Structural Division*, 98, 949–954.

Nour-Omid, B. and Rankin, C., (1991), Finite Rotation Analysis and Consistent Linearization Using Projectors, *Computer Methods in Applied Mechanics and Engineering*, 93, 353–384.

Oden, J.T., (1972), *Finite Elements of Nonlinear Continua*, McGraw-Hill, New York, pp. 318–321.

Ogden, R.W., (1972), Large Deformation Isotropic Elasticity—On the Correlation of Theory and Experiment for Incompressible Rubberlike Solids, *Proceedings of the Royal Society of London: Series A, Mathematical and Physical Sciences*, 326, 565–584.

Ortiz, M. and Popov, E.P., (1985), Accuracy and Stability of Integration Algorithms for Elastoplastic Constitutive Relations, *International Journal for Numerical Methods in Engineering*, 21, 1561–1576.

Parisch, H., (1978), Geometrical Nonlinear Analysis of Shells, *Computer Methods in Applied Mechanics and Engineering*, 14, 159–178.

Parisch, H., (1979), A Critical Survey of the Nine-Node Degenerated Shell Element with Special Emphasis on Thin Shell Application and Reduced Integration, *Computer Methods in Applied Mechanics and Engineering*, 20, 323–350.

Parisch, H., (1981), Large Displacement of Shells Including Material Nonlinearities, *Computer Methods in Applied Mechanics and Engineering*, 27, 183–214.

Pecknold, D.A., Ghaboussi, J. and Healey, T.J., (1985), Snap-Through and Bifurcation in a Simple Structure, *Journal of Engineering Mechanics*, 111, 910–922.

Rajasekaran, S. and Murray, D.W., (1973), On Incremental Finite Element Matrices, *Journal of the Structural Division*, 99, 7423–7438.

Ramm, E., (1981), Strategies for Tracing the Nonlinear Response Near Limit Points, in *Nonlinear Finite Element Analysis in Structural Mechanics*, eds. W. Wunderlich, E. Stein, and K.-J. Bathe, Springer Verlag, Berlin, Germany, pp. 63–89.

Ramm, E., (1982), The Riks/Wempner Approach—An Extension of the Displacement Control Method in Non-Linear Analysis, in *Non-Linear Computational Mechanics*, ed. E. Hinton, Pineridge Press, Swansea, UK, pp. 63–86.

Rankin, C. and Nour-Omid, B., (1988), The Use of Projectors to Improve Finite Element Performance, *Computers and Structures*, 30, 257–267.

Rankin, C.C. and Brogan, F.A., (1986), An Element Independent Corotational Procedure for the Treatment of Large Rotations, *Journal of Pressure Vessel Technology*, 108, 165–174.

Riks, E., (1972), The Application of Newton's Method to the Problem of Elastic Stability, *Journal of Applied Mechanics*, 39, 1060–1065.

Riks, E., (1979), An Incremental Approach to the Solution of Snapping and Buckling Problems, *International Journal of Solids and Structures*, 15, 529–551.

Rivlin, R.S. and Saunders, D.W., (1951), Large Elastic Deformations of Isotropic Materials. VII. Experiments on the Deformation of Rubber, *Philosophical Transactions of the Royal Society of London: Series A, Mathematical and Physical Sciences*, 243, 251–288.

Rodrigues, O., (1840), Des Lois Geometriques qui Regissant les Deplacements d'un Systeme Solide dans Espace, et de la Variation des Coordonnees Provenant de ses Deplacements Consideres Independamment des Causes qui Peuvent les Produire, *Journal de Mathematiques Pures et Appliques*, 5, 380–440.

Seide, P., (1975), *Small Elastic Deformations of Thin Shells*, Noordhoff International Publishing, Leyden, the Netherlands.

Sidarta, D.E. and Ghaboussi, J., (1998), Modeling Constitutive Behavior of Materials from Non-Uniform Material Test, *International Journal of Computer and Geotechnics*, 22(1), 53–71.

Simo, J.C. and Pister, K.S., (1984), Remarks on Rate Constitutive Equations for Finite Deformation Problems: Computational Implications, *Computer Methods in Applied Mechanics and Engineering*, 46, 201–215.

Simo, J.C. and Taylor, R.L., (1985), Consistent Tangent Operators for Rate-Independent Elastoplasticity, *Computer Methods in Applied Mechanics and Engineering*, 45, 101–118.

Simo, J.C. and Vu-Quoc, L., (1986), A Three-Dimensional Finite-Strain Rod Model. Part II. Computational Aspects, *Computer Methods in Applied Mechanics and Engineering*, 58, 79–116.

Skallerud, B. and Haugen, B., (1999), Collapse of Thin Shell Structures—Stress Resultant Plasticity Modelling within a Co-Rotated ANDES Formulation, *International Journal for Numerical Methods in Engineering*, 46, 1961–1986.

Struik, D.J., (1961), *Lectures on Classical Differential Geometry*, 2nd ed., Addison-Wesley, Reading, MA.

Sture, S., Runesson, K. and Macari-Pasqualino, E.J., (1989), Analysis and Calibration of a Three-Invariant Plasticity Model for Granular Materials, *Ingenieur-Archiv*, 59, 253–266.

Surana, K.S., (1983), Geometrically Nonlinear Formulation for the Curved Shell Elements, *International Journal for Numerical Methods in Engineering*, 19, 581–615.

Szwabowicz, M.L., (1986), Variational Formulation in the Geometrically Nonlinear Thin Elastic Shell Theory, *International Journal of Solids and Structures*, 22, 1161–1175.

Taig, I.C., (1961), Structural Analysis by the Matrix Displacement Method, Report SO 17, English Electric Aviation, London.

Thurston, G.A., Brogan, F.A. and Stehlin, P., (1986), Post-Buckling Analysis Using a General-Purpose Code, *Journal of the American Institute of Aeronautics and Astronautics*, 24, 1013–1020.

Treloar, L.R.G., (1944), Stress-Strain Data for Vulcanised Rubber under Various Types of Deformation, *Transactions of the Faraday Society*, 40, 59–70.

Truesdell, C., (1955), The Simplest Rate Theory of Pure Elasticity, *Communications on Pure and Applied Mathematics*, VIII, 123–132.

Wempner, G., (1969), Finite Elements, Finite Rotations and Small Strains of Flexible Shells, *International Journal of Solids and Structures*, 5, 117–153.

Willam, K.J. and Warnke, E.P., (1975), Constitutive Models for the Triaxial Behavior of Concrete, *Proceedings of the International Association for Bridge and Structural Engineering*, 19, 1–30.

Wu, X., (1991), Neural Network Based Material Modeling, PhD thesis, Department of Civil and Environmental Engineering, University of Illinois at Urbana–Champaign.

Wu, X., Ghaboussi, J. and Garrett, J., (1992), Use of Neural Networks in Detection of Structural Damage, *Computers and Structures*, 42(4), 649–659.

Yun, G.-J., (2006), Modeling of Hysteretic Behavior of Beam-Column Connections Based on Self-Learning Simulations, PhD thesis, Department of Civil and Environmental Engineering, University of Illinois at Urbana–Champaign.

Yun, G.-J., Ghaboussi, J. and Elnashai, A.S., (2008a), A New Neural Network Based Model for Hysteretic Behavior of Materials, *International Journal for Numerical Methods in Engineering*, 73, 447–469.

Yun, G.-J., Ghaboussi, J. and Elnashai, A.S., (2008b), Self-Learning Simulation Method for Inverse Nonlinear Modeling of Cyclic Behavior of Connections, *Computer Methods in Applied Mechanics and Engineering*, 197(33–40), 2836–2857.

Zhang, M., (1996), Determination of Neural Network Material Models from Structural Tests, PhD thesis, Department of Civil and Environmental Engineering, University of Illinois at Urbana–Champaign.

Zienkiewicz, O.C., (1977), *The Finite Element Method*, 3rd ed., McGraw-Hill, London.

Zienkiewicz, O.C. and Nayak, G.C., (1971), A General Approach to Problems of Plasticity and Large Deformation Using Isoparametric Elements, *Proceedings of the Third Conference on Matrix Methods in Structural Mechanics*, AFFDL-TR-71-160, 881–928.

Index

Milton Keynes UK
Ingram Content Group UK Ltd.
UKHW052021071024
449327UK00027B/2370

9 780367 875244